AF335407

The Enzyme Catalysis Process

Energetics, Mechanism, and Dynamics

NATO ASI Series

Advanced Science Institutes Series

A series presenting the results of activities sponsored by the NATO Science Committee, which aims at the dissemination of advanced scientific and technological knowledge, with a view to strengthening links between scientific communities.

The series is published by an international board of publishers in conjunction with the NATO Scientific Affairs Division

A	**Life Sciences**	Plenum Publishing Corporation
B	**Physics**	New York and London
C	**Mathematical and Physical Sciences**	Kluwer Academic Publishers Dordrecht, Boston, and London
D	**Behavioral and Social Sciences**	
E	**Applied Sciences**	
F	**Computer and Systems Sciences**	Springer-Verlag
G	**Ecological Sciences**	Berlin, Heidelberg, New York, London,
H	**Cell Biology**	Paris, and Tokyo

Recent Volumes in this Series

Series A: Life Sciences

The Enzyme Catalysis Process

Energetics, Mechanism, and Dynamics

Edited by

Alan Cooper

Glasgow University
Glasgow, Scotland, United Kingdom

Julien L. Houben and Lisa C. Chien

Istituto di Chimica Quantistica ed Energetica Molecolare
CNR
Pisa, Italy

Plenum Press
New York and London
Published in cooperation with NATO Scientific Affairs Division

Proceedings of a NATO Advanced Study Institute on
The Enzyme Catalysis Process:
Energetics, Mechanism, and Dynamics,
held July 11–23, 1988,
at Il Ciocco, Barga, Italy

Library of Congress Cataloging in Publication Data

NATO Advanced Study Institute on the Enzyme Catalysis Process: Energetics, Mechanism, and Dynamics (1988: Barga, Italy)
  The enzyme catalysis process: energetics, mechanism, and dynamics / edited by Alan Cooper, Julien L. Houben, and Lisa C. Chien.
      p.   cm—(NATO ASI series. Series A, Life sciences; v. 178)
  "Published in cooperation with NATO Scientific Affairs Division."
  Includes bibliographical references.
  ISBN 0-306-43331-1
  1. Enzyme kinetics—Congresses. 2. Enzymes—Structure—activity relation-
ships—Congresses. I. Cooper, Alan, 1945-   . II. Houben, Julien L. III. Chien,
Lisa C. IV. North Atlantic Treaty Organization. Scientific Affairs Division. V. Title.
VI. Series.
QP601.3.N38   1988   89-37195
574.19′25—dc20   CIP

© 1989 Plenum Press, New York
A Division of Plenum Publishing Corporation
233 Spring Street, New York, N.Y. 10013

Printed in the United States of America

PREFACE

This volume represents the proceedings of a NATO Advanced Studies Institute held near Barga (Italy), July 11-23, 1988, involving over 90 participants from more than twelve countries of Europe, North America and elsewhere. It was not our intention at this meeting to present a complete up-to-the-minute review of current research in enzyme catalysis but rather, in accord with the intended spirit of NATO ASIs, to give an opportunity for advanced students and researchers in a wide variety of disciplines to meet together and study the problem from different points of view. Hence the lectures cover topics ranging from the purely theoretical aspects of chemical reaction kinetics in condensed matter through practical experimental approaches to enzyme structure, dynamics and mechanism, including the new experimental opportunities arising from genetic engineering techniques. Our approach was unashamedly physical, both because the more biochemical aspects of enzymology are amply covered elsewhere and because progress in our understanding and application of the molecular basis of enzymic processes must ultimately come from advances in physical knowledge. We tried to cover as wide a spectrum as possible, and succeeded in gathering an expert and enthusiastic team of speakers, but there are some inevitable omissions. In particular, and with hindsight, our discussions might have been enriched by more detailed coverage of general aspects of chemical catalysis - but readers requiring this background should find adequate references herein.

It is difficult to judge the effectiveness of a meeting such as this, but perhaps one measure of success is that at least three international research collaborations were initiated here. Other benefits are less tangible, but we hope that a greater understanding of the capabilities, and failings, of various experimental and theoretical techniques will further progress in this important area. Perhaps most important of all, the opportunity for younger scientists to meet with established workers, warts and all, in a friendly and informal setting, shows the next generation that science is an intensely human activity, with all its attendant frailities, but is no less exhilarating for that.

We would like to thank the NATO Scientific Affairs Division for their financial and organizational support. Additional financial support and assistance with local organization, in most congenial surroundings, was generously provided by the Comitato Scienze Fisiche and the Gruppo Nazionale Cibernetica e Biofisica of the Consiglio Nazionale delle Recerche (CNR, Italy).

Alan Cooper   Glasgow and Pisa
Julien L. Houben and   1988
Lisa Chien

CONTENTS

(C) Reactivity and Catalysis

(A) <u>Introductory Lectures</u>

*"If a man will begin with certainties, he shall end
in doubts; but if he will be content to begin with
doubts, he shall end in certainties."*

Francis Bacon

*"First listen, my friend, and then you may shriek
and bluster."*

Aristophanes

ENZYME CATALYSIS: AN OVERVIEW FROM PHYSICS

Giorgio Careri

Dipartimento di Fisica
Università di Roma I
00185 Roma, Italy

INTRODUCTION

The organizers of this workshop asked me, as the first speaker, to present "a broad overview/introduction/ philosophy-like oriented lecture, as it appears that many of our collegues still have problems accepting the idea that a theoretical model can be of any use". I have accepted their invitation gladly, but let me say at the start that I share some of the perplexities of my colleagues, and that I am not so naive as to believe that "our ultimate goal must be to develop a general and practically useful model of the total enzyme catalysis process", as it is mentioned among the aims of this NATO-ASI. In the following, I will try to convince you that instead of a model we would do better to focus on a few general physical features of the process in question, each of these features coming into play with different emphasis in specific cases met in the laboratory.

Before entering into business, let me comment on the notion of "general physical features" for a process like enzyme action, which is only one part of the larger one called the living process. There is little doubt that some of these general physical features cannot be accounted for by physics alone, but require the wider viewpoint of biology. Let us take a look at a couple of examples:
1 - Why are enzymes such large and sophisticated macromolecules? An answer to this question can come only from biology: because the living process needs highly reliable kinetic units (e.g. the enzymes) to accomplish its operative tasks, leaving only DNA to introduce appropriate changes to its operative program.
2 - Why do enzymes work at the highest speed compatible with physical laws? Again an answer from physics is impossible, while biology suggests that an intrinsic property of the living process is the continuous urge to change itself and to control itself, and the quicker these operations are carried out the better for survival.

We may ask ourselves if there are general features of enzyme action that can be described from the viewpoint of physics only, even if the enzymes we deal with today result from a large evolution towards some biological goals. My opinion on this matter is optimistic, not because I am a physicist, but because I believe that what is going on in enzyme catalysis is similar to what happens in any physical process involving molecular media.

Therefore let me outline at least 3 physical features which seem

intrinsic to catalytic action, leaving to the following part of this lecture and even more to other lecturers, the task of expanding these statements:
a) While fully accepting the body of structural data offered by biochemistry, we must add that the enzyme macromolecule must have the capability to display some _useful fluctuations_ in its space-charge distribution when coupled to a random medium.
b) In order to operate as a single system, the macromolecule must display some _long-range connectivity_ among its subsystems.
c) Under stresses imposed by the environment or by other subsystems, each subsystem changes its space-charge distribution. Eventually, some of these _induced changes_ can became cross-correlated in time.

SPONTANEOUS FLUCTUATIONS

Although globular proteins have long been considered "floppy bodies", meaning systems capable of displaying a large set of conformational fluctuations, the description of these fluctuations and of their biological significance is a comparatively recent acquistion. Because of the small size of a single macromolecule, transient fluctuations are inevitable, even at thermodynamic equilibrium. This is not a unique property of globular proteins, but here the occurrence of these large fluctuations can have pertinent implications upon their function, for instance as enzymes.

The first evidence that rapid structural fluctuations are present in a large number of globular proteins and enzymes comes from experiments on fluorescence quenching performed in 1973 by Lakowicz ad Weber [1]. Almost contemporaneously I suggested [2] that the ability to time-correlate these fluctuations could be an essential kinetic property of the macromolecule if it is to work as an enzyme. Since then, the notion of the enzyme as a fluctuating unit has gained credit; its experimental and theoretical basis has been progressively strengthened [3) (4].

In order to proceed in our physical description of biomacromolecules near equilibrium in their thermal bath, we shall first take a closer look at the interconnection between three major physical concepts, namely "events" "statistical macro-variables", and "statistical correlations". Let us begin with a proper definition of a statistical macrovariable Y. By this we mean a macroscopic statistical variable that is necessary for describing function at a molecular level. Examples of these macro-variables are the average charge on specific residues at the active site of an enzyme; the concentration of a particular ligand; and so on. Clearly, Y quantities are themselves functions of several microscopic quantities: for instance, the average charge on an active-site residue is a function of the charges and distances of nearby atoms, and so forth. Note that these Y macro-variables are of a truly statistical nature, because the small size of the system necessarily implies a certain number of fluctuations.

Events can now be defined as the measurable change dY of the Y macrovariables. Of course such a measurement is in practice possible only if amplitude and duration can be calculated using available experimental techniques; this means that the energy must be well above average thermal energy. One such typical event is the transition over a free-energy barrier, the measurable quantity being the rate of chemical species produced during the barrier transition. As a rule, only rare large amplitude fluctuations can be considered as measurable events.

Once a set of relevant macro-variables is identified, their cross-correlation can be mathematically expressed as the non-vanishing average of these quantities over a duration time. It stands to reason that in order to be cross-correlated, macrovariables must display close values of self-correlation time, so that some form of coupling can become effective among them. The notion of "correlated events" then emerges in a simple and direct way. As a matter of fact, large amplitude fluctuation of the macro-variables can be described as relevant spontaneous events, that can become cross-correlated if their duration is similar and if their coupling is sufficiently strong. Thus it becomes quite evident that when too many such events become cross-correlated, the complexity limit within this description level may be reached.

Some aspects of enzyme action are truly general; they will be the focus of our concern hereafter. For instance physics aims to describe how free energy is exchanged in order to overcome free energy barriers, and since such barriers operate in a time sequence, physics hopes to understand how this planned sequence of internal events can be achieved in a thermal bath. This implies that an enzyme should be described not only by its spatial but also by its temporal structure, the two merging together in the spacetime fluctuation spectrum of the macromolecule. This ambitious programme has not been achieved as yet. Since the structural aspects of enzymes have been widely considered in the literature, only some of the temporal aspects will be considered below.

Let us start by trying to increase the precision of our notion regarding the statistical time event relevant for catalysis. According to the picture of the fluctuating enzyme, the reaction coordinate ought to be a function of only a few statistical macrovariables, which by high-amplitude spontaneous fluctuations create biological function. Therefore, we must first consider the relationship between the time scale of the spontaneous flutuations of the macrovariables involved and accordingly we must identify the main classes of these relevant macrovariables as a function of the time-data considered above. If this class is relevant for catalysis it must provide the active fluctuation; namely, the high-amplitude fluctuation that occurs rarely and well away from the mean value of the Gaussian curve. In our picture, this active fluctuation is the bottleneck for catalysis. We believe that a free-energy increase of one order of magnitude above the average thermal energy (say, up to 5 Kcal/mole) should be considered sufficient, because this increase must be lower than the net free energy of structure stabilization of a globular protein, which is the order of 10 Kcal/mole. Therefore, active fluctuations towards catalysis must occur with a probability factor around 10 exp -5, and if we take 10 exp -3 as a representative value for the enzyme turnover, we anticipate that some active fluctuations must originate from that part of the frequency spectrum which is centered around 10 exp -8[5].

As this stage we should combine the two main points: on the one hand, the free-energy transducing property of the enzyme between the thermal bath and the active site and, on the other, the value of the common-time constant of the spontaneous fluctuations involved in the transducing processes. This will help us identify the specific classes of fluctuating macrovariables which are relevant for catalysis. Expressed in different and more concrete terms, we should identify some specific process occurring on the macromolecule surface and others occurring at the active site. There is reason to assume these processes can couple, because both have a time constant close to 10 exp-8 sec. This search was performed some time ago [5], and among suitable candidates which could account for processes occurring at the protein surface, we have already pointed out the proton-transfer reaction of bound water, the charge fluctuations of the ionic medium and the side-chain motions.

Once this set of relevant macrovariables is identified, we should proceed to calculate their statistical cross-correlations, in order to derive the natural laws which describe the behaviour of this macromolecule. Such is the final aim of this description level, and it is obviously quite remote because of the non-linear coupling which probably exists among the macrovariables themselves. At this point different coupling regimes may occur, according to the degree of displacement of the system from thermodynamic equilibrium. In practice, the enzyme will use the regime that fits best in the overall course of the functional order. There is no preconceived scheme according to which every enzyme must work; rather, in each instance, nature has evolved the best way to achieve its goal, based upon the laws of statistical physics.

LONG RANGE CONNECTIVITY

A general statistical-physical approach, called the percolation model, has been developed for a wide range of processes where spatially random events and topological disorder are of intrinsic importance. A typical physical application of the percolation theory [6] is to the electrical conductivity of a network of conducting and non-conducting elements. One of the most appealing aspects of the percolation process is the presence of a sharp transition, where long-range connectivity among the elements of a system suddenly appears at a critical concentration of the carriers. My aim here is to show that this model can successfully describe the emergence of catalytic function in a simple enzyme. This is because proteins are often disordered at the microscopic scale; then long-range connectivity between subunits must be established according to statistical laws, and the presence of a threshold should be expected according to the percolation model.

Recent work by our group has shown that powders of lysozyme at low hydration display protonic conductivity and that the conduction process follows the percolation model [7] (8). In this picture, the conductivity reflects motion of protons along threads of hydrogen-bonded water molecules adsorbed on the surface of the macromolecule, with long-range proton movement developing along with the extended network at the percolation threshold.

For lysozyme-saccharide complexes a higher value of the percolation threshold has been found, suggesting that the presence of a "foreign body", where the water bridges may not be favorable for proton transfer, must affect the long-range connectivity on the protein surface. This hydration level, hc=0.25, is close to the critical level for the onset of enzymatic activity in Lysozyme catalysis. The value of the critical exponent found for the saccharide complex suggests that the protonic conduction remains a surface process. This observation is in agreement with the suggestion that preferred paths of proton movement pass through the active site; substrate would be expected to block these without changing the surface character of the percolation.

The percolation transition, although it exhibits some characteristics of a phase transition, differs importantly in that it reflects change in the connectivity within a system without there being also a change in the chemistry of the elements of the system. A percolative transition can be found even if there is no interaction between elements, as in the case of a mixture of conducting and non-conducting spheres. For this example, one can identify in principle two phases, one consisting of those conducting elements which are part of the infinite thread, the other of conducting elements not part of it; however, the chemistry of the conducting elements in the two putative phases is clearly indistinguishable. The percolative transition is detected as a discontinuity in a process, a property of the entire system, for example conduction, the onset of which reflects the onset of long-range connectivity. This is developed stochastically in a randomly-structured system of many elements, as a result of a change in system composition (in the example, change in the fraction of conducting elements). We note that one may develop an alternative description of the conductive regime of hydrated lysozyme powders, as a series of correlated single proton transfers along a random thread of water molecules. Since each single proton transfer can be considered a small-scale event due to a local fluctuation, the large-scale conductivity results from the correlation of very many of these small-scale fluctuations.

We have extended the application of the experimental techniques developed through study of lysozyme to measurement of the dielectric properties of the purple membrane of <u>Halobacterium</u> <u>halobium</u> [9]. The data demonstrate a cooperative transition for proton movement, with a threshold hydration level close to the onset of photoresponse of this system.

The percolation model bears on the transport of protons and small molecules across membranes. This movement is thought to involve, generally, a water channel, although for this discussion it is sufficient to imagine some collection of conducting elements. Several points can be made: (1) The channel or surface need be only partly filled with conducting elements. The percolation threshold, above which there is conduction, is 0.45 fraction filled for a surface, and 0.16 for a 3-dimensional region. (2) Gating might be associated directly with a simple chemical event, such as binding of a ligand or change in state of ionization of a protein group, without its being mediated by a change in protein conformation. If the system is poised near the critical percolation concentration, a small change in the fraction filled would serve to switch the system. A two percent decrease in water content shifts the lysozyme system from a proton conductive to a non-conductive mode. (3) The central focus of percolation theory is the randomness of the arrangement of conducting elements. A membrane channel or conducting surface that operates through a percolation mechanism would need no structure extending over the full thickness of the hydrocarbon core of the membrane. For example, in the case of the proton pump of the purple membrane there need be no proton wire of hydrogen-bonded groups leading from the membrane surface to the active site. Instead, it is possible that protons diffuse to and from the gate along the threads of a fluctuating random percolation network within the protein-lipid interface. The photoreaction at the active site would serve to give a vectorial kick to the proton movement. In this picture, the requirement for a special non-random structure would be limited to the gate, and the size of the active site would be that found to be typical for other biochemical processes such as enzyme catalysis.

INDUCED CHANGES

Broadly speaking, we can call "induced changes" all kinds of changes that one site of a macromolecule displays because of the action of the surroundings. Typically, this is the case of a change induced by binding of a substrate, or by the mutual interactions among nearby sidechains at the active site. All these effects are widely considered in the biochemical literature and will not be a matter of concern hereafter. Instead, what is worth mentioning here is the possibility that the change at one site of the macromolecule can result from the action of one which is far away. Although experimental evidence for this kind of long-range effect is still lacking, it seems quite possible that such effects can take place in some specific cases. The best-known example is the proton transfer along chains of water molecules adsorbed on the protein surface, along a pathway which may also include some sidechains. Of course, in this way some distant side chains relevant for acid-base catalysis can become time-correlated to develop concurrent factors for catalysis. A similar but much more sophisticated long range process can be assisted by such long living excitations as the so called Davydov solitons. However, in spite of the existence of vibrational trapped states in a model system for the amide group of proteins, there is no evidence as yet that this kind of process can be helpful in real enzymes [10].

It seems worthwhile to point out the great relevance of the hydrogen bond in most of the inductive effects considered here. As is well known, the active sites of several enzymes are just an embroidery of hydrogen bonds connecting substrate to side chains, and even the protein surface is but a network of hydrogen bonds connecting absorbed water molecules to the amide backbone. It seems that the hydrogen bond frame which encompasses the whole protein is the best system occuring in nature to produce large amplitude fluctuations (because of the low energy of this bond), long-range connectivity paths (because of the nature of the water molecules and of the amide group), and strong inductive effects (because of the high polarizability of this bond). In other words, most of the general physical features occurring in enzymatic action are rooted in the properties of the hydrogen bond itself.

CONCLUDING REMARKS

i - In the previous sections I have used a dynamical picture of the enzyme which is still tailored to its structure. The design of the macromolecule was assumed to have been optimized by evolution in order to make good use of both thermal disorder and of strucutral order, the first playing the role of a chance carrying factor and the second of a selecting factor. And the flexibility of the enzyme structure we tacitly assumed to allow for the temporal series of space-charge changes. An _irreducible_ _complexity_ of the enzymes's structure thus appears, because the system is made of several interacting subsystems, each capable of spontaneous fluctuations and of mutual inductions assisted by random interconnecting links.
    On the other hand, one can use a description grounded directly on the process itself, so that the _functional_ _complexity_ of the enzymatic action becomes expressed in terms of a series of cross-correlated temporal events [11]. Of course the two above description are complementary to each other, because an event is by definition a change occurring in a structure (or, better, a transition among two states of that structure). Yet the description stressing the process rather than the structural character of the enzymatic action is more abstract and general, and for these reasons should be better used when comparing the enzymatic process to other processes occurring in the living cell.

ii - Let me add something about the future work which seems more interesting from the personal viewpoint developed in this overview. In my opinion, there are at least three lines which are worth mentioning:
a - detecting cross-correlation among fluctuating macrovariables by non-destrctive probes.
b - using computer simulation to reach the fluctuation spectrum centered near 10 exp -8[5].
c - producing by bio-engineering simpler enzymes, to be used as model system ameneable to theoretical treatments.
    I am glad to see that these topics will be a matter of concern for several of the next speakers. The progress we shall make here along these avenues will be a good indication of the success of this ASI.

REFERENCES

1.   J.R. LAKOWICZ, & G. WEBER, G. Biochemistry. _12_, 411 (1973).

2.   G. CARERI, The Fluctuating Enzyme. In: B. KURSUNOGLU, S.L. MINTZ & S.M. WIDMAYER (eds.) "Quantum Statistical Mechanics in the Natural Sciences". Plenum, New York. pp. 15-35 (1974).

3.   G. CARERI, P. FASELLA & E. GRATTON, Ann. Rev. Biophys. Bioeng. _8_, 69 (1979).

4. G.R. WELCH, B. SOMOGY & S. DAMIANOVICH, Progr. Biophys. Mol. Biol. 39, 109 (1982).

5. G. CARERI, P. FASELLA & E. GRATTON, Crit. Rev. of Biochem. $\underline{3}$, 141 (1975).

6. R. ZALLEN, "The Physics of Amorphous Solids" John Wiley & Sons, New York, (1983).

7. G. CARERI, A. GIANSANTI & J.A. RUPLEY Proc. Nat. Acad. Sci. U.S.A. $\underline{83}$, 6810, (1986).

8. G. CARERI, A. GIANSANTI, J.A. RUPLEY Phys. Rev. A ($\underline{37}$, 2703) (1988).

9. J.A. RUPLEY, L. SIEMANKOWSKI, G. CARERI & F. BRUNI Proc. Nat. Acad. Sci. U.S.A. (in press) (1988).

10. G. CARERI, U. BUONTEMPO, F. GALLUZZI, A.C. SCOTT, E. GRATTON, E. SKYAMSUNDER, Phys. Rev. B $\underline{30}$ 4685 (1984).

11. G. CARERI "Order and Disorder in Matter" Benjamin Cummings, Palo Alto. Ch. 5, (1983).

ENZYME CATALYSIS: THE VIEW FROM PHYSICAL CHEMISTRY

Alan Cooper and Julien L. Houben+

Department of Chemistry, Glasgow University
Glasgow G12 8QQ, Scotland, U.K., and +Istituto
di Chimica Quantistica ed Energetica Molecolaré
CNR, Via del Risorgimento 35, I-56100 Pisa, Italy

The role of an introductory lecturer is a difficult one. There is a quite natural human temptation to attempt to appear erudite and profound about the subject, but this must be balanced by an awareness of the diverse backgrounds and different areas of specialisation of the audience. Even though we all share a common scientific language (expressed in various degrees of "fractured English"), there are many dialects associated with the various sub-disciplines (physics, chemistry, biology...) which can make communication difficult. This introductory lecture, therefore, is merely an attempt to set the scene for what what follows and to establish some common ground. But we should caution from the start that the title might mislead one into the belief that there is an agreed consensus concerning the understanding of enzyme catalysis from the physico-chemical point of view. This is not the case, and no such global understanding exists. Indeed, if it did, then there would be no need for a meeting such as this. This is a healthy situation indicative of the intellectual excitement and practical importance of the topic in current times when biotechnology offers so much promise. So, in what follows, we do not pretend to give any comprehensive view of the physical chemistry of enzyme catalysis, rather we will present a series of basic ideas - some well established and others rather less so, some abstruse and some trivial - which will form a backcloth for the more detailed lectures that follow. Few, if any, of these observations will be original, and further details and background references may be found in the Bibliography at the end of this chapter.

Enzymes are, of course, a product of evolution and natural selection, and a consequence of the biological need for specific, efficient and controllable catalysis within complex chemical systems. They are large dynamic molecules made up of polypeptide chains* that

---

* Though some RNAs, together with other synthetic and natural polymers, have also been shown to have enzyme-like properties.

fold into more or less unique structures which provide specific binding
sites for substrates, cofactors, and so forth, within which certain
chemical reactions may be catalysed with rates and specificities that
are the envy of any synthetic chemist. How and why this is so raises
many questions which should, in principle, be answerable in physico-
chemical terms. For example:-

<u>Protein Structure</u>: What is it, and what does it have to do with
    catalysis ? How can we modify it usefully ? How and why does it come
    about ? Can we predict structure from amino acid sequence ?

<u>Protein Dynamics</u>: How big and how fast are the motions that take place
    within a protein molecule ?   And so what ?   Do these motions
    have anything to do with any biological function ?

<u>Ligand Binding</u>: What forces govern the affinity and specificity of
    protein binding sites ?   Are there any basic rules of molecular
    recognition that would allow us to design new enzymes or new drugs ?

<u>Enzyme Catalysis</u>: This is the major problem - why are enzymes so
    efficient ? But, is there really a problem and, if so, whose is it ?
    In other words: are any new physical principles involved in enzyme
    catalysis, or do we just need a better understanding of chemistry in
    general ? (We incline to the latter view.)

<u>Catalytic Control</u>: Why are enzymic reactions usually so stereospecific ?
    And what is the origin of other important control functions such as
    allostery and cooperativity ?

These questions illustrate the sorts of topics that will be tackled in
detail in later lectures. But before we get onto that we must first
review some of the basics.

1. ENZYME BINDING and KINETICS

No enzyme-mediated catalysis can take place without binding of
substrate molecules (S) to specific sites on the enzyme (E). This leads
to the classic (1913) model of Michaelis and Menten for the description
of enzyme kinetics which, in simplest terms, may be written as follows:

$$E \; + \; S \; \xrightleftharpoons{K_M} \; ES \; \xrightarrow{k_{cat}} \; E \; + \; P$$

where the assumption of a rapidly-established binding equilibrium gives
the usual hyperbolic expression for the initial rate of the enzyme
catalysed reaction, v, in terms of total enzyme and substrate
concentrations, $[E]_0$ and $[S]$ respectively:

$$v \; = \; d[P]/dt \; = \; k_{cat}[E]_0[S]/(K_M+[S])$$

Most enzymes appear to follow this mechanism, with possible
modifications for multiple substrates, and K_M is usually a reasonable
indicator of substrate dissociation constant (K_S), i.e.

$$K_M \; \approx \; K_S \; = \; [E][S]/[ES]$$

but there are exceptions[1]. For example, in cases where the enzymic
mechanism involves formation of several enzyme-bound intermediate

species in significant amounts, K_M represents the overall dissociation constant of all such species. Similarly, in some very efficient enzymes the reactions are rate limited by the initial substrate binding step (that is k_{cat} approaches the diffusion limited value usually estimated to be about 10^8-10^9 s^{-1}M^{-1}) so that, although these systems still appear empirically to follow Michaelis-Menten kinetics, K_M is not directly related to the substrate dissociation constant. Non-productive substrate binding may also lead to spurious estimates of K_S.

Various kinds of inhibition due to competitive or non-competitive binding of other, non-substrate molecules can be easily incorporated into these simple kinetic schemes and this can lead to useful information about enzyme-inhibitor dissociation constants (K_I) and so forth.

The specificity or selectivity of an enzyme for catalysis of different competing substrates depends on both K_M and k_{cat} and is usually expressed in terms of the specificity constant k_{cat}/K_M. Thus enzyme specificity is enhanced both by increasing k_{cat} and by lowering K_M (i.e. tighter binding). But there are obvious limits, imposed essentially by thermodynamics, as to how far this may go. Bearing in mind that enzymes are strictly reversible catalysts which must work equally well in both directions without affecting the overall equilibrium of the free reactants and products:

$$S \rightleftharpoons P \quad ; \quad K_{Eq} = [P]/[S] \; ,$$

though the equilibrium ratio in the enzyme active site can, and frequently is affected. This leads to the classic Haldane relationship between kinetic parameters for forward and reverse reactions:

$$k_{cat,S} K_{M,S}/k_{cat,P} K_{M,P} \;\; = \;\; K_{Eq}$$

Thus these parameters may not be varied independently. In particular, making $K_{M,S}$ lower, for example to improve specificity in the forward direction, may be counterproductive as it might force tighter binding of product ($K_{M,P}$ lower) and result in an overall reaction rate limited by release of product.

2. INTERMOLECULAR FORCES

Protein folding and ligand interactions are dominated by a range of quite weak, non-covalent interactions: hydrogen bonds, Van der Waals (dispersion) forces, hydrophobic and electrostatic interactions. But, whilst the names are now familiar, the essential thermodynamic nature of these forces and the extent to which they are dominated by solvent (especially water) effects are not always appreciated. When considering, for example, the stability of a folded protein or its complex with some other molecule, we must look at the overall process and consider the thermodynamic free energy change ($\Delta G = \Delta H - T.\Delta S$) for the transition. This free energy change will contain contributions not only from the energy (strictly enthalpy, ΔH, at constant pressure) changes that might be computed from intermolecular potential energy diagrams, but also the possibly comparable and frequently dominating entropy changes (ΔS) arising from solvation and configuration effects. This is most familiar for the hydrophobic interaction which arises not from any particular attractive force between non-polar groups, but rather from their mutual repulsion from solvent water, and which is a consequence of entropic effects in this highly structured solvent. But similar complications arise even for apparently simpler interactions. Take, for example, the hydrogen bond and imagine the process of folding a polypeptide chain or the binding of a substrate molecule to an enzyme active site. In both cases, molecular groups which are initially exposed to water will become buried and make interactions with other protein groups. H-bonding groups will, therefore, simply change from making bonds with the solvent to similar bonds within the protein. The overall number of hydrogen bonds involved will not change and, if the bond strengths are similar, there will be no nett energy change, and any stabilizing free energy must come from favourable entropic effects due to release of solvated water, for instance. Similar considerations apply equally to electrostatic or dispersion forces where the balance of effects during the transition from exposure to water to burial within the protein must be taken into account. As a consequence of these solvent effects we have only a poor quantitative understanding of the thermodynamic forces central to protein and enzyme interactions. Even more disturbing, we have no rational way of writing down realistic pair potentials for any of the forces in a suitable analytical form that might be of use in molecular mechanics or computer simulations.

Simple pair-wise interaction potentials may not be appropriate anyway in some cases because of cooperative, long-range interactions. Studies on homopolypeptides have demonstrated that stereoregular structures, α-helices for example, are unstable at low degrees of polymerization (n < 20 to 30), even with favourable amino acids. This is due to the fact that the first three residues in such a configuration have parallel permanent dipoles. After the first turn of the helix, some head-to-tail interactions and hydrogen bonding take place and this progressively stabilizes the structure. In proteins, stereoregular structures are often short (about 5-20 residues) and often conform to the sequence/composition rules suggested from homopolymer studies. But even here there are many exceptions. The long-range forces within folded proteins (i.e. those forces exerted between the side chains of residues which are located great distances apart along the polypeptide backbone, but physically close because of folding) can apparently be sufficient in some cases to oppose the short range forces (i.e. between neighbouring residues) and frustrate predictions.

Despite recent advances in the rational incorporation of solvent (H_2O) into molecular mechanics and dynamics calculations of protein-ligand interactions, for example, the understanding of intermolecular forces in proteins remains primarily an empirical problem. The rigorous and systematic study of the thermodynamics of such interactions is only now getting under way.

3. A TEXTBOOK EXAMPLE: THE SERINE PROTEASES

Mechanistic studies are beginning to provide a clearer picture of the chemical steps involved in catalysis by a range of various enzymes. Perhaps the best documented case is that of the serine protease family (trypsin, chymotrypsin and elastase) responsible for catalysis of hydrolysis of peptide bonds, where structural considerations and analogies with small-molecule organic chemistry have led to a picture, at least, of what might be going on.

Some idea of the basis for molecular recognition by these proteins can be gained from examination of the available structures. These three proteins are quite similar in overall conformation, and share common features in the active site - primarily an active serine residue (number

195 in the chymotrypsin sequence) thought to be central to the hydrolysis mechanism in each case. But there are significant differences in the region of the active site cleft, particularly residues 189, 216 and 226, which can at least qualitatively explain their different substrate specificities:

In chymotrypsin these residues are Ser-189, Gly-216 and Gly-226, all relatively small groups which give a fairly deep and wide active site cleft (1.0 to 1.2nm deep and 0.35-0.4 by 0.55-0.65nm across) which can thus accommodate aromatic amino acid residues of the substrate.

In elastase the glycine residues are replaced by the larger valine-216 and threonine-226 respectively, so that the site may only accommodate substrates with smaller amino acids.

In trypsin the glycine residues are unchanged, but the neutral serine is replaced by an aspartate-189 group, thereby introducing a strong electrostatic interaction that must be matched by a appropriate lysine or arginine on the substrate.

Comparisons such as these between naturally occurring homologous proteins can rationalize differences in substrate binding affinities, and are now being complemented by a range of site-directed mutagenesis studies, but we are still a long way from being able to predict the magnitudes of these affinities in any systematic manner. Much more quantitative work is clearly needed.

Following formation of the non-covalent enzyme-substrate complex, the catalytic process in these serine proteases may be described in two discrete steps. The first involves a concerted proton double transfer (or "charge relay") involving the active serine-195 and the residues His-57 and Asp-102 of the protein in the base-catalysed nucleophilic attack of the serine hydroxyl group on the carbonyl carbon of the substrate. This results in the formation of an unstable tetrahedral intermediate which collapses to give a covalent acyl-enzyme complex, releasing the C-terminal portion of the peptide substrate. The second step involves a reversal of this process involving, this time, a water molecule acting as the nucleophile to hydrolyse the acyl intermediate *via* another base-catalysed, tetrahedral intermediate transient process which releases the remaining portion of the substrate and returns the enzyme to its original state.

This is of course only a description of what probably happens and, by no means, an explanation of the source of catalytic rate enhancement in these enzymes. (It is even incomplete as a description because there are several more protein groups which might be involved in the process.) But it does illustrate some of the features that might be common to the enzymic process in general. For example, there is no new chemistry involved - all the steps are familiar to physical organic chemists from small molecule reaction mechanistic studies. However, the mechanism is different from that which would be likely to occur during the uncatalysed hydrolysis of a peptide bond involving direct nucleophilic attack by water. Moreover, the reactions in the enzyme take place in an essentially non-aqueous environment, paradoxically for a hydrolysis process, but this might be a significant factor in explaining the source of rate enhancement and, in many enzymes, is crucial to the specificity of the catalysed reaction. (In the case of the kinases, for example, which involve the transfer of phosphate groups from one molecule to another, it is essential that solvent water be rigorously excluded during the catalytic act otherwise hydrolysis and inorganic phosphate production, rather than transfer, would predominate.)

4. SOURCES OF RATE ENHANCEMENT

As indicated by the examples above, enzymic processes require that certain functional groups are in the right place at the right time. But this is equally true of un-catalysed reactions. Yet enzymic reactions are very fast, with k_{cat} typically 10^8-10^{12} times faster than the equivalent non-enzymic reaction - though this can be hard to quantify since the un-catalysed reactions are often too slow to measure under comparable conditions and the mechanisms frequently differ. So, what is the source of this vast enhancement of catalytic efficiency in enzymes ? There is probably no single answer, and many apparently different suggestions have been made over the years - though many of them are really just alternative expressions of the same basic concept[3-5].

Perhaps the most significant contribution to rate enhancement comes from what have been variously and loosely called "entropy", "proximity", "propinquity", "approximation" effects, and so forth. The basic idea here is that, whereas in a non-enzymic reaction various groups must come together (either simultaneously or in some specific sequence, and in some specific orientations) every time a molecular reaction occurs, in an enzyme active site most, if not all the functional groups are already present in the appropriate stereochemical arrangement for the reaction to proceed (within the constraints imposed by the dynamics of the protein structure). In simple terms this means that the entropy of activation for the reaction in the enzyme is vastly reduced, and this is "paid for" by the free energy of folding of the polypeptide chain. Initial estimates for the magnitude of this effect suggested a rate enhancement by a factor of 55 per catalytic group (arising from the molarity of water and imagining every water molecule replaced by a functional group), but careful later analysis based on the study of intramolecular reactions in small molecules[4,5] has shown that this is a naive and severe underestimate, and that rate enhancement factors of upto 10^8 are feasible.

But simply arguing that reactive groups are optimally placed in an enzymic reaction does not constitute a description of the catalytic process. The actual chemical process may involve intra-molecular strain or transition state stabilisation, both of which may contribute to rate enhancement by altering activation energy parameters for example. Electrostatic and other environmental effects must surely also play a part through their effects on pK_A's of reactive groups, bond polarization, and so forth - but in a sense these are simply a consequence of the initial proximity of active site groups.

A class of theories based on somewhat different principles has developed in recent years[8]. These are based on the known dynamic properties of protein molecules and suggest that, in some way, thermal fluctuations in the system might be focused or concentrated in the region of the enzyme active site to provide the extra energy required to overcome activation energy barriers. The problem with such theories is that they are difficult to quantify and test experimentally. Moreover, they appear, at first sight, to violate the second law of thermodynamics. Perhaps only when it has been shown that there is no alternative explanation of enzyme behaviour should we take such a dramatic step. Although it is reasonable to expect that the rate and form of the thermal fluctuations in an enzyme active site might be influenced by the structure of the surrounding protein, and that this might affect the catalytic functions (see below), any such effects must conform to the conventional rules of physical chemistry.

5. ENZYME CATALYSIS AS A CHEMICAL REACTION

Up to this point the actual chemical event - i.e. the point at which covalent bonds are changed - has hardly been touched upon, even though it is in fact the key event. Later lectures will be dedicated to the general theory of chemical reactions in the condensed phase, but a few introductory remarks are necessary here. Let us consider, for the sake of illustration, a co-linear chemical reaction:

$$XY + Z \rightleftharpoons X + YZ$$

where X,Y and Z are heavy groups so that tunnelling effects may be ignored. The phase space can then be divided into two subspaces: one for the reactants XY and Z, and one for the products X and YZ. When the two species are sufficiently far apart, any mutual interactions can be neglected and the potential surface of the transferable particle, Y, may take the form of two independent and deep potential wells, separated by a large barrier. But as the distance between the species is progressively reduced, the mutual interactions increase and, in the case of mutual attractions, the potential functions merge to two wells with a relatively small barrier between them. In such simple cases the chemical reaction can be seen as the result of a particle passing from one potential well to another. Any satisfactory description of this process must be based upon the following points:-

(a) An analysis of the collision process, and in particular of its steric and temporal aspects. Do the interactions tend to predispose the system to certain relative orientations of the species during collision or not? Is the collision time sufficient to permit any molecular dynamics, such as energy exchanges or structural reorganizations, to take place ?

(b) Reliable potential surfaces, since these directly influence the fate of the reactants during the collision. This requires a very good knowledge of: the inter- and intra-molecular forces which fix the frequencies of the modes active at the bottom of the potential wells; the interactions between the two species in the transition region, since these fix the exact form of the saddle and the transmission coefficient at the curve crossing; and the time response of the surrounding matrix to the evolution of the species during the course of the reaction, since this will govern the contribution that viscosity and damping effects may make to the process.

(c) The mechanism and efficiency of the energy exchange between the various modes: since the potential barrier is normally significantly higher than thermal energy it is necessary to concentrate a fair amount of energy on the reaction coordinate (as a first approximation this can be associated with the chemical bond involved) for the reaction to occur.

In addition to these problems of chemical kinetics in general, in the enzyme catalysis event the reactions are taking place within a rather specific matrix, and a satisfactory theoretical model must take account of the additional factors that this implies:

(d) The collision process and the chemical reaction may be divided into two distinct phases. The former may be a fairly slow process, first requiring some stereospecific interactions between the macromolecule and its substrate and, in many cases, a significant restructuring of both

the enzyme and substrate after complex formation. The chemical reaction
then occurs in this predisposed and long-lived, but dynamic, state.

(e) The potential surface: enzyme rigidity and flexibility can, in
principle at least, allow the correct interplay between the best
possible stabilisation of the intermediates at the saddle point (that
is, the greatest possible decrease in the potential barrier) and the
continuous adaptation of the active site to the geometry and polarity of
the reacting species during the course of the reaction.

(f) The energy exchange: enzymes have an anisotropic structure which
can, maybe, be used to "focus" energy on precisely one spot. Since the
magnitudes of energy fluctuations are determined by the size of the
enzyme, these fluctuations may well have the most suitable amplitude and
length to optimize the rate constants. (But note that the thermodynamic
validity of such energy focusing schemes has yet to be established.)

6. ENZYME STRUCTURAL FLUCTUATIONS

Protein molecules are relatively small ("mesoscopic") systems in
physical terms, subject to quite intense thermal fluctuations arising
from collisions and interactions with the surrounding solvent
environment[6,7]. Consequently, when considering detailed structural
mechanisms in enzyme catalysis it must not be forgotten that these
structures are dynamic. Despite the closely packed structures that
emerge from crystallographic studies, the folded forms of protein
polypeptides are only marginally stable and a range of experimental and
theoretical work has shown that biological macromolecules fluctuate over
a large number of readily accessible conformations under normal
conditions. The magnitudes of these fluctuations are such that
individual groups within a protein may undergo relative movements of
quite large distances (1-2Å, or more) over times ranging from
picoseconds to seconds. The forces that hold enzymes together - hydrogen
bonds, hydrophobic, electrostatic and dispersion interactions, etc. -
are mostly fairly weak, especially in a polar solvent like water, and
frequent local fluctuations in these "bonds" are to be expected. But
even small deformations of just a few dihedral angles (an almost
isoenergetic process even in stereoregular regions such as α-helix and
β-sheet) can induce large conformational changes, and interdomain
movements are not expected to be opposed by any significant potential
barrier. Such movements are, however, expected to be strongly coupled to
the environment because of the changes and rearrangements of surface
groups and viscous, frictional interactions with the surrounding
solvent. The overall magnitudes of thermal fluctuations in these systems
can be determined from thermodynamic properties[6], and the detailed
motion at the molecular level can be explored using molecular dynamics
simulations[7].

The functional relevance of conformational fluctuations in proteins
has not yet been fully explored, but it is difficult to imagine that
evolution has not taken advantage of the possibilities that this allows
to develop additional facilities that are not available in small
molecules. One area where conformational fluctuations must surely play a
major role is in cases where ligand binding sites are fully or partially
buried within the protein matrix. Such situations are well known in some
proteins where it is essential to isolate the ligand from surrounding
water to avoid competing reactions, for example in the oxygen binding
sites of haemoglobin or myoglobin where oxidation ("rusting") of the
haeme iron would take place in the presence of water or, again, in the

kinases to avoid nucleophilic competition by water for the transferrable phosphate group. Indeed, exclusion of water may contribute to catalytic rate enhancement in enzymes in general. But all these cases raise the problem of how the substrates and products can gain easy access to or from these site without compromising their isolation from the bulk solvent. Transient conformational fluctuations, which can open up gates or channels within the protein, are clearly the answer here. Less clear is the possible role of dynamic processes in other enzyme functions. One serious possibility is at the level of control, that is allostery and cooperativity, where it has been established in theory, at least, that dynamic conformational fluctuations can act as a means of transmitting information across macromolecules *via* a mechanism complementary to the conventional scheme involving static conformational changes[6]. But more controversial is the postulated involvement of fluctuations in the catalytic act itself, where it has been suggested that the relatively large energy fluctuations within enzyme molecules may somehow become focused onto the active site and thereby help overcome the kinetic activation barrier for the reaction[8]. The problem with such schemes is that they appear, superficially at least, to defy the second law of thermodynamics by requiring spontaneous formation of local hot spots or temperature fluctuations in an equilibrium, isothermal system.

The magnitudes and molecular nature of protein fluctuations will be considered in detail in later lectures, but it might be interesting to take note of a few rough numbers at this stage in order to remind ourselves of the global physical properties of the average enzyme. An enzyme may be considered in many cases as a multi-domain structure with the active site cleft located in an interdomain region. Within the body of a single domain the atoms and groups are packed with a density approaching that of typical crystalline organic solids – the packing density is about 0.75 for proteins, compared to the theoretical limits of 0.71 and 0.91 for close-packed spheres and cylinders respectively. This implies that the dynamic motions of chains or groups within these regions cannot be pictured as a smooth sliding motion because of the intimate intertwining and interdigitation of the groups involved. In fact, it more closely resembles the motion of "Lego-like" structures whose single elements, in order to move, must first disentangle themselves from the general framework. Such movements might be described as a series of discrete elementary steps which must occur in a precise time sequence. On the other hand, interdomain movements are probably more like the smooth but unpredictable Brownian diffusion of two heavy particles which are only very loosely bound: the major limitation limitation on their movement being that they must remain confined within a certain volume of the configuration space. Some estimate of the likely timescale of these processes may be made as follows.

Consider an enzyme (or globular domain) having a molecular weight of ≈ 25000 daltons, assuming it to be a sphere with a density (d) of about 1.3 g/ml, a typical heat capacity (c_P) of about 1.3 $JK^{-1}g^{-1}$, and a thermal conductivity (λ) of about 2×10^{-3} J $s^{-1}cm^{-1}K^{-1}$ at room temperature. Its radius (r) is about 2 nm and it contains ≈ 250 amino acids (corresponding to ≈ 3000 atoms), up to 60% of which might be typically found in stereoregular regions (α-helices or β-sheet structures). $kT \approx 26$ meV (2.5 kJ mol^{-1}) at room temperature, and the internal thermal energy content of the enzyme can be estimated to be about 50 eV (5000 kJ mol^{-1}), distributed according to the Bose-Einstein law almost entirely among the various vibrational modes and pseudo-vibrational motions of the polypeptide. Due to interactions with the matrix which solvates this enzyme, the energy actually fluctuates about this mean value in an approximately Gaussian manner with a root mean-square fluctuation of about ± 1.7 eV (165 kJ mol^{-1}). But how fast do these fluctuations occur ? Using classical thermal diffusion theory, the fluctuation length (relaxation time) may be calculated as:

$$\tau \quad = \quad r^2 c_p d/\lambda \quad \approx \quad 30 \text{ psec}$$

for uncorrelated thermal motions in the protein. Another estimate, more relevant to harmonic modes within the protein, comes from realising that $kT \equiv 200$ cm^{-1} at room temperature. Consequently, normal modes of frequencies greater than 200 cm^{-1} will rarely be excited and the majority of molecular motions will occur at lower frequencies, that is on time scales slower than about 0.1 psec. Large scale conformational fluctuations involving correlated inter-domain motions ("hinge-bending") will be dominated by solvent viscosity and damping effects and are likely to be of order 10 psec, or slower.

It is easy to show, using simple statistical theory, that these numbers are reasonably consistent. The number of enzyme-water collisions per second is given by $N_{coll} = k_{coll}MR$, where k_{coll} is the collision rate constant, M is the molarity of water and R is the ratio between the surface areas of the enzyme and a water molecule. This gives a mean value of about 100 collisions per psec as a reasonable estimate for this model enzyme. As each water molecule has an energy of about 3kT, the rate of energy exchange can be assumed to be 3kT/2 per collision. Assuming that the energy exchange does not depend too much on the relative energy content, the energy fluctuations in 10 psec - that is, the difference between the energy produced and the energy taken away - is directly given by the fluctuations in the difference between the collisions of each type, i.e. 50 collisions/10 psec giving energy fluctuations of about 2eV, consistent with the thermodynamic estimate above.

7. CONCLUSION

Description of enzyme catalysis is a multidisciplinary problem. Reduced to its bare essentials, it can be said that enzyme catalysis is a chemical reaction catalysed by a macromolecule. This macromolecule is structurally organized, but it is not rigid on the time scale of the crucial steps of the catalytic event and, due to its mesoscopic dimensions, it is the seat of relatively large and long-lasting energy fluctuations. Consequently a range of disciplines, including the following, must be involved:-

(a) Chemistry: in order to understand the chemical reaction itself in terms of dynamics, reactivity and catalysis. It should be noted that some of these aspects are still very empirical - for example, chemical reactivity is not very amenable to quantitative treatment on a theoretical basis - so any model for enzyme catalysis is bound to be semi-empirical in nature.

(b) Spectroscopies and X-ray crystallography: to determine the structure and dynamics of the enzyme and the enzyme-substrate complex. The simultaneous use of various spectroscopic techniques is essential here, as no single technique is able to give a complete and unambiguous picture.

(c) Molecular mechanics and dynamics: to employ the power of modern computational techniques in simulating and describing the processes of protein folding and dynamics, substrate recognition, binding and catalysis. But care must be taken not to be too seduced by the apparent graphic detail available from such simulations - they are only as good as the empirical or intuitive force fields used and the numerical approximations enforced by computational constraints.

(d) Statistical physics: to explore the fundamental basis of kinetics and thermodynamics of mesoscopic and condensed systems. In particular, to determine the limits imposed by the fundamental physical laws on rate processes in macromolecular systems and, equally important, to confirm the thermodynamic validity of any empirical models of enzyme catalysis.

So, to conclude, what is the physical chemists' view of enzyme catalysis ? As indicated at the very beginning, there is no consensus as yet and a range of views has been indicated here. But it is probably fair to say that majority opinion would hold that there is really nothing particularly novel about enzymes from a fundamental point of view. That is not to say that we understand them, because we don't, but that the processes that take place within enzyme molecules do not involve any new fundamental physical laws. They are, rather, highly refined examples of chemical rate processes in condensed matter in general, and this degree of refinement and catalytic efficiency has come about in the course of evolution to take full advantage of whatever the fundamental laws allow. This only pushes the problem one stage back, however, since we still lack a comprehensive theory of chemical kinetics in condensed matter that would give us a quantitative description of the special case of enzymes and that would allow us to design novel catalysts that could match them. Paradoxically, it may well be that the study of enzymes might be rather easier, both for the experimentalist and the theoretician, than the more general problem since, by virtue of the known (or, at least, knowable) structure of proteins, the number of potential degrees of freedom is much reduced compared to, say, a simple liquid phase reaction. Moreover, the diversity of available enzymes and, even more, the availability now of site-directed mutagenesis techniques allows much greater scope for testing of theoretical hypotheses.

REFERENCES

1. A.R. Fersht, "Enzyme Structure and Mechanism", W.H. Freeman, San Francisco (1985).

2. W.P. Jencks, "Catalysis in Chemistry and Enzymology", McGraw-Hill, New York (1969).

3. J. Kraut, "How Do Enzymes Work ?", Science $\underline{242}$, 533-540 (1988).

4. M.I. Page (editor), "The Chemistry of Enzyme Action", Elsevier, Amsterdam (1984).

5. M.I. Page & A. Williams, "Enzyme Mechanisms", Royal Society of Chemistry, London (1987).

6. A. Cooper, "Protein Fluctuations and the Thermodynamic Uncertainty Principle", Progr.Biophys.Mol.Biol. $\underline{44}$, 181-214 (1984).

7. J.A. McCammon & S.C. Harvey, "Dynamics of Proteins and Nucleic Acids", Cambridge University Press, Cambridge (1987).

8. B.Somogyi, G.R.Welch & S.Damjanovich, "The Dynamic Basis of Energy Transduction in Enzymes", Biochim.Biophys.Acta $\underline{768}$, 81-112 (1984).

(B) <u>Protein Structure, Spectroscopy and Dynamics</u>

*"When you can measure what you are speaking about
and express it in numbers, you know something about
it; but when you cannot measure it, when you cannot
express it in numbers, your knowledge is of a meagre
and unsatisfactory kind. It may be the beginning of
knowledge, but you have scarcely in your thought
advanced to the stage of science."*

William Thomson (Lord Kelvin)

"E pur si muove."

Galileo Galilei (attributed)

PROTEINS: INTERACTIONS AND DYNAMICS

R.L. Somorjai

Division of Biological Sciences
National Research Council of Canada
Ottawa, ON Canada

INTRODUCTION

Enzymes (proteins) have characteristic, stable three-dimensional structures. Despite their apparently miraculous feats of discrimination and catalytic power, enzymes are stabilized and functionally primed by the same physical forces that operate on less glamorous systems such as liquids and
solids. Nevertheless, it is important to characterize these forces as they
appear to act in proteins since first-principle quantum mechanical calculations on such large systems are entirely impractical. We are faced with
inevitable approximations and simplifications when attempting to calculate
protein structure and dynamics.

The first essential approximation replaces quantum mechanical atom-atom
interactions by classical and empirically derived atom-atom potentials. Typically, a convenient functional form for the potential is chosen and its parameters optimized to fit a wide variety of experimental data that were determined for small systems. The pious hope is that these potentials arc transferable to proteins. (Whether such transfer is successful is debatable.)
Implicit in the switch to classical mechanics is the assumption that quantization effects are negligible at biological temperatures. (This is probably a
safe enough assumption, except when we try a detailed computation of the
enzymatic action. Vibrational quantization, tunnelling etc may then become
critical and have to be dealt with directly.)

Although we do not really know how to calculate many-body potentials
accurately and effectively, in protein structure and dynamics calculations
not only 2- but also 3- and 4-body nonbonded interactions are used. Of
course, the pair-potentials are effective ("dressed") pairs: many-body
effects are included in an averaged form by fitting the potential parameters
to experimental data (which obviously contain all interactions). However, the
explicit 3- and 4-body effective interactions are computationally useful.
They include bond-angle and torsional potentials and will be discussed later.

The interactions that we typically encounter and compute in proteins
can be classified in a number of ways. Which of these classifications is more

useful or appropriate is obviously context-dependent. The following classification scheme is particularly relevant for macromolecules, considering the extent and nature of the various simplifications one has to resort to in the simulation studies.

A. CLASSIFICATION OF INTER- AND INTRAMOLECULAR INTERACTIONS IN PROTEINS

The interaction is

    I . Solvent-independent (intrinsic to protein)
    II. Solvent-influenced
    III. Solvent-dominated ("hydrophobic")
    IV. Direct solvent-protein.

A.I. Solvent-Independent Interactions

These can be further classified into

    1. covalent
    2. noncovalent ("nonbonded").

A.I.1. According to this sub-classification all chemical bonds that do not dissociate under normal biological conditions belong to the covalent class. If we want to preserve the primary protein structure, we have to assume that these chemical bonds never break.

We shall not consider covalent bonds in any great detail, except to list the covalent atom-atom interactions one encounters in protein studies:

C – C	C – N	C – O	C – H	
N – H	O – H	S – H	C – S	S – S

The covalent disulfide bridge S – S, which connects two half-cysteines, plays a particularly important role in the stabilization of many proteins.

The vibrational (stretching) potential functions generally used for covalent bonds are harmonic:

$$V(r) = K(r - r_0)^2,$$

where r_0 is the equilibrium distance for the bond. Of course, all covalent bonds are assumed to be universal in that they are independent of the environment in which they occur.

A.I.2. The most important empirical nonbonded pair potentials, the van der Waals atom-atom potentials are characterized by the fact that the overall interaction is made up of two opposing tendencies as a function of interatomic distance r: a repulsive part at short distances and an attractive portion at longer ranges of r.

The total van der Waals attraction between neutral molecules arises
from several contributions, depending on the nature of the molecules. These
are:

1. Dipole – dipole (Keesom)
2. Dipole – induced dipole (Debye)
3. Induced dipole – induced dipole (London).

Of course, only the London dispersion forces are operative on the atom-
atom level, although all three have dominant r^{-6} dependence. The London interaction energies depend on the polarizabilities and ionization potentials of
the atoms involved. However, in simulations one uses empirically fitted interaction strengths A:

$$E_{atr} = -A / r^6, \qquad \text{negligible for } r > 8 \text{ A.}$$

When the two atoms approach each other, the overlap between the atomic
charge densities generates a rapidly increasing repulsion. This is much more
difficult to calculate theoretically. One frequently assumes exponential- or
inverse power-dependence:

$$E_{rep} = B / r^n, \qquad 5 \leq n < 13$$

$$E_{rep} = b\, e^{-cr}, \qquad 4.30 \leq c \leq 4.75.$$

The total van der Waals potential is the sum of the attractive and repulsive
parts:

$$E_{vdW} = E_{atr} + E_{rep}.$$

Due to its computational convenience, the <u>Lennard-Jones</u> or 6-12 potential is
the potential of choice in protein simulation studies.

Whatever the pair potential, its parameters will depend on the nature
of the interacting atoms. The parametrizations are done on a finite number of
<u>like</u> atom pairs (e.g. C ⋯ C, H ⋯ H, O ⋯ O, etc). The mixed atom-atom
interactions (e.g. O ⋯ N) are calculated from the constituent like-like
terms, using their geometric means (this is the best one can do without knowing and using the polarizabilities). A common refinement is to extend the
original primitive list of atom pairs (H ⋯ H, O ⋯ O, N ⋯ N, C ⋯ C and
S ⋯ S) to a dozen or more. Thus the nature of the adjacent atom(s) to which
the atom in question is <u>covalently</u> bound can be taken into account. As an example, Momany & al. distinguish 4 types of H ⋯ H interactions : aliphatic,
aromatic, hydroxy or carboxylic and secondary amide or amine.

Typical equilibrium separations for the pair potentials range from 2.7
to 4.7 A, while energies at the minima of the potentials are in the 30 – 200
cal/mole range.

The total van der Waals interaction of the protein molecule in a given
conformation is calculated as the sum over all distinct pair interactions.

The <u>bond bending</u> potential is the simplest of the many-body potentials
that are traditionally approximated empirically. It involves three adjacent

atoms connected by two covalent bonds. It is commonly represented in a quadratic approximation:

$$E(\theta) = K_\theta (\theta - \theta_0)^2,$$

where θ_0 is the equilibrium bending angle and K_θ can be determined by IR spectroscopy.

Internal rotation about covalent bonds is not completely free; the two polyatomic groups linked by the bond repel each other due to varying degrees of charge density overlap. As a consequence, barriers to rotation develop. Depending on the number of bonds emanating from the linked atoms (usually three, occasionally two), these periodic torsional barriers have dominant three- or two-fold symmetry. The maxima correspond to the eclipsed positions of the substituents, while the minima occur at the staggered positions. In hydrocarbons, such as ethane, there are three equivalent minima and maxima for rotation about the C - C single bond joining the tetrahedral C atoms. The general form of the torsional potential for such rotation is

$$E = (E_1/2) (1 - \cos 3\theta) + (E_2/2) (1 - \cos 6\theta) \quad \ldots$$

where E_1 , E_2 etc are the barrier heights for 3-, 6-, etc fold symmetry. In proteins the most frequently considered rotations are about the N - C_α (phi), C_α - C' (psi) and C $\cdots$ N (Ω) peptide bonds of the backbone, and various single bond rotations in the side chains. The barrier heights are low for the first two (0.6 kcal/mole for the phi-rotation, 0.2 kcal/mole for the psi-rotation), while the partial double-bond character of the peptide bond produces a barrier of 20 kcal/mole for the Ω-rotation. The minima occur when psi = $\pm$ 60° and 180° and phi = 0° and ±120°.

The side-chain rotation barriers are typically in the 0-3 kcal/mole range.

The torsional potential around the S - S bridge has two minima at $\pm$ 90° and a barrier of 12 kcal/mole.

There is considerable uncertainty about the strength, periodicity and even location of the minima in torsion potentials. The currently used values are dictated by compromises and "guestimates". In some cases experimental techniques, particularly IR and NMR, can provide reasonable values for barrier heights. It is important to realize that at room temperature both the phi- and psi-barriers are "transparent" to thermal motion.

A.II. Solvent-Influenced Interactions

1.The Hydrogen Bond

The ubiquity of the so-called hydrogen bond in biological structures makes it imperative that we discuss its properties separately. (This despite the contention of some researchers that van der Waals plus electrostatic interactions are both necessary and sufficient to fully simulate its traits, without the need for special hydrogen bond potentials.) It is the most characteristic interaction that is influenced by the presence and nature of the solvent. It is one of the strongest among nonbonded interactions, comparable

in strength to salt bridges. (Whether one classifies hydrogen bonds as strong
nonbonded interactions or weak chemical bonds is almost a matter of taste,
and depends on which current quantum mechanical calculations one gives more
credence to.)

Phenomenologically speaking, hydrogen bonds may form whenever two elec-
tronegative atoms are linked by a hydrogen atom which is covalently bound to
one of them:

$$D - H \cdots A$$

where D is called the donor and A the acceptor atom. In proteins both D and A
are N or O. The D $\cdots$ A distance is shorter than the sum of van der Waals
radii for A and D; strong attractive electrostatic interactions are thought
to be the cause. H bonds are strongest when the three atoms D - H $\cdots$ A are
collinear. Optimally, one of the lone electron pairs of A should be directed
toward H. However, the angular preferences are not very strong; e.g. in the
peptide bond N - H $\cdots$ O = C the angle H - O - C may vary from the ideal 180°
all the way to 90°. Hydrogen bond energies vary between 2 - 8 kcal/mole.

The following types of H bonds are found in proteins:

1) Between two peptide groups
2) Between peptide group and side-chain
3) Between two side-chains:

    a) neutral - neutral
    b) neutral - charged
    c) charged - charged

The charged-charged H bond is often classified as an ion pair or salt bridge
(e.g. as in   - NH_3^+ $\cdots$ ^-OOC -).

The most common H bonds in proteins are N - H $\cdots$ O and O - H $\cdots$ N. Very
weak H bonds may be formed with a donor S - H group. The π-electron cloud of
an aromatic ring may behave as an acceptor for hydrogen bonding (e.g. in aromatic clusters).

The N - H $\cdots$ O = C H bond between peptide groups of the backbone is
the bond most frequently observed in proteins. It stabilizes α-helices, β-
sheets, turns and even irregular folds. Its potential energy function has the
shape of a typical interatomic potential: repulsive for short ($\leq$ 2.3 Å)
distances, turning attractive as r increases, reaching its minimum in the
range of 2.9 - 3.1 Å, then increasing again towards zero. The minimum value
is 4-5 kcal/mole, but this depends on the H bond angle: any deviation from
collinearity will decrease the strength of the H bond (at 50° deviation the
minimum energy is only about 1 kcal/mole). Of course, many of these values
depend somewhat on the particular model one assumes for the H bond potential.
A common choice is a Lennard-Jones type 10-12 potential.

Note that the binding energy of any H bond depends on its surroundings:
accessibility to water will change the effective strength due to competition
with water H bonds:

$$> N - H \cdots O = C < + H_2O \cdots H_2O \longleftrightarrow > N - H \cdots OH_2 + HOH \cdots O = C <$$

If _polar_ side-chains are at the _interior_ of the protein, they cannot interact with the solvent. This energetically unfavorable situation is alleviated by the formation of internal H bonds, either with another buried polar side-chain or with the peptide backbone.

Hydrogen bonding may modify the physical properties of groups involved in the bond. The most commonly used experimental techniques that can be used to detect the presence of H bonds include UV, IR, fluorescence spectroscopy, and especially NMR.

2. Electrostatic Interactions

Under native conditions, proteins may have a net charge. Even at the isoelectric point, where the net charge of the molecule is zero, there are numerous positively and negatively charged amino acid side-chains. These charges interact with each other and with free ions in the solvent. Clearly, these electrostatic interactions depend on the number and distribution of the charges; consequently they depend on the conformational state of the protein.

In addition to the fully charged residues, there are partial charges on the atoms. These arise because chemical bonds between unlike atoms possess a partial electrostatic character. This reflects the electronegativity differences between the bonded atoms. In principle, charge density distribution can be calculated by solving the Schrödinger equation. In practice this is too cumbersome. Instead, _partial_ _point_ _charges_, located on the atoms, are assigned. These are usually calculated by semiempirical quantum mechanical methods (e.g. by the so-called Mulliken population analysis or some variant of CNDO or PSILO), with their attendant uncertainties. They should at least reproduce the known bond moments and the overall dipole moment of the entire molecular group or fragment they are part of.

Whatever their inadequacies, electrostatic interactions between partial charges play an important role in representing and computing nonbonded interactions in proteins.

Two charges q_i and q_j, separated by a distance r_{ij}, have an interaction energy governed by Coulomb's law:

$$E_{elec} = q_i\, q_j \,/\, \epsilon\, r_{ij} \,,$$

where $\underline{\epsilon}$ is the _dielectric_ _constant_ of the medium. In vacuum ϵ is unity. One practical difficulty in protein structure and dynamics calculations is how to choose ϵ. For _intramolecular_ calculations popular choices range from 2 to 4. More recently, _distance-dependent_ ϵ's have been advocated. These somewhat ad hoc choices are based on the argument that 1) polarizable media weaken the _net_ interactions ($\epsilon > 1$) and that 2) the larger r_{ij} the more the charges are _screened_. On the other hand, Lifson argued that when the charge-carrying atoms _are_ the medium, ϵ should be 1.

In the case of _intermolecular_ electrostatic interactions (e.g. between molecules or parts of a large molecule dissolved in a solvent, the dielectric constant of the solvent should only be used if it can be regarded as a continuous medium. It is a matter of controversy whether it is legitimate to use continuum models at microscopic distances.

Instead of the monopole model just discussed, the dipole approximation could be implemented. Known or calculable bond or group moments would then be used. The dipole-dipole interaction energy of two dipoles is given by the following equation:

$$E_{dipole} \;=\; -(\mu_1 \mu_2 / \epsilon r^3)(2 \cos \theta_1 \cos \theta_2 - \sin \theta_1 \sin \theta_2),$$

where r is the distance between the centers of the two dipoles. The dipole moments are μ_1 and μ_2, and the angles θ_1 and θ_2 define the orientations of the two dipoles with respect to the line that passes through the dipole centers. It is further assumed that $r \gg L$, the length of the dipoles. Note that unlike the long-range (r^{-1}) Coulomb interaction, the dipole-dipole energy is of shorter range. For atom-atom separations of a few A's, the monopole approximation is a better model. Furthermore, for the sake of consistency, other interactions, such as charge-dipole, would have to be considered. In practice, the monopole model is used predominantly, even though the peptide bond has a large dipole moment of 3.7 debyes and is known to contribute significantly to the cooperative stabilization of α-helices. (An excellent approximation to the macro line-dipole of an α-helix can be obtained by placing half-charges of opposite signs at the beginning and end of the helix.)

3. Solvent-Dominated Interactions

It is generally recognized that the so-called hydrophobic interaction plays an essential role in the folding and conformational stability of proteins. The term implies "dislike" of the aqueous environment and points to the need to understand the structure and dynamics of liquid water. But water is a highly unusual system, due mainly to the ability of H_2O molecules to form relatively strong H bonds with their neighbors. Each water molecule can form four hydrogen bonds, the two H's serving as protons donors, the two lone pairs as proton acceptors. As a consequence, a reasonably regular 3-dimensional network exists in ice; liquid water is also structured, although it is much less regular.

A number of theoretical models have been proposed for water. Some of these models have been tested by molecular dynamics and Monte Carlo simulations. The qualitative picture that emerges is described by Stillinger as "a macroscopically connected random network of hydrogen bonds with frequent strained and broken bonds continually undergoing topological reformation." It should be pointed out that none of the simulations are capable of reproducing all of the experimental data on water. This is particularly notable for macroscopic properties, such as the bulk dielectric constant ($\epsilon = 78$ debyes).

Given the propensity of water molecules to form H bonds with each other, how does that influence their interaction with nonpolar molecules that are incapable of forming H bonds? When water molecules approach such a molecule, they are faced with the following dilemma: whichever way they orient, one or more of their tetrahedrally placed partial charges will be lost to possible H bond formation, because they point towards the inert solute molecule. What is the remedy?

If the inert molecule is not too large, it is possible for water molecules to pack around it without giving up <u>any</u> of their H bonding sites. The main effect of bringing water molecules and nonpolar molecules together is the <u>reorientation</u> of the former so that they can participate in H bond formation as in bulk water.

The sizes and shapes of nonpolar solute molecules appear to be fairly critical in determining the structure that water assumes around them. This is referred to as <u>hydrophobic hydration</u>. Both theoretical and experimental studies indicate that the restructuring of water around nonpolar solutes is <u>entropically unfavorable</u>, since it disrupts the existing water structure and replaces it with a new, more ordered one.

This is the reason for the low solubilities of nonpolar substances in water. The mainly entropic origin of the unfavorable free energy of solubilization is known as the <u>hydrophobic effect</u> (Kauzmann, Tanford).

In globular proteins the tendency of the residues with nonpolar side-chains to be found <u>inside</u> the protein globule has been ascribed to this effect. Its stabilizing role for proteins (where 40-50% of all residues have at least partially nonpolar side-chains) implied a strong interaction between these <u>when in an aqueous environment</u>. Because of the strength, it was thought that some sort of "hydrophobic bond" was responsible. But it should be clear that there is no bond associated with this mainly entropic phenomenon, which has its origin in the structural rearrangement of water molecules in the overlapping hydration zones as two hydrophobic residues come together. (If we were to simulate hydrophobic interactions as specific bonds, they would have to be of much longer range than any typical bond.)

The general importance of hydrophobic interactions both for protein stabilization and in directing the folding process lead to a great deal of experimental and theoretical investigation.

The experimental work focused mainly on determining the free energy change that accompanies the transfer of amino acid side-chains as well as backbone peptide units from aqueous to non-aqueous media. It was found that there is a linear relationship between the free energy of transfer and the accessibility of residues to the solvent in folded proteins. This lead to a classification of the 20 amino acids as to their propensity to be found <u>inside or outside</u> a protein.

Over 40 different experimental and theoretical hydrophobicity scales have been constructed over the years. The theoretical scales were obtained mostly from statistical studies on known protein structures. Their usefulness derives from the fact that a number of interesting properties of proteins can now be predicted from them, <u>without knowing the tertiary structure</u>. Among these are the prediction of <u>reverse turns</u> from minima in the hydrophobicity plots, <u>amphipatic structures</u> (using hydrophobic periodicities and moments), <u>antigenic determinants</u> and even possible contacts between secondary structural elements through hydrophobic <u>patches</u>.

4. <u>Direct Solvent-Protein Interactions</u>

This is a computational rather than a conceptual sub-classification. That is to say, the previous two classifications were based on the implicit assumption that although the presence and nature of the solvent is critical, its influence on protein properties and behavior appears only indirectly, through parametrization. With the increase in computing power, it is now feasible to include in simulation studies a large number of water molecules. As a consequence, specific solvent-solute interaction potentials could be modelled and used directly. This, in principle, would automatically take into account the hydrophobic interaction and eliminate the use of both an effective dielectric constant or function and explicit H bond potentials.

Although there are in existence specific water-protein potentials, they lack the self-consistent calibration that intra-protein potentials have. In addition, our theoretical knowledge of simulating _pure_ water structure and dynamics is still in its infancy and hence cannot be used with confidence on the protein-solvent interaction problem.

B. PROTEIN DYNAMICS AND ITS SIMULATION

We have classified the types of forces that operate at the microscopic level between atoms of the protein-solvent system. Given reasonably accurate force field representations, and the ever-increasing computer power, it is natural to try to simulate various properties of the protein. X-ray techniques were the first to give us an essentially "frozen", structural view of a protein molecule and earlier calculations reflect this. Static minimization algorithms were in the vogue in attempts to compute the native structure. There are many reasons, both technical and conceptual, why such an approach is bound to fail; they will be discussed in connection with the protein folding problem. In this lecture we focus on one aspect: the empirically derived force field is extremely complex, with hundreds of parameters.

The various types of interaction terms we discussed are combined into an overall energy function or interaction potential

$$V(r) = V(r_1, r_2, \ldots , r_N) =$$

$$\sum_{bonds} (K_b/2)(b - b_0)^2 \quad + \quad \sum_{angles} (K_\theta/2)(\theta - \theta_0)^2 \quad +$$

$$\sum_{torsions} (K_\omega/2)(\Omega - \Omega_0)^2 \quad + \quad \sum_{dihedrals} K_\phi [1 + \cos(n\Phi - \delta)] \quad +$$

$$\sum_{pairs (i,j)} [A(i,j)/r^{12}_{ij} - B(i,j)/r^6_{ij} + q_i q_j/\epsilon r_{ij}] \quad +$$

$$\sum_{H\ bonds} [C(i,j)/r^{12}_{ij} - D(i,j)/r^{10}_{ij}]$$

where the position vector r_i labels the ith atom. The first term represents the covalent bond-stretching interaction along bond b. The second term describes the bond-angle interaction. Two forms are used for the torsional or dihedral-angle interactions: the harmonic terms for angles that are forbidden to make transitions, and sinusoidal terms for angles that can rotate by 360

degrees if they can surmount the n-fold barrier. The next term is a sum over
all pairs of atoms and represents the effective non-bonded interaction, consisting of the van der Waals and the Coulomb interaction between atoms i and
j with charges q_i and q_j at a distance r_{ij}. The last term is over all H bonds.
Frequently these include angular terms. The subscripted constants K_b etc are
the force constants, the A's, B's, C's and D's are Lennard-Jones parameters,
and n, b_0 etc are geometrical reference values.

As the consequence of this complexity, the overall potential hypersurface contains an astronomically large number of local minima. A local minimizer can, and inevitably will get trapped in any of these, usually the one
closest to the starting point. Clearly, this is an unphysical situation;
local minimizers need external help.

Protein molecules are not rigid, static entities. They operate in solution, fluctuate about their equilibrium position at finite temperatures,
pressures etc and under the influence of the solvent. They need to be flexible in order to carry out their biological function. In short, they are
dynamical systems. (Even in the crystalline state proteins are not static, as
evidenced by X-ray-determined B-factors.) At any given time a protein molecule exhibits a wide variety of motions. These range from local atom fluctuations, side chain oscillations etc, through rigid body motions of helices,
subunits, domains, to large-scale displacements such as the folding/unfolding
motion of the whole molecule. Collective, correlated events are of particular
importance. The large variety is accompanied by equally large scales of motions. Thus at room temperature the amplitudes have been estimated to range
from 0.1 - 1000 nm, the energies from 0.1 to 100 kcal/mole, while the times
cover a 10^{-15} to 10^3 sec span. There is a hierarchy of scales of motion.

A realistic equilibrium statistical mechanical simulation of protein
molecules in solution demands that we consider the influence of external,
macroscopic parameters such as temperature or pressure as well as the internal, microscopic interactions. We do this by using "dynamical" minimizers
(DM's).

The most important characteristics of DM's is that they are neither
local, nor terminating. At finite temperature they will sample continuously
the available configuration space.

There are two major types of DM's currently in use. The Monte Carlo
(MC) method is explicitly stochastic. It is not truly dynamical, because time
does not enter into its formulation. Microscopic properties are ensemble
averages. In contrast, the Molecular Dynamics (MD) approach is deterministic.
It simulates the time evolution of the system. Microscopic properties are
calculated from time averages over their dynamical trajectories. The equivalence between ensemble averages and time averages is guaranteed by the
validity of the ergodic hypothesis.

1. The Monte Carlo Method

The experimentally observed equilibrium properties of physical systems
are ensemble averages. The canonical (T,V,N) ensemble operates under constant
temperature, volume and particle number. Monte Carlo simulations are usually
carried out in the canonical ensemble, although it is relatively simple to
change to other ensembles.

The classical version of MC simulation was introduced by Metropolis. It is implemented as follows. One of the N atoms is chosen (randomly or cyclically) and is displaced in a random direction. The energy change δE due to this displacement is calculated. If it is negative, the move is accepted and the process is repeated with another atom. If δE is positive, a uniformly distributed random number in (0-1) is generated and compared to $\exp(-\delta E/kT)$. If the latter is greater, the move is accepted; if not, it is rejected and the configuration reset to its previous state. Thus _descent_ to the minimum is _not_ _monotonic_: at finite T there is a finite probability that an _uphill_ move is accepted. This means that the system can easily surmount barrier heights of the order of kT. Consequently, much larger portions of the low-energy conformation space can be sampled. The penalty is that MC simulations are much more expensive than simple static minimizations.

It is important to select the average random displacement carefully. If it is too small, most moves will be accepted but progress will be painfully slow; if too large, most uphill moves will be rejected and the system will tend to stay in place. Experience with simulating liquid and solid properties suggests that an acceptance ratio of about 0.25 - 0.50 is reasonable. However, since proteins are polymers with very anisotropic potential surfaces, it is not clear that these acceptance ratios are appropriate.

There are both advantages and disadvantages in using the MC method. Among the advantages is the relative ease with which one can change the particular ensemble in which one wishes to carry out the simulation. Thermodynamic variables, such as T,P, enter only parametrically. (This is not an unequivocal gain; one does not have a ready "thermometer" during the simulation to verify that the requested temperature is in fact well-maintained.) Another benefit is that derivatives of the potential are not needed. Computer coding is very simple and the statistical averages of interest can be calculated as the simulation proceeds.

A serious problem with MC simulations is that equilibration is slow. Typically, a third to a half of the initial run is discarded because the system took that long to "forget" the initial conditions. The problem can be traced back to the Markov chain generation process. This persistence of correlation in long MC Markov chains plays an essential role in the estimation of errors. This is best done in term of a correlation length τ. For any given property F of interest, this can be calculated from the MC run, using expectation values and variances of F only. τ is the number of trial moves for which subsequent values of F are to be regarded as correlated. When successive F's are independent, $\tau = 1$, while a $\tau = m$, say, means that the Markov chain samples an independent conformation only after every m "macro" steps (a macro step corresponds to moving each of the N particles of the system once).

More powerful MC methods are known under the names of _force_ _bias_ and _smart_ Monte Carlo. They are designed to descend more rapidly towards the low-energy regions of the system. Improvements in efficiency by factors of 2-4 have been claimed, even though forces (derivatives) had to be computed.

A combination of static minimizers with MC seems promising. The idea is to utilize the speed of local minimizers to descend to a minimum and then use MC to explore the vicinity of that minimum in a statistically meaningful way. Care must be exercised in calculating averages: if MC helps the system escape

from the basin of attraction of the last minimum, configurational averagings
have to be restarted.

A relatively recent MC-inspired development is _simulated annealing_. It
can be regarded as a _nonequilibrium_ statistical mechanical procedure. System
simulation by MC is started at very high temperatures, allowing most of the
conformation space to be sampled. The search is guided towards low-energy
regions by gradually lowering the temperature. As T approaches zero, the system is "frozen" out into the global minimum. Thus the method may be viewed as
one particular member of a class of stochastic global minimizers: they are
guaranteed to find the global minimum with probability one, provided the
computation time is long enough. The _cooling schedule_ is critical: too fast
cooling and one gets trapped in a local minimum; too slow, and the search
becomes exhaustive (dumb) random search, exorbitantly expensive. The X-PLOR
package of Brünger has an annealing option.

2. Molecular Dynamics

In recent years, molecular dynamics (MD) simulations have gained in
popularity. The reason is pragmatic: it is believed that for the same computational effort a larger portion of conformation space can be sampled than by
MC. (There is evidence, at least in solids and liquids, that the relative
sampling efficiencies of the two methods are comparable.) Furthermore, if
time-dependent properties are needed, only MD can produce them.

In an MD simulation Newton's classical equations of motion are solved
for the N-particle system (protein). Thus we need the _forces_, the partial
derivatives of the interaction potentials. The equations of motion are

$$d^2 r_i / dt^2 \; = \; F_i \, [r(t)] \, / \, m_i \, , \qquad i = 1, 2, \, . \, . \, . \, , N$$

where the _force_ exerted on atom i is

$$F_i(t) \; = \; - \, \partial V[r(t)] \, / \, \partial r_i(t).$$

In solving the equations numerically, time is discretized. Sufficiently
small time increments have to be chosen so that numerical instabilities are
avoided. Time steps are typically in the femtosecond (10^{-15} sec) range. As in
the MC method, an initial equilibration is needed. The equilibration period
is terminated when no systematic changes in T are observed over a period of
about 10 psec or 1000 time steps; The temperature T is simply related to the
average kinetic energy of the system on N atoms,

$$\sum_{i=1}^{N} m_i \, \langle v_i^2 \rangle = 3 \, N \, kT,$$

where m_i and $\langle v_i^2 \rangle$ are the mass and average velocity squared of the ith atom
and k is the Boltzmann constant. The atomic momenta ought to obey a Maxwel-
lian distribution and different regions of the protein should have the same
local average temperature. Once the system equilibrates, time averages of the
properties of interest over portions of the trajectories are taken. These

averages should be the same as the MC ensemble averages, according to the ergodic theorem.

The classical MD method operates in the microcanonical (E,N,V) ensemble. Consequently, the temperature fluctuates during the simulation and has to be periodically readjusted. The readjustment is done by scaling the velocities. Recent developments in theory lead to a number of different canonical or constant-T MD algorithms; unfortunately, the calculated dynamical (nonequilibrium) properties may no longer be correct. These algorithms also tend to be more time-consuming. Canonical MD algorithms have not been used extensively in protein dynamics simulations.

A definite disadvantage of MD methods is that there is an additional uncertainty about the accuracy of the simulation. This arises because of differences in the numerical integrator algorithms. MD simulators of protein dynamics tend to rely on constant-stepsize, constant-order ordinary differential equation (ODE) solvers. The early Verlet algorithm has generally been superseded by the Beeman algorithm (Levitt) or Gear's 4th order predictor-corrector method (Karplus). We give, as an example, the integration scheme recommended by Beeman. We write Newton's equations in the condensed form

$$d^2 r/dt^2 = a(r) .$$

Here r, $v = dr/dt$, and a are vectors with 3N components, representing the x, y, and z components of the position, velocity, and acceleration of each of the N atoms. When solving these equations, starting from an initial set of coordinates and velocities at $t = 0$, time is discretized. The time step h is the constant amount by which time is advanced during simulation. Then the discrete version of Beeman's third order predictor-corrector is

$$r_{n+1} = r_n + hv_n + (h^2/6)(4a_n - a_{n-1}) \qquad \text{(predictor)}$$

$$hv_{n+1} = r_{n+1} - r_n + (h^2/6)(2a_{n+1} + a_n) \qquad \text{(corrector)},$$

and these iterations are carried out in the sequence $P(r)E(a)C(v)$, where P, E, and C stand for "predict", "evaluate" and "correct".

It is frequently argued that there is no need to use anything more sophisticated or accurate, because the trajectories tend to follow tortuous paths in phase space which would interfere with the efficiency of methods that assume smoothness of higher derivatives. But Newton's equations for a protein with its widely different time scales of motion are stiff. (This is the dynamical manifestation of the extreme anisotropy of the potential we have mentioned in connection with MC.) Thus the maximum fixed time step that can be tolerated while maintaining numerical stability is very small: it must gracefully handle fast "rattlings" of some of the variables in very steep-walled potentials. In the meantime, many other variables experience much weaker forces; they could have tolerated much larger steps. Thus the tail wags the dog.

There are several possible remedies.

1) Variable-stepsize variable-order stiff ODE solvers could be tried. They are widely available in standard mathematical libraries such as IMSLIB.

2) Alternatively, Bennett's suggestion of replacing the kinetic energy function by a more general quadratic form could be used to equalize motions

of widely dissimilar time scales. The dynamics will be different but the equilibrium properties will not change. An efficient choice for the "mass tensor" is not a trivial problem.

3) I prefer changing the independent variable t to $\sigma(t)$, where σ is the _arclength_ of the trajectory. Overhead is negligible, with one additional first order ODE to solve for $dt/d\sigma$. Note that the advantage of this new system of equations is that for a _fixed_ _stepsize_ _in_ σ the system itself would automatically adjust the stepsize h, to reflect local changes in the curvature of the trajectory. If we rewrite Newton's equations of motion as a set of first order equations, they are of the form

$$dz_i/dt = f_i(z_1, z_2, \ldots, z_M).$$

The implicit change of variable $dt/d\sigma = 1/G$ would give the equations

$$dz_i/d\sigma = g_i[z(\sigma)] \, , \, i = 1, 2, \ldots, M$$

$$dt/d\sigma = 1 / G \, ,$$

where the $g_i = (f_i/F)/ G$, $F = \max_i |f_i|$ and a useful functional form for G is

$$G = C \, [(1/F)^2 + (f_1/F)^2 + (f_2/F)^2 + \ldots + (f_M/F)^2]^{1/2}$$

with C some preselected constant. A simple analysis of the changed expressions shows clearly how the variable t-steps are produced. (Division by F ensures that all derivatives with respect to σ are ≤ 1 in magnitude.)

We have to mention the importance of boundary conditions in statistical mechanical simulation work. The choice of _periodic_ boundaries helps mimic an infinite system through the use of replicas of one's finite number of particles. Its usefulness for simulating proteins in solution appears to be limited. This is because periodicity would simulate an overly high protein concentration. In solids or liquids, for which it was originally devised, periodic boundary conditions work well because of the relatively uniform distribution of the atoms or molecules across box interfaces.

From the computational point of view, neither method is clearly superior _if_ one is only interested in equilibrium averages. At present, there are only a few time-dependent experimental observations (e.g. inelastic neutron scattering or Mössbauer spectroscopy) that would require MD. (Microscopic simulation of kinetic results would need computer times far longer than practicable at present.)

Both MC and MD share an important advantage over static minimization: they can be used to simulate relative _free_ _energy_ _changes_. This is the foundation of the powerful and useful _thermodynamic_ _perturbation_ _theory_ that is gaining popularity in theoretical protein engineering studies.

We can readily calculate statistical equilibrium averages both from MC and MD simulations, provided the averaged properties are _mechanical_ quantities, such as kinetic and potential energies, structural properties, fluctuations, etc. However, it is virtually impossible to simulate two important _extensive_ properties of a thermodynamical system: the absolute values of the

entropy and the free energy. Yet many important biochemical quantities of
interest are directly related to free energy differences and changes. In
recent years however, several promising methods have been developed or redis-
covered for evaluating relative free energy changes. Although computationally
intensive, these methods are of great significance in protein engineering and
design.

Several methods exist for calculating relative free energy differences.
The two currently most useful in protein engineering are called thermodynamic
integration and thermodynamic perturbation methods, respectively. Both rely
on the fact that free energy changes produced by perturbing the system can be
determined during simulation. The free energy difference between two states A
and B, denoted by $\delta\, G_{BA}$, can be determined from a simulation, provided the
potential function $V(\mathbf{r})$ or even the total Hamiltonian, $H(\mathbf{p},\mathbf{r})$ is changed so
gradually that the system itself changes slowly along a reversible path from
A to B.

The essence of the procedure is to introduce a coupling constant, say
σ, into that part of the potential that we want to change. This could be any
and all of the parameters of the potentials we are planning to alter. One
commonly allows σ to range between 0 and 1: $\sigma = 0$ may correspond to state A
and $\sigma = 1$ to state B. Let the parameter we change be P. Then

$$P(\sigma) \;=\; (1-\sigma)\, P_A \;+\; \sigma\, P_B \;.$$

The Gibbs free energy of the system becomes a function of σ, $G(\sigma)$:

$$G(\sigma) \;=\; -kT \ln Z(\sigma) \;, \tag{*}$$

where $Z(\sigma)$ is the partition function in the appropriate ensemble. The free
energy difference $\delta\, G_{BA}$ is

$$\delta\, G_{BA} \;=\; G(\sigma_B) - G(\sigma_A) \;=\; -kT \ln \{Z(B)/Z(A)\}$$

and this can be expressed as an ensemble average:

$$= -kT \ln \{\langle \exp[V(B) - V(A)]/kT\rangle_{\sigma_A}\},$$

where the brackets $\langle \ldots \rangle_\sigma$ denote an ensemble average at the value σ. This
is a perturbation formula because it is accurate only when state A and B are
not too different. Normally this is not the case and then one computes the
difference between A and B as a sum of small free energy increments. Differ-
entiating (*) with respect to σ gives

$$\partial\, G(\sigma)/\partial\sigma \;=\; -(kT/Z(\sigma))\, \partial\, Z(\sigma)/\partial\sigma \;=\; \langle \partial V(\sigma)/\partial\sigma\rangle_\sigma.$$

This then leads to the standard statistical mechanical integration formula
for the free energy difference

$$\delta\, G_{BA} \;=\; \int_A^B \langle \partial V(\sigma)/\partial\sigma\rangle_\sigma \; d\sigma.$$

For slow enough change in σ this integration can be carried out during the
course of the simulation. MD seems to be the method of choice, although MC
would work just as well. It should be mentioned that the mathematical basis

of the thermodynamic perturbation method is known as the <u>continuation</u> method, and has been used extensively in applied mathematics. The rate of changing σ is critical for accuracy and elaborate precautions are taken in mathematical applications. This does not seem to be the case yet for simulations in bio-physics.

REFERENCES (PART A)

Creighton, Th. E., 1983 "Proteins", W.H. Freeman & Co. N.Y.
Israelachvili, J.N., 1985 "Intermolecular and Surface Forces", Academic Press, N.Y.
Franks, F. (ed.), 1982 "Biophysics of Water", Wiley-Interscience, N.Y.
McCammon, J.A. and Harvey, S.C. (1987) "Dynamics of Proteins and Nucleic Acids", Cambridge University Press, Cambridge.
Schulz, G.E. and Schirmer, R.H. (1979) "Principles of Protein Structure", Springer-Verlag, N.Y.
Baker, E.N. and Hubbard, R.E. Prog. Biophys. Molec. Biol. (1984) $\underline{44}$, 97-179
Hydrogen Bonding in Globular Proteins
Cornette, J.L., Cease, K.B., Margalit, H., Spouge, J.L., Berzofsky, J.A. and DeLisi, Ch., J. Mol. Biol. (1987) $\underline{195}$, 659-685
Hydrophobicity Scales and Computational Techniques for Detecting Amphipatic Structures in Proteins
Eisenberg, D, Wilcox, W. and McLachlan, A.D. J. Cellular Biochem. (1986) $\underline{31}$, 11-17
Hydrophilicity and Amphiphilicity in Protein Structure
Hermans, J., Berendsen, H.J.C., van Gunsteren, W.F. and Postma, J.P.M., (1984), Biopolymers $\underline{23}$, 1513-1518
A Consistent Empirical Potential for Water-Protein Interactions
Hvidt, A., Ann. Rev. Biopys. (1983) $\underline{12}$, 1-20
Interactions of Water with Nonpolar Solutes
Némethy, G., Peer, W.J. and Scheraga, H.A., Ann. Rev. Biophys. Bioeng. (1981), $\underline{10}$, 459-497
Effect of Protein-Solvent Interactions on Protein Conformation

REFERENCES (PART B)

Beeman, D. (1976) J. Comp. Phys. $\underline{20}$, 130-139
Some Multistep Methods for Use in Molecular Dynamics Calculations
Bennett, Ch. (1975) J. Comp. Phys. $\underline{19}$, 267-279
Mass Tensor Molecular Dynamics
Binder, K. (ed.) (1979) "Monte Carlo Methods in Statistical Physics", Springer-Verlag, Berlin
van Gunsteren, W.F. (1988) Protein Engineering $\underline{2}$, 5-13
The Role of Computer Simulation Techniques in Protein Engineering
Jacucci, G. and Rahman, A. Il Nuovo Cimento (1984) $\underline{4D}$, 341-356
Comparing the Efficiency of Metroplis Monte Carlo and Molecular-Dynamics Methods for Configuration Space Sampling
Wood, W.W. and Erpenbeck, J.J. (1976) Ann. Rev. Phys. Chem. $\underline{27}$, 319-348
Molecular Dynamics and Monte Carlo Calculations in Statistical Mechanics

THEORIES OF PROTEIN FOLDING

R.L. Somorjai

Division of Biological Sciences
National Research Council of Canada
Ottawa, ON Canada

INTRODUCTION

It is generally recognized that one of the most important and most challenging unsolved problems of molecular biology is the _protein folding_ problem. What is this problem and why is it so important?

We can formulate it simply as follows: given the primary amino acid sequence of any protein, find its native, three-dimensional structure. For _enzymes_, this is the _active_ conformation. Such deceptively simple-sounding formulation hides a wealth of deep and difficult conceptual problems. Its solution is an essential stepping stone to the ultimate goal of protein engineering: the _ab initio_ design of an enzyme with a novel function.

However, to reach this goal, first we have to understand how structure is related to function. But the elucidation of _structure-function_ relationship demands that we develop predicting capabilities that would lead reliably from sequence to configuration - i.e. that we solve the protein folding problem.

BASIC CONCEPTS

It is important to realize that the phrase _protein folding_ means different things to different people. I shall try to show that this is still a conceptual stumbling block, a red herring that prevented progress. What is then a useful definition?

Here we have to make an essential distinction between the _experimentalist_'s view and that of the _theoretician_. The former, quite properly, regards protein folding as a _process_. After all, he induces folding/unfolding by changing the external parameters, the _environment_. He monitors the effect of these changes indirectly and the conclusions he reaches are inferential. As a consequence, the interpretation of the experiments may be ambiguous.

In contrast, the theoretician has been guided (and beguiled) by the basic _hypothesis_ that the linear sequence contains the full code necessary to arrive at the unique native structure. Hence the continuing quest for the Holy Grail: the ultimate _folding algorithm_. At present, no method may claim even the resemblance of success. To understand the failure, and to put it in proper perspective, we shall have to know the motivation and the arguments that lead to the different prediction methods. Each of these methods has its own conceptual and/or practical foundation and weakness; we shall discuss these in turn.

The essential experimental facts of <u>in vitro</u> folding/unfolding (F/U) are:

E1) F/U is a <u>reversible</u>, equilibrium process;
E2) F/U is <u>fast</u>;
E3) The reversibility (and often the speed) are <u>qualitatively</u> <u>independent</u> of the environmental factors that are used to induce F/U.

The observation of reversibility leads quite naturally to the so-called <u>Thermodynamic Hypothesis</u>. This asserts that under native external conditions (e.g. temperature, pH, ionic strength etc) the protein + environment system is stable and in a <u>global</u> <u>free</u> <u>energy</u> <u>minimum</u>. Implicit is the assumption that F/U is an equilibrium process.

However, because a protein molecule (coupled to its solvent surroundings) has many degrees of freedom, its <u>conformation</u> <u>space</u> (half of the phase space of statistical mechanics) is vast. The complete sampling of this space would take eons. This conclusion is contradicted both by the in vitro experimental facts and by biological exigency: folding times are in the seconds-minutes range. The obvious inference is that only a very limited region of conformation space is sampled by the folding protein. The observed rapid folding is the consequence of specific <u>constrains</u>, encoded in the linear sequence. Thus the native state is the <u>kinetically</u> <u>most</u> <u>accessible</u> outcome of the folding process; it is <u>not</u> a thermodynamic imperative that the biologically relevant conformation be the global free energy minimum and it probably isn't. What these constraints are we don't know, but they provide the <u>kinetic</u> <u>control</u> that helps confine the folding molecule to a manageable number of <u>preferential</u> <u>pathways</u>. (The original, crude folding rate calculations no doubt overestimate the true rates and have been modified by Dill. Based on his model calculations, he argues against the need for kinetic control. His folding time estimates are almost realistic. However, the necessity of constraints is as compelling as before, as we shall see.)

The need to reconcile apparently contradictory experimental facts thus lead to the view that to successfully predict the native conformation we have to find a set of relevant constraints. We are then faced with the problem: how are these constraints expressed in nature and how are we to realize them in our simulation studies?

An important clue is the observation that protein structures are organized. Furthermore, this <u>structural</u> <u>organization</u> is <u>modular</u> or pseudo-<u>hierarchical</u>. The hierarchical units are easily perceived in the X-ray structures and range in size from domains (compact, often functional regions) through supersecondary structural elements down to secondary structures and/or hydrophobic or aromatic clusters. (Several algorithms exist that sequentially decompose a protein into smaller and smaller subunits.)

Given the structural hierarchy, it is tempting to postulate that protein folding itself is a hierarchical (or more accurately, multi-stage) process. Furthermore, it seems reasonable to implicate the various structural subunits in the kinetics/dynamics of folding. In fact, several variants of hierarchical condensation as the dominant mechanism of protein folding have been advanced. Such a mechanism is conceptually attractive for numerous reasons. It rationalizes the speed of folding; it also provides an efficient editing/correcting scheme that reduces the time spent in dead-end pathways. In addition, nucleation is a natural concept in a multi-stage self-assembly, as is the existence of (possibly marginally stable) intermediates. (It is possible that the transient intermediate stages differ considerably from their counterpart in the final organized structure. If true, this implies that some type of feedback mechanism must play an important role in the folding process. This points to the relevance of classifying intramolecular interactions into ranges (short, medium, long), defined according to separation along the main chain. Feedback would operate through the long-range interactions and would help reorganize and stabilize transient substructures.)

What is then the current view of the folding process?

It is now commonly accepted that folding starts, independently and more or less simultaneously, in several regions of the polypeptide chain by the formation of short segments of secondary structural elements. Such small subunits (nucleation/seed sites) can appear in microseconds because each consists of only a few, neighboring amino acids. They are metastable at best, again because only a few residues are involved.

The next stage establishes stereospecific interactions between neighboring subunits that happened to diffuse together (a "diffusion-collision" model). As a result of these long-range interactions, the appropriate local secondary structures are selected and stabilized. Furthermore, precise relative positioning of these subunits occurs, thus creating local supersecondary units. These, in turn, coalesce (and rearrange) into domains. The domains, already folded into a native-like conformation, then self-assemble into a globular form. It has been postulated that a molten globule state is reached before the side-chains finally acquire their most stable state through conformational fine-tuning. This then leads to the final stabilization of the native structure.

An important complication of this qualitative picture needs mentioning. Studies of refolding kinetics indicate that the 'unfolded' state of a protein may consists of two 'substates'. In one, every proline residue is in the same cis or trans conformation as in the native protein; in the other, at least one proline is the wrong isomer. Brandts suggested that proteins with the wrong isomer in the unfolded state would refold slowly; the cis-trans isomerization of proline would be the rate-limiting step. Experimental evidence supports this view.

In retrospect, a quantitative assessment of these qualitative ideas on the folding process would have been next logical step, especially in view of the availability of ever-increasing computing power. This is not what had happened. The main reason is that for most theorists "simulation of protein folding" became synonymous with "prediction of the 3-dimensional structure from the linear sequence". This change of emphasis from process to structure resulted in a predominantly algorithmic approach. Its goal is the development of a (hopefully simple) set of unambiguous folding rules. Their systematic application to the linear sequence would generate the correct native structure. Naturally, such folding algorithms need not mimic the actual physical process of folding. They are most unlikely to do so.

PREDICTION METHODS

A. <u>Statistical Methods</u>

The impetus for such a rule-oriented program was provided by the steadily growing crystallographic information on detailed 3-dimensional protein structures. The direct observability of the predicted regular structural features (α-helices, β-sheets) naturally invited <u>statistical</u> prediction schemes. Such methods aim to correlate the intrinsic properties of individual amino acid residues with their observed <u>propensities</u> to form or belong to particular secondary structural units. In fact, several of the early algorithms (Chou-Fasman, Robson-Pain, Lim) predict secondary structures with only 50-55% success (3-state prediction), even though the influence of near neighbors in the linear sequence are included. (Note that the hierarchical classification of protein structure is implicitly included in these methods: they focus on the lowest non-trivial level of the hierarchy.) This notable lack of success has been attributed to the fact that the methods cannot incorporate long-range information. That this is only partly correct is demonstrated by the sharply increased (15-20%) prediction success achieved if the proteins are first classified into the four major types (all-α, all-β, $\alpha+\beta$ and α/β) and for each of these the appropriate <u>optimum</u> decision constants are used in the Garnier-Robson algorithm. Note again the general utility of hierarchical <u>build-up</u> principles.

Another method of enhancing the accuracy of predictions is to make use of <u>homologous</u> sequences. Homology need not involve the whole protein. Furthermore, homology could be considered with respect to some property (say, hydrophobicity). The success-rate is variable, from poor to excellent.

A particularly interesting physico-chemical attribute, <u>hydropathy</u>, derives its importance form the fact that it reflects the propensity of the given residue to be found inside the folded protein. Thus hydro<u>phobic</u> residues with nonpolar aliphatic or aromatic side-chains favor the interior, while the hydro<u>phillic</u> residues tend to appear near the surface, often in reverse <u>turns</u>. This suggested the use of smoothed hydropathy profiles in which <u>minima</u> indicate turn locations. Additional statistical information, exemplified by the Rose-Wetlaufer regression equation:

$$T = 0.125 \, N + 2.28,$$

where T is the number of turns in a protein of N residues, can help in minimizing <u>false</u> <u>positives</u> (prediction of turns where there are none). Success may be very high, especially when additional constraints can be used. As an example, Cohen and coworkers, using a hierarchical pattern-matching algorithm, correctly assigned 43 of 45 turns in 7 α/α proteins, 138 of 145 turns in 8 α/β proteins and 117 of 127 turns in 11 β/β proteins. They also predicted secondary structural elements in 10 α/β proteins with high accuracy: reverse turns (98%), helices (83%), sheets (93%) and coils (74%). All of this points to a prediction future which relies more and more heavily on extended data bases and the increasing adaptation of sophisticated concepts and techniques from <u>knowledge</u> <u>engineering</u> and <u>artificial</u> <u>intelligence</u>.

Despite notable successes, even for supersecondary structure predictions, there are inherent limitations, especially when it comes to the assembly of these predicted structural elements into the final three-dimensional
protein. Because no explicit energetics can be incorporated into these rule-
driven prediction methods, and because it is improbable that the rules will
select a _unique_ most likely structure, the final choice from a (hopefully
small) number of 'possibles' will have to be based on other criteria. An
obvious candidate is structural 'fine-tuning', based on _energy minimization_.

B. Static Minimization Methods

Historically, minimization methods were intended to solve directly the
sequence - structure prediction problem. The rationale for these attempts is,
of course, the Thermodynamic Hypothesis, reinforced by the successful derivation of reasonable empirical atom-atom potentials. Thus a classical many-par-
ticle potential function can be constructed, that is applicable to any conformation of a protein; such a conformation is represented as a point on the
multidimensional energy surface. The overall potential energy is then a function of all explicitly considered degree of freedom. The global minimum is
assumed to correspond to the native structure. "Folding" is identified with
"motion" on this hypersurface from some high-energy region (e.g. the extended
conformation) to the lowest minimum.

There are a number of practical and conceptual problems with minimization methods. Let us consider the most serious difficulties.

Practical problems
1. Large number of variables

Even when we ignore the solvent, the number of atoms in a protein of N
residues is ~ 12 N, i.e there are ~ $M = 36 N$ Cartesian coordinates. A small
protein of 100 residues has ~ 3600 variables. The number of pair-interaction
calculations for any given conformation is $O(M^2)$; this can be improved to
$O(kM)$, where $k < M$, if we use a cutoff distance. However, the long-range electrostatic interactions would require corrections.

2. Large number of local minima

This is known as the _multiple-minimum_ problem. Its acuteness can be
demonstrated by taking a simplified representation of a protein molecule: we
assume that there are only rotational degrees of freedom. This reduces the
number of variables from 36 N to 5 N. Assume that there are 3-fold rotational
barriers about each bond. Then the number of minima are $O(3^{5N})$, a huge number
even for moderate N. A variety of ingenious methods have been devised to cir-
cumvent or bypass searching this bumpy surface (somehow the protein does it
effectively). Unfortunately, global minimization is notoriously difficult and
extremely time-consuming (exhaustive search of high-dimensional space).

3. Validity and accuracy of the empirical potentials

A legitimate worry is that the potentials are not calibrated properly
for proteins. (Simulation methods in general, and minimization methods in
particular, use empirically determined potential functions that were fitted
to small-molecule crystal data.) This can be tested by starting a local minimizer from the experimental coordinates and determining how far the calcula-

ted minimum "drifts" from the starting point. Most major molecular modelling
packages such as AMBER have been carefully calibrated so that this should not
be a serious problem. The calibration involves the creation of well-balanced
parameter sets for the various types of potentials the package uses. Energy
refinement on porcine insulin with the AMBER parametrization gave an rms
deviation of 0.28 A for the backbone, attesting to the success of the parameter tuning.

Conceptual problems

Native structure ≡ Global minimum ?

Let us assume that a brilliant computational breakthrough occurs and we
can calculate, with certainty, the global free energy minimum of the protein.
Can we assert that this is the native conformation? The depressing answer is
that we cannot! This is because the native structure is a constrained free
energy minimum, and we do not know what the constraints are and how they act.
Notice the emphasis on free energy. Current minimization approaches are confined to internal energy (enthalpy) calculations; entropic contributions are
generally ignored. (Attempting to estimate entropy by the logarithm of the
determinant of the Hessian matrix at the minimum of the potential is a crude
and expensive local approximation.)

An important reason for this predestined failure of minimizers is that
progress along the potential hypersurface (via minimization or otherwise) is
not equivalent to the folding process. Of course, if the simulation of the
proper folding pathway were not necessary, then minimization would be relatively efficient computationally; it could bypass the simulation of the time-
consuming reequilibration that the folding protein undergoes as external conditions are changing. Unfortunately, compelling arguments can be advanced in
favor of process-oriented folding strategies . Until we know what constraints
have to be imposed, minimization is not likely to succeed.

It should be emphasized that the multiple-minimum problem is an arte-
fact of the algorithmic philosophy that characterizes minimization methods.
That is, encountering the plethora of local minima and their intervening
barriers during minimization is the consequence of equating the folding process with traversing an invariant multidimensional "landscape" while descend-
ing from high- to low-energy regions of conformation space. But in reality,
this multidimensional landscape changes with changing folding conditions.
This means, in particular that different environments require different parametrization. This in turn implies that the location, size and even existence
of local minima varies with varying external conditions. (Recall that the
potentials are parametrized to produce good agreement with the experimentally
determined native structure. They do not and cannot reflect changes that different unfolding conditions would induce.)

Physical and biological processes, such as protein folding, take place
at finite temperatures, pressures etc. Minimization would mimic a process
that occurred at 0°K, or one that was very rapidly "quenched" to absolute
zero. At room temperature barriers of the order of kT (0.6 kcal/mole) are
transparent to conformational motion; local minimizers could get trapped by
any of them.

Of course, most researchers are aware of at least some of these conceptual difficulties and there is a definite trend in protein structure simulations toward "dynamical" minimizers. Although this is a step in the right direction, dynamics per se is no remedy for our major conceptual malady: the hypersurface is still invariant with respect to external conditions (except for temperature and possibly pressure, of course). Currently, one uses "dynamics" only to avoid getting trapped in local minima. At least, this particular technical problem is solved in a physically meaningful way.

C. Dynamical Minimization Methods

Unlike the static (zero temperature) minimizers, which are purely algorithmic, these methods are based on statistical mechanical concepts. They operate at finite temperatures (and pressures etc). The temperature is either an externally imposed parameter or is made to behave as such. Two methods are currently in use. The Monte Carlo (MC) method is explicitly stochastic in nature, while its chief competitor, the Molecular Dynamics (MD) approach, is deterministic. Statistical-stochastic aspects enter into the latter indirectly, by invoking the equivalence of ensemble averages of physical observables to their time averages over dynamical trajectories. (Ergodic theorem.) Let us review the salient features of the two methods.

1. The Monte Carlo Method

The experimentally observed equilibrium properties of physical systems, such as protein molecules, are ensemble averages. Which of the ensemble averages one measures depends on the thermodynamic variables one holds constant. The canonical (T,V,N) ensemble assumes constant temperature, volume and particle number. Monte Carlo simulations are predominantly canonical ensemble simulations, although it is relatively simple to change to other ensembles.

The traditional version of MC simulation was introduced by Metropolis. Its statistical sampling characteristics arise from random moves that produce a memoryless Markov process. For the purposes of this discussion the essential point is that the Metropolis sampling procedure is preferentially biased towards generating conformations with lower and lower energy. (Details of the implementation can be found in the article on Interactions and Dynamics.) The essential point is that descent to the minimum is not monotonic: at finite T there is a finite probability that an uphill move is accepted. This means that the system can easily surmount barrier heights of the order of kT. Consequently, much larger portions of the low-energy conformation space are sampled than by a static minimizer. The penalty is that MC simulations are much more expensive than simple static minimizations.

A relatively recent MC-inspired development is simulated annealing. One starts the system simulation by Monte Carlo at very high temperatures, allowing thorough sampling of the conformation space. The temperature is then gradually lowered, guiding the search towards the low energy regions. As the temperature approaches zero, the system is "frozen" out into the global minimum. The cooling schedule is critical: too fast cooling and one gets trapped in a local minimum; too slow, and the search becomes exhaustive (dumb) random search, exorbitantly expensive. The X-PLOR package of Brünger has an annealing option.

2. <u>Molecular Dynamics</u>

In recent years, molecular dynamics simulations have gained in popularity. The reason is practical: it is believed that for the same computational effort a larger portion of conformation space can be sampled. Furthermore, if time-dependent properties are needed, only MD can produce them.

In an MD simulation Newton's classical equations of motion are solved for the N-particle system (protein). Thus we need the <u>forces</u>, the partial derivatives of the interaction potentials. In solving the equations, sufficiently small time increments are chosen so that numerical instabilities are avoided. (This is equivalent to selecting the random stepsizes in MC simulations just large enough so that 25-30% of the moves made are accepted.) Once the system equilibrates, time averages of the properties of interest over portions of the trajectories are taken. These averages should be the same as the MC ensemble averages, according to the ergodic theorem.

The classical MD method operates in the microcanonical (E,N,V) ensemble. Consequently, the temperature is not constant during the simulation and has to be periodically readjusted. Since the temperature is simply related to the average kinetic energy of the system, the readjustment is done by scaling the velocities. Recent developments in theory lead to constant-T MD algorithms; they have not been used extensively in protein dynamics simulations.

From the computational point of view, neither method is clearly superior <u>if</u> one is only interested in equilibrium averages. At the moment there are only a limited number of experimental techniques (notably inelastic neutron scattering and Mössbauer spectroscopy) that would require MD.

Both MC and MD share an important advantage over static minimization: they can simulate <u>free energy changes</u>. This is the foundation of the powerful and useful <u>thermodynamic perturbation theory</u>. In fact, with enormous computer resources (not now available!), one could envision a fully microscopic simulation of the protein folding process in its entirety. The simulation would start with the denatured protein + environment system. External conditions (e.g. denaturant concentration, pH etc) would then be gradually changed and after each change the new system would be allowed to reequilibrate. This process of change and equilibration would be continued until native conditions were reached.

Would the protein acquire its native structure at the conclusion of such massive simulation? Possibly. And yet, one suspects that since <u>all</u> interactions, especially with solvent molecules, had to be treated <u>explicitly</u>, modelling these interactions accurately would become much more critical. Since a fully quantum mechanical treatment is out of the question, the need for a much more detailed representation (hence understanding) of the various types of empirical interactions we use, becomes that much more acute. This is particularly true with regard to their dependence on macroscopic, environmental parameters.

It should be evident that neither the minimization methods (static or dynamic) nor the statistically based model-building approaches are capable of solving the protein folding problem, especially on their own. We need more than just new, efficient solutions to the computational problems. The more

critical conceptual difficulties require a new approach, as was mentioned
earlier.

An integral part of any new approach ought to be the search for the
physical manifestations of those constraints that are absolutely necessary
for the proper control of the folding process. An important element of this
search is the requirement that we focus on the essential features of the
folding process, bypassing brute force, _unstructured_ simulations. We have to
understand how evolution had produced the self-organizing, _directing_ aspects
of folding.

The consequence of this requirement is that judicious _simplifications_
have to be introduced as we simulate the folding process. Furthermore, the
simplifications should depend on the particular stage of folding we are in
the process of simulating.

I do not mean to imply that either the model building or minimization
approaches are to be rejected. They each play important roles and should form
integral parts of an overall simulation strategy. Thus _model building_, with
its strong hierarchical foundation, should provide us with good initial conditions for the simulation (imitating the early stages of folding). On the
other hand, the _dynamic minimizers_ will be required both for physically relevant equilibrations along the folding pathways, and for the concluding fine-
tuning, once the main-chain topology has been established (molten globule ->
native structure).

Let us summarize the relevant concepts of protein folding.

ESSENTIAL CONCEPTS OF FOLDING: THE 3 C'S

    I.   C H A N G E :   This is induced by altering the protein's
    environment: F/U is a _process_

    II. C H O I C E :   The folding protein needs the CHANCE to
    1) explore alternative pathways
    2) _edit_ non-viable choices and
    3) _correct_ sub-optimal ones
    These are possible through _random fluctuations_:

    ==> "DYNAMICS"

    III. C O N T R O L :   Via both _structural_ and _kinetic_ CONSTRAINTS,
    in order to _limit_ excessive random search

The basic PREMISE is that an understanding of the folding process will
be achieved when we decode the hidden, as yet unknown _constrains/controls_.
Furthermore, we postulate as a reasonable working hypothesis that a first
step toward the physical simulation of the controls can be achieved by impos-
ing _hierarchical constraints_ on folding.

The notion of hierarchical folding is not new. It has been proposed
before, explicitly and implicitly. However, the earliest descriptions were

plausibility arguments against a random-search nucleation model, with qualitative reasoning for the existence and importance of nuclei/seeds in early
folding. The expression "hierarchical condensation" was coined later, based
on the observation of well-defined substructures in X-ray structures.

The multi-stage model of folding was treated in a more quantitative way
by two separate research groups. The diffusion-collision coalescence (DCC)
model of folding was introduced by Karplus and Weaver and applied to simplified protein models. Go and coworkers in Japan have carried out statistical
mechanical simulations on two-dimensional models of proteins on lattices.
They also introduced and studied the noninteracting local-structure model of
folding, which has several common features with the Karplus-Weaver model.

The DCC model is based on the premise that short segments (generally
the ultimate secondary structural elements) of the unfolded polypeptide chain
start folding into microdomains, and that this proceeds more or less independently. The microdomains are usually unstable, but if they diffuse together,
they may stabilize. Eventually they coalesce into larger domains, which may
become even more stable. With the assumption that the diffusional collision
is rate limiting, kinetic folding rates were calculated. In early work a
large number of simplifying assumption were made, to the extent that much of
the possible relevance to proteins was lost (two spheres joined by a feature-
less string is hardly a good representation of the folding of two subdomains
separated by an unfolded coil segment). The most recent calculations however
are much more realistic: the folding dynamics of two helical segments, connected by a flexible polypeptide chain, were simulated by Brownian dynamics.
This attempts to mimic a possible elementary step in early folding.

It is difficult to assess how realistic the results are. The assumption
that diffusion is the rate limiting step was questioned for the overall folding rate of RNase A, which was found experimentally to be independent of
solvent viscosity. The analysis of the simulation results is not trivial, and
is based on additional assumptions.

Go calls the preliminary unstable microdomains embryos. He designates
the general multi-stage folding process as the embryo-nucleus mechanism. In
this framework, he postulates that the growth in size of an embryo may follow
one of two typical models:

1) growth-merge mechanism: an embryo will grow in size by absorbing
contiguous random coil segments or by merging with another nearby embryo.

2) diffusion-collision mechanism: this is essentially the DCC model of
Karplus-Weaver, according to which embryos grow as a result of collision between two or more embryos that are not contiguous along the chain.

The two mechanisms are not mutually exclusive; both may be operative
at different stages of the folding process. Go claims that the growth-merge
mechanism of folding is more likely than folding by the DCC process; the
former leads to a smaller entropy loss because of its contiguous character.
DCC would become operative only when the growing embryo encounters a very
unstable section of the polypeptide chain. This section would then fold only
when flanking embryos diffuse close to it and help its stabilization.

Go introduced the concept of consistency among the various energy terms
that stabilize the native structure of proteins. He restated the hierarchical
condensation model of folding by emphasizing that the range of interactions
and the order at which they become operative or "dominant" is the principal
determining factor for proper folding. There is evidence that short-range

interactions are important; there are also indications that long-range interactions are critical. The consistency principle is introduced then to reconcile these conflicting points of view.

Simply stated, it asserts that <u>both</u> types of interactions may appear
dominant because both are <u>consistent</u> with the native structure. This means
that in minimizing the various energy terms, none of them should be singled
out, i.e. they should be individually minimized. If consistency were perfect,
there would be no local stress anywhere in the protein molecule in its native
state. There is no perfect consistency in proteins. This is because proteins
did not evolve purely for maximum stability. The fact that they are <u>function-
al</u> implies that other factors besides stability are important. Nevertheless,
the consistency argument enabled him to propose a model of folding which is
not hierarchical. This he called the <u>nonspecific globule model</u>. He defines
nonspecific interactions as those interactions that are <u>not</u> in their operative range in the native structure. The prime candidates for nonspecific
interactions are the hydrophobic interactions. The postulate implies then
that in the early stages of folding hydrophobic interactions give rise to a
globule in which the hydrophobic residues are packed randomly in the interior
of the molecule. This "nonspecific globule" should have a rather loose structure, with good prospects for large fluctuations. Two possible mechanisms can
be envisaged for the transition from the nonspecific to the native globule.
The "slithering snake" mechanism assumes that because of the loose structure,
the polypeptide chain can somehow move within the random globule until more
specific interactions stabilize it in the native structure. If such motions
are restricted, then the native structure could only be acquired by several
folding-unfolding attempts until the right near-native conformations can be
approximated. When this occurs, the specific interactions will lock in the
native structure.

Go speculates about the conditions under which the nonspecific globular
state could be observed either as an equilibrium state or as a kinetic intermediate. He argues that under conditions favoring refolding the nonspecific
globular state is likely to be observed kinetically. The molten globule state
has been tentatively identified with the nonspecific globule state.

The hierarchical embryo-nucleus model of folding is expected to be
dominant under conditions which do not favor refolding.

What is the most fruitful approach to the folding problem? I believe
that the hierarchical model captures the essence of the process, even if
details will have to be modified. In particular, "hierarchy with feedback and
rearrangement" should be a reasonable operating principle. It is rather more
difficult to suggest specific simulation strategies. However, recent computer
simulation work on the folding of small proteins <u>under the guidance of experimentally observed short-distance constraints</u> (NMR) indicates that the early
acquisition of native secondary structural elements is a necessary folding
protocol; if not encouraged and reinforced, even small proteins will fold
incorrectly. This argues for the importance of "turning on" of the interactions in a nonuniform manner, with short-range interaction preceding and
possibly dominating the influence of longer-range ones. The numerous advantages of a multi-stage organization of the folding process support the
likeliness that the concepts are correct. The advent of site-directed mutagenesis should enable the theoretician to test various folding theories much
more effectively.

REFERENCES

Anfinsen, C.B. and Scheraga, H.A. (1975) Adv. Prot. Chem. 29, 205-300
Experimental and Theoretical Aspects of Protein Folding
Brünger, A.T., Clore, G.M., Gronenborg, A. and Karplus, M. (1986)
Proc. Natl. Acad. Sci. USA 83, 3801-3805
Three-Dimensional Structure of Proteins Determined by Molecular Dynamics
with Interproton Distance Restraints: Application to Crambin
Dill, K.A. (1985) Biochemistry 24, 1501-1509
Theory for the Folding and Stability of Globular Proteins
Fetrow, J.S., Zehfus, M.H. and Rose, G.D., (1988) Biotechn. 6, 167-171
Protein Folding: New Twists
Ghelis, C. and Yon, J., 1982 "Protein Folding", Academic Press, N.Y.
Go, N., (1976) Adv. Biophys. 9, 65-113
Statistical Mechanics of Protein Folding, Unfolding and Fluctuation
Go, N. and Abe, H. (1981) Bioplymers 20, 991-1011
Noninteracting Local-Structure Model of Folding and Unfolding Transi
tions in Globular Proteins. I. Formulation
Go, N. (1984) Adv. Biophys., 18, 149-164
The Consistency Principle in Protein Structure and Pathways of Folding
Go, N., (1983) Ann. Rev. Biophys. Bioeng. 12, 183-210
Goldberg, M.E. (1985) TIBS 388-391
The Second Translation of the Genetic Message: Protein Folding and
Assembly. Theoretical Studies of Protein Folding
Harrison, S.C. and Durbin, R. (1985) Proc. Natl. Acad. Sci. USA 82,
4028-4030
Is there a Single Pathway for the Folding of a Polypeptide Chain?
Karplus, M. and Weaver, D.L., (1976) Nature 260, 404-406
Protein-folding Dynamics
Karplus, M. and Weaver, D.L., (1979) Biopolymers 18, 1421-1437
Diffusion-Collision Model for Protein Folding
Kim, P.S. and Baldwin, R.L. (1982) Ann. Rev. Biochem. 51, 459-489
Specific Intermediates in the Folding Reactions of Small Proteins and
the Mechanism of Protein Folding
King, J., (1986) Biotechnology 4, 297-303
Genetic Analysis of Protein Folding Pathways
Kuntz, Jr., I.D. in "The Protein Folding Problem" 1983, D.B. Wetlaufer,
ed., Westview Press, Inc., Boulder. AAAS Selected Symposium 89, pp 65-81
Stability and Dynamics of Globular Proteins
Lee, S., Karplus, M., Bashford, D. and Weaver, D.L., (1987), Biopoly-
mers 26, 481-506
Brownian Dynamics Simulation of Protein Folding: A Study of the
Diffusion-Collision Model
Li, Z> and H.A. Scheraga (1987) Proc. Natl. Acad. Sci. USA 84, 6611-6615
Monte Carlo-Minimization Approach to the Multiple-Minima Problem in
Protein Folding
Miyazawa, S. and Jernigan, R.L. (1892) Biochemistry 21, 5203-5213
Most Probable Intermediates in Protein Folding-Unfolding with a Non-
interaction Globule-Coil Model
Némethy, G. and Scheraga, H.A. (1977) Q. Rev. Biophys. 10, 239-352
Ohgushi, M. and Wada, A., (1984) Adv. Biophys. 18, 75-90
Liquid-like State of Side Chains at the Intermediate Stage of Protein
Denaturation

Oxender, D.L. (ed.) 1987 "Protein Structure, Folding and Design 2"
Alan R. Liss Inc, New York
Ptitsyn, O.B. and Finkelstein, A.V. (1979) Int. J. Quantum Chem. $\underline{16}$,
407-418
Mechanism of Protein Folding
Robson, B. and Garnier, J. 1986 "Introduction to Proteins and Protein
Engineering", Elsevier, Amsterdam
Scheraga, H.A. (1983) Biopolymers $\underline{22}$, 1-14
Recent Progress in the Theoretical Treatment of Protein Folding
Tanaka, S. and Scheraga, H.A. (1975) Proc. Natl. Acad. Sci. USA $\underline{72}$,
3802-3806
Model of Protein Folding: Inclusion of Short-, Medium-, and Long-Range
Interactions
Weaver, D.L., (1982) Biopolymers, $\underline{21}$, 1275-1300
Microdomain Dynamics in Folding Proteins
Wetlaufer, D.B. (1973) Proc. Natl. Acad. Sci. USA $\underline{70}$, 697-701
Nucleation, Rapid Folding, and Globular Intrachain Regions in Proteins
Wetlaufer, D.B. and Ristow, S. (1973) Ann. Rev. Biochem. $\underline{42}$, 135-158
Acquisition of Three-Dimensional Structure of Proteins
Wetlaufer, D.B. in "The Protein Folding Problem" 1983, D.B. Wetlaufer,
ed., Westview Press, Inc., Boulder. AAAS Selected Symposium 89, pp 29-45
Modular Processes and Rapid Selection for Rapid Folding

STRUCTURAL AND FUNCTIONAL PROPERTIES OF CONSECUTIVE ENZYMES IN THE GLYCOLYTIC PATHWAY

Herman C. Watson

Biochemistry Department
School of Medical Sciences
The University
Bristol, BS8 1TD, England

Introduction

In this article I will outline the basic mechanistic features of triosephosphate dehydrogenase (GPD), phosphoglycerate kinase (PGK) and phosphoglycerate mutase (PGM); enzymes whose structures have been studied by me and by my co-workers over many years. Emphasis will be placed on the properties of these enzymes which appear to be functionally important as seen from the static - snap shot - type pictures observed using the analysis of X-ray diffraction data. The information provided will show that great strides have been made towards our understanding of the enzymic processes involved but that much remains to be done to confirm the proposed reaction mechanisms.

Two of the selected enzymes are tetramers (GPD and PGM) whilst the linking enzyme (PGK) is a monomer as is shown in figure 1. In reality the yeast mutase is essentially a dimer of dimers as indicated by the fact that the enzyme from sources other than yeast is invariably composed of only two subunits (Fothergill-Gilmore and Watson, 1989). These three enzymes represent three very different reaction mechanisms (for details see below) and, for those interested in subunit interactions, the three of the most prevalent quaternary structure classes. At the tertiary structure level all three enzymes are of the alpha-beta-alpha type (Levitt and Chothia, 1974), two of the enzymes (GPD and PGK) having two domains one of which - the nucleotide binding domain - corresponds closely to that commonly referred to as the Rossmann fold (Rossmann et al., 1976). The other two domains of these subunits, as is the single domain of PGM, are similar to that of the Rossmann fold but differ in the order of the strands of the parallel beta sheet relative to their position in the linear sequence of amino acids (see Buehner et al., 1974 and Banks et al., 1979). The mutase subunit also has one of its six beta sheet strands running in the opposite (anti-parallel) direction to the remainder (for details see Campbell et al., 1974). It is perhaps not surprising therefore to find that this enzyme has no known nucleotide binding function.

In the direction of glycolysis the dehydrogenase reaction requires one substrate - glyceraldehyde 3-phosphate - and, following abstraction of a proton and a hydride ion, forms a covalent intermediate - the hemithioacetyl (for chemical reaction scheme see Trentham, 1971). The release of the co-valent intermediate is brought about by the reaction with inorganic phosphate but is also aided by the replacement of the reduced nucleotide cofactor by NAD^+. The kinase mechanism is conceptually less

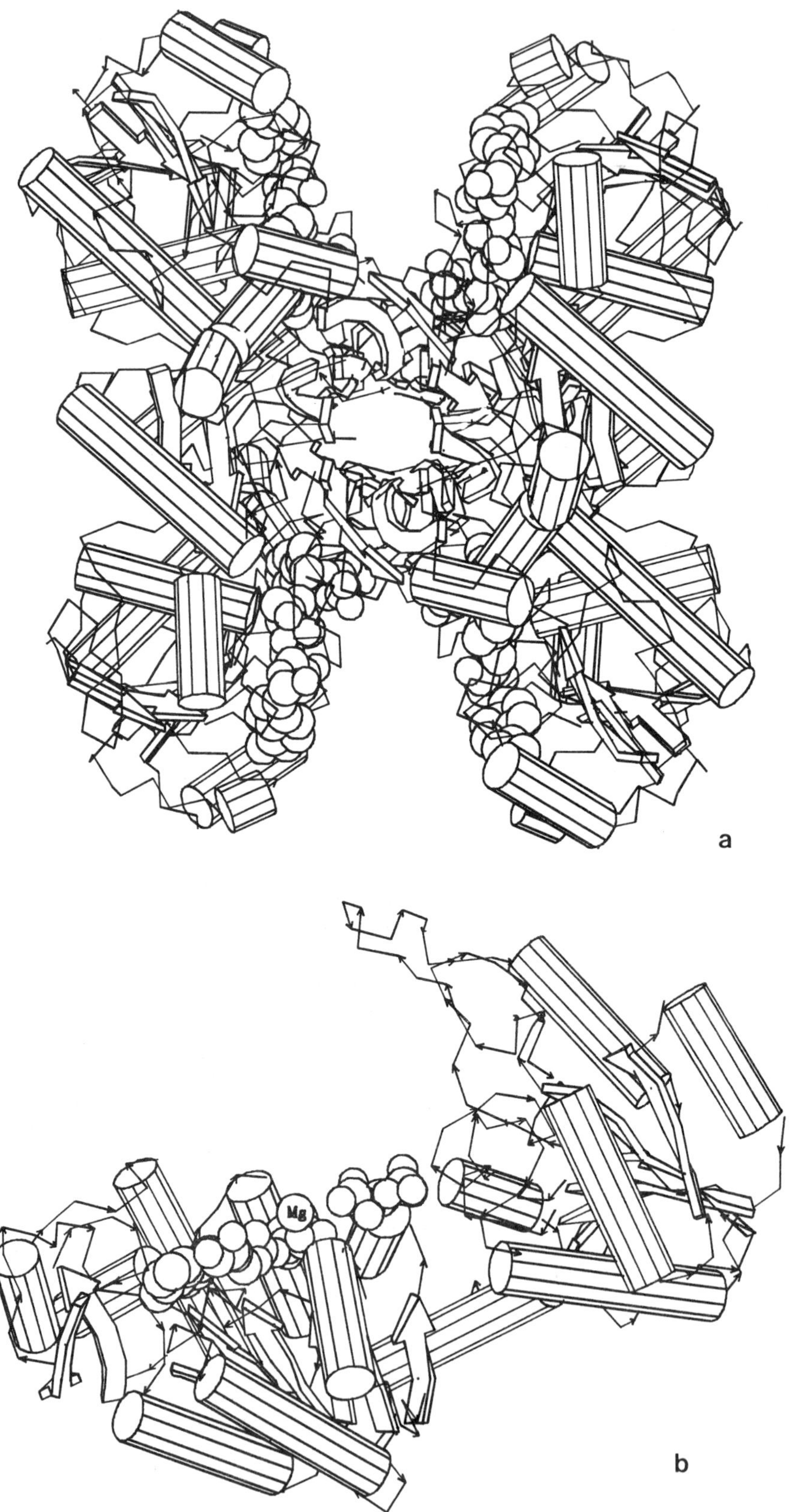

a
Mg
b

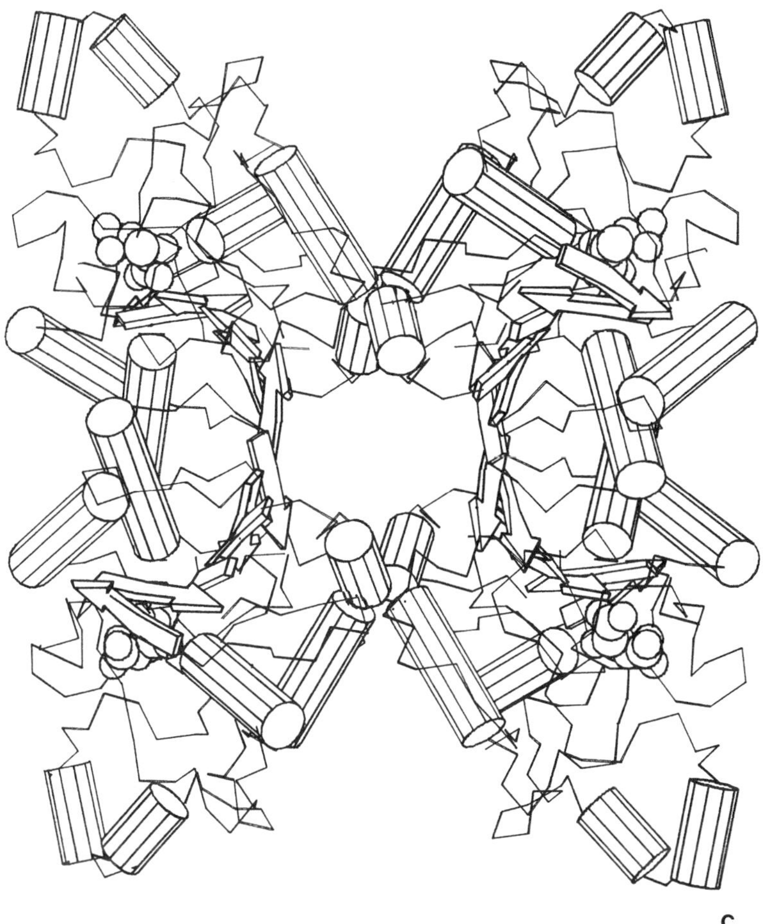

c

Figure 1. Computer drawings showing the overall secondary, tertiary and quaternary structures of the triosephosphate dehydrogenase from human muscle tissue (a), yeast phosphoglycerate kinase (b), and yeast phosphoglycerate mutase (c). As explained in the text the mutase is in essence a dimer of dimers. The functional dimer would be represented by either the right or left hand half of (c). These two halves of the molecule are related by a two fold symmetry axis passing through the centre of the molecule. In this figure the alpha helices are represented as barrels, the beta sheet as laths and the connecting regions by arrows indicating a single peptide units. The substrates as they are bound to these enzymes are represented by atom spheres. These drawings were prepared using a computer program written by Lesk and Hardman (1982).

complicated than that of the dehydrogenase in that no proton abstraction is required (for chemical reaction scheme see Scopes, 1973) and that the ion involved (Mg^{2+}) plays only a passive role in the reaction. The mutase reaction would appear to be the simplest of the three enzymes under consideration (for overall chemical reaction scheme see Rose, 1982). Unlike the kinase the mutase requires only one substrate and unlike the dehydrogenase it does not require a cofactor or additional ion. For the mutase reaction to proceed, however, the enzyme must be in its phosphorylated form and a proton must be abstracted from the attacking hydroxyl group.

From this brief introduction, and the information contained in the review references quoted, it will be seen that "our" three enzymes differ markedly. In particular they differ in subunit molecular weight by increments of 10,000 daltons varying from 25,000 daltons for the mutase to 45,000 daltons for the kinase. They differ in subunit number and, more relevant to this article, in reaction type and complexity. In muscle glycolysis for example it is the tetramer's task to add inorganic phosphate to the triose sugar at the expense of reducing the energy rich co-factor NAD^+. The "dimer's" role is to move a phosphate group between hydroxyl oxygens associated with adjacent carbons. This phospho transfer facilitates energy removal from the triose substrate following the abstraction of a water molecule by the enzyme enolase. The smallest of "our" three enzymes - the kinase - transfers the phosphate group added to the triose sugar during the dehydrogenase reaction to form energy rich ATP.

In order to ensure that energy is derived from the reduction of NAD^+ the dehydrogenase has to solve the problem that the thioester formed during the enzymic reaction must be phosphorylated rather than hydrolysed. Also both the kinase and mutase must ensure that phosphoryl transfer occurs to the proper substrate and not to water. In the following sections we will see how each of the three enzymes achieves its allotted task. In the penultimate section I will deal with the way that techniques other than protein crystallography are being used to substantiate the various mechanistic proposals and also to help develop a dynamic rather than a static picture of these reaction processes. In the concluding section I will detail reaction similarities and differences whilst emphasizing the simple but effective way in which these enzymes reach their objectives.

The Dehydrogenase

Glyceraldehyde 3-phosphate dehydrogenase is a tetrameric enzyme containing four identical chains. There has been considerable controversy relating to the symmetry of ligand binding. The binding of NAD^+ to the enzyme is definitely co-operative but glyceraldehyde 3-phosphate (G-3-P) can be shown to bind independently to all four subunits. A comparison of the bacterial and human muscle apo- and holoenzymes has indicated that the binding of NAD^+ causes large movements in the coenzyme binding domain. Unfortunately a meaningful discussion of co-operativity, particularly as it relates to the structural changes seen in the high resolution studies of the bacterial enzyme, are beyond the scope of this article. For further information on this feature of the enzymic process the reader is referred to papers by Wonacott and his colleagues (see Skarzynski *et al.*, 1987 and the references therein).

The first step in the GPD reaction sequence is the formation of a hemithioacetal between a cysteine and the substrate G-3-P. This has the effect of converting the carbonyl, which is not easy to oxidize directly, into a hydroxyl that is readily dehydrogenated. During this reaction step a hydride ion is transferred from the C_1 carbon to form a reduced nucleotide cofactor which then leaves the active site to be replaced by NAD^+. The thioester formed then reacts with inorganic phosphate to produce the acylphosphate as is shown in figure 2b. It can be shown that the acyl transfer is very slow unless NAD^+ is bound to the enzyme. The replacement of NADH by NAD^+ is therefore a necessary part of the reaction sequence.

The crystal structures of the holoenzymes from lobster (Buehner *et al.*, 1974) and human muscle (Reed *et al.*, 1989) have been determined to a resolution of 2.9 and 2.4Å resolution and that of the *Bacillus stearothermophilus* holoenzyme to 1.8Å (Skarzynski *et al.*, 1987). The structures of enzyme-substrate complexes have been deduced from model-building experiments on both the muscle and bacterial enzymes. The following description is a composite of these studies.

Two binding sites for sulphate ions have been identified at the active sites of all GPD enzymes that have been crystallised from ammonium sulphate. A stereochemically reasonable model for the course of the GPD reaction may be constructed by assuming that these are the binding sites for the phosphate group associated with the thioester's C_3 phosphate and the inorganic phosphate (but for a 'complication' see below). Prior to the acylation of the thioester the covalent intermediate makes hydrogen bonds with the hydroxyl of a threonine, the positively charged side chain of an arginine and the 2'-hydroxyl of the ribose ring that is attached to the nicotinamide of NAD^+. The C_2 hydroxyl of the substrate can then form a hydrogen bond with a serine when the C_1 hydroxyl forms a similar bond with an imidazole nitrogen of a histidine. The transition state for the attack of inorganic phosphate on the thioester is stabilized by hydrogen bonds to the phosphate oxygens from a number of residues including a serine, two threonines and a main chain nitrogen. The existence of this extensive and species invarient binding site for phosphate presumably explains why the thioester is phosphorylated rather than hydrolyzed during the reaction sequence. The sulphur atom of the thioester is close enough to the C_4 carbon of the nicotinamide ring of NAD^+ to be polarized by its positive charge presumably explaining the activation of the acyl transfer reaction when NADH replaces NAD^+.

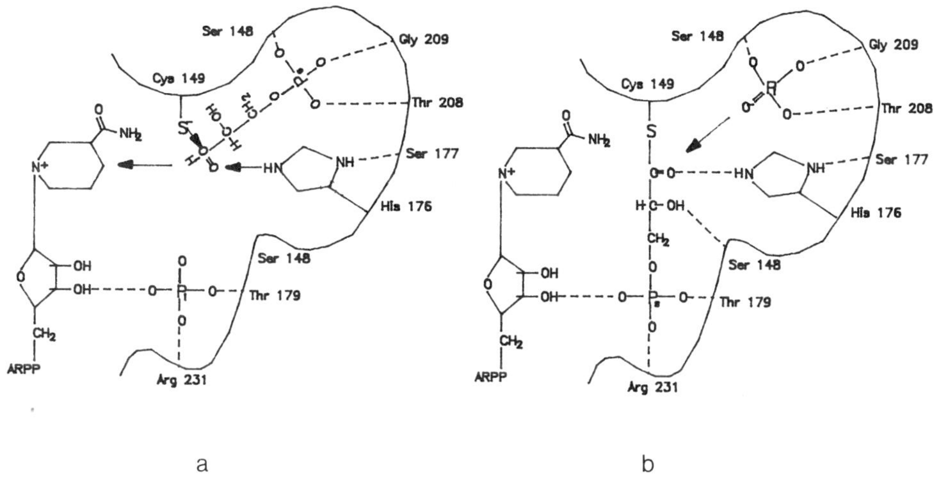

Figure 2. A schematic drawing of the active site of the dehydrogenase showing :-

a) the substrate G-3-P with the C_3 phosphate bound at the 'P$_i$' site as prior to the formation of the hemithioacetyl intermediate (for explanation see below).

b) the thioester intermediate with its C_3 phosphate in the 'P$_s$' position.

The right hand side of this figure is essentially that produced by Wonacott based on his 2.7Å studies of the *Bacillus* enzyme. (a) is the same as (b) except that it has been modified by the author (placement of G-3-P) to include the information implicit in the 1.8Å structure determination and referred to in the text (for further details see Skarzynski *et al.*, 1987). Broken lines indicate hydrogen bond interactions. The residues indicated in this figure are invariant in all known GPD sequences.

Several minor but important changes have recently been made to the original interpretation of the sequence of events which takes place at the active site as G-3-P is converted to 1,3-DPG. These changes take into account the greater accuracy of atomic parameters implicit in the high resolution structural study of the holo enzyme from *Bacillus stearothermophilus*. This study indicates that the hydrogen bond between the ND atom of His-176 and the carbonyl oxygen of Ser-177 (*Bacillus stearothermophilus* numbering) serves to anchor the imidazole ring throughout the reaction sequence. With the revised orientation of the side chain of Histidine 176 it is no longer possible to build a stereochemically reasonable model of the hemithioacetal with its C_3 phosphate group in the site designated P_s by Moras *et al.*, (1975). Although it is still believed that the P_i site must be the location for the inorganic phosphate in the phosphorylation step of the reaction, the P_i site is now also thought to play a role in binding the C_3 phosphate of the substrate (G-3-P) and the hemithioacetyl intermediate. This implies that the inorganic phosphate and much of the co-valent intermediate must be relocated during the course of the reaction in the manner indicated in figure 2.

The Kinase

Phosphoglycerate kinase is a monomer enzyme which, in its substrate binding form, has two very well separated domains (see figure 1b). One domain is composed of residues from the carboxyl end of the polypeptide chain but the last 15 or so residues are required to complete the amino terminal domain. The active site of the enzyme involves a major portion of the interface between the amino and carboxyl domains. This region of the molecule contains a very large proportion of those residues which have been found to be invariant when comparing the twenty one sequences which have been derived for PGKs from prokaryotic and eukaryotic organisms. This feature of the molecule appears to confirm the prediction made originally by Blake and his colleagues (Banks *et al.*, 1979) that the two domains move relative to each other during the reaction process. Indeed this feature of the reaction mechanism has led to some confusion as to the interpretation of experimental results and has served to show us that earlier concepts of enzyme mechanism, as derived from studies of extra cellular enzymes, are perhaps over simplified. This is a point I shall come back to in the penultimate section of this article. Unless otherwise stated all the PGK residues referred to are invariant in all species for which sequence information is available.

An examination of the cleft region of PGK reveals a relatively confined surface patch containing six closely grouped basic residues. X-ray structure studies have shown that the nucleotide substrates bind to the section of the cleft opposite to the basic patch area. The sugar substrate also binds between the two domains with its non transferable phosphate hydrogen bonded to the side chains of at least two amino domain residues as is shown in figure 3a. In this position, deep in the inter domain interface, the carboxyl group of 3-phosphoglycerate (3-PGA) is located in a position, and at the correct distance from the terminal phosphate of ATP, to effect phosphoryl transfer (Watson *et al.*, 1982).

From the X-ray structure results summarised in figure 3a, it can be seen that both alpha and beta phosphate oxygens interact with the terminal groups of lysine residues. Two of the three gamma phosphate oxygens interact with the amino nitrogens at the end of an alpha helix. The third phosphate oxygen is exposed to the solvent in the open form of the enzyme in a position where protein ligands could not possibly form charge or hydrogen bond interaction with this potentially electro negative atom. The magnesium ion associated with the nucleotide substrate carries two positive charges but one of these charges is presumably balanced by the negative charge of an aspartic acid side chain which is positioned in the structure in such a way as to suggest that its primary role is to help desolvate the metal ion (see Watson and Gamblin, 1985). Taking into account the 1/2+ charge associated with helical dipole (see Hol, 1980), and making appropriate allowance for the charges associated with interacting groups, the charge of the bound nucleotide substrate appears to be -1/2. Similar arguments relating to 3-PGA and its binding site lead to the conclusion that the charge associated with the triose substrate is also -1/2. Thus we see in charge terms why phosphoryl transfer is unlikely to occur when the appropriate substrates are properly bound to the enzyme. This leads us to believe that a structural rearrangement occurs following the formation of the true ternary complex and that it is this feature of the enzyme mechanism which ensures that there is virtually no phosphatase activity. Presumably the necessary structural change alters the charge situation associated with the substrate carrying the transferable phosphate making it transiently positive.

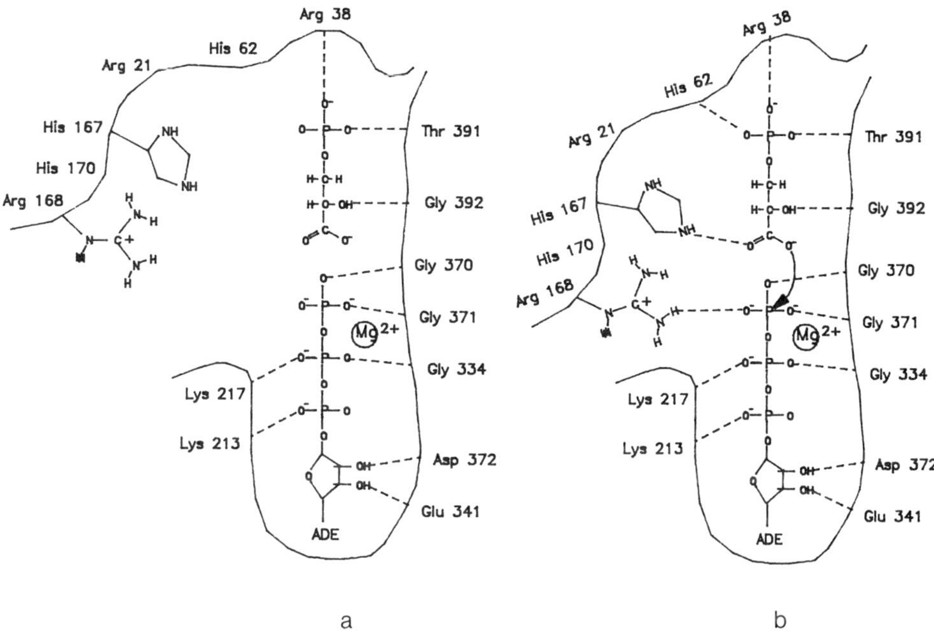

Figure 3. A schematic drawing of the active site of the kinase (PGK) showing substrate protein hydrogen bond (broken lines) interactions in both the 'open' (a) and 'closed' forms (b) of the enzyme. The residues associated with the amino terminal domain form the upper 'half' of the active site. These residues have been rotated 15^{0} relative to those representing the carboxyl domain (lower 'half') to produce the closed form structure. The centre of rotation was assumed to be located close to the positive component of the salt link between arginine 38 and phosphate of 3-PGA.

In addition to providing a possible understanding of the lack of appreciable phosphatase activity the arguments detailed above also prompt us to ask questions about the nature of the molecular rearrangement which affects the charge situation surrounding the transferable phosphate. Inspection of the open or substrate binding form of the yeast enzyme indicates that if the domains were to move relative to each other following the formation of the true ternary complex, then the 'basic patch' residues could be positioned with their centre of charge in close proximity to the transferable phosphate site. Crystal soaking experiments (for details see Watson *et al.*, 1982) have shown that substrates bind with two of the transferable phosphate oxygens bonded to the main chain nitrogens associated with the amino terminal end of an alpha helix. With the transferable phosphate located in this way, the remaining non-linking oxygen is exposed to solvent in a position where it could interact with a 'basic patch' residue following domain movement. If this is the case then in this closed or catalytically active form of the enzyme a positively charge residue could act as a transition state stabilising residue. The situation drawn in figure 3b shows arginine 168 adopting this role and effectively altering the charge of the substrate carrying the transferable phosphate.

Efforts to crystallise the closed form structure have not proved successful. In the absence of a closed form structure we have been forced therefore to use non crystallographic methods in our attempts to test this interesting hypothesis. These methods and the results that have been obtained are described in the penultimate section.

The Mutase

There are at least four different types of phosphoglycerate mutase each of which are kinetically and structurally distinct (Fothergill-Gilmore and Watson, 1989). The enzyme we are considering is the mutase which catalyses the inter-conversion of 3-phosphoglycerate (3-PGA) and 2-phosphoglycerate (2-PGA). This enzyme is activated by catalytic amounts of the cofactor/primer 2,3-bisphosphoglycerate (2,3-DPG). Depending on the conditions prevailing in the cell (eg. ions, pH) the enzyme has to be re-primed approximately once in every 10^5 enzyme catalysed reactions. It will be see therefore that 2,3-DPG is not a cofactor in the accepted sense of that word and will be referred to subsequently as "the primer".

The high resolution structure of the yeast phosphoglycerate mutase has been determined and the active site located using crystals soaked in 3-PGA (Winn *et al.*, 1981). The active site lies at the bottom of a deep hollow and is formed entirely by residues of one subunit. In the tetrameric structure (see figure 1c) the four active sites are well separated and appear to be freely accessible to the solvent. Like the situation found in the dehydrogenases two sulphate ions bind in the crystal structure at positions assumed to be occupied by the phosphoryl groups associated with the bound substrate. The most prominent feature of the active site is two histidine side chains which are parallel and approximately 4Å apart. One sulphate ion binds close to histidine 8 and is located where it can make hydrogen bonds to a serine and a threonine (see figure 4). Groups located at this position are referred to as binding at the transferable phosphate site. The second active site sulphate appears to be stabilized by the hydrogen bonds formed (and the dipole effect) at the amino end of an alpha helix. Phosphoryl groups bound in this position are said to be bound at the non-transferable binding site.

Modelling studies based on the observed binding site for 3-PGA indicate that the primer 2,3-DPG can bind at the active site in two possible orientations. One of these orientations has its 3 phospho group located at the transferable phosphate position and the 2 phospho group at the non-transferable position. The second orientation has the primer's two phosphate groups reversed. In both orientations the carboxyl group associated with the 1 carbon is able to form a salt bridge with an arginine side chain buried deep in the active site pocket. Not surprisingly it has been found that substrates for this enzyme have a mandatory requirement for a negatively charged group associated with the C_1 position (Pizer and Ballou, 1959).

The C-terminal 14 residues (-Ala-Ala-Gly-Ala-Ala-Ala-Val-Ala-Asn-Gln-Gly-Lys-Lys) are not observed in the electron density map of the un-phosphorylated enzyme presumably because they form a flexible tail. These residues are particularly susceptible to proteolysis and enzyme which has lost "tail" residues, but is otherwise unmodified, shows complete loss of mutase activity. It has been suggested therefore (see Winn *et al.*, 1982) that the terminal part of this flexible tail moves into the active site cleft during the catalytic process as is show schematically in figure 4b. The function of this tail is probably threefold:-

1) to provide charge balance when negatively charged substrates bind at the active site (the tail has an effective charge of $+1$).

2) to fill the space in the active site pocket which would otherwise be occupied by water molecules.

3) to provide the ligand (one of the lysine epsilon nitrogens) which stabilises the transition state species.

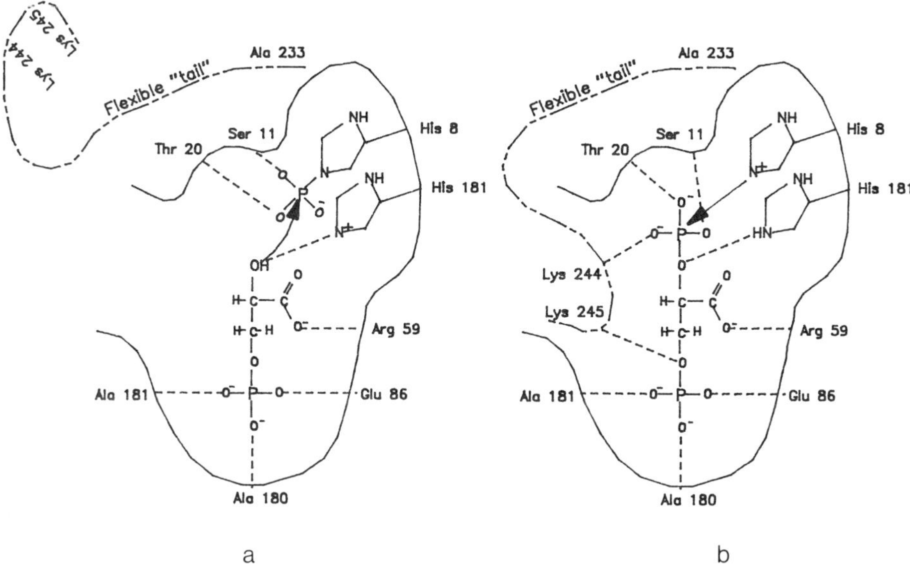

Figure 4. A schematic drawing of the active site of the mutase showing :-

a) the substrate 3-PGA bound with its C_3 phosphate located at the non -transferable phosphate position. The C-terminal tail is drawn in the tail out position as if the triose were just about to enter the substrate binding pocket.

b) the 'primer' 2,3-PGA bound with its C_3 phosphate located at the non -transferable phosphate position. The C-terminal tail is drawn in the tail in position as it would be if phosphoryl transfer were about to proceed.

Broken lines indicate hydrogen bond interactions. The arrows indicate the direction of proton and phosphoryl transfer.

Using all the available chemical, kinetic and structural data a scheme has been proposed (see Fothergill-Gilmore and Watson, 1989) for the overall mutase reaction which is summarised in figure 5. In this scheme the priming of the reaction occurs when 2,3-DPG binds in one of two possible orientations to the unliganded form of the enzyme. In the orientation shown in figure 4b the phosphoryl group associated with the 'primers' C_2 atom is positioned at the transferable phosphate site and has one of its oxygens free to form a transition state stabilising reaction with the C-terminal lysine residue. In this position the other phosphate group is located at the non transferable phosphate binding site where it is stabilised by bonds formed at the amino terminal end of an alpha helix. As the transition state develops the second active site histidine donates its proton to the bridging oxygen and the phosphorylated enzyme is formed. Following the release of the mono phosphoglycerate the enzyme is primed for its role in glycolysis - the interconversion of 2 and 3 phosphoglycerates. It will be appreciated that in order to carry out its proton donating/abstracting role the second active site histidine must be positioned very precisely with respect to the bridging/attacking oxygen as was the case for the similarly employed histidine in the dehydrogenase reaction (see figure 2).

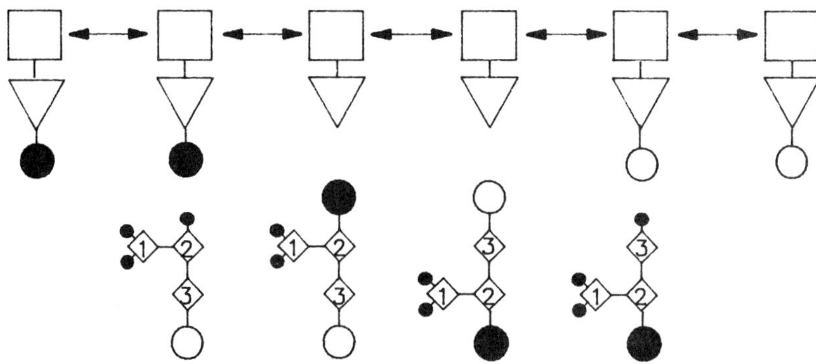

Figure 5. Proposed reaction sequence for the mutase. The open squares represent the enzyme subunit and the open triangles histidine 8. The numerals 1, 2, 3 label the carbon atoms of the glycerate moiety. The large circles (both open and filled) represent phosphoryl groups and the small circles oxygen atoms associated with the substrates' hydroxyl and carboxyl groups.

A round of catalysis would be initiated when either 3-PGA or 2-PGA binds at the active site. The relative concentrations of the mono phosphoglycerates presumably influences whether the mutase acts as in glycolysis or as in gluconeogenesis. Note that the 2,3-DPG intermediate in the reaction scheme referred to above ((b) and (c) in figure 5) are the same as for the initial priming reaction. In other words the primer of the mutase reaction is in reality a reaction intermediate. An important feature of this scheme which is illustrated in figure 5 is that there is a requirement for 2,3-DPG to re-orient during each round of catalysis. Modelling studies show that the physical characteristics of the active site pocket are suitable for this and manoeuvre if it is assumed that the reaction intermediate moves away from the phosphorylatable and proton accepting imidazole groups depicted schematically in figures 4. The available data would seem to suggest that the reaction intermediate rarely leaves the active site cleft during catalysis (Warwicker and Watson, 1982).

Protein Engineering

At the start of a enzyme structure study the reaction catalysed is usually known and information is therefore available about the cofactors involved, about the nature of the natural substrates and, in some cases, the reaction intermediates. If the enzyme is one that has been subject to detailed biochemical study then information is also available concerning the amino acid sequence and the involvement in the enzymic reaction of specific amino acid side chains. With some of the extra cellular enzymes (most notably those of lysozyme and chymotrypsin) it was a relatively straight forward task to use the mechanistic information implicit in the structure determination to provide a much more detailed description of the overall reaction than that based on chemical information alone. The enzymes described above follow the lysozyme/chymotrypsin example but in each case the active site, as seen in the structure determination, cannot be used to provide a complete description of the enzymic reaction without making major assumptions. These assumptions involve movement of one or more of the reaction components and as such cannot be fully evaluated using the standard methods of chemical, kinetic and crystallographic enzymology. In order to test detailed hypotheses based on the structure determinations about the relative movement of substrates and protein it is necessary to adapt conventional techniques in order to examine protein modified by the use of site-directed mutagenesis.

The use of site directed mutagenesis to study enzyme mechanism is still in its infancy. Some workers, particularly Fersht and his colleagues (e.g. Leatherbarrow and Fersht, 1986), have already used this method to confirm the basic mechanistic features of specific enzymes. Below I shall describe rather briefly our most recent attempt (but see also Wilson et $al.$, 1987) to use this technique to prove that the PGK domains move during catalysis. It should perhaps be noted that similar experiments are being carried out to confirm the movement of the thioacetyl intermediate in the GPD reaction (Mougin et $al.$, (1988) and with PGM (Fothergill-Gilmore et $al.$, (1989)) to confirm that the flexible tail stabilises the transition state complex as has been predicted from the sequence data.

An arginine residue remote from the PGK phosphoryl transfer site has been implicated as being involved in stabilising the transition state intermediate following domain movement as is shown in figure 3. In designing the relevant site-directed mutagenesis experiments it was assumed that the effect on activity of modifying this residue would provide definitive evidence for domain movement during catalysis. It was anticipated, for example, that replacing such a vital residue with one of similar charge would have a measurable effect on the enzyme's catalytic efficiency (V_{max}/K_m) but that a change involving the substitution of a neutral residue would eliminate transition state stabilisation and thus abolish enzymatic activity. The kinetic results summarised in Table 1 indicate that, even though residue 168 is some 13Å from the phosphoryl transfer site in the X-ray structure, this arginine manages to influence the binding of both substrates. These results can be taken therefore as direct evidence that this residue is close to the phosphoryl transfer site following the binding of substrates. The experiments show, however, that the effect of substituting residue 168 has a greater effect on the binding of the triose substrate than on the nucleotide which carries the transferable phosphate. It seems reasonable to conclude therefore that arginine 168 does not form a bond with the transferable phosphate but that it forms a salt link with one of the two oxygens associated with the triose's C_1 carbon. Such a bonding arrangement would explain the effect of the lysine and methionine substitutions on the K_ms for both the nucleotide and triose substrates. In the case of the lysine mutant, for example, the 'displaced' proton donating atom would alter the binding position of the attacking carboxyl group and thus affect the location of the charge at the transferable phosphate site. The methionine mutant would result in the loss of a salt link with the triose substrate and would lead to the retention of the destabilising charge interaction between the negatively charged carboxyl group and the transferable phosphate following domain closure.

Table 1 Michaelis-Menten kinetic parameters for R168K and R168M mutant and wild-type yeast PGK. The enzyme activity (EU) is expressed as μmol NADH converted/min at 25^{0}C.

PGK Enzyme	V_{max} (EU/mg)		K_m (mM)	
Substrate	ATP	G-3-P	ATP	G-3-P
Wild-type	500	499	0.33	0.78
Arg --> Lys	201	203	0.90	2.11
Arg --> Met	102	93	1.18	6.10

In summary these carefully designed and executed experiments have provided some evidence that domain movement does occur but they have failed to corroborate the proposal implicit in figure 3 (a and b) that domain movement is an essential feature of the PGK enzymic reaction process. Further experiments of this type, possibly involving the introduction of a strategically placed tryptophan residue for use in fluoresence experiments, must now be undertaken to provide information about the dynamics of the domain movement.

Summary Observations

In the preceding sections I have shown how the proposed reaction mechanisms for three consecutive enzymes in a single biochemical pathway have been modified to include information implicit in the respective high resolution structure determinations. Based on structural observations, and also many related biophysical studies, the molecular movement is see to be of critical importance in any detailed understanding of biological function. The first of our three enzymes (the dehydrogenase) moves its substrate during the reaction probably following the formation of NADH but before its replacement with NAD$^+$. The second enzyme (the kinase) moves its domains relative to each other following the binding of substrates in order to complete the active site. The third enzyme (the mutase) moves part of its subunit during the reaction and, like the dehydrogenase, moves the reaction intermediate as well. All three processes are quite different although they are used to solve related chemical problems.

For both the dehydrogenase and mutase reactions to proceed a proton has to be abstracted from a hydroxyl before phosphorylation occurs. With the kinase this is not necessary as the oxygen which joins the transferable phosphate to form the trigonal bipyrimidal intermediate is unprotonated both before and after the reaction. In other words the dehydrogenase and the mutase reactions involve a protonatable sidechain (a histidine) whereas the kinase does not. For the same histidine to both abstract and donate a proton from the thioacetyl group (for the dehydrogenase) and

2,3-DPG (for the mutase) requires that both intermediates are relocated with respect to the protein during the overall reaction sequence. These two enzyme reaction mechanisms have another common feature which is not immediately obvious. Both use moveable groups to stabilise, by introducing a positive charge, their reaction intermediates. The dehydrogenase presumably uses the cofactor NAD^+ rather than the protein because, unlike the mutase, it has a requirement to 'loose' a hydride ion.

The major problem overcome by the kinase and mutase enzymes is the possible transfer of the phosphoryl group to a water molecule (phosphatase activity). Both enzymes appear to solve this problem by completing their respective active sites with water excluded from the region associated with phosphoryl transfer. The question therefore arises as to why one enzyme relies on the hinge bending of a monomeric enzyme to provide the necessary transition state ligand when the other uses the tail wagging property of a multimeric protein? Perhaps the reason why such different strategies have evolved relates to the requirement that the mutase reaction intermediate - 2,3-DPG - has to turn over in its active site for the enzymic reaction to proceed. In the absence of a verifiable explanation it seems reasonable to assume that moving the tail partway towards the solvent restricts access to water whilst also helping to contain the non-covalently linked reaction intermediate within the substrate binding pocket. Loss of the reaction intermediate would of course reduce enzymic efficiency. The kinase reaction on the other hand does not involve reaction intermediates and the hinge bending motion of this enzyme would certainly be sufficient to restrict access of water to the active site and also effect the appropriate equilibrium of substrates.

Finally we shoud ask the question why is the mutase multimeric. The dehydrogenase shows co-operative effects although these are not thought to be of any specific biological importance. Skarzynski *et al.*, (1987) have indicated , however, that an invarient arginine residue contributes to the formation of the active site of a symmetry related subunit. For the mutase the active site appears to be formed exclusively from residues of one subunit suggesting that it should be active as a monomer. Arguments have been put forward which suggest that the formation of multimeric enzymes helps stabilise the monomer structure and provides some degree of protection against proteolysis. The fact that the kinase has evolved to be fully functional as a monomer would appear to suggest that other biologically advantageous reasons exist for subunit association probably.

Acknowledgements

Over the past twenty years many people have passed through the Bristol Protein Structure Laboratories and all have been involved in some way with our study of the structural and mechanistic properties of the enzymes of the glycolytic pathway. Some have made major contributions during a brief stay while others have persisted over many years with difficult and often intractable problems. To mention individuals would therefore be quite invidious. I would like to express my thanks to all those who have helped with the Bristol based work on the the glycolytic enzymes and also to acknowledge the continued support of the Science and Engineering Research Council.

References

Banks, R.D., Blake, C.C.F., Evans, P.R., Haser, R., Rice, D.W., Hardy, G.W., Merrett, M. & Phillips, A.W. (1979) Nature (Lond.) *279*, 773-777.
Buehner, M., Ford, G.C., Moras, D., Olsen, K.W. & Rossmann, M.G. (1974) J. Mol. Biol. *90*, 25-49.

Campbell, J.W., Watson, H.C. & Hodgson, G.I. (1974) Nature *250*, 301-303.
Fothergill, L.A. & Watson, H.C. (1989) in "Advances in Enzymology" (Meister, A. ed.), *Vol.62*, 227-313. Wiley: New York.
Hol, W.G.J., van Duijnen, P.T. and Berendsen, H.J.C. (1978) Nature Lond., *273*, 443-446.
Lesk, A.K. and Hardman, K.D. (1982) Science, 216, 539-540.
Leatherbarrow, R.J. and Fersht, A.R. (1986) Protein Engineering, *1*, 7-16.
Levitt, M.R. and Chothia, C. (1976) Nature *261*, 552-556.
Moras, D., Olsen, K.W., Sabesan, M.N., Buehner, M., Ford, G.C. and Rossmann, M.G. (1975) J. Biol. Chem., *250*, 9137-9162.
Mougin, A., Corbier, C., Soukri, A., Wonacott, C., Bralant, C. and Bralant, G. (1988) Protein Engineering, *2*, 45-48.
Pizer, L.F. and Ballou, C.E. (1959) J. Biol. Chem., *234*, 1138-1142.
Reed, R., Fujinaga, M., Littlechild, J.A., and Watson, H.C. (1989) to be published.
Rose, Z.B. (1982) in "Methods in Enzymology", (Colowick, S.P. and Caplan, N.O. eds.) *87*, 42-51. Academic Press: New York.
Rossmann, M.G., Liljas, A., Branden, C.-I. and Banaszak, L. (1975) in "The Enzymes" (Boyer, P.D. ed.) Vol 2, 61-93. Academic Press: New York.
Scopes, R.K. (1975) in "Methods in Enzymology", (Colowick, S.P. and Caplan, N.O. eds.) *42*, 134-137. Academic Press: New York.
Skarzynski, T., Moody, P.C.E. & Wonacott, A.J. (1987) *193*, 171-187.
Trentham, D.R. (1971) Biochem. J., *122*, 59-61.
Warwicker, J. & Watson, H.C. (1982) J. Molec. Biol., *174*, 671-679.
Watson, H.C., Walker, N.P.C., Shaw, P.J., Bryant, T.N., Wendell, P.L., Fothergill, L.A., Perkins, R.E., Conroy, S.C., Dobson, M.J., Tuite, M.F., Kingsman, A.J., & Kingsman, S.M. (1982) E.M.B.O. J., *1*, 1635-1640.
White, M.F., and Fothergill,-Gilmore, L.A. (1988) FEBS Lett., *229*, 383-387.
Wilson, C.A.B., Hardman, N., Fothergill-Gilmore, L.A., Gamblin, S.J. and Watson, H. (1987) Biochem. J., *241*, 609-614.
Winn, S.I., Watson, H.C., Harkins, R.N. & Fothergill, L.A. (1981) Phil. Trans. R. Soc. Lond. B, *293*, 121-130.

THE SPECTROSCOPY OF ENZYMES: INTRODUCTORY REMARKS

J. L. Houben

Istituto di Chimica Quantistica ed Energetica Molecolare
Consiglio Nazionale delle Ricerche
Via Risorgimento, I-56100 Pisa (Italy)

The problems of the determination by molecular spectroscopies of the enzyme structure and dynamics as related to their catalytic activity appear, at least at first sight, to be fairly trivial and easy to resolve; a close collaboration between enzymologists and spectroscopists and an awareness that one must begin by posing the problem in functional terms, seem to be all that are needed. The appropriate questions may well be formulated in the following terms: "How does an enzyme actually work? How is it structured? What are the internal dynamic components relevant to the catalytic event? How can the relevant structural and dynamic parameters be measured?"

In practice, the problems are much less easily resolved. The first reason is the obvious difficulty which scientists from widely different backgrounds have in communicating with one another: asking the right question requires an in-depth knowledge of enzymology and spectroscopy. The second reason is more complex as it is related to the description of a chemical reaction. To calculate the potential surface along the reaction coordinate, it is necessary to know the structure of the matrix around the reactants and its dielectric time response in the time scale of the event (1). This may involve extremely short time intervals which are not always directly measurable for various reasons: the time resolution of the available techniques and/or the lack of a local probe. The third reason, which will be discussed below, is the difficulty of defining the relevant variables.

Before entering into a discussion of the application of some spectroscopy techniques, it appears timely and appropriate in the present context to give a general introduction to molecular spectroscopy (2).

THE SEMI-CLASSICAL DESCRIPTION OF A ONE-MOLECULE - PULSED LIGHT SPECTROSCOPY EXPERIMENT

In principle, the description of a real pulsed beam of light propagating along the z axis requires the introduction of the concept of the wave packet to account for its time, spatial and frequency properties. In the following semi-classical description of the interactions between molecules and light, a standing monochromatic wave can be used without any loss of general applicability and since this description greatly simplifies the equations without threatening the validity of the conclusions, it will be used in the following discussion. Light is described by an electromagnetic wave to which correspond the electric, E, and magnetic, H, fields (in the following, italic letters describe vectors) given by an equation of the type:

$$F(x,y,z; t,\upsilon) = F(x,y) \{\exp i(-kr + 2\pi\upsilon t)\} \tag{1}$$

where: $k = (0,0,k)$ is the wave vector, $F(x,y)$ gives the wave amplitude and polarization (the fields are orthogonal and located in the x-y plane), and υ is the wave frequency; note that $k = 2\pi\upsilon/c = 2\pi/\lambda$, where c is the velocity of light and λ the wavelength.

In this approximation, a simple one-molecule spectroscopy experiment can be described as the measurement of the signal resulting from the interactions of the electromagnetic wave with a molecule situated at the origin of the axis. The interaction energy, H, between an electromagnetic wave and a molecule is given by:

$$H = E \cdot \mu_T + H \cdot m_T \tag{2a}$$

where μ_T and m_T are the total (permanent plus induced) electric and magnetic dipoles. In the case of organic molecules, with the sole exception of magnetic spectroscopies (NMR, ESR), the electric term is generally orders of magnitude larger than the second term, therefore equation 2a may be reduced to:

$$H = E \cdot \mu_T \tag{2b}$$

At low light intensities, the induced dipole, μ, is directly proportional to the field strength; i.e. it may be described by:

$$\mu = \alpha E \tag{3}$$

where α is the polarization tensor. In the molecular axis (X,Y,Z), such a tensor is diagonal so that the dipole is given by:

$$\mu = (\alpha_{XX} E_X, \alpha_{YY} E_Y, \alpha_{ZZ} E_Z) \tag{4}$$

Since generally the components of the tensor are not equal (for example: in the near-UV and visible, the in-plane polarizability of an aromatic molecule is larger than the out-of-plane one and the two in-plane components are strongly frequency-dependent), the induced dipole and the incident field are not generally parallel to each other. If $|0\rangle$ and $|j\rangle$ are the total molecular wave functions for the ground and excited states of the molecule, out of resonance, this tensor is given by:

$$\alpha(\nu) = cte \cdot \nu \cdot \sum \cdot \frac{\langle o|r|j\rangle \cdot \langle j|r|o\rangle}{\left[\Delta E_{o,j}^2 - (h\nu)^2\right] + O(\Gamma)} \tag{5}$$

where Γ is the bandwidth and $\Delta E_{o,j}$ is the difference in energy between the states 0 and j. At resonance, this equation reduces to :

$$\alpha(\nu) = cte \cdot \sum \frac{\langle o|r|j\rangle \cdot \langle j|r|o\rangle}{\left(\Delta E_{o,j} - h\nu\right) + i\Gamma} \tag{6}$$

The electromagnetic field due to the induced dipole at a distance, r, and an angle, γ, between the dipole and the direction of observation is given by an equation of the type:

$$D(x,y,z; t',\upsilon') = Cte \cdot (\sin\gamma)/r \tag{7}$$

A photodetector situated at a distance r and an angle θ between the direction of light propagation and the direction of observation, can measure:

- in the direction of light propagation ($\theta = 0$), the fields due to the incident wave and to the dipole, i.e.:

$$F\ (x,y,z;\ t,\upsilon) + D\ (x,y,z;\ t',\upsilon') \qquad (8a)$$

- in all the other directions ($\theta \neq 0$), only the second term:

$$D\ (\ x,y,z;\ t',\upsilon') \qquad (8b)$$

The light intensity, I, is given by:

$$I = \{F\ (x,y,z;\ t,\upsilon) + D\ (x,y,z;\ t',\upsilon')\ \}^2 \qquad (9a)$$

$$I = \{D\ (x,y,z;\ t',\upsilon')\ \}^2 \qquad (9b)$$

It is easy to demonstrate that the imaginary part of the polarization tensor describes the absorption process and the real part describes the refraction index (3).

Consequently all the information yielded by a one-molecule experiment reflect the properties of the induced dipole. Since they can be interpreted in terms of the molecular eigen-functions through the polarization tensor, all of this spectroscopic information is related directly to the electronic structure of the molecule. When such a quantum mechanical analysis is not feasible, a semi-empirical treatment based on model systems must be applied.

In this ideal experiment, where only one molecule interacts with the incident wave, the physical observables are:
- the spectral distribution (which can be measured as a function of time): a spectrum is characterized by the number of bands present and their relative position, width and intensity. It is from these characteristics that structural and dynamical information may be obtained. For example, the linewidth reflects both the dynamic of the observed eigenstates and the interactions of the molecule with the matrix; these phenomena are called homogeneous and inhomogeneous line broadening, respectively.
- the time evolution of the intensity of the signal at a given frequency: in a pulsed experiment, the time evolution of the induced dipole can be followed as regards the polarization (orientation), energy (changes in frequency) and state lifetime. From such a measurement, it is often possible to obtain information on the dynamics of the interactions between the matrix and the induced dipole and between the various excited states of the molecule.

It should be noted that, in principle these two sets of measurements are equivalent: they merely represent two different sections in the time-frequency plane.

FROM A ONE- TO AN N-MOLECULE SPECTROSCOPY EXPERIMENT

Compared to the one-molecule case discussed above, a real experiment presents only one substantial difference: the samples generally contain anywhere from a large, up to a very large, number of molecules. As a consequence, the electromagnetic field associated with the induced dipoles is the sum of the individual components; if D_j and F_T indicate the fields associated with the jth dipole and the total field, respectively, then equations 8a and 8b become:

$$F_T = F\ (x,y,z;\ t,\upsilon\ ) + \Sigma\,D_j\ (x,y,z;\ t',\upsilon') \qquad (10a)$$

$$F_T = \Sigma\,D_j\ (x,y,z;\ t',\upsilon') \qquad (10b)$$

while equations 9a and 9b now read:

$$I = \{F\,(x,y,z;\,t,\upsilon\,) + \Sigma\,D_j\,(x,y,z;\,t',\upsilon')\,\}^2 \qquad\qquad (11a)$$

$$I = \{\Sigma\,D_j\,(x,y,z;\,t',\upsilon')\,\}^2 \qquad\qquad (11b)$$

respectively. In a disordered system with incoherent light or in any system at low beam intensities, no phase relationship may be expected between the induced dipoles since they oscillate independently. Therefore, equation 11b does not contain any cross terms; that is since:

$$\Sigma\,D_j\,D_k = 0$$

equation 11b reduces to:

$$I = \Sigma\,\{D_j\,(x,y,z;\,t',\upsilon')\,\}^2 \qquad\qquad (11c)$$

If an intense coherent wave is focused on the sample, all the individual dipoles are phased and thus a third observable appears: the phase relaxation time. This corresponds to the time which the prepared coherent population takes to decay to a random population; the process of dephasing results from the fact that, due to molecular non-homogeneity, the molecules experience different interactions with the matrix. In NMR for example, this implies different local magnetic fields, and therefore the dipoles precess at different Larmor frequencies around the external magnetic field. From the phase relaxation time it is thus possible to collect information on the matrix inhomogeneity.

Finally, the number of molecules, n, in a sampling volume fluctuates around its mean value as $\sqrt{n}$. In emission spectroscopies of highly "diluted" samples (i.e., for small values of n), since the intensity of the signal is directly proportional to the number of molecules sampled, the analysis of the noise can yield information on the number of particles (through the signal to noise ratio) and their dynamics (through the noise frequency). This is done by plotting the function of correlation, F(t), defined as:

$$F(t) = <I(t)\,.\,I(t+\tau)> \qquad\qquad (12)$$

where I(t) and I(t+τ) are the intensities of the signal at t and t+τ, respectively. This is a simple and powerful means of obtaining information on the rate of exchange between various structures if they are distinguishable by emission spectroscopy.

POPULATION AND TIME AVERAGING

A direct consequence of the nearly infinite number of molecules which constitutes a real sample is that spectroscopic measurements give population and time averages. As discussed below, this may be added in some cases to another time averaging - the one due to the non-negligible scattering time.

<u>Population averaging</u>

Since what is measured in absorption spectroscopies is the ratio between the intensities of the reference and sample beams, the number of molecules in the optical path must always be high (about 10^{15} molecules per cc in UV and visible absorption spectroscopies) to obtain a satisfactory signal to noise ratio. As a consequence, what is measured is a population-averaged property. It should be noted that this is equivalent in non-rigid matrices to the process of time averaging on one molecule over a sufficiently long period of time. Therefore, due to the high number of molecules involved and to the fact that they are not phased, there is no dynamic information contained in the noise. Nevertheless, the spectra contain information on the interstate dynamics. Unfortunately, however,

especially in the UV-visible absorption and emission spectroscopies of
biomolecules, this data is often buried in the unresolved Franck-Condon en-
velopes and/or under inhomogeneous line broadening.

Time averaging

Two examples of time averaging can be made to illustrate the point: (a) in
fluorescence where averaging is observed for processes occurring in times
shorter than about 1 ns., and (b) in NMR, when samples equilibrating at a rate
faster than about 100 Hz are studied.

The fluorescence of any molecule can be regarded as being due to two instantaneous processes (i.e., with associated times shorter than 10^{-15} s): the absorption and the emission of a photon separated by a certain time interval, the
mean value of which is the lifetime of the excited state. A molecule can decay
through various mechanisms (see ref. 4 for more details): emission (em), intersystem crossing (isc), internal conversion (ic) or quenching (quen). The fate
of an excited molecule can be analyzed in terms of the lifetime (the "when" - see
eq.13) and the quantum yields (the "how's" - see eq.14). They depend on the
rates of the decay mechanisms: k_{em}, k_{isc}, k_{ic} and k_{quen} through the equations:

$$\tau = 1 / (k_{em} + k_{isc} + k_{ic} + k_{quen}[q]) \qquad (13)$$

$$F_i = k_i /(k_{em} + k_{isc} + k_{ic} + k_{quen}[q]) \qquad (14)$$

where [q] is the quencher concentration. Since these rate constants depend on
the vibrational state of the molecule, j, and on the space coordinates, x, of the
molecule and the quencher, they should be written as:

$$k_i = k_i(j,x) \qquad (15)$$

They are thus time dependent. At the molecular level, the "when" and the "how"
are stochastic parameters and therefore only the mean lifetime and the yields
are physically relevant. The question of time averaging can be formulated as:
"What are the parameters which are measurable and/or relevant in the description of the process?" (see ref. 5 for a detailed discussion of the concept of relevant and irrelevant variables).

In the condensed state the vibrational states decay on a time scale which
is much shorter (2 to 3 orders of magnitude) than that of the electronic excited
state. Similarly, in low viscosity liquids every space variable varies on a time
scale which is at least one order of magnitude shorter than its fluorescence lifetime. Thus, both of them may be termed fast; these variables are said to be
irrelevant since the excited molecules will be allowed to experience the whole
range of their values many times during their lifetime. If all the space variables of any importance in equation 15 are fast, the fluorescence intensity decay will be mono-exponential. In such cases the fluorescence lifetime and yield
do not convey any direct information about the molecular dynamics; the lifetime and yield give only indirect information regarding rigidity, permeability,
etc. The emission spectra, being associated with an instantaneous process, will
reflect this multitude of individual states from which emission can occur
through a certain line broadening.

Variables which are either slower than, or on the same order of magnitude as, the excited state lifetime are instead termed relevant since they can influence both the lifetime, the yields and the spectral properties: non-monoexpo-
nential decays may be observed and the emission spectra may be time dependent.

To illustrate the problem presented by NMR, consider a system oscillating between configurations A and B at an exchange rate k and characterized by lines with chemical shifts of 100 Hz. If the rate of exchange between the two structures is slower than 100 Hz, two distinct lines are observed. If, on the contrary, it is much higher, only one narrow line at an intermediate frequency is seen; the exact position of this line being dependent upon the dwelling time of the molecule in each configuration. In the intermediate cases, a broad line is observed between the two specific frequencies.

In NMR, and in particular in 2D-NMR, this averaging process will outweigh the short distance contribution. The consequences of this are twofold: the first is that NMR tends to suggest structures which are more densely packed than is actually the case. The second is that any process faster than about 1 ms is cancelled out; that is, the molecular structure given by the spectra appears to be rigid.

INFORMATION CONTENT AND DATA ANALYSIS - ENZYME STRUCTURE AND DYNAMICS AND COMPLEMENTARY TECHNIQUES

Raw spectroscopic data contains a certain quantity of information which may be revealed by data analysis. In more complex cases, the analysis cannot discriminate between various interpretations. In this case, some other techniques must be used. This is one of the reasons why the determination of enzyme structure and dynamics is often carried out using complementary techniques; not a very difficult task since most of the experimental techniques are computer-assisted. In such situations, however, it is difficult to know exactly what are the physical models underlying the computer programmes used, and the empirical, semi-empirical or theoretical ab-initio nature of the models; that is, to be aware of exactly where the information content begins and ends, and thereby to avoid the danger of over-analyzing the experimental data. Sometimes, the interpretations of the data appear contradictory. A crucial point should always be kept in mind: computers are powerful tools, but do not actually generate data.

Circular Dichroism (CD) and Vibrational Spectroscopies

The CD spectra of a chromophore is directly related to its structure. At resonance, the interaction with the light beam is given by equations 2 and 6. In absorption spectroscopy, due to the small dimension of the chromophore compared to the wavelength of the incident light, the space property of the electric field can be neglected; i.e., for a molecule situated at the origin of the axis, equation 1 reduces to:

$$E\,(0,0,0;\,t,\upsilon) = E\,(0,0)\,\{\exp i(2\pi\upsilon t)\} \qquad (16)$$

The physical origin of optical activity being the space property of the molecule, it is necessary to introduce the space property of the electric field; i.e., to introduce into equation 16 the second term in the development in series:

$$E\,(x,y,z;\,t,\upsilon) = E\,(0,0,0;\,t,\upsilon) + \{(z-z_0)/\lambda\,\}.\{\,\partial E\,(0,0,0;\,t,\upsilon)/\,\partial z\,\} + \ldots \qquad (17)$$

The Maxwell equations show that the second term in the preceding equation is equivalent to the time derivative of the magnetic field. Consequently, the polarization tensor (equation 6) contains cross terms between the magnetic and electric dipoles; i.e.:

$$<0\,|\mu|\,j> . <0\,|m|\,j> \qquad (18)$$

It is easy to show that such a term, and consequently the CD spectrum, is non-zero only for some asymmetric deformations of the chromophore. An ab-initio calculation of the CD spectra requires a much more precise knowledge of the excited state wavefunctions than the actual state of quantum chemistry computerization permits. It is therefore rather difficult to calculate ab initio a CD spectrum.

Protein CD spectra analyses are thus based on empirical techniques using reference spectra from model polypeptides or reference proteins for which high resolution X-ray structures are available (6-8). Furthermore, since the CD spectra have few structural features and a rather poor signal to noise ratio, the number of significant reference spectra must therefore be kept small; otherwise the problem of data over-analysis as discussed above results, with the data analysis yielding, at least apparently, more information than is actually contained in the spectra. Another aspect to be noted is that the CD spectrum, like much spectroscopy data, does not discriminate between similar chromophores: all the peptide bonds are observed in the same wavelength range, for example. Therefore, CD spectra can only yield space averaged properties.

A vibrational spectrum contains a series of bands characterized by their position, polarization, intensity and width. In proteins, however, the width is often simply the result of non-homogeneous line broadening so that, unless correlation spectroscopy is added, it can hardly be used in conformation or dynamics analysis. The other three parameters reflect more directly the nature of the vibrational mode; i.e., the geometry and the interactions of the group of atoms involved. The vibrational spectroscopies (infrared absorption, Raman and neutron scattering) generate a great amount of information as the data analysis is relatively simple (the spectrum can be analysed assuming simple band shapes). Nevertheless, even when combined they cannot provide an ab-initio analysis of protein structure. Indeed, if for a molecule consisting of N nuclei the number of force constants is N(N-1)/2, the number of vibrational modes is (3N-6) so that the number of unknowns very rapidly becomes larger than the number of equations. Vibrational data analyses of enzymes are thus also based on empirical techniques using reference spectra from model polypeptides or reference proteins for which high resolution X-ray structures are available (6).

<u>Fluorescence lifetime measurements</u>

Even if one supposes a "perfect" experimental set-up (delta pumping and stable electronic characteristics), and a simple chemical system (assuming no kinetic complications due, for example, to excimer formation, etc.) such that the data can be gathered and analysed without any particular problem, the information content in a fluorescence lifetime experiment is very often too limited to discriminate between various models. Indeed, such a measurement only yields a monotonously decaying curve, which either can or cannot be convoluted by a mono-exponential decay function; i.e.:

$$I(t) = I_0 \exp(-at) + b \qquad (19a)$$

or

$$I(t) \neq I_0 \exp(-at) + b \qquad (19b)$$

Every other analysis, such as the multi-exponential or continuous distribution of lifetimes, has no value per se, other than to offer eventually a "best fit". Since multi-exponential decay suggests a discrete series of sites (the number of which must be at least equal to the number of terms in the decay function) and a continuous distribution suggests a high density of discrete sites or even a continuum distribution of sites, these two analyses correspond to very different physical descriptions of the system. In the former, the discrete sites are in slow thermal equilibrium among themselves (in this case "slow" means $k < 10^8$ sec^{-1}). In the latter there is a slow space variable whose characterisitic time for moving between experimentally distinguishable states is on the order of magnitude of, or shorter than, the fluorescence lifetime. The quality of the fit cannot be used to discriminate between the two models; that is, the physical model sustaining these complex decay functions must be verified by another set of experimental data.

A FEW CAUTIONARY REMARKS

It is useful at this point to draw attention to some difficulties which may be encountered in enzyme spectroscopy.

Relevant and irrelevant variables

In the discussion of relevant and irrelevant variables, the condition "in liquid solutions" was specified. In such cases, it is possible, and indeed easy, to define time characteristics - simply on the basis of the state lifetime, for example - such that each space variable can be classified easily within one of these two groups. In solution, it is also evident that only those space variables which can be associated with the solvation shell of the chromophore are of interest. In systems such as proteins, the situation is far more complex, however. For example, if two groups must come into close contact for fluorescence quenching to occur, this may require a series of well-defined configurations to exist for a certain length of time. As a consequence in fluorescence spectroscopy, variables much faster than fluorescence lifetime may become relevant. As far as regards the concept of the solvation shell, similar caution should be taken. Due to the enzyme's structural organization and rigidity, perturbations may be transmitted even if they are applied at a site quite distant from the spectroscopic probe.

Perturbations of the system due to the wave probe

Spectroscopic probes are used to determine protein structure and dynamics but the measurement induces a perturbation of the system under analysis. For example, tryptophan residues in the ground and first excited states have very different dipole moments; therefore, they have quite different solvation properties in these two states. When measuring the fluorescence behavior of tryptophan, one should always be aware that what is actually being probed is the properties of a biologically irrelevant species: the tryptophan excited state.

The extent to which local heating during excitation in high vibronic state has really any measurable influence on the spectroscopic properties of the probe has still to be studied even though it is certainly an important factor in rigid matrices. The problem stated simply is that due to vibrational relaxation, the energy difference between the prepared state and the emitting one is dissipated in about 1 ps so that the local energy content - to speak very loosely, the temperature - is very high for a short period of time, too short probably to induce any conformational changes in the structured region, but possibly long enough to perturb less rigid regions.

CONCLUSIONS

The material presented here is not intended to offer any startling new ideas. It is only an attempt to indicate to non-spectroscopists by use of a few selected examples, the limitations of some of the more common spectroscopic techniques and certain problems which may arise either out of the basic principles of physics underlying these techniques or from the limitations of the techniques themselves. At this point it should be obvious that:

a. no single spectroscopic technique can give a complete, or even satisfactory, picture of the conformation of proteins in solution;
b. structures appear "rigid" or "flexible" depending on the technique's time scale for averaging or, in other words, on its capacity to distinguish between fluctuating structures and equilibrium populations;

c. care should be taken when transferring results from one technique to another, since certain fundamental physico-chemical aspects (even so banal a factor as concentration) may change drastically in the conversion.

REFERENCES

(No attempt has been made in this chapter to present a complete bibliography of the subject since it is intended only to be a general introduction to the other chapters of the present volume. A few basic references not expected to be cited elsewhere have been included.)

(1) J.L. Houben: Introduction to the basic concepts in reaction dynamics, in the present volume.
(2) N. Mataga, T. Kubota: "Molecular interactions and electronic spectra." M. Dekker, New York (1970).
(3) J.L. Houben, N. Rosato: Light migration in biological systems. p. 15 in: "Molecular models of photoresponsiveness." G. Montagnoli, B.F. Erlanger (eds.) Vol. 68, NATO ASI Series A, Life Sciences , Plenum Press, New York (1983).
(4) for a more detailed discussion, see: A. Szabo: The fluorescence properties of aromatic amino acids: their role in the understanding of enzyme structure and dynamics, in the present volume.
(5) P. Grigolini, F. Marchesoni: Basic description of the rules leading to the adiabatic elimination of fast variables, p. 29 in: "Memory approaches to stochastic processes in condensed matter," M.W. Evans, P. Grigolini and G. Pastori Parravicini (eds.), in Adv. Chem. Phys., Vol. 62, I. Prigogine and S. Rice (Gen. eds.), J. Wiley, New York (1985).
(6) Y.H. Chen, J.T. Yang, H.M. Martinez: Determination of the secondary structures of proteins by circular dichroism and optical rotatotry dispersion, Biochem. 11, 4120, 1972.
(7) S.W. Provencher, J. Gloeckner: Estimation of globular protein secondary structure from circular dichroism, Biochem. 20, 33, 1981.
(8) J.P. Hennessey, W.C. Johnson Jr.: Information content in the circular dichroism of proteins, Biochem. 20, 1085, 1981.

INTRODUCTION TO VIBRATIONAL SPECTROSCOPY - FROM THE NORMAL MODE TO

THE LOCAL MODE: INFRARED, RAMAN AND INELASTIC NEUTRON SCATTERING

Francois Fillaux

Laboratoire de Spectrochimie Infrarouge et Raman
Centre National de la Recherche Scientifique
2 rue Henri-Dunant, 94320 Thiais, France

INTRODUCTION

Atoms in molecular entities and in crystals are oscillating around their equilibrium position. This fact has been recognized many decades ago when it appeared that it is possible to observe vibrational spectra with specific excitation and detection devices. Measurements of vibrational frequencies is one of the most important goals in molecular spectroscopy. They provide valuable information concerning the structure and dynamics of atoms and molecules. The absorption or emission spectra arising from the vibrational motion of a molecule which is not electronically excited is mostly in the infrared region. The internal vibrations, which are mainly related to the nature of the chemical bonds, are generally observed between 200 and 4000 cm^{-1}, i.e. their frequencies are between 6×10^{12} and 120×10^{12} Hz and their periods between $\sim 0.3 \times 10^{-12}$ and 0.8×10^{-15} sec. Most of the vibrations due to much weaker intermolecular interactions, sometimes denoted external vibrations, appear below 200 cm^{-1}. In some cases, it is not possible to distinguish internal and external vibrations. This is the case for large systems such as enzymes where very low frequency vibrations related to the protein folding may occur.

The vibrational frequencies for molecules depend on several parameters (atomic masses, nature of the bond; molecular geometry, interactions, environment...) and it is not straightforward to sort out of the spectra the relevant information concerning the structure and the dynamics of a particular system. Theoretical models are thus necessary and the idea has emerged rapidly that molecular vibrations can be represented by particles held together by certain forces. These forces were crudely thought of as weightless springs which only approximately obey Hooke's law (i.e., the restoring force is proportional to the atomic displacement relative to its equilibrium position) and which hold the atoms in the neighborhood of certain configurations relative to one another. This picture of forces as springs is useful for visualization, but it is not sufficiently general for all cases and much more sophisticated models including higher terms in the series expansion of the forces are needed (anharmonic vibrations). In any case, the nature of these forces and the search for potential functions which involve a small number of parameters and which at the same time permit good agreement with experimental data are some of the chief problems in vibrational spectroscopy.

This lecture is an introduction to vibrational spectroscopy and the following aspects are presented: i) the normal mode description of molecular vibrations; ii) the local mode approach for high vibrational excited-states; and iii) the transition moments and band intensities in infrared, Raman and inelastic neutron scattering.

NORMAL COORDINATES AND NORMAL VIBRATIONS

Normal mode definition and calculation are extensively presented in a number of textbooks [1-3]. In this lecture we do not intend to enter into all the details; only a brief overview of the problem and of its implications in the interpretation of vibrational spectra is given, avoiding most of the mathematical aspects.

In a diatomic molecule the vibration of the nuclei occurs along the line connecting the two nuclei. Its frequency is $v = (1/2\pi) \sqrt{k/\mu}$ where K is the force constant associated with the harmonic potential $V=(1/2)Kq^2$, q being the internuclear distance, and μ is the reduced mass $1/\mu = (1/m_1 + 1/m_2)$ for nuclei 1 and 2. In polyatomic molecules, the situation is much more complicated because all the nuclei perform their own oscillations. It can be shown, however, that any of these extremely complicated vibrations may be represented as a superposition of a number of normal vibrations, provided certain limitations for the shape of the potential function.

Let the displacement of each nucleus be expressed in terms of rectangular coordinate systems with the origin of each system at the equilibrium position of each nucleus. Then the kinetic energy of an N-atom molecule would be expressed as:

$$T = \frac{1}{2} \sum_{N} m_N \left[\left(\frac{d\Delta x_N}{dt}\right)^2 + \left(\frac{d\Delta y_N}{dt}\right)^2 + \left(\frac{d\Delta z_N}{dt}\right)^2 \right]$$

(1)

If generalized coordinates such as:

$$q_1 = \sqrt{m_1}\,\Delta x_1 \;;\; q_2 = \sqrt{m_1}\,\Delta y_1 \;;\; q_3 = \sqrt{m_1}\,\Delta z_1 \;;\; q_4 = \sqrt{m_2}\,\Delta x_2 \;; \ldots$$

(2)

are used, the kinetic energy is simply written as:

$$T = \frac{1}{2} \sum_{i=1}^{3N} \dot{q}_i^2$$

(3)

The potential energy of the system is a complex function of all the coordinates involved. For small values of the displacements, it may be expanded in a Taylor series such as:

$$V(q_1, q_2, \ldots) = V_0 + \sum_{i=1}^{3N} \left(\frac{\partial V}{\partial q_i}\right)_0 + \frac{1}{2} \sum_{i,j=1}^{3N} \left(\frac{\partial^2 V}{\partial q_i \partial q_j}\right)_0 q_i q_j + \ldots$$

(4)

where the derivatives are evaluated at $q_i=0$, the equilibrium position. The constant term V_0 can be taken as zero if the potential energy at $q_i=0$ is taken as a standard. The first derivative terms also become zero, since V must be a minimum at $q_i=0$. Thus V may be represented by:

$$V(q_1, q_2, \cdots) = \frac{1}{2} \sum_{i,j}^{3N} b_{ij} q_i q_j \tag{5}$$

neglecting higher-order terms. This is the harmonic approximation.

If the potential energy did not include any cross products such as $q_i q_j$, the problem could be solved directly by using Newton's equation:

$$\frac{d}{dt}\left(\frac{\partial T}{\partial \dot{q}_i}\right) + \frac{\partial V}{\partial q_i} = 0 \quad i=1, 2, \ldots 3N \tag{6}$$

or:
$$\ddot{q}_i + \sum_{j=1}^{3N} b_{ij} q_j = 0 \qquad i = 1, 2, \ldots 3N \tag{7}$$

and if $b_{ij} = 0$ for $i \neq j$:

$$\ddot{q}_i + b_{ii} q_i = 0 \quad i = 1, 2, \ldots 3N \tag{8}$$

and the solution is given by:

$$q_i = q_{i0} \sin\left(\sqrt{b_{ii}}\, t + \delta_i\right) i = 1, 2, \ldots 3N \tag{9}$$

where q_{i0} and δ_i are the amplitude and the phase constant, respectively.

Since, in general, this simplification is not applicable, the coordinate q_i must be transformed into a set of new coordinates Q_i through the relationship:

$$q_k = \sum_i B_{ki} Q_i \tag{10}$$

The Q_i are called normal coordinates for the system. By an appropriate choice of the coefficients B_{ki}, both the potential and the kinetic energies can be written as:

$$T = \frac{1}{2} \sum_i \dot{Q}_i^2 \tag{11}$$

$$V = \frac{1}{2} \sum_i \lambda_i Q_i^2 \tag{12}$$

without any cross products.

If these equations are combined with Newton's equation (6), there results:

$$Q_i = Q_{i0}\sin\left(\sqrt{\lambda_i}\, t + \delta_i\right) \tag{13}$$

The frequency is:

$$\nu_i = \frac{1}{2\pi}\sqrt{\lambda_i} \tag{14}$$

Such a vibration is called a normal vibration.

For the general N-atom molecule, it is obvious that the number of normal vibrations is only 3N-6, since 6 coordinates are required to describe the translational and rotational motion of the molecule as a whole. (Linear molecules have 3N-5 normal vibrations.) Thus, the general form of the molecular vibration is a superposition of the 3N-6 (3N-5) normal vibrations given by equation (13).

The physical meaning of the normal vibration may be demonstrated in the following way. As shown in equation (10), the original displacement coordinate is related to the normal coordinate by:

$$q_k = \sum_i B_{ki}Q_i \tag{15}$$

Since all normal vibrations are independent of each other, consideration may be limited to a special case in which only one normal vibration, indicated by subscript 1, is excited (i.e., $Q_{10} =\!\!/\, 0$, $Q_{20} = Q_{30} = \ldots = 0$). It then follows from equation (13) and (15) that:

$$q_k = B_{k1}Q_{10}\sin\left(\sqrt{\lambda_1}\, t + \delta_1\right) \tag{16}$$

This relation holds for all k. Thus, it is seen that the excitation of one normal vibration of the system causes vibrations, given by equation (16), of all the nuclei in the system. In other words, in the normal vibration all the nuclei move with the same frequency and in phase.

This is true for any other normal vibration. Thus equation (16) may be written in the more general form:

$$q_k = A_k\sin\left(\sqrt{\lambda}\, t + \delta\right) \tag{17}$$

which combined with equation (7) gives:

$$-\lambda A_k + \sum_j b_{kj}A_j = 0 \tag{18}$$

This is a system of first-order simultaneous equations with respect to A. In order for all the A's to be non-zero,

$$\begin{vmatrix} b_{11}-\lambda & b_{12} & b_{13} & \cdots \\ b_{21} & b_{22}-\lambda & b_{23} & \cdots \\ b_{31} & b_{32} & b_{33}-\lambda & \cdots \\ \cdots & \cdots & \cdots & \cdots \end{vmatrix} = 0 \tag{19}$$

The order of this secular equation is equal to 3N. Suppose that one root, λ, is found for equation (19). If it is inserted in equation (18), A_1, A_1,... are obtained for all the nuclei. The same is true for the other roots of equation (19). Thus, the most general solution may be written as a superposition of all the normal vibrations:

$$q_k = \sum_l B_{kl} Q_{l0} \sin\left(\sqrt{\lambda_1}\, t + \delta_1\right) \tag{20}$$

Practically, the kinetic energy is easily set up in terms of cartesian displacement coordinates of the atoms. Changes in the interatomic distances or in the angle between chemical bonds, on the other hand, can be used to provide a set of 3N-6 (3N-5) internal coordinates which are unaffected by translations or rotations of the molecule as a whole. They provide the most physically significant set for use in describing the potential energy of the molecule. The simple types of coordinates in terms of which the potential functions for most molecules are usually expressed are stretching (υ), bending (δ or γ) and torsion (τ).

The kinetic energy of vibration can be written in terms of internal coordinates:

$$T = \frac{1}{2} \sum_{tt'} (G^{-1})_{tt'} \dot{S}_t \dot{S}_t \tag{21}$$

where $(G-1)_{tt'}$ is a matrix element, and S'_t and $S'_{t'}$ are the derivatives of internal coordinates S_t and $S_{t'}$. The potential energy is then:

$$V = \frac{1}{2} \sum_{tt'} F_{tt'} S_t S_{t'} \tag{22}$$

Then, the secular equation is:

$$\left| G\,F - \lambda E \right| = 0 \tag{23}$$

The G matrix is totally determined by the molecular structure. The F matrix contains the force constants which are adjustable parameters. E is the unity matrix. Diagonalization of the GF matrix gives the eigen values $\lambda^2 = 4\pi^2\nu^2$, i.e., the normal frequencies of the system. The associated eigen vectors describe the potential energy distribution of the normal mode among the internal coordinates. This can be converted into the mean square amplitude for each atom. Once normal frequencies have been computed for a given set of force constants, iterative procedures are available to adjust the desired subset of force constants in order to fit the experimental frequencies.

THE LOCAL MODE APPROACH

Owing to the development of tunable lasers in the near infrared region, experimental data on high vibrational molecular states have been accumulated over the past ten years. The realization has developed that the conventional normal mode description of molecular vibrations is inefficient at high vibrational energies. An alternative approach involving the concept of "local" vibrational modes has been applied mainly to bond-stretching vibrations involving hydrogen and deuterium [4-15].

The formal theory has been given by Wallace [4]. For a simple system like YXY, if the bending motions are neglected, then the Hamiltonian is written as:

$$H = T(p_1, p_2) + V(r_1) + V(r_2) \qquad (24)$$

where r_1 refers to one XY bond-stretching coordinate and r_2 to the other.

$$T = -\hbar^2 \left[\left(\frac{1}{m_1} + \frac{1}{m_2} \right) \left(\frac{\partial^2}{\partial r_1^2} + \frac{\partial^2}{\partial r_2^2} \right) + \frac{1}{m_1} \cos \Theta \, \frac{\partial^2}{\partial r_1 \partial r_2} \right] \qquad (25)$$

where Θ is the average bond angle.

Therefore, in the local mode approach it is supposed that the potential energy is well localized on the chemical bond and that kinetic couplings are mainly responsible for the coordinate mixing in the vibrational states.

The transition from normal mode to local mode behavior as a function of energy can be visualized through the potential surface of such an XY_2 molecule (Fig.1).

In the region of small oscillations, the contours of the potential energy surface are elliptical, with principal axes along the symmetric and antisymmetric stretching directions. At high energy, the surface is dominated by the possibility of dissociation of the XY bonds, and for a description of the complete surface the two XY bond extensions are the natural coordinates. Usually, the entire potential energy surface can be very reasonably represented by a simple sum of Morse (or other one-dimensional) functions, one for each local oscillator.

The results obtained with this approach fit well the details of overtone spectra for simple molecules with light atoms like H_2O, NH_3, CH stretching, ... They therefore provide a rationale for understanding the values found for the

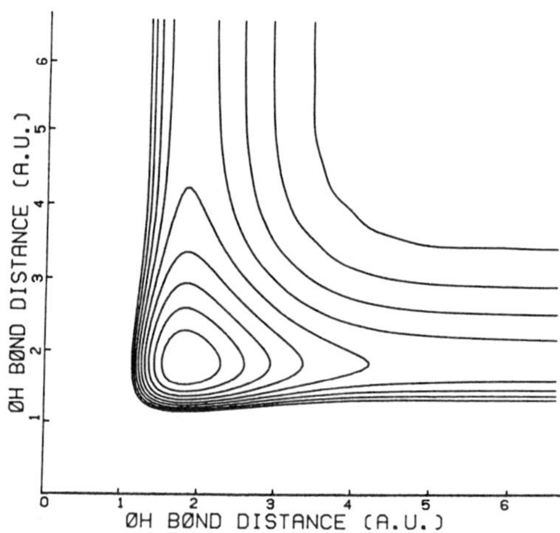

Fig. 1. Potential surface for the HOH molecule. Contours shown are -9.0, -8.0, ..., -1.0eV. After J. S. Wright, D. J. Donaldson and R. J. Williams, J. Chem. Phys., 81, 397 (1984).

anharmonicity constants when the spectra are analyzed via any of the conventional normal mode approaches based either on expansion of the potential energy in powers of the normal coordinates or of the total energy in powers of the vibrational quantum numbers. However, the local mode picture becomes less efficient for heavy atoms. For example, the local mode description is inferior to the normal mode analysis for SO_2.

TRANSITION MOMENTS AND BAND INTENSITIES

Three major techniques are nowadays available in vibrational spectroscopy: infrared (IR), Raman scattering and inelastic neutron scattering (INS). These techniques involve different physical properties and the spectra provide different information on the atomic and molecular vibrations. The description of the experimental set up for these techniques is beyond the scope of this lecture. Here we will be concerned with a brief description of the physical information which can be sorted out of the spectra obtained by IR, Raman and INS.

Infrared

Infrared spectra can be obtained through absorption, reflection or emission. Within the field of biological systems, absorption IR spectroscopy is the most important. It occurs when an incident radiation of appropriate energy interacts with the molecular dipole moment, which may be expanded in a Taylor series of the normal coordinates:

$$M_\rho = M_{\rho 0} + \sum_{k=1}^{3N-6} M_{\rho k} Q_k + \sum_{k,l=1}^{3N-6} M_{\rho kl} Q_k Q_l + \dots \quad \rho = x, y, z \tag{26}$$

The dipole moment associated with the change of state m -> n for the kth normal coordinate Q_k is:

$$M_{\rho mn} = \int \Psi_m (Q_k) M_{\rho k} \Psi_n (Q_k) dQ_k \quad \rho = x, y, z \tag{27}$$

The absorbed intensity is:

$$I_{\rho mn} \alpha\, I_{\rho 0} \nu_{mn} \left| M_{\rho mn} \right|^2 \quad \rho = x, y, z \tag{28}$$

It is perhaps worth emphasizing that it is not the amplitude of the vibration that is primarily responsible for the intensity of an absorption band. The amplitude is fixed by the force constants, masses and geometry; in fact, those transitions in which the change of amplitude is greatest (overtones) are generally quite weak. It is primarily the difference in electrical character of the atoms and the amount of asymmetry in the form of the vibration that determine the intensity, as shown by the fact that the expression for the intensity of a vibration involves derivatives evaluated at the equilibrium position (equation (26)).

Raman

Raman scattering may be observed when the electromagnetic field of the exciting radiation can induce a dipole moment in the molecules under investigation. The induced dipole ($\vec{\mu}$) is given in terms of the applied field by:

$$\vec{\mu} = |\alpha| \, \vec{E} \tag{29}$$

where $\{\alpha\}$ is a symmetrical tensor of the second rank:

$$|\alpha| = \begin{bmatrix} \alpha_{xx} & \alpha_{xy} & \alpha_{xz} \\ \alpha_{yx} & \alpha_{yy} & \alpha_{yz} \\ \alpha_{zx} & \alpha_{zy} & \alpha_{zz} \end{bmatrix} \tag{30}$$

It is apparent that a dipole moment can be produced in the x direction by an electrical field along the y axis or the z axis. The components of the polarizability tensor will vary as a bond is oscillating and each component can be expanded as a Taylor series of the normal coordinates:

$$\alpha_{\rho\sigma} = \alpha_{\rho\sigma 0} + \sum_k \alpha_{\rho\sigma k} Q_k + \sum_{kl} \alpha_{\rho\sigma kl} Q_k Q_l + \cdots \qquad \rho, \sigma = x, y, z \tag{31}$$

Raman scattering is a two photon process involving virtual absorption from the initial state (m) to the entire manyfold of excited vibronic states (r), followed by virtual emission to the final state (n).

$$I_{mn} \, \alpha \, (\nu_0 + \nu_{mn})^4 \sum_{\rho\sigma} |(P_{\rho\sigma})_{mn}|^2 \qquad \rho, \sigma = x, y, z \tag{32}$$

$$(P_{\rho\sigma})_{mn} = (\alpha_{\rho\sigma})_{mn} E = \frac{1}{h} \sum_r \left[\frac{(M_\rho)_{rn}(M_\sigma)_{mr}}{\nu_{rm} - \nu_0} + \frac{(M_\rho)_{mr}(M_\sigma)_{rn}}{\nu_{rn} + \nu_0} \right] E \tag{33}$$

ν_0 is the frequency of the incident light; ν_{rm}, ν_{rn} and ν_{mn} are the frequencies corresponding to the energy differences between subscripted states; $(M_\rho)_{mr}$ are the cartesian components of the transition moments. It should be noted that the states n, m and r are vibronic states.

Since the electric dipole operator acts only on the electronic wave functions, equation (33) can be rewritten as:

$$(\alpha_{\rho\sigma})_{mn} = \frac{1}{h} \int \Psi_m^g (\alpha_{\rho\sigma})_{gg} \Psi_n^g dQ \tag{34}$$

where:

$$(\alpha_{\rho\sigma})_{gg} = \sum_e \left[\frac{\langle g|\mu_\sigma|e \rangle \langle e|\mu_\rho|g \rangle}{\nu_{eg} - \nu_0} + \frac{\langle g|\mu_\rho|e \rangle \langle e|\mu_\sigma|g \rangle}{\nu_{eg} + \nu_0} \right] \tag{35}$$

ν_{eg} corresponds to the energy of a pure electronic transition between the ground and excited states.

To discuss the Raman scattering, we expand this term as a Taylor series with respect to the normal coordinate Q_k:

$$(\alpha_{\rho\sigma})_{gg} = (\alpha_{\rho\sigma})_{gg0} + (\alpha_{\rho\sigma})_{ggk} Q_k + \cdots \tag{36}$$

and from equation (34):

$$(\alpha_{\rho\sigma})_{m\,n} = \frac{1}{h}(\alpha_{\rho\sigma})_{gg0}\int\Psi_m^g\Psi_n^g\,dQ + \frac{1}{h}(\alpha_{\rho\sigma})_{ggk}\int\Psi_m^g Q_k\Psi_n^g\,dQ_k \tag{37}$$

The first term on the right-hand side is zero unless $i = j$. This term is responsible for Rayleigh scattering. The second term determines the activity of fundamental vibrations in Raman scattering; it vanishes for a harmonic oscillator unless $j = i\pm1$. Higher order terms in equation (36) give non-linear effects which are not discussed here.

Inelastic Neutron Scattering

Unlike electromagnetic-radiation scattering processes, the scattering of neutrons can be a coherent or incoherent phenomenon, depending upon the presence of spin and isotopic effects [16-21]. We will be primarily concerned here with the study of hydrogenous substances, where incoherent scattering is dominant. An important difference of INS compared to Raman scattering is that neutrons can exchange both energy (E_f-E_i) and momentum ($h(\vec{k}_i\text{-}\vec{k}_f)=h\vec{Q}$) with the scattering particles whereas the latter is always negligible in Raman ($\Delta\vec{k}\sim0$) For neutrons, the probability per unit time that a scattering event can take place is:

$$W\left(m,\vec{k}_i\,;\,n,\vec{k}_f\right) = \frac{2\pi}{h}\delta[E_n\text{-}E_m\text{-}(E_f\text{-}E_i)]\left\langle n,\vec{k}_f|U|m,\vec{k}_i\right\rangle \tag{38}$$

where n_0 and n denote the scatterer initial and final states, and U is the neutron-nuclear interaction potential inducing the transition. This is a first-order process, namely a transition arising from direct interaction between neutron and scatterer. For potential scattering at low energies, it is customary to replace the short-range potentials by point interactions:

$$U\left(\vec{r},\vec{R}_1,\vec{R}_2,\ldots\vec{R}_N\right) = \frac{2\pi\bar{h}^2}{m_0}\sum_{l=1}^{N}a_l\,\delta\left[\vec{r}\text{-}\vec{R}_l\right] \tag{39}$$

where m_0 is the neutron mass, $\vec{r}$ and $\vec{R}$ are the neutron and nuclear positions, and the scatterer is taken to be a system of N nuclei, each labeled by the subscript l. This is the well-known Fermi pseudopotential, which contains as an experimental parameter the bound-nucleus scattering length a. The latter is a measure of the strength of the interaction; equivalently, it gives the magnitude of the spherically outgoing scattered wave. For scattering systems of identical nuclei having zero spin, a is simply a constant. However, fluctuation in the scattering length can occur if the scatterer is an isotopic mixture or if the nuclei have a non-zero spin. In such cases the interactions must be treated as isotopic or spin-dependent.

The neutron wave function far from the interaction region (the scatterer) can be represented as plane waves. This means that the matrix element in equation (37) can be written as:

$$\left\langle n,\vec{k}_f|U|m,\vec{k}_i\right\rangle = \frac{2\pi\bar{h}^2}{m_0}\left\langle n\left|\sum_l a_l\exp\left(i\vec{Q}.\vec{R}_l\right)\right|m\right\rangle \tag{40}$$

where $\vec{Q}=\vec{k}_i-\vec{k}_f$ is proportional to the momentum transfer. It is understood that if the nuclei have non-zero spin, $|m\rangle$ and $|n\rangle$ must be expanded to include the neutron and nuclear spin states. The transition probability is next converted into a cross-section by summing over all the final states and averaging over the initial states:

$$\frac{d^2\sigma}{d\Omega dE} = \left(\frac{E_f}{E_i}\right)^{1/2} \sum_{mn} P(m)\, \delta\left[E_m - E_n + \bar{h}\omega\right] \frac{1}{N} \left| \left\langle n \left| \sum_l a_l \exp(i\vec{Q}\cdot\vec{R}_l) \right| m \right\rangle \right|^2 \tag{41}$$

$P(m)$ is the probability that the initial scatterer state is m.

The distinction between coherent and incoherent scattering lengths is essential to the understanding of any INS study. Experimentally, the relative importance of each type of process can be estimated from a knowledge of the scattering lengths of the given scatterer. It is well known that neutron scattering by a proton is predominantly incoherent because of the spin effects. Theoretically, the analyses of coherent and incoherent spectra involve different considerations. The incoherent problem has proved to be more tractable and the molecular substances investigated usually contain significant amounts of hydrogen. The total cross section of hydrogen ($81.5 \times 10^{-24}\,cm^2$) is an order of magnitude greater than the cross-section of other elements. For deuterium, the total cross-section is $7.5 \times 10^{-24}\,cm^2$. This is the reason why one normally considers only the incoherent contribution from hydrogen in analyzing INS spectra of hydrogenous compounds.

The experimental data are usually given in terms of the coherent and incoherent scattering laws.

$$\frac{d^2\sigma}{d\Omega dE} = \left(\frac{E_f}{E_i}\right)^{1/2} \left[a_{inc}^2\, S_{inc}(Q,\omega) + a_{coh}^2\, S_{coh}(Q,\omega) \right] \tag{42}$$

$$S_{coh}(Q,\omega) = \frac{1}{2\pi N} \int dt \exp(-i\omega t) \sum_{ll'} \left\langle \exp\left(-i\vec{Q}\cdot\vec{R}_l(t)\right) \exp\left(-i\vec{Q}\cdot\vec{R}_{l'}(0)\right) \right\rangle \tag{43}$$

$$S_{inc}(Q,\omega) = \frac{1}{2\pi N} \int dt \exp(-i\omega t) \sum_{l} \left\langle \exp\left(-i\vec{Q}\cdot\vec{R}_l(t)\right) \exp\left(-i\vec{Q}\cdot\vec{R}_l(0)\right) \right\rangle \tag{44}$$

For an isotropic harmonic oscillator the scattering law can be evaluated analytically:

$$S_{inc}(Q,\omega) = \exp(-2w) \sum_m \exp(-mz_0)\, I_m\left(\frac{\bar{h}Q^2}{2M\omega_0} \operatorname{csch}(z_0)\right) \delta\left(E_i - E_f + m\bar{h}\omega_0\right) \tag{45}$$

where:

$$2W = \frac{\bar{h}Q^2}{2M\omega_0} \coth(z_0) \tag{46}$$

and:

$$z_0 = \frac{\bar{h}\omega_0}{2k_BT} \tag{47}$$

k_B being the Boltzmann's constant. I_n is the modified Bessel function of the first kind.

The scattered neutron spectrum is a series of equally spaced lines, each corresponding to a transition in which a definite number of oscillator levels are excited. The $n = 0$ term gives the elastic scattering component, and positive (negative) n terms represent excitation (de-excitation) of the system. The factor $\exp(-2W)$ is the Debye-Waller factor familiar in X-ray diffraction studies. Its presence implies that at equilibrium each oscillator generates a thermal cloud by virtue of its vibratory motions. The extent of this cloud is, of course, temperature dependent, and it does not vanish at $T = 0$ because of the zero-point vibrations.

In molecular spectroscopy, under certain experimental conditions the scattering law may be related to the normal modes [20 - 24]:

$$S_M(Q, \omega) = \sum_d \exp(-2W_M^d) \prod_\lambda \left[\sum_{n\lambda} \exp(-nz_\lambda) I_{n\lambda} \left(\frac{\bar{h}\,|\vec{Q}.\vec{C}_d^\lambda|^2}{2m_d\omega_\lambda \sinh(hz_\lambda)} \right) \delta\left(\omega - \sum_\lambda n_\lambda \omega_\lambda \right) \right] \tag{48}$$

where C_d^λ is the displacement vector for the dth atom in the λth normal mode, and n_λ defines the number of quanta of energy lost or gained by the neutron. The Debye-Waller factor only includes the mean square displacement due to intramolecular vibrational modes.

For a single molecular mode, one finds:

$$S_M(Q, \omega) = \sum_d \exp(-2W_M^d) \left[1 + n_B(\omega_\lambda) \right] \frac{h\,|\vec{Q}.\vec{C}_d^\lambda|^2}{2m_d\omega_\lambda} \delta(\omega - \omega_\lambda) \tag{49}$$

where $n_B(w_\lambda)$ is the Bose occupation factor.

The measured scattering law is a convolution of the intra- and intermolecular functions:

$$S_{inc}(Q, \omega) = S_M(Q, \omega) \times S_L(Q, \omega) \tag{50}$$

Therefore, significant structure due to the lattice density of the states should be observed around the internal mode transitions. However, in spite of this complicated aspect of the problem, the important point is that it is possible to relate the atom displacements of an atom in a given normal mode to the observed INS intensity.

SUMMARY AND CONCLUSION

This brief overview of the main characteristics of IR, Raman and INS techniques leads to an intuitive understanding of their main specificities.

i) The wavelength used in IR and in Raman are very large compared to the interatomic and intermolecular distances. Therefore, only in phase vibrations of a large number of atoms or molecules are observed. The wavelength used in INS (a few Å) is similar to the internuclear distances and single particle motions are observed.

ii) A vibrational mode is not always active in IR and Raman. Its activity depends on whether the transition moment has a non-zero value for the particular atomic displacements associated with this mode. On the contrary, there is no selection rule in INS since the interaction is localized on a particular atom, or on a very limited number of atoms.

iii) As already underlined, force field calculations for large molecular systems is a mathematically undetermined problem. For N degrees of freedom, the number of unknown force constants, $N(N+1)/2$, greatly exceeds the number, N, of observable frequencies. Therefore, no unique force field can be extracted from frequency data alone. In IR and Raman, the dipole moment and the polarisability tensor are usually unknown for polyatomic systems and it is not possible to rely on the observed intensities to discriminate between several force fields giving good agreement with the observed frequencies. The intensities of the INS bands, on the other hand, can be calculated exactly for any force field, and dynamical models which produce very poor fits to the INS data can be quickly eliminated as irrelevant. In this sense force fields consistent with INS intensities must be considered as the most firmly based. However, for hydrogenous systems, the intensity fit is only consistent with a good description of the proton dynamics, but the description of normal modes involving negligible proton displacement remains largely arbitrary.

REFERENCES

1. E.B. Wilson Jr., J.C. Decius and P.C. Cross, "Molecular Vibrations, The Theory of Infrared and Raman Vibrational Spectra", McGraw-Hill Book Company Inc., London (1955).
2. P. Gans, "Vibrating Molecules. An Introduction to the Interpretation of Infrared and Raman Spectra", Chapman and Hall, London (1971).
3. K. Nakamoto, "Infrared and Raman Spectra of Inorganic and Coordination Compounds", John Wiley and Sons, New York (1986).
4. R. Wallace, Chem.Phys., 11:189 (1975).
5. B.R. Henry, in:"Vibrational Spectra and Structure," J. Durig, ed., Elsevier, New York, Vol.10 (1981).
6. I.A. Watson, B.R. Henry and I.G. Ross, Spectrochim.Acta, 37A:857 (1981).
7. H.L. Fang and R.L. Swofford, in:"Advances in Laser Spectroscopy," Heyden, Philadelphia (1982).
8. M.L. Sage and J. Jortner, Adv.Chem.Phys., 47:293 (1981).
9. G.J. Scherer, K.K. Lehmann and W. Klemperer, J.Chem.Phys., 78:2817 (1983).
10. M.J. Berry and D.F. Heller, J.Chem.Phys., 76:2814 (1982).
11. J.S. Hutchison, W.P.Reinhart and J.T. Hynes, J.Chem.Phys., 79:4247 (1983).
12. E.L. Sibert III, W.P.Reinhart and J.T.Hynes, J.Chem.Phys., 81:1115 (1984).
13. E.L.Sibert III, W.P. Reinhart and J.T. Hynes, J.Chem.Phys., 81:1135 (1984).
14. J.S. Hutchinson, E.L. Sibert III and J.T. Hynes, J.Chem. Phys., 81:1314 (1984).

15. L.A. Findsen, H.L. Fang, R.L. Swofford and R.R. Birge, <u>J.Chem Phys.</u>, 84:16 (1986).
16. P.A.Egelstaff, ed., "Thermal Neutron Scattering," Academic Press, New York (1965).
17. G.E. Bacon, "Neutron Scattering in Chemistry", Butterworths, London (1967).
18. H. Boutin and S. Yip, "Molecular Spectroscopy with Neutrons," The M.I.T. Press, Cambridge, Massachusetts (1968).
19. W.Marshall and S.W.Lovesey, eds.,"Theory of Thermal Neutron Scattering," Clarendon Press, Oxford (1971).
20. A. Griffin and H.Jobic, <u>J.Chem.Phys.</u>, 75:5940 (1981).
21. M. Warner, S.W. Lovesey and J. Smith, <u>Z.Physik</u>, B 51:109 (1983).
22. H. Jobic and H.J. Lauter, <u>J.Chem.Phys.</u>, 88:5450 (1988).
23. G.J. Kearley, <u>J.Chem.Soc. Faraday Trans. II</u>, 82:41 (1986).
24. F. Fillaux, J. Tomkinson and J. Penfold, <u>Chem.Phys.</u>, in press.

NONLINEAR COUPLING AND VIBRATIONAL DYNAMICS

Francois Fillaux

Laboratoire de Spectrochimie Infrarouge et Raman
Centre National de la Recherche Scientifique
2 rue Henri-Dunant, 94320 Thiais, France

INTRODUCTION

In the previous lecture, we have seen how the molecular vibrations in a quadratic potential can be described in terms of normal modes, i.e., an ensemble of classical harmonic oscillators. The dissociation energy of most of the chemical bonds is much larger than in the lower vibrational states and the corresponding vibrational modes are well described by the harmonic approximation. However, there are some modes which show very large anharmonicity and more sophisticated approaches are necessary to account for their dynamics. These modes are usually due to oscillations of particles with very large amplitudes (namely CH_3 torsional modes, nitrogen inversion, four member ring-puckering, five-member ring pseudo-rotation, etc.) [1]; to chemical bonds with a dissociation energy close to the vibrational energy level (hydrogen bond) [2,3]; or to weak inter-molecular interactions (Van der Waals, ionic) [4]. Although these vibrational motions are rather rare compared to the total number of vibrational modes in a molecule, a thorough analysis of their dynamics provides an accurate description of the potential functions governing large amplitude displacements of atoms in a molecule or in a complex. These are related to important phenomena in physics and chemistry, such as the transport of atoms or charges, dynamical exchanges between molecular conformations, and the preliminary steps in chemical reactions.

The theoretical approach relevant for the description of the anharmonic vibrations in molecular entities is quanto-mechanical in nature and the purpose of this lecture is to present briefly different aspects of this problem.

THE HARMONIC OSCILLATOR

The harmonic potential [5,6] is limited to the quadratic term:

$$V(x) = \frac{1}{2} kx^2 \tag{1}$$

or:

$$V(x) = \frac{1}{2} m\omega^2 x^2 \quad ; \quad \omega = \sqrt{\frac{k}{m}} \tag{2}$$

where m is the reduced mass of the oscillator, k is the force constant and ω is the harmonic frequency multiplied by 2π. The energy levels and the wave functions for this simple system are:

$$E_n = \left(n + \frac{1}{2}\right)\bar{h}\omega \tag{3}$$

$$\Psi_n(x) = \left(\frac{mw}{\pi\bar{h}}\right)^{1/4} \frac{1}{\sqrt{2^n n!}} H_n\left(x\sqrt{\frac{m\omega}{h}}\right) \exp\left(-\frac{m\omega}{2h}x^2\right) \tag{4}$$

where $\bar{h}$ is the Planck's constant divided by 2π and H_n is the Hermite polynomial of degree n. The normalisation condition for the wave functions gives:

$$\langle n \mid n'\rangle = \delta_{nn} \tag{5}$$

$\delta nn'$ being the Kronecker symbol. According to equation (3) the vibrational levels are equidistant:

$$E_n - E_{n'} = \bar{h}\omega \tag{6}$$

Moreover, the Hermite polynomial and the wave functions are easily calculated using the following equations:

$$2x\,H_n(x) = H_{n+1}(x) + 2n\,H_{n-1}(x) \tag{7}$$

$$2x\,\Psi_n(x) = \sqrt{2(n+1)}\,\Psi_{n+1}(x) + \sqrt{2n}\,\Psi_{n-1}(x) \tag{8}$$

The general expression for the first-order transition moment is:

$$\langle n \mid x \mid n'\rangle = \frac{1}{2}\sqrt{2(n+1)}\,\langle n \mid x \mid n'+1\rangle + \frac{1}{2}\sqrt{2n}\,\langle n \mid x \mid n'-1\rangle \tag{9}$$

which gives the well-known selection rule: $\Delta n = \pm 1$. Transitions involving one quantum of energy are observed.

It is important to note that this selection rule may be cancelled by high order terms of the transition moment operator (the dipole moment or the polarisability tensor element). For example $<n|x^2|n'>$ is non-zero for $\Delta n = 0, \pm 2$. For inelastic neutron scattering (INS) the transition moment is:

$$\langle n \mid \exp(-iQ.x) \mid n'\rangle \tag{10}$$

In this case there is no selection rule and transitions involving several quanta may be observed. (However, the observed intensities decrease rapidly at high energies because of the Debye-Waller factor.)

94

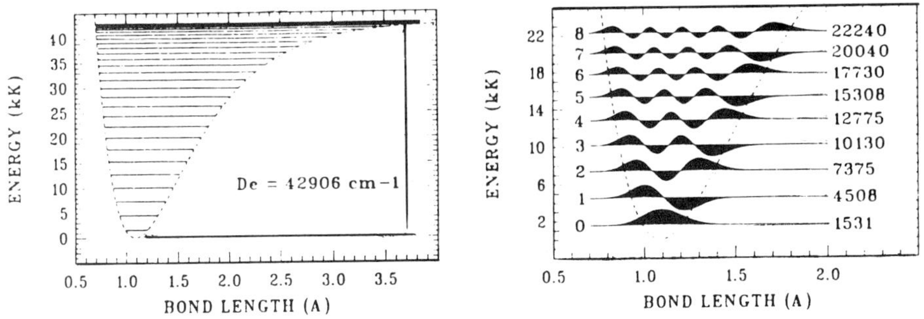

Fig. 1. Morse potential (left) and wave functions (right) for the CH bond in acetaldehyde, after L. A. Findsen and co-workers, J.Chem.Phys., 84:16 (1986).

THE MORSE POTENTIAL

The Morse potential is [5,6]:

$$V(x) = A + B\left[e^{-2cx} - 2\,e^{-cx}\right] \tag{11}$$

This potential is asymmetric and anharmonic (Fig. 1). It is convenient for the stretching mode of a diatomic molecule or for local modes in more complicated molecules. A is the dissociation threshold and A-B is the depth of the well. The Schrödinger equation can be solved analytically for this potential:

$$E_n = A - B\left[1 - \frac{c\overline{h}}{\sqrt{2mB}}(n+1)\right]^2 \tag{12}$$

and the level spacing is :

$$E_{n+1} - E_n = 2\,B\,\frac{c\overline{h}}{\sqrt{2mB}}\left[1 - \frac{c\overline{h}}{\sqrt{2mB}}(n+1)\right] \tag{13}$$

The energy levels are not equidistant. Moreover, the selection rule for the harmonic potential (equation (9)) is no longer valid and transitions involving several quanta may be observed. It is important to notice that two transitions are sufficient for a complete determination of the potential parameters.

The potential function:

$$V(x) = \frac{-A}{\cosh\left(\alpha\,x^2\right)} \tag{14}$$

represents a symmetric well with a dissociation threshold A. Its vibrational level pattern is very similar to that of the Morse function [6]. It is sometimes used for the description of bending vibrations of weakly bonded complexes.

In both cases, the discrete level progressions merge into continua for transitions above the dissociation threshold.

THE GENERAL PROBLEM

Most of the very anharmonic molecular vibrations which show complicated spectra with several transitions cannot be represented exactly with simple Morse-type potentials. It is necessary to consider the Taylor series expansion of the potential around the equilibrium position of the oscillator:

$$V(x) = \sum_i a_i x^j \tag{15}$$

The energy levels and the wave functions are computed using the variational method [7-11]. The wave function for the eigen state, p, of the system is expanded over a basis set of harmonic wave functions:

$$\Phi_p = \sum_n c_{np} \Psi_n \tag{16}$$

The matrix elements are then computed and the matrix is diagonalised. This gives the eigen values E_p and the eigen vectors $\{C\}$. Such calculations are now quite easy. Basis sets with more than 40 harmonic wave functions are easily handled on a microcomputer and the final result may be obtained within one minute. The accuracy is better than $0.01 cm^{-1}$ for the 5 lowest levels and downgrades for higher levels. Automatic procedures are available to adjust the coefficients of the potential expansion (equation (15)) in order to fit the observed frequencies.

DOUBLE MINIMUM POTENTIALS AND THE TUNNEL EFFECT

When a molecular system can exist in two different conformations with the same energy, its potential function has two symmetric minima separated by a potential barrier. At the zero approximation level, there are two identical states localized in each well. However, owing to the vanishing part of the wave function in the classically forbidden region, these two states interact and the degeneracy is removed. Finally, the lower state is split into two states which correspond to symmetric (lower) and anti-symmetric (upper) combinations of the wave functions (Fig. 2). The particle is not localized in a particular well and the tunnel splitting is inversely proportional to the residence time in each site. Similar state mixings occur for higher vibrational states.

Different theoretical approaches have been proposed to calculate the tunnel splitting and the wave functions in double minimum potentials. The most general is the variational method described previously (see above). The potential is expanded as:

$$V(x) = -a x^2 + b x^4 \tag{17}$$

and the wave functions are expanded on a harmonic basis set [7-9]. Another type of potential with a gaussian barrier has been considered by several authors [10]:

$$V(x) = \sum_i a_i x^i + b e^{-cx} \tag{18}$$

The NH_3 molecule was one of the first examples for which a vibrational tunnel effect has been observed. This arises from the pyramidal structure of

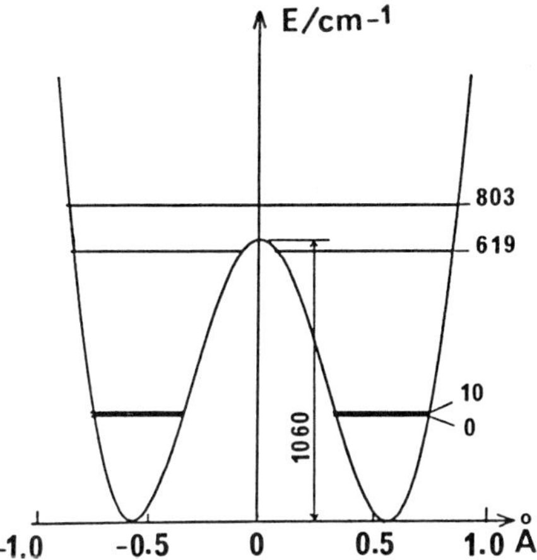

Fig. 2. Double minimum potential for the out-of-plane NH bending
mode of N-methylacetamide at low temperatures. After ref. 13.

the molecule, and the potential function for the bending mode shows two
minima separated by a barrier of about 2000 cm^{-1}. More recently, double
minimum potentials were proposed to account for the dynamics of the out-of-
plane vibrations in N-methylacetamide, which is one of the simplest molecules
containing a peptide unit. Raman [12] and IR data [13] show that the molecule
is not planar and two different modes, namely the torsion around the central
C-N bond and the NH out-of-plane bending, are governed by double minimum
potentials (Fig. 3). The potential barrier for the nitrogen inversion is about
half of that observed for the ammonia molecule and the minima correspond to
an angle of about ± 30° for the NH bond, with respect to the mean molecular
plane (Fig. 2). The decrease of the potential barrier is consistent with the par-

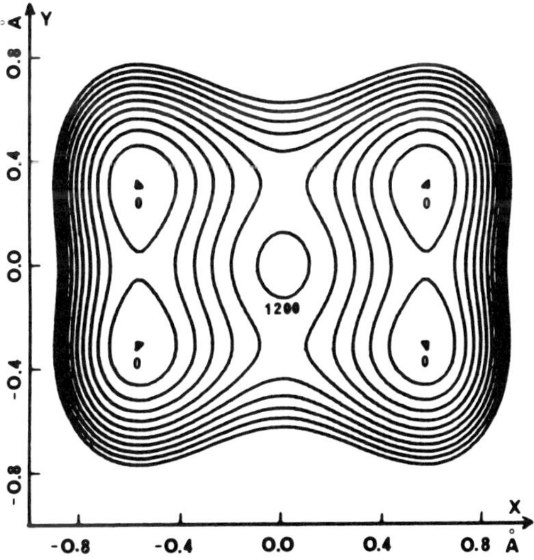

Fig. 3. Potential surface for the out-of-plane NH bending (x) and for the
C-N torsion (y) modes of N-methylacetamide at low temperature. The
isoenergy contours are at 0, 200, 400, ...cm^{-1}. After ref. 13.

tial electronic delocalisation in the CONH group. For the C-N torsional mode, on the other hand, the potential barrier is only 200 cm^{-1} and the minima are at about ±10° with respect to the mean plane. These results suggest that the vibrational dynamics of the peptide unit are different from the planar model of Pauling [14]. These large amplitude atomic displacements could play a role in the dynamics of proteins and enzymes [15].

Another example of the double minimum potential which may be of importance in enzyme catalysis is the proton transfer between a donor AH and an acceptor B. Potassium hydrogen carbonate (KHCO$_3$) forms cyclic dimers in the crystalline state and neutron diffraction studies have evidenced disorder due to proton transfers along the hydrogen bond [16,17]. Quantitative analyses of the OH stretching band-shapes in Raman [18] and INS spectra in the 200cm^{-1} region [19] are consistent with a double minimum potential for the proton displacements along the O...O bond (Fig. 4). This is the most accurate evaluation of the potential barrier for the proton transfer in an OH...O hydrogen bond of 2.61 Å. The barrier is about 5000 cm^{-1} (~15 kcal/mole) and the tunnel splitting in the ground state is very small. However, because of the crystal environment the potential symmetry is broken and the two sites do not have exactly the same energy. The observed splitting (~200 cm^{-1}) thus is mainly due to dimer-dimer interactions. The wave functions are no longer symmetric or anti-symmetric; rather, they are localized in each well.

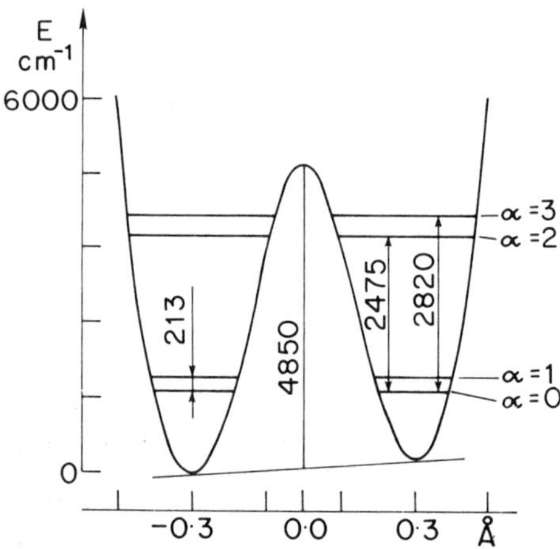

Fig. 4. Double minimum potential for the proton transfer along the hydrogen bond in the potassium hydrogen carbonate crystal at low temperatures. After ref. 18.

NONLINEAR COUPLINGS IN VIBRATIONAL SPECTROSCOPY

So far, we have been considering either an ensemble of coupled harmonic oscillators or independent anharmonic potentials. In some cases, however, complicated spectra must be interpreted in terms of strongly coupled anharmonic oscillators. Usually, the problem is restricted to two degrees of freedom (x and y) and the potential is expanded as:

$$V(x, y) = a x^2 + b y^2 + c x y^2 + d x^2 y + \ldots \tag{19}$$

The Schrödinger equation leads to a system of coupled equations which can be solved analytically for different terms of the Taylor series [20-25]. Franck-Condon progressions are calculated and, in some cases, it is possible to adjust the potential parameters to fit the observed frequencies and intensities. However, for double minimum potentials and for very large anharmonic terms, equation (19) does not converge rapidly to the desired potential shapes and it is useful to consider the equivalent of the Born-Oppenheimer (adiabatic) approximation for the electronic transitions. The system is thus separated into a fast (for example, x) and a slow (for example, y) coordinate which are dynamically separable. It is assumed that the fast coordinate adjusts its frequency instantaneously when the slow coordinate oscillates. For each state, n, of the fast coordinate the dynamics of the slow motion are governed by the adiabatic potential Vn(y). Although the frequency ratio in vibrational spectroscopy (fast/slow ~ 10) is much lower than that for electronic transitions (fast/slow ~ 100), it is widely accepted that this type of dynamical separation, first proposed by Stepanov for hydrogen bonded systems [26,27], is relevant [28]. This approach has been used to analyse the spectra of N-methylacetamide [29] and of several hydrogen bonded systems [25].

In the NH stretching region, N-methylacetamide shows several bands which are assigned to Franck-Condon type progressions due to changes of the potentials for the out-of-plan bending modes (namely γNH and τC-N)) in the νNH excited state [29]. The molecule becomes planar and the bending vibrations soften dramatically (Fig. 5). It appears that the peptide unit becomes looser when the νNH mode is excited. Another strong coupling between the τC-N mode and the C-N stretching is observed in the Amide I region (~1630 cm^{-1}) as well (Fig. 6) [30]. The initial hypothesis of Davydov, which assumes strong couplings of the νN...O and Amide I modes to stabilize the solitons in polypeptides and proteins [31-33], thus might be reconsidered.

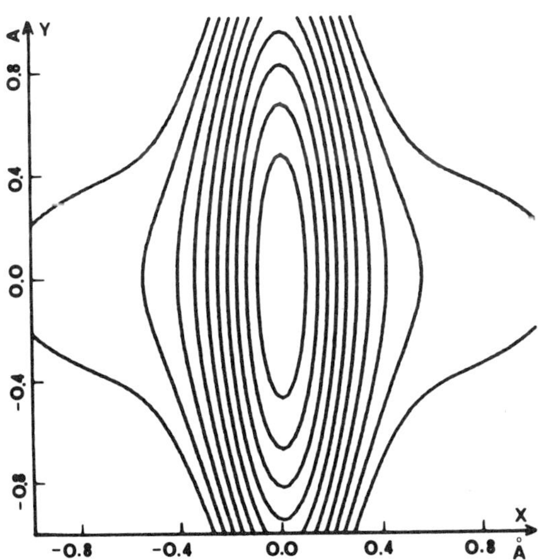

Fig. 5. Potential surface for the out-of-plane NH bending (x) and for the C-N torsion (y) modes in the NH stretching excited state of N-methylacetamide at low temperatures. The isoenergy contours are at 0, 40, 80, ..cm^{-1}. After ref. 29.

The most important class of systems showing very strong anharmonic couplings includes strongly hydrogen-bonded complexes [34-38]. In some cases, a detailed analysis of the band-profiles observed in IR, Raman and INS spectra leads to a complete description of the potential surface governing the dynamics of the stretching modes of the hydrogen bond. In the case of $K_3H(SO_4)_2$ [38], which forms $(SO_4...H...SO_4)^{3-}$ dimers in the crystalline state, this potential surface (Fig. 7) shows interesting features: for very short inter-nuclear distances ($R_{O...O}$), the proton experiences a very flat potential and its

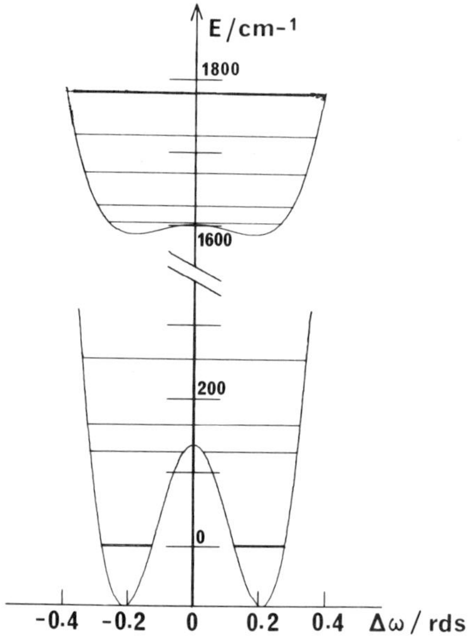

Fig. 6. Adiabatic potentials for the C-N torsional mode in the ground
(bottom) and first excited (top) states of the Amide I mode of
N-methylacetamide at low temperatures.

frequency is lower than 1000 cm^{-1}. When $R_{O...O}$ increases, the double minimum appears and the potential barrier increases very rapidly. The proton position with respect to the center of the bond remains approximately the same, whereas it is usually supposed that the hydrogen should follow the oxygens. It seems that the proton position is primarily determined by the crystal environment, while the proton dynamics are strongly affected by the O...O distance. This last example shows that vibrational spectroscopy is likely to provide unique information on the dynamical aspects of simple chemical reactions.

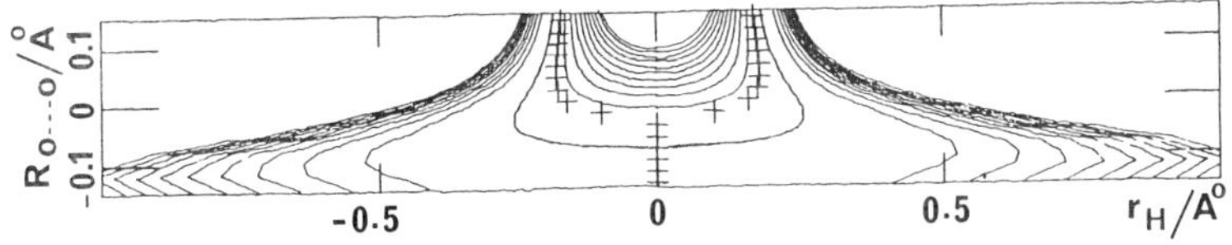

Fig. 7. Potential surface for the OH stretching (z) vibration of the symmetric O...H...O bond in $K_3H(SO_4)_2$. The mean O...O (R) is ~2.47Å. The isoenergy contours are at 0, 500, 1000, ..cm^{-1}. After ref. 38.

REFERENCES

1. J.R. Durig, ed., "Vibrational Spectra and Structure," vol. 1, Marcel Dekker Inc., New York (1972).
2. G. Zundel and C. Sandorfy, eds., "The Hydrogen Bond. Recent Developments in Theory and Experiments," vols. 1-3, P. Schuster, North-Holland, Amsterdam (1976).
3. A. Novak, Struct.Bonding, 18:177 (1974).
4. H. Ratajczak and W.J. Orville-Thomas, in "Molecular Interactions," vol.1, Wiley and Sons, New York (1980).
5. G. Herzberg, "Infrared and Raman Spectra of Polyatomic Molecules," Van Nostrand, New York (1960).
6. L. Landau and E. Lifchitz, "Mécanique Quantique," Mir, Moscow (1967).
7. E. Heilbroner, H. Rutishauser and F. Gerson, Helv.Chim.Acta, 42:2285 (1959).
8. E. Heilbroner, H. Rutishauser and F. Gerson, Helv.Chim.Acta, 42:230 (1959).
9. R.L. Somorjai and D.F. Hornig, J.Chem.Phys., 36:1980 (1962).
10. S.I. Chan and D. Stelman, J.Chem.Phys., 39:545 (1963).
11. Y. Guissani and H. Ratajczak, Chem.Phys., 19:361 (1981).
12. F. Fillaux, M.H. Baron, C. de Lozé and G. Sagon, J.Raman Spectr., 7:244 (1978).
13. F. Fillaux and M.H. Baron, Chem.Phys., 62:275 (1981).
14. L. Pauling, "The Nature of the Chemical Bond", Cornell, Ithaca (1960).
15. G.N. Ramachandran, A.V. Lakshminarayanan and A.S. Kolaskar, 303:8 (1973).
16. J.O. Thomas, R. Tellegren and I. Olovsson, Acta Cryst., B30:1155 (1974).
17. J.O. Thomas, R. Tellegren and I. Olovsson, Acta Cryst., B30:254 (1974).
18. F. Fillaux, Chem.Phys., 74:405 (1983).
19. F. Fillaux, J. Tomkinson and J. Penfold, Chem.Phys. (in press).
20. A. Witkowski and Y. Maréchal, J.Chem.Phys., 48:3697 (1968).
21. C.A. Coulson and G.N. Robertson, Proc.Roy.Soc., A337:167 (1974).
22. C.A. Coulson and G.N. Robertson, Proc.Roy.Soc., A342:289 (1975).
23. Y. Bouteiller and E. Maréchal, Mol.Phys., 32:277 (1976).
24. Y. Bouteiller and Y. Guissani, Mol.Phys., 38:617 (1979).
25. J.C. Lassègues and J. Lascombe, in: "Vibrational Spectra and Structure," p. 51, J.R. Durig, ed., Elsevier (1982).
26. B.I. Stepanov, Zh.Fiz.Khim., 19:507 (1945); and 20:408 (1946).
27. B.I. Stepanov, Nature, London, 157:808 (1946).
28. S.A. Barton and W.R. Thorson, J.Chem.Phys., 71:4263 (1969).
29. F. Fillaux, Chem.Phys., 62:287 (1981).

30. F. Fillaux, unpublished results.
31. A.S. Davydov, <u>Stud.Biophysics</u>, 62:1 (1977).
32. A.S. Davydov, <u>Phys.Scr.</u>, 20:387 (1979).
33. A.S. Davydov, <u>Phys.Stat.Sol.(b)</u>, 138:559 (1986).
34. F. Fillaux, <u>Chem.Phys.</u>, 74:395 (1983).
35. F. Fillaux, B. Marchon and A. Novak, <u>Chem.Phys.</u>, 86:127 (1984).
36. L. Soulard and F. Fillaux, <u>Chem.Phys.</u>, 100:355 (1985).
37. P. Bernardet and F. Fillaux, <u>Chem.Phys.</u>, 106:447 (1986).
38. F. Fillaux, A. Lautié and J. Tomkinson, unpublished results.

LOW FREQUENCY DYNAMICS OF PROTEINS STUDIED BY INELASTIC NEUTRON

SCATTERING

Stephen   Cusack

Grenoble   Outstation
European   Molecular   Biology   Laboratory
c/o ILL, 156X, 38042 Grenoble, France

1. CONFORMATIONAL CHANGE AND EQUILIBRIUM FLUCTUATIONS IN PROTEINS

In order to perform their biological function many proteins need to be able to adopt two or more different conformations and are induced to change from one to another by ligand or substrate binding or by a change in environmental conditions such as pH [1,2,3]. In several cases the nature of such conformational changes has been revealed by X-ray crystallography and shown to range from subtle reorientations of a few sidechains, to displacements of loops (e.g. phosphorylase, triose phosphate isomerase), pseudo-rigid body inter-domain motions (e.g. hexokinase, phosphoglycerate kinase [4]), subunit reorientations (heamoglobin, aspatate-transcarbamylase) and drastic rearrangements of the whole protein (e.g. influenza virus hemagglutinin). Conformational change or functional flexibility of this kind clearly involves internal protein motion. However there is overall direction to the motion and it is usually reversible.

In whichever conformation a protein might be, it will also be undergoing continual random thermal motion. From fundamental thermodynamic arguments it is possible to estimate the overall magnitude of internal energy and volume fluctuations in proteins [1,5]. Because of their complex macromolecular structure, proteins possess a wide spectrum of possible equilibrium fluctuations [6-8]. They range from very fast covalent bond vibrations (10^{-13}-10^{-14}s) to slower more global vibrational motion (10^{-11}-10^{-12}s) and surface side-chain motion (10^{-10}-10^{-11}s), inter-domain motion (10^{-7}-10^{-11}s) and on down to internal aromatic ring flips (1-10^{-4}s) and partial denaturing (10-10^{-5}s). Whereas it is clear that conformational change is of evident biological importance, the role of thermal fluctuations in protein function is often less obvious. Are they just an inevitable background jiggling of the protein atoms which permit, but do not positively promote, functionally important motions to occur? Or is the pattern of thermal fluctuations tailored by evolution to enhance protein function, for instance in favouring motions which are required along a catalytic reaction pathway? In other words, do proteins function because of or in spite of thermal fluctuations? In some cases the situation is apparently clear: without thermal fluctuations, oxygen would not be able to penetrate into the heme pocket in myoglobin and in the trypsin/trypsinogen system proteolytic activity is regulated by a disorder-order transition of the activitation domain. These and other more speculative ideas about the role of fluctuations in protein function are discussed in references [1-3,8-11], but in general it is clear that a better experimental and theoretical characterisation of protein structural fluctuations is required before their importance can be fully established.

An alternative picture of protein dynamics involves the concept of confromational substates. Because of the very large number of degrees of freedom in a protein molecule there is not one single protein conformation, but a distribution of a very large number of closely related structures or substates separated by energy barriers of 0-10's kJ/mol. At physiological temperatures there can be rapid transitions between different substates whereas at low temperatures (typically below 200K) there is the possibility that an individual molecule becomes 'frozen' into a particular substate. Evidence for this picture comes from low temperature flash photolysis studies of ligand binding in myoglobin where it was found that the observed non-exponential rebinding kinetics could only be explained by postulating the existence of a distribution of substates with different rebinding rate constants [12,13].

2. SCOPE OF THIS CHAPTER

The object of this chapter is to show how inelastic neutron scattering can be used as a tool to probe the equilibrium fluctuations of proteins on a timescale of faster than 10^{-10}s. This overlaps with the timescale now accessible to computer simulations of protein dynamics and an important part of this work will be to show how experimental results match up to theoretical predictions. Particular emphasis will be given to the measurement of the density of states of low frequency vibrational modes in proteins and a comparison with various normal mode analyses in which it is assumed that the protein vibrates as a harmonic system. Also a preliminary description of measurements on myoglobin as a function of temperature will be given which clearly show that above 200K proteins have much enhanced mobility due to the excitation of fast non-harmonic motions. This motion may be associated with the transitions between conformational substates. Earlier reviews of the contribution of inelastic neutron scattering to the understanding of protein dynamics are to be found in [14,15].

3. INTRODUCTION TO NEUTRON SCATTERING

Neutrons are ideal probes of thermal fluctuations in condensed systems. This is because neutrons with wavelengths on the scale of atomic distances have energies comparable to that of typical thermal excitations. Thus a neutron of wavelength 1.8 Å has an energy of 25 meV equivalent to $k_B T$ at 300K ($\lambda^2 = h^2/(2m_n k_B T) = 950/T$), whereas an X-ray photon of the same wavelength has an energy of 6.9 KeV correponding to a temperature of 8.10^7 K ($\lambda = hc/k_B T = 1.43 \times 10^8/T$). Neutrons interacting with molecular systems at normal temperatures can therefore exchange a significant proportion of their energy with thermal excitations and these energy changes can be readily measured by a variety of techniques. Current neutron spectrometers are capable of probing a wide range of diffusional, rotational and vibrational motion with characteristic times in the range 10^{-7}-10^{-13}s. The technique overlaps at the high frequency end with optical spectroscopy (e.g. Raman spectroscopy) and at the lower end with NMR, fluoresence depolarisation [16] and Mössbauer spectroscopy [17]. The only drawback to the use of neutron scattering is that intense neutron beams are only available at major facilities such as the High Flux Reactor at the Institut Laue-Langevin in Grenoble,France or the pulsed neutron source ISIS in England. Even at these centres the neutron fluxes are relatively low ($\sim 10^7$ neutrons/cm^2/s) compared to for instance X-ray fluxes at a synchrotron ($\sim 10^{12}$ photons/cm^2/s), requiring that samples are large (100's mgs) and counting times long (hours to days). This combined with the competition for beamtime severely limits the number of experiments that can be done. Nevertheless, because neutrons have a distance and energy scale matched to thermal fluctuations in atomic systems, the method can provide information unattainable by other techniques. Good textbooks on neutron scattering are [18-20].

In a typical neutron scattering experiment, a beam of monoenergetic neutrons is incident on a sample and the number of neutrons scattered at various angles θ_S to the incident beam are measured. In the case of diffraction or scattering experiments, the detector does not discriminate the energy of the scattered neutrons and from the resultant diffraction pattern or scattering curve, structural information can be derived [21,22]. In inelastic neutron scattering experiments, the energy changes occurring when the incident neutrons interact with moving nuclei in the sample are also measured such that an energy spectrum of scattered neutrons is obtained at each scattering angle. The fundamental quantities measured are the *static structure factor* $S(q)$ in the case of diffraction, where $\hbar q$ is the momentum transfer between neutron and system $(q=4\pi/\lambda \sin(\theta_S/2))$ and the *dynamic structure factor* $S(q,\omega)$ in the case of inelastic neutron scattering, where $\hbar\omega$ is the energy transfer from neutron to system. In inelastic neutron scattering a variety of energy/frequency units are employed, $1\text{meV} = 8.07 \text{ cm}^{-1} = 0.24\text{THz} = 11.61\,^{\circ}\text{K}$.

4. INELASTIC NEUTRON SCATTERING AND TYPES OF MOTION

Figure 1 shows a schematic representation of neutron scattering experiments and the main features of the spectra obtained in the case of macromolecules such as proteins. What will concern us here is the way in which different kinds of protein motion are reflected in the measured $S(q,\omega)$. In general, three features may be distinguishable in an

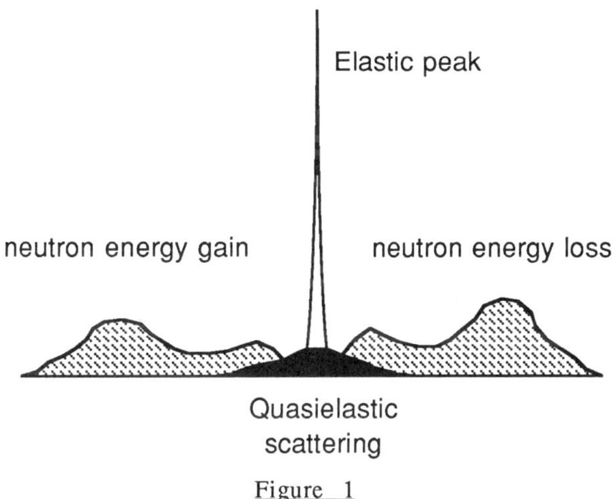

Figure 1

inelastic neutron spectrum. These are (a) the *elastic peak* $S(q,\omega=0)$ which arises from neutrons which are scattered with no change in energy and is generally the most intense feature in the spectrum in the case of solid (as opposed to liquid or gaseous) samples, (b) the *quasielastic scattering* which appears as a more or less broad, continuous feature centred at $\omega=0$ [18,20]. This arises from rotational or translational diffusive motion (for example the self diffusion of a particle in a liquid gives rise to a Lorentzian quasielastic scattering with width Dq^2 where D is the translational diffusion coefficient) or activated motions, for example translational or rotational motion of a particle between discrete sites requiring the surmounting of an energy barrier (see section 13). Transitions between conformational substates of proteins should manifest itself in the quasielastic spectrum. (c) the *inelastic scattering* which arises from transitions of the system between well-defined energy levels and thus gives rise to features shifted from $\omega=0$. This is the case for a system with harmonic normal modes of vibration in which for instance an incident neutron can excite or de-excite a normal mode of angular frequency ω, giving rise to peaks in the inelastic neutron spectrum at $+\omega$ or $-\omega$ respectively. The separation between low frequency inelastic scattering and quasielastic scattering is sometimes difficult to make in practice as the features due to different kinds of motion may overlap. Also the distinction between the two is not always clear: for a damped harmonic (or Brownian) oscillator, there is a continuity between a well defined inelastic peak and a broad quasielastic feature centred at $\omega=0$ depending on the degree of under- or over-damping [15]. It must be remembered that a measured spectrum will always be convoluted with the instrumental energy resolution function such that for example the elastic peak will not be appear as a true delta function but as having a finite width. Thus on a given instrument it may not be possible to detect a very slow diffusive motion which gives rise to quasielastic broadening with a width less than the resolution. Different neutron spectroscopic techniques are adapted to different energy resolutions [20,23] e.g. time-of-flight spectroscopy with resolution $\Delta E \sim 50\text{-}100\mu eV$, suitable for studying vibrational motion and very fast diffusive motion, backscattering spectroscopy with $\Delta E \sim 1\text{-}10\mu eV$, suitable for diffusive and jump processes and spin-echo spectroscopy with $\Delta E \sim 10 neV$, suitable for slow diffusive motions. By comparison, Mössbauer spectroscopy has an energy resolution of $\Delta E = 4.6 neV$ [17].

5. COHERENT AND INCOHERENT NEUTRON SCATTERING

The scattering of neutrons by an individual fixed nucleus is isotropic and characterised by a single parameter b, called the *scattering length* , which gives the amplitude of the spherical scattered wave per unit incident flux. For an assembly of N nuclei at positions $\mathbf{R}_L$ and scattering length b_L the static structure factor $S(\mathbf{q})$ is given by the expression:

$$S(\mathbf{q}) = \sum_{L,L'}^{N} e^{i\mathbf{q}\cdot(\mathbf{R}_L - \mathbf{R}_{L'})} < b_L b_{L'} > \qquad (1)$$

where $<\ >$ denotes an ensemble average.

However there is an important complication. In general b depends on which particular isotope the nucleus is and, if it has nonzero spin, on whether the neutron and nuclear spins are parallel or antiparallel. Provided spin orientation and isotope distribution is uncorrelated with position (the usual situation), the ensemble average can be evaluated as:

$$< b_L b_{L'} > = < b_L >< b_{L'} > + \delta_{LL'} (< b_L^2 > - < b_L >^2) \qquad (2)$$

and equation (1) becomes:

$$S(\mathbf{q}) = \sum_{L,L'}^{N} b_L^{coh} b_{L'}^{coh} e^{i\mathbf{q}\cdot(\mathbf{R}_L - \mathbf{R}_{L'})} + \sum_{L=1}^{N} \frac{\sigma_L^{inc}}{4\pi} \qquad (3)$$

The first term in (3) is the *coherent* scattering contribution and depends on the mean scattering length $b_L^{coh} = < b_L >$ for each kind of nucleus (averaged over spin orientation and isotope). Its intensity is given by the interference of scattered waves from each nucleus with phase differences determined by the nuclear position and it therefore contains structural information. In contrast, the second term is the *incoherent* scattering and depends on the incoherent cross-section of the nucleus, $\sigma_L^{inc} = 4\pi (< b_L^2 > - < b_L >^2)$. It results from spin (or isotope) disorder and is the sum of non-interfering individual nuclear contributions and contains no explicit structural information. Table 1 gives coherent and incoherent scattering lengths for the most frequently occurring nuclei in proteins. The most striking fact is the anomalously large incoherent cross-section of the ordinary hydrogen nucleus ^{1}H. Since 50% of the atoms in proteins are hydrogens, the incoherent scattering from them is the overwhelming contribution to the total cross-section. The isotope deuterium ^{2}H behaves very differently from ^{1}H and this is used in many aspects of neutron scattering from biological systems [21,22]. In the following discussions on inelastic neutron scattering and dynamics, we shall only be concerned with the incoherent scattering as this is the major contribution from normal (i.e. undeuterated) proteins. Experiments on proteins are usually done in the presence of D_2O rather than H_2O, to avoid the very large incoherent scattering from H_2O.

TABLE 1. Neutron scattering lengths (b) and cross-sections ($\sigma=4\pi b^2$) of the principle nuclei found in biological molecules.

Nucleus	Spin	b_{coh} (10^{-12} cm)	σ_{coh} (10^{-24} cm^2)	σ_{inc} (10^{-24} cm^2)
^{1}H	1/2	-0.374	1.76	79.7
^{2}H	1	0.667	5.59	2.0
C	0	0.665	5.56	0.0
N	1	0.94	11.1	0.2
O	0	0.580	4.23	0.01

A general expression for the incoherent dynamic structure factor is:

$$S_{inc}(\mathbf{q},\omega) = \frac{1}{2\pi}\int_{-\infty}^{\infty} dt\, e^{-i\omega t}\sum_{L=1}^{N}\frac{\sigma_L^{inc}}{4\pi} < e^{-i\mathbf{q}\cdot\hat{\mathbf{R}}_L(0)}\, e^{i\mathbf{q}\cdot\hat{\mathbf{R}}_L(t)} >_T \qquad (4)$$

where $\hat{\mathbf{R}}_L(t)$ is the time-dependent Heisenberg position operator of nucleus L and the brackets $< >_T$ indicate an ensemble average at temperature T. This again shows that the incoherent scattering is a sum of individual nuclear contibutions (no interference), but that it does contain explicit dynamical information because of the dependence on the atomic trajectories $\mathbf{R}_L(t)$.

6. SIMULATING PROTEIN DYNAMICS

Models of protein dynamics which give explicit expressions for each atomic trajectory can be used to calculate inelastic neutron spectra and can therefore be tested against experimental data. However proteins are extremely complex, inhomogeneous many-body systems whose structure and dynamics are governed by a variety of interatomic forces whose nature is known in principle but not accurately in detail. Faced with this situation, simplifying assumptions have to be made. Much current work in theoretical protein dynamics is based on the use of *empirical potential energy functions* (or force-fields) to describe the most important interactions in proteins (and with the solvent). These enable the potential energy $V(\{\mathbf{R}_L\})$ of a protein to be estimated as a function of the co-ordinates $\{\mathbf{R}_L\}$ of the component atoms. A fuller discussion of protein potential energy functions is given elsewhere in this book [24] and in [25]. Suffice to say here that most protein force-fields include terms to describe covalent bond stretching, angle bending and torsion as well as non-bonded terms interactions arising from van der Waals forces (short-range) and electrostatics (long-range). Hydrogen bonds may or may not be included explicitly. Given such a force-field and the existence of todays powerful computers, it is now possible to perform a *molecular dynamics simulation* of the protein dynamics. This involves numerically solving the classical equation of motion simulataneously for each atom in the protein as it evolves under the influence of the potential energy function V. The result is a time series $\{R_L(t_n)\}$ giving for each atom at discrete times $t = n\tau$, $n = 0,N$ where τ is typically 0.01ps and $N\tau$ is up to 100ps or more. From these trajectories it is possible to calculate directly the inelastic neutron scattering spectrum using a modification of equation (4), but for technical reasons this is somewhat difficult [15]. Despite the fact that many simulations of different proteins have now been performed (e.g. on BPTI [26], lysozyme [27] and myoglobin [28]) comparisons of calculated and experimental spectra are only just beginning to be done [29].

Molecular dynamics simulations give an extremely detailed description of the motion of each atom in a protein (although the accuracy of this detail is another matter!) and ultimately this is what is important in understanding what goes on in an enzyme active site. However without considerable analysis it is difficult to get from them an overall picture of protein dynamics and furthermore few experimental techniques are capable of furnishing equivalently detailed data for comparison. Thus it is of interest to look at even simpler models of specific kinds of motion that may occur in proteins.

7. VIBRATIONAL MOTION IN PROTEINS

Proteins like all molecular systems are expected to undergo vibrational motion and indeed optical spectroscopic studies of high frequency (100-3000 cm^{-1}) local vibrations characteristic of the particular chemical groups in proteins have for a longtime been used as a probes and monitors of both structure and function [30]. However because of their large size and the importance of soft forces in protein structure it is probable that they also have lower frequency, more global modes of vibration which were originally conceived of as 'breathing' modes or other distortions of a protein considered as an elastic body [31]. These low frequency modes would be themally excited at room temperature ($h\nu < k_B T = 200$ cm^{-1} at 300K) and would contribute significantly to the atomic mean-square displacements. Collective motions of this type have been identified by analysis of molecular dynamics simulations [32] but a more direct way to look theoretically at vibrational motion is to perform a *normal mode analysis*. In this procedure we begin as before with a protein and its empirical potential energy function, but make the further assumption that the protein conformation is at a minimum of the energy function and that motion corresponds to small displacements δR_L of each atom about its equilibrium position R_L^0. We can then expand the potential energy in a Taylor series about the minimum configuration giving:

$$V(\{R_L\}) = V(\{R_L^0\}) + \frac{1}{2}\sum_{ij}^{3N}\left(\frac{\partial^2 V}{\partial R_i \partial R_j}\right)_{\{R_L^0\}}\delta R_i \delta R_j + O(\delta R)^3 \qquad (5)$$

If the displacements are small enough we can make the harmonic approximation and only consider the terms quadratic in the displacements. Then using the force constants given by the second derivative matrix in the equations of motion it can be shown that the resultant vibrational motion of each atom can be expressed as a superposition of 3N-6 harmonic oscillations or *normal modes*. Each normal mode is characterised by a nonzero angular frequency ω_λ and a set of N orthonormal eigenvectors c_L^λ giving the relative magnitude and direction of displacement of each atom according to the normal mode λ. A more detailed discussion of the theory of normal modes is given in [33].

8. CALCULATIONS OF NORMAL MODES IN PROTEINS

Since proteins contain hundreds of atoms, the number of degrees of freedom and hence the number of normal modes is extremely large and calculating their frequencies and eigenvectors is a formidable computational problem. However the result is in principle a complete vibrational spectrum from the highest frequency bond stretching modes to the lowest global collective modes (i.e 3-3000 cm^{-1}). To reduce the scale of the problem, simplifications are often made. For instance it is possible to include non-polar hydrogens (e.g. methyl hydrogens) with the heavy atom to which they are attached (extended atom approximation) or to restrict attention to the softer torsional degrees of freedom by constraining bond lengths and angles. In the following we shall focus on $N(\omega)$ the *frequency distribution* of normal modes (or *density of vibrational states*) where $N(\omega)d\omega$ gives the number of modes with frequency in the range ω to $\omega+d\omega$. This function is directly accessible to measurement by inelastic neutron scattering (see below) and is also sufficient, within the harmonic approximation, to calculate vibrational contributions to thermodynamic functions such as entropy and specific heat [15].

TABLE 2. Details of various normal mode calculations on BPTI.

	Model	Number of degrees of freedom	Electrostatics (see text)	Lowest mode	Authors	Ref.
A	Torsion	208	?	4.6 cm^{-1}	Levitt et al (1983)	[36]
B	Polar H	1740	Shift	6.2 cm^{-1}	Brooks et al. (1983)	[37]
C	Polar H	1740	Switch	11.0 cm^{-1}	Tidor et al. (1988)	[38]
D	All H	2712	Switch	9.6 cm^{-1}	Tidor et al. (1988)	[38]
E	Torsion	241	?	5.7 cm^{-1}	Go et al. (1983)	[39]
F	Vibrational force field	1040	-	15.2 cm^{-1}	Derreumeux (1987)	[40]

Normal mode calculations have been performed on bovine pancreatic trypsin inhibitor (BPTI), lysozyme, crambin and ribonuclease [34] as well as on a DNA fragment [35]. Table 2 gives some details of the six different normal mode calculations on BPTI, a small protease inhibitor of 58 residues much studied in the protein dynamics field. The reader is referred to [34,36,39] for pictorial representations of some of the calculated normal modes of BPTI.

A comparison of the frequency distributions of the first four calculations in the above table are shown in Figure 2. This shows that models A and B predict far more very low frequency modes (<50 cm^{-1}) than models C and D. The differences between models B,C and D are very informative as they were performed with basically the same potential (the CHARMM potential [41]), but with a change in only one feature of the calculation. Calculations B and C used exactly the same number of degrees of freedom but differ in the method of truncating the long-range electrostatic forces: they used respectively shift and switch methods [38,41]. This clearly has a drastic effect on the number of low frequency modes and demonstrates the importance of an accurate representation of the electrostatic forces. Calculations C and D both use switch electrostatics but differ in that C uses the extended atom approximation with only explicit polar hydrogens, whereas in D, all hydrogen atoms were included explicitely. The differences between C and D are rather less dramatic, but D has more modes at higher frequency due to the considerably greater number of degrees of freedom. This will be discussed further below as part of the comparison with experimental results.

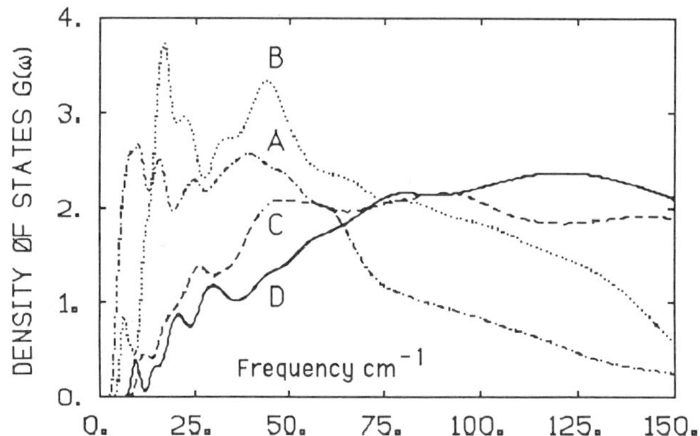

Figure 2. Comparison of the frequency distributions of four of the normal mode models of BPTI described in Table 2. The curves have been convoluted with a typical instrumental resolution function.

9. NORMAL MODES AND INELASTIC NEUTRON SCATTERING

For a harmonically vibrating molecule, expression (4) for $S_{inc}(\mathbf{q},\omega)$ can be evaluated exactly in terms of normal mode frequencies and eigenvectors. The general result is quite complicated and discussed in detail in [15,42]. However when the momentum transfer q is small (as is generally the case in the experiments to be described) only two scattering processes are important. These are the elastic scattering $S_{inc}(\mathbf{q},0)$ given by the expression:

$$S_{inc}(\mathbf{q},0) = \sum_{L=1}^{N} \frac{\sigma_L^{inc}}{4\pi} e^{-2W_L(\mathbf{q})} \qquad (6)$$

which only depends on the Debye-Waller factor $\exp[-2W_L(\mathbf{q})]$ and cross section of each atom L and results in an intense elastic peak. The true inelastic scattering is dominated by the *one-phonon scattering* in which neutrons exchange one quantum of vibrational energy with a particular mode λ. This gives rise to a spectrum of peaks at $\pm\omega_\lambda$ (corresponding to neutron energy gain or loss) with intensities given by the expression:

$$S_{inc}(\mathbf{q},\pm\omega_\lambda) = e^{\pm\hbar\omega_\lambda/k_BT} \sum_{L=1}^{N} \frac{\sigma_L^{inc}}{4\pi} e^{-2W_L(\mathbf{q})} \frac{\hbar\,|\mathbf{q}.\mathbf{c}_L^\lambda|^2}{4\omega_\lambda m_L \sinh(\hbar\omega_\lambda/2k_BT)} \qquad (7)$$

110

The last term in (7) can be interpreted as the mean-square amplitude of motion of atom L undergoing mode λ projected onto the direction of $\mathbf{q}$. This equation shows that low frequency modes which have larger amplitudes give stronger peaks. The Debye-Waller factor exponent $W_L(\mathbf{q})$ is given by:

$$W_L(\mathbf{q}) = \sum_{\lambda=1}^{3N-6} \frac{\hbar |\mathbf{q}.\mathbf{c}_L^\lambda|^2}{4\omega_\lambda} \coth\left(\frac{\hbar\omega_\lambda}{2k_BT}\right) \tag{8}$$

Expressions (6) and (7) are correct to order q^2; for high q, multiphonon scattering processes can become important [42]. If measurements are made for range of values of q, the following extrapolation can be made:

$$G(\omega) = \lim_{q\to 0} \frac{6\omega}{\hbar q^2}(e^{\hbar\omega/k_BT}-1)S_{inc}(q,\omega) = \sum_{\lambda=1}^{3N-6}\sum_{L=1}^{N}\frac{\sigma_L^{inc}}{4\pi m_L}|\mathbf{c}_L^\lambda|^2\delta(\omega-\omega_\lambda) \tag{9}$$

Here $G(\omega)$ is the amplitude-weighted (or generalised) frequency distribution and the result is written for an isotropic system (e.g. protein solution or powder). Since neutron inelastic scattering uses the hydrogen atoms as probes of vibrational motion, and these are rather evenly distributed over the protein, it can be imagined that $G(\omega)$ has a very similar form (apart from a scale factor) to the true vibrational frequency distribution $N(\omega)$ and this has been verified by a model calculation [42]. In practice, great care has to be taken in performing the extrapolation implied by equation (9) on real data. This is because coherent scattering can affect the results at low q, multiphonon scattering at high q and also multiple scattering has to be taken into account as has been shown recently [43].

10. EXPERIMENTAL METHODS

Experimental inelastic neutron scattering spectra have now been obtained on a number of proteins including hexokinase [44], lysozyme [45], BPTI [38,46], cytochrome C [46] and myoglobin [47]. In most cases the time-of-flight spectrometer IN6 at the Institut Laue-Langevin has been used to make the measurements [23]. This instrument combines high intensity with resonable energy resolution ($1cm^{-1}$ at the elastic peak, but deteriorating with increasing energy transfer). The instrument uses incident neutrons of wavelength 5.1Å and measures energy changes upon scattering by measuring the delay (neutron energy loss) or advance (neutron energy gain) in neutron arrival times at detectors placed a fixed distance away from the sample. Typically time-of-flight spectra at 18 different scattering angles in the range $10^0<\theta_s<110^0$ can be measured simultaneously, corresponding to elastic momentum transfers of $0.2 < q_{el} < 2.1$ Å^{-1} (in general q depends on both scattering angle and energy transfer). Most useful data is obtained on the neutron energy gain side and the spectrum extends up to 200-300 cm^{-1}. Standard corrections for background, transmission and detector efficiency enable the experimental dynamic structure factor to be derived with good accuracy.

Samples typically require several hundreds of milligrams of material which is contained in thin-walled aluminium cells 0.1-0.5 mm thick and up to 5 cm in diameter (neutrons beams are large!). Normally the thickness is chosen to acheive a neutron transmission of about 90%, less than which could introduce significant multiple scattering effects. Proteins have been measured in various states: in concentrated D_2O solution [46], in crystalline form (again with D_2O as the solvent) and in powder form hydrated to various degrees with either D_2O or H_2O [38,47]. Most early experiments were done on solutions often with the aim of being able to change the state of the protein easily during the experiment (e.g. to bind a ligand as in the case of hexokinase [44], or to change oxidation state in the case of cytochrome C [46]). However the very large scattering contribution from the solvent (even when D_2O) and the broadening of the elastic peak due to the translational diffusion of the molecule meant that these early attempts at difference experiments were not very reliable. More recent experiments have been mainly on hydrated protein powders in which the solvent scattering is minimised while maintaining the protein in a near native state e.g. a myoglobin sample hydrated to 0.35 gD_2O/g protein is essentially fully hydrated, but the solvent scattering contribution is only 4%.

11. EXPERIMENTAL RESULTS

Figure 3a shows a series of time-of-flight spectra at different scattering angles obtained on spectrometer IN6 at room temperature from a fairly dry powder of BPTI (0.12 $gD_2O/gprotein$). Note that the time-of-flight scale is non-linear in energy shift and that neutron energy gain increases to the left of the elastic peak. Figure 3b shows the *predicted* time-of-flight spectrum of BPTI at 66° calculated from the normal mode models B and D in Table 2 (and convoluted with the IN6 resolution function). Figure 3c shows spectra of myoglobin at two different hydrations: a dry sample at 0.1 gD_2O/g protein and a fully hydrated sample at 0.35 gD_2O/g protein.

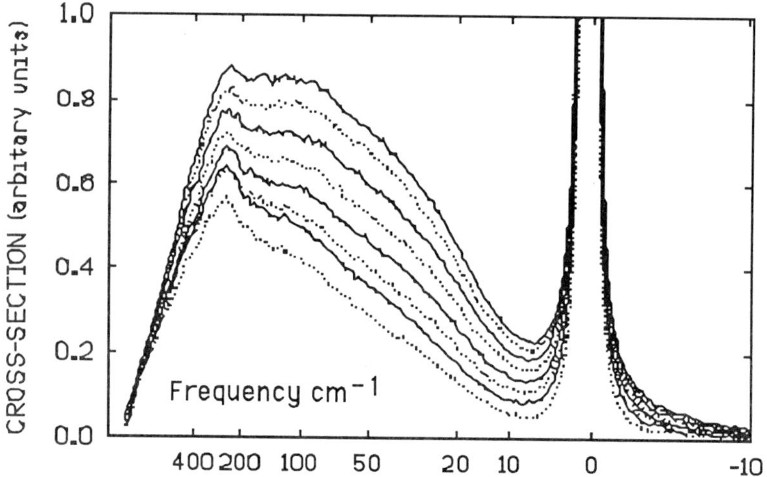

<u>Figure 3a.</u> Time-of-flight spectra of dry BPTI obtained at several different scattering angles.

From these Figures we can remark the following. Experimental time-of-flight spectra of proteins are smoothly varying and show no marked structure, but have a characteristic shape (intense elastic peak, minimum at about 10 cm⁻¹, then intensity rising steadily to 200-250 cm⁻¹ before dropping to zero). The sharp peak at roughly 220 cm⁻¹ probably arises from torsional vibrations of methyl groups of which there are usually several in a protein [48]. The observed shape is roughly reproduced for BPTI by normal mode analysis D (all hydrogens, switch electrostatics), but not at all by normal mode analysis B (polar hydrogens only, shift electrostatics). Even so, the distinct peaks predicted by the normal mode analyses (which arise not from single modes, but from clusters of modes in the frequency distribution [42]) are not observed. Possible reasons for the smoothness of the distribution are discussed further below. Spectra of different proteins (e,g, BPTI, lysozyme, myoglobin) are found to be remarkably

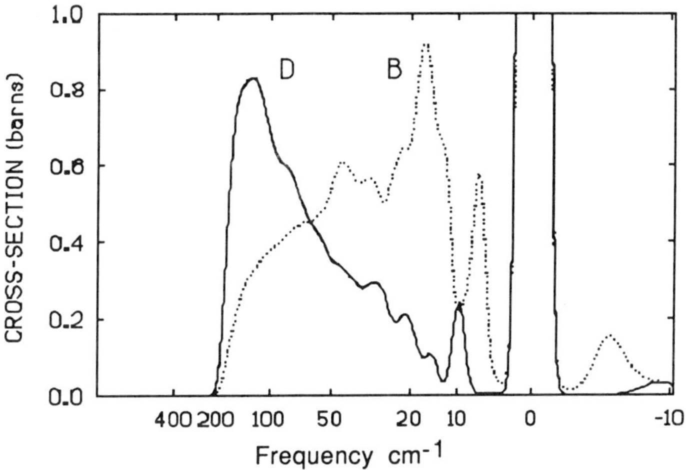

Figure 3b. Time-of-flight spectra of BPTI (convoluted with the instrumental resolution function) calculated from normal mode models B and D.

similar at comparable hydrations. This suggests that the low frequency spectrum is determined by general rather than detailed features of protein structure. The effect of hydrating a sample of myoglobin is shown in Figure 3c. It can be seen that there is very little effect on high frequency motion above about 20 cm^{-1} whereas below this frequency there is enhanced scattering which appears to arise from extra mobility in the hydrated protein. Very similar results are obtained on comparison of data of dry BPTI and BPTI crystals (with deuterated mother liquor). Figure 3d compares a linear combination of the spectra of dry BPTI and pure D_2O with the spectrum of BPTI crystals. The crystal sample was estimated to contain 50% D_2O and it is the latter which gives rise to the hump in

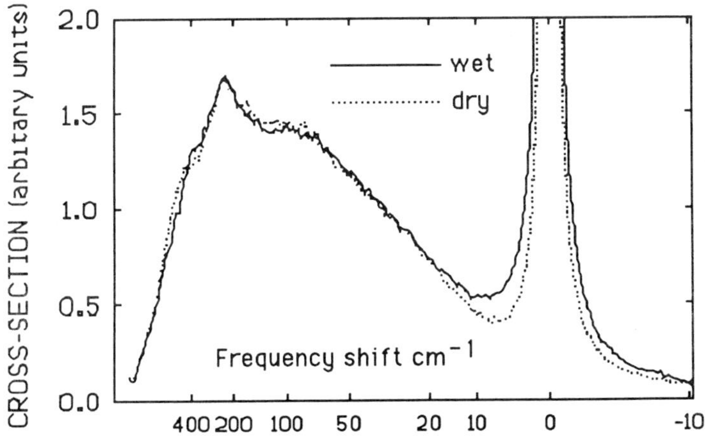

Figure 3c. Comparison of time-of-flight spectra of dry and hydrated myoglobin.

the spectrum at 50 cm^{-1}. The linear combination explains well the data above 25 cm^{-1}, again suggesting a limited effect of hydration above this frequency. At lower frequencies there is again additional scattering from the hydrated protein. In section 13 below, we show by examining the temperature dependence of this extra low frequency scattering that it is unlikely to arise from additonal low frequency vibrational motion, but is quasielastic scattering arising from fast transitions.

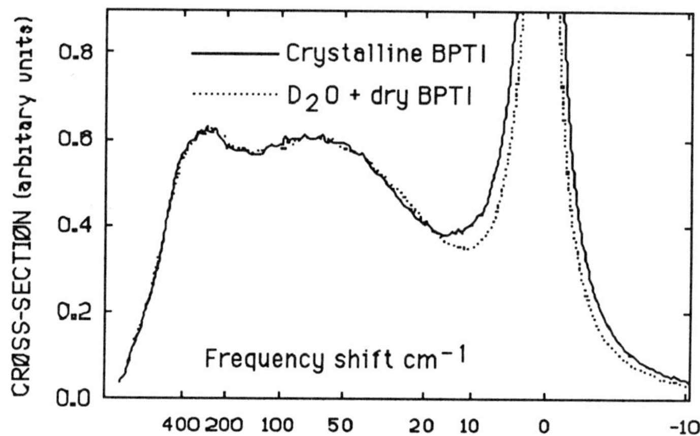

Figure 3d. Comparison of time-of-flight spectra of BPTI crystals (with deuterated mother liquor) and a linear combination of dry BPTI and pure D$_2$O.

12. COMPARISON OF EXPERIMENTAL AND CALCULATED FREQUENCY DISTRIBUTIONS

Figure 4 shows the experimental frequency distribution $G(\omega)$ derived for dry BPTI from the data of Figure 3a and compared with the frequency distributions of normal mode models B and D [38]. The results again show that the envelope of model D (and indeed model C) accords reasonably with the experimental curve whereas model B predicts far more very low frequency modes (this is also true of model A illustrated in Figure 2). Thus inelastic neutron scattering is able to give a very direct test of the quality of various models and potential functions in predicting dynamic phenomena. However it cannot be claimed that model D is in good accord with the experimental data, principally because it predicts far more structure in $N(\omega)$ than is observed. This is presumably due to the inadequacies inherent in the normal mode method where it is assumed that there is a well-defined minimum energy configuration, that displacements away from the minimum are small enough to be harmonic and that there are no frictional effects (e.g. from solvent). None of these assumptions are really valid for proteins. Anharmonic effects are known to exist at room temperature particularly for low frequency motions [37,39,49] and the concept of multiple conformational substates mentioned in the introduction also complicates the simple notion of normal modes. Preliminary evidence that molecules in different substates may have slightly different vibrational motion has come from recent molecular dynamics simulations of myoglobin

114

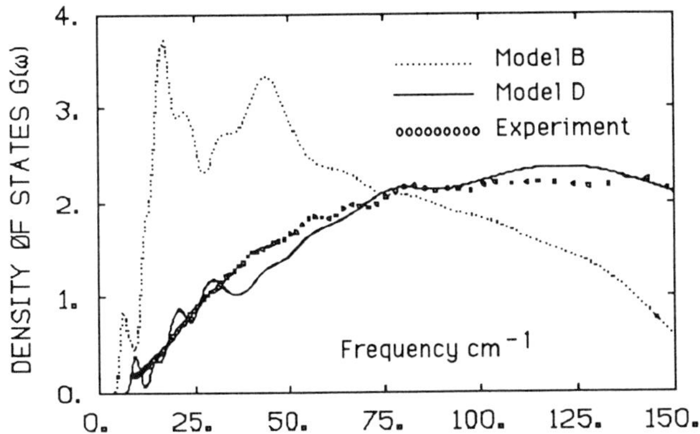

Figure 4. Comparison of experimentally determined vibrational frequency distribution for BPTI with normal mode models B and D.

quenched to 80K [29].   All these factors as well as frictional effects will act towards smoothing of the vibrational frequency distribution. It may be supposed that going to low enough temperatures one could freeze out transitions between substates, and reduce the amplitude of vibrational motion such as to eliminate anharmonic effects. Indeed this appears to be the case; evidence from Mössbauer spectroscopy [17] and inelastic neutron scattering (see below) show that protein dynamics is essentially harmonic below 200K. However this does not change significantly the vibrational frequency distribution. Figure 5 shows $G(\omega)$ derived for a myoglobin powder hydrated with D_2O at 180K (at 100K the result is very similar). The curve is still smooth and similar in form to that in Figure 4.

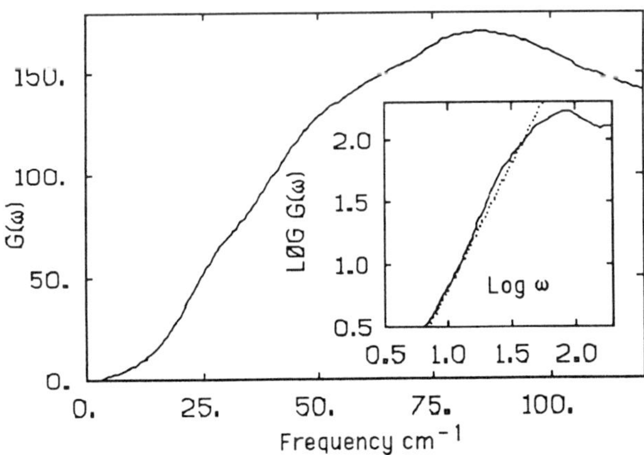

Figure 5. Experimental vibrational frequency distribution of hydrated myoglobin. Inset: log/log plot showing deviations from Debye behaviour (the dotted line has slope 2).

The inset log/log plot in Figure 5 shows that at very low frequencies the density of states is Debye-like i.e. $G(\omega) \propto \omega^2$, but that there is a distinct enhancement at about 25 cm^{-1}. (Note that this $G(\omega)$ is derived from the myoglobin data at 180K shown in Figure 8; the enhancement in $G(\omega)$ corresponds to the marked peak in $S_{inc}(q,\omega)$ at 25 cm^{-1}). It is remarkable that very similar curves are obtained with simple amorphous materials [50]. This together with the similarity between different proteins suggests that simple models of close-packed but disordered systems may be sufficient to explain the density of states of proteins [51]. Finally it should be remarked that it is impossible experimentally to observe isolated molecules (as treated in most protein dynamics simulations) and in real protein systems (e.g. powders and crystals) at very low frequencies one expects to see evidence of inter-molecular modes of motion (this could explain the low frequency Debye-like behaviour).

13. THE TEMPERATURE DEPENDENCE OF PROTEIN DYNAMICS

In studies of dynamics it is very important to observe the variation with temperature as this can give important clues on the type of motion involved. For example the temperature dependence of harmonic motion is determined essentially by the occupation factor $n(\omega)=[\exp(h\omega/k_B T) -1]^{-1}$ for the mode ω, whereas for an activated motion, there is usually an Arrhenius-type temperature dependence $\exp(-\Delta E_a/RT)$ where ΔE_a is the activation energy.

The temperature dependence of a number of properties of myoglobin have been studied over a wide range of temperatures, yielding a number of important ideas. The concept of conformational substates arose from the analysis of low temperature flash photolysis experiments which measure ligand (e.g. oxygen) binding rates to the heme iron [12,13]. Elsewhere in this book Mössbauer spectroscopic and X-ray crystallographic studies of myoglobin as a function of temperature are described in detail [17]. Mössbauer experiments have revealed a striking dynamic transition at about 200K, above which temperature the mean-square displacement of the heme iron shows a much faster increase with temperature than expected for a harmonic system. This has been interpreted as due to the excitation of new slow modes of motion above the transition temperature [17,52].

Recently it has been shown that the dynamic transition in myoglobin at 200K can also be clearly observed in inelastic neutron scattering experiments and analysis of the data gives some new insights into the nature of fast motions in proteins [47]. A brief description of this work will now be given.

A schematic view of the observed incoherent dynamic structure factor $S_{inc}(q,\omega)$ of myoglobin at fixed q and three different temperatures is shown in Figure 6. At very low temperatures (Figure 6a), there is essentially no motion and the spectrum consists only of the sharp elastic peak. As the temperature is raised (to say 150K, Figure 6b), the amplitude of harmonic vibrations increases giving rise to an inelastic vibrational spectrum (according to equation (7) and a corresponding decrease of the elastic peak (according to equation (6)). Figure 6c shows the situation above the transition temperature at say 300K. In addition to the increased scattering from the vibrations, new non-vibrational modes of motion have been excited and give rise to a quasielastic spectrum, with a corresponding sharper decrease in the elastic peak. Note that this discussion illustrates well the *zeroth-order sum rule*:

$$\int_{-\infty}^{\infty} S_{inc}(\mathbf{q},\omega)\, d\omega = \sum_{L=1}^{N} \frac{\sigma_L^{inc}}{4\pi} = \text{constant} \qquad (10)$$

which is an exact result and shows that the overall scattered intensity is conserved but can be redistributed between elastic, quasielastic and inelastic scattering.

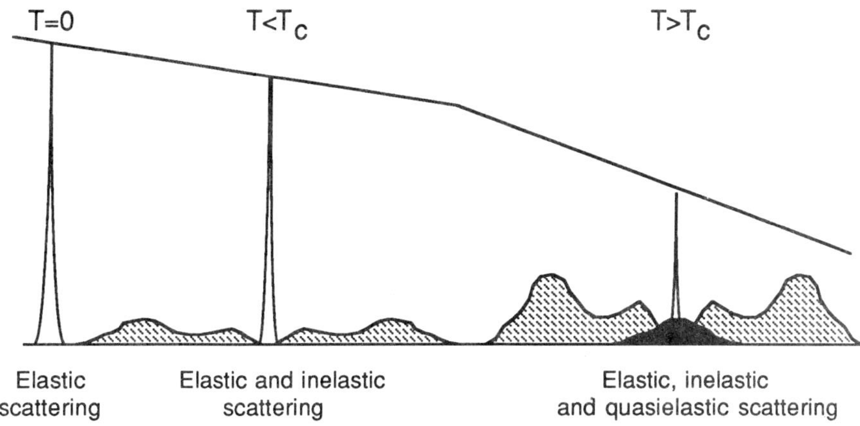

<u>Figure 6.</u> Schematic diagram of the temperature variation of the inelastic neutron scattering spectrum of myoglobin.

Myoglobin behaves in precisely the way illustrated in Figure 6, although because different spectrometers were used to collect the data it is not possible to show the exactly corresponding experimental spectra. However, Figure 7 shows the temperature variation between 4 and 320K of the elastic intensity $S(q,0)$ at a number of different values of q measured on the backscattering instrument IN13

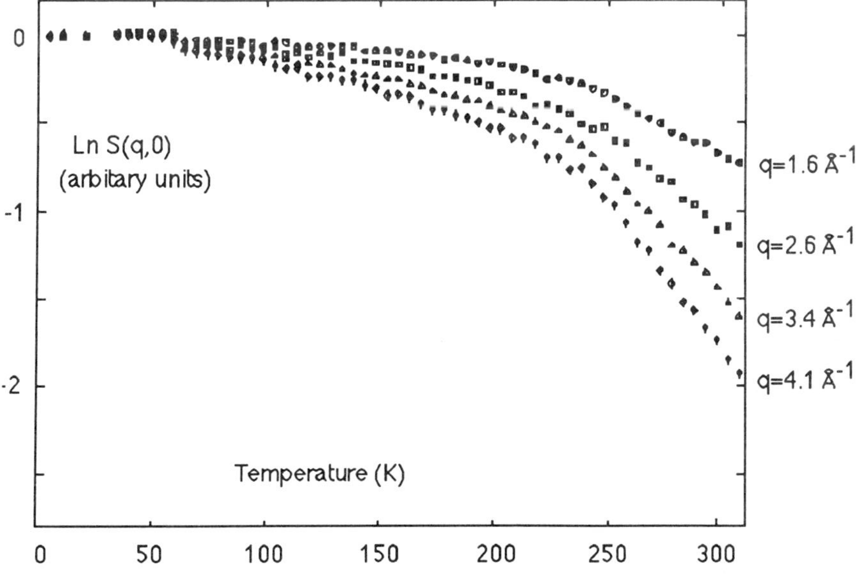

<u>Figure 7.</u> Normalised elastic intensity $S(q,0)$ for hydrated myoglobin as a function of temperature for four different values of q.

117

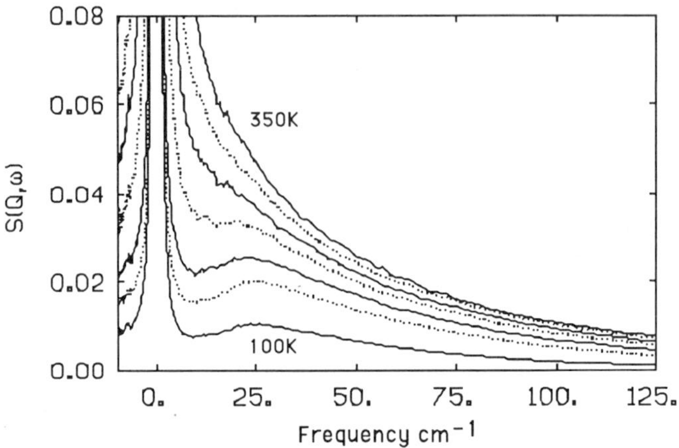

Figure 8. Inelastic neutron scattering spectra (at a fixed scattering angle) of hydrated myoglobin at temperatures of 100,180,220,270,300,320 and 350K.

at the Institut Laue-Langevin. This instrument has a resolution of about 10mev, so that motions slower than 10^{-10}s are not resolved. The transition at about 200K is clearly seen. Figure 8 shows quasielastic and inelastic spectra obtained on instrument IN6 (resolution 50μev) at temperatures between 100 and 320 K. These show that there is little change in the form of the scattering between 100 and 180K but that above this temperature the inelastic peak prominent at 25 cm⁻¹ is increasingly submerged below a broad quasielastic line.

How do we interpret these data? Myoglobin behaves as a harmonic system below 180K. The evidence for this is firstly that below 180K the data of Figure 7 fits the expected form for a vibrational Debye-Waller factor, i.e. $S(q,0) \sim exp(-q^2<\Delta x^2>_{vib})$, where $<\Delta x^2>_{vib}$ is the average mean-square displacement of the hydrogen atoms in the protein and $<\Delta x^2>_{vib} \propto T$. Secondly, the spectra at 100K and 180K in Figure 8, show no quasielastic scattering, and their different intensities are consistent with the temperature dependence of harmonic motion. In other words, the vibrational frequency distribution derived from the 180K data and shown in Figure 5 is a temperature-independent property of the system at these temperatures. The extra motion excited above 180K has been modelled by assuming that hydrogen atoms are able to jump between discrete sites a distance d apart and separated by free energy ΔG. This simple asymmetric double-well model (see inset of Figure 9) has been widely used, for instance in the elegant neutron work on hydrogen-bond dynamics in carboxylic acids [53]. For this model it can be shown that the powder-averaged elastic intensity $S(q,0)$ is given by:

$$S_{inc}(q,0) = e^{-q^2<\Delta x^2>_G}\left\{1-2p_1p_2\left(1-\frac{sin(qd)}{qd}\right)\right\} \qquad (11)$$

where $<\Delta x^2>_G$ is the Gaussian contribution to the mean-square displacement (at low temperatures $<\Delta x^2>_G = <\Delta x^2>_{vib}$). The second term describes the non-Gaussian jump process and p_1 and p_2 are respectively the probabilities of being in the ground or excited state, with $p_2/p_1 \propto exp(-\Delta G/RT)$. Least squares fits of equation (11) to the data of Figure 7 allows the parameters of the model to be derived. A plot of $Ln\{p_2/p_1\}$ against $1/T$ gives an energy asymmetry of $\Delta H = 12$

(± 2) kJ/mole and entropy $\Delta S/R = 3.0$. The value of the jump distance d is found to be 1.5(± 0.1)Å. The magnitude of this value would indicate the involvement of torsional degrees of freedom and indeed large amplitude, fast dihedral angle fluctuations are observed in molecular dynamics simulations of proteins [26]. Figure 9 shows the variation with temperature of the different contributions to the average mean-square hydrogen displacement which is given by the expression:

$$<\Delta x^2> = -\left(\frac{d\,\{LnS(q,0)\}}{d(q^2)} \right)_{q=0} = <\Delta x^2>_G + \frac{p_1 p_2 d^2}{3} \tag{12}$$

$<\Delta x^2>_G$ is principally given by the vibrational contribution $<\Delta x^2>_{Vib}$ but there is an additional slower process apparent above 240K. The second term in (12) gives rise to a very strong temperature dependence above 180K. Figure 9 is very reminiscent of results obtained by Mössbauer spectroscopy [17,52] but the time-scale of the motions implicated is very different.

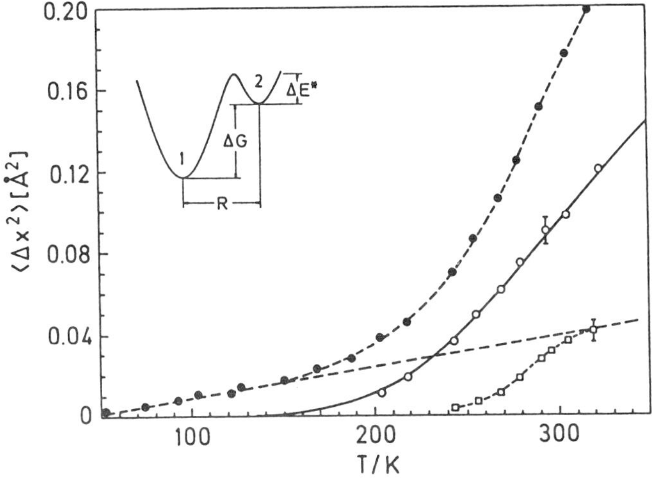

Figure 9. Hydrogen mean-square displacements in hydrated myoglobin as a function of temperature derived by fitting an asymmetric double potential well model (inset) to the data of Figure 7. Full circles: total mean-square displacement (equation 12); open circles: second term in equation (12); dashed straight line: vibrational contribution extrapolated from low temperatures; open squares: additional Gaussian process above 240K.

To get more idea of the time-scales of these motions, one has to look at the complete spectrum rather than just the elastic peak. The spectra of Figure 8 have been analysed on the assumption that the vibrational component at a given temperature is simply given by scaling with the correct harmonic temperature dependence the spectrum at 180K where quasielastic scattering is absent. Figure 10 shows the resulting quasielastic spectra for T>180K, obtained by subtracting the scaled vibrational contribution. The quasielastic scattering is essentially zero beyond 3meV and shows an increase in intensity with temperature consistent with the asymmetric double well model discussed above. The data can be Fourier transformed with simultaneous deconvolution of the instrumental resolution function to give time-correlation functions (Figure 11). These clearly show two distinct processes, a fast motion with correlation time 0.3-0.5 ps which is associated with the jumping, and a slower relaxtion with estimated mean

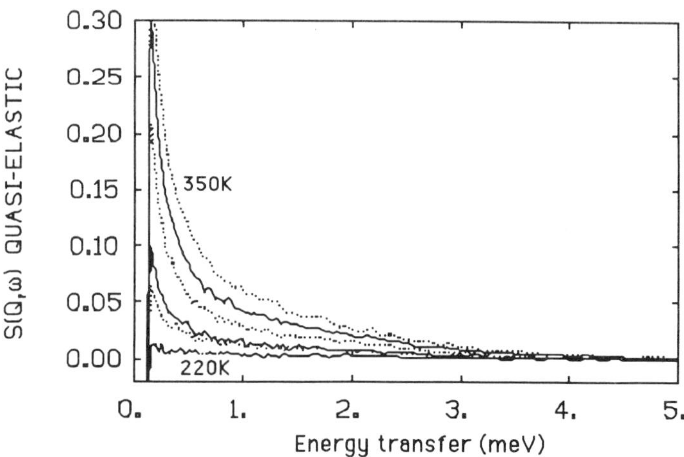

Figure 10.   Quasielastic scattering from hydrated myoglobin at temperatures 220,250,270,300,320 and 350K, obtained by subtracting the vibrational component from the spectra of Figure 8.

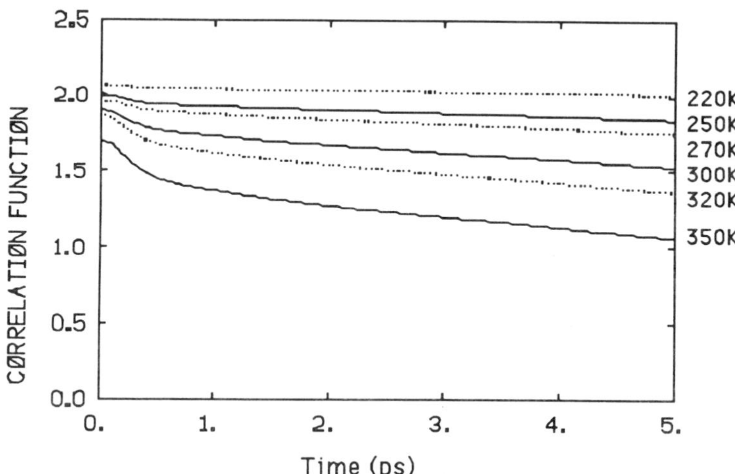

Figure 11. Correlation functions of hydrated myoglobin obtained by Fourier transformation of the quasielastic spectra in Figure 9 (arbitary units).

correlation time of 20 ps which is associated with the additional Gaussian process above 240K [47].

The exact atomic nature of these motions is still unclear. However as with the vibrational frequency distribution, the effects just described appear to be general for globular proteins. Furthermore this dynamic behaviour has many features expected of a liquid-glass transition [47,54] and similar effects have been observed by neutron scattering in other glassy systems [55,56]. One thing is clear, only by constant interaction between theorists and experimentalists will further progress be made in understanding the dynamics of such complex systems. In this context one should mention recent work based on molecular dynamic

120

simulations of myoglobin which analyse transitions between multiple conformational substates [57] and begin to compare the temperature dependence with that obtained experimentally [29].

ACKNOWLEDGEMENTS

The ideas and results discussed here are to a large extent based on collaborations with Dr. Jeremy Smith and Professor Martin Karplus (Chemistry Department, Harvard) and Dr. Wolfgang Doster (Technischen Universität München). I would also like to thank Drs. A. Dianoux and W. Petry (Institut Laue-Langevin) for assistance with the experiments and frequent discussions.

REFERENCES

[1] Cooper, A., Sci.Prog. Oxf. 66:473 (1980).

[2] Huber,R. and Bennett,W.S. Bioploymers 22:261 (1983).

[3] Ringe, D. and Petsko,G.A. Prog. Biophys. Mol. Biol. 45:197 (1985).

[4] Watson,H. Elsewhere in these proceedings.

[5] Cooper,A. Elsewhere in these proceedings.

[6] McCammon, J.A. Rep. Prog.Phys. 47:1 (1984).

[7] Karplus, M. in "Structural Molecular Biology" eds. Davies,D.B., Saenger,W. and Danyluk,S.S. NATO ASI Series A: Life Sciences 45, Plenum, London (1982).

[8] McCammon,J.A. and Harvey,S.C. "Dynamics of proteins and nucleic acids." Cambridge University Press, Cambridge.

[9] Sturtevant,J.M. PNAS 74:2236 (1977).

[10] Careri,G., Fasella,P. and Gratton,E. Ann.Rev.Biophys.Bioeng. 8:69 (1979).

[11] Careri,G. Elsewhere in these proceedings.

[12] Austin, R.H., Beeson,K.W., Eisenstein,L., Frauenfelder,H. and Gunsalus,I.C. Biochem. 14:5355 (1975).

[13] Ansari,A., Berendson,J., Bowne,S.F., Frauenfelder,H., Iben,I.E.T., Sauke,T.B. Shyamsunder,E. and Young,R.D. PNAS 82:5000 (1985).

[14] Middendorf, H.D. Ann.Rev.Biophys.Bioeng. 13:425 (1984).

[15] Cusack,S. Comm.Mol.Cell.Biophys. 3:243 (1986).

[16] Gratton,E. Elsewhere in these proceedings.

[17] Parak,F. Elsewhere in these proceedings.

[18] Springer,T. "Quasielastic neutron scattering for the investigation of diffusive motions in solids and liquids" Springer Tracts in Modern Physics 64, Springer-Verlag, Berlin (1972).

[19] Lovesey,S.W. "Theory of neutron scattering from condensed matter." Vol 1, Clarendon Press, Oxford (1984).

[20] Bee,M. "Applications of quasielastic neutron scattering to solid-state chemistry, biology and material science." Adam Hilger, London (1988).

[21] Zaccai,G. and Jacrot,B. Ann.Rev.Biophys.Bioeng. 12:139 (1983).

[22] "Neutrons in Biology" ed. Schoenborn, B.P. Basic Life Sciences Vol 27. Plenum Press, New York (1984).

[23] Neutron research facilities at the ILL high flux reactor. Institut Laue-Langevin, Grenoble, France.

[24] Somorjai,R. Elsewhere in these proceedings.

[25] Lifson,S. in "Structural Molecular Biology" eds. Davies,D.B., Saenger,W. and Danyluk,S.S. NATO ASI Series A: Life Sciences 45, Plenum, London (1982).

[26] Levitt,M. J.Mol.Biol. 168:595 and 168:621 (1983).

[27] Post,C.B., Brooks,B.R., Karplus,M., Dobson,C.M., Artymiuk,P.J., Cheetham,J.C. and Phillips,D.C. J.Mol.Biol. 190:455 (1986).

[28] Levy,R.M., Sheridan,R.P., Keepers,J.W., Dubey,G.S., Swaminathan,S. and Karplus,M. Biophys.J. 48:509 (1985).

[29] Smith,J., Kuczera,K., Tidor,B., Karplus,M., Doster,W. and Cusack,S. Physica B 156-157:437 (1989).

[30] Peticolas,W. in "Structural Molecular Biology" eds. Davies,D.B., Saenger,W. and Danyluk,S.S. NATO ASI Series A: Life Sciences 45, Plenum, London (1982).

[31] Suezaki,Y. and Go,N. Int.J.Pept.Prot.Res. 7:333 (1975).

[32] Swaminathan,S., Ichiye,T., van Gunsteren,W. and Karplus,M. Biochem. 21:5230 (1982).

[33] Fillaux,F. Elsewhere in these proceedings.

[34] Levitt,M., Sander,C. and Stern,P.S. J.Mol.Biol. 181:423 (1985).

[35] Tidor,B., Irikura,K.K., Brooks,B.R. and Karplus,M. J.Biomol.Struc.Dyn. 1:231 (1983).

[36] Levitt,M., Sander,C. and Stern,P.S. Int.J.Quant.Chem. 10:181 (1983).

[37] Brooks,B.R. and Karplus,M. PNAS 80:6571 (1983).

[38] Cusack,S., Smith,J., Finney,J., Tidor,B. and Karplus,M. J.Mol.Biol. 202:903 (1988).

[39] Go,N., Noguti,T. and Nishikawa,T. PNAS 80:3696 (1983).

[40] Derreumeux,P. Thesis, University of Lille II (1988).

[41] Brooks,B.R., Bruccoleri,R.E., Olafson,B.D., States,D.J., Swaminathan,S. and Karplus,M. J.Comp.Chem. 4:187 (1983).

[42] Smith,J. , Cusack,S., Pezzeca,U., Brooks,B.R. and Karplus,M. J.Chem.Phys. 85:3636 (1986).

[43] Cusack,S. and Doster,W. Submitted to J. Chem Phys.

[44] Jacrot,B., Cusack,S., Dianoux,A.J. and Engelman,D.M. Nature 300:84 (1982).

[45] Bartunik,H.D., Jolles,P., Berthou,J. and Dianoux,A.J. Biopolymers 21:43 (1982).

[46] Cusack,S., Smith,J., Finney,J., Karplus,M. and Trewhella,J. Physica 136B:256 (1986).

[47] Doster,W. , Cusack,S. and Petry,W. Nature 337:754 (1989).

[48] Drexel,W. and Peticolas,W.L. Bioploymers 14:715 (1975).

[49] Mao,B., Pear,M.R., McCammon,J.A. and Northrup,S.H. Biopolymers 21:1979 (1982).

[50] Dianoux,A.J., Page,J.N. and Rosenberg,H.M. Phys.Rev.Lett. 58:886 (1987).

[51] Elber,R. and Karplus,M. Phys.Rev.Lett. 56:394 (1986).

[52] Parak,F., Knapp,E.W. and Kucheida,D. J.Mol.Biol. 161:177 (1982).

[53] Stöckli,H., Furrer,A., Schoenenberger,Ch., Meier,B.H., Ernst,R.R. and Anderson,I. Physica 136B:161 (1986).

[54] Doster,W., Bachleitner,A., Dunau,R., Hiebl,M. and Lüscher,E. Bioploymers 50:213 (1986).

[55] Frick,B., Richter,D., Petry,W. and Buchenau,U. Z.Phys. B70:1 (1988).

[56] Fujara,F. and Petry,W. Eurphys.Lett. 4:921 (1987).

[57] Elber,R. and Karplus,M. Science 235:318 (1987).

THE FLUORESCENCE PROPERTIES OF AROMATIC AMINO ACIDS: THEIR ROLE IN THE

UNDERSTANDING OF ENZYME STRUCTURE AND DYNAMICS

Arthur G. Szabo

Division of Biological Sciences
Protein Biochemistry and Spectroscopy
National Research Council
Ottawa, Canada K1A OR6

INTRODUCTION

Fluorescence spectroscopy has been extensively used in studies of
enzymes and proteins[1-5]. It has been shown to provide insights into
important aspects of the interrelationships of enzyme structure, function
and dynamics. The reasons for the wide applicability of fluorescence
spectroscopy are due to several factors. Fluorescence has a high degree
of sensitivity allowing one to work at low concentrations typical of "in
vivo" conditions. Since the process depends on the absorption and
emission of light energy it is selective, probing only those molecular
subunits which have appropriate chromophoric properties. A wide vartiety
of information is available including the study of inter and intramolecular interactions; the local environment of the fluorescent chromophore;
conformational heterogeneity; the rates of diffusional processes; the
dynamics of the flexibility of the protein segments; and the kinetics of
enzymatic processes. These studies have been facilitated by the relative
ease of obtaining high quality fluorescence measurements using
commercially available instruments.

In this review some of the fundamentals relating to the fluorescence
process are outlined. The fluorescence of the aromatic amino acids are
discussed and examples of the use of these properties in studies of
proteins and enzymes are presented.

FLUORESCENCE PARAMETERS

When a molecule absorbs a photon of light an electron is promoted
from the highest occupied molecular orbital to one of the lowest energy
unoccupied molecular orbitals. There are a number of well documented
selection rules which describe the efficiency of this absorption process
[6]. The important point is that a new electronic distribution in the
molecule is achieved and the new electronic state, usually a singlet
state, is a metastable state of higher energy than the original ground
electronic state. Now a number of competing processes are turned on, by
which this metastable state can lose its excess energy and return to the
original ground state species or to a structurally altered ground state.
These processes are not normally available in the original ground state.

Fluorescence, or the emission of a photon of light is only one of these deactivation pathways. After the molecule absorbs a photon of light the excited state rapidly undergoes an internal conversion (10^{-14}s) to relax to the lowest vibrational level of the lowest energy excited state. The various deactivation processes are summarized below where $^1M^*$ is the excited state, $^3M^*$, the excited triplet state, and M the ground state of the chromophore of interest.

Fluorescence

$$^1M^* \longrightarrow M + h\nu \qquad\qquad k_F$$

Non-radiative deactivation

$$^1M^* \longrightarrow M \qquad\qquad k_{NR}$$

Intersystem crossing to the triplet state

$$^1M^* \longrightarrow {}^3M^* \qquad\qquad k_{ISC}$$

Photoproduct formation

$$^1M^* \longrightarrow P \qquad\qquad k_P$$

Quenching by diffusion (Stern Volmer)

$$^1M^* + Q \longrightarrow M + Q \qquad\qquad k_Q[Q]$$

Resonance energy transfer

$$^1M^* + A \longrightarrow M + {}^1A^* \qquad\qquad k_{ET}$$

Proton transfer

$$^1M^* + BH \longrightarrow {}^1(MH)^* + B^- \qquad\qquad k_H$$

These processes are described in greater detail in monographs and articles (7).

One can write a differential equation describing the deactivation rate of the singlet excited state:

$$-d\,[^1M^*]/dt = -I_0M + (k_F + k_{NR} + \ldots)\,[^1M^*]$$

where I_0 is the intensity in einsteins 1^{-1} s^{-1} of absorbed photons. Under photostationary conditions (Steady State) one can define the efficiency of fluorescence or fluorescence Quantum yield,

$$\Phi_F = k_F\,[^1M^*]/I_0M$$

which is the rate of fluorescence emission compared to the rate of absorption of photons. It can be shown that

$$\Phi_F = k_F/(k_F + k_{NR} + \ldots)$$

If one excites the sample with an infinitely narrow light pulse, a δ excitation pulse, the function describing the rate of fluorescence is:

$$F(t) = F_0 \exp(-k_M t)$$

where

$$k_M = k_F + k_{NR} + \cdots$$

The singlet lifetime is then defined as

$$\tau_S = 1/k_M$$

and the radiative lifetime is

$$\tau_R = 1/k_F$$

Hence a relationship exists between the quantum yield, Φ_F, and the singlet lifetime, τ_S, and the radiative lifetime, τ_R:

$$\Phi_F \tau_R = \tau_S$$

FLUORESCENT CHROMOPHORES IN ENZYMES

In proteins the selectivity of the fluorescent method is advantageous since there are only three fluorescent amino acids, phenylalanine (PHE), tyrosine (TYR), and tryptophan (TRP),

which serve as intrinsic probes. One can make derivatives of the protein using fluorescent labels such as Dansyl[2] and these extrinsic probes can also provide useful information. However, the presence of the intrinsic aromatic amino acids offers considerable advantages since one does not have to be concerned with the effect derivatization may have on the protein structure or function. When the absorption and fluorescence spectroscopic properties of PHE, TYR, and TRP are considered then it is realized that usually TYR and TRP are the most often used fluorescent probes in protein studies. The absorption spectra of the zwitterions of these three aromatic amino acids are shown in Figure 1. Table 1 summarizes the position of their absorption maxima and extinction coefficients. The fluorescence spectra are shown in Figure 2 and the data on their fluorescence quantum yields are also included in Table 1. The low extinction coefficient of PHE coupled with its relatively low quantum yield precludes its utility as an informative fluorescent aromatic amino acid. Furthermore, the abundance of PHE in proteins and enzymes does allow selective information using this residue as a fluorescent probe. On the other hand the extinction coefficient of TYR is appreciable at 275 nm and its fluorescence quantum yield indicates that its fluorescence can readily be monitored. Its fluorescence spectral maximum in proteins is usually found near 303 nm. However, the absorption spectrum of TRP, an

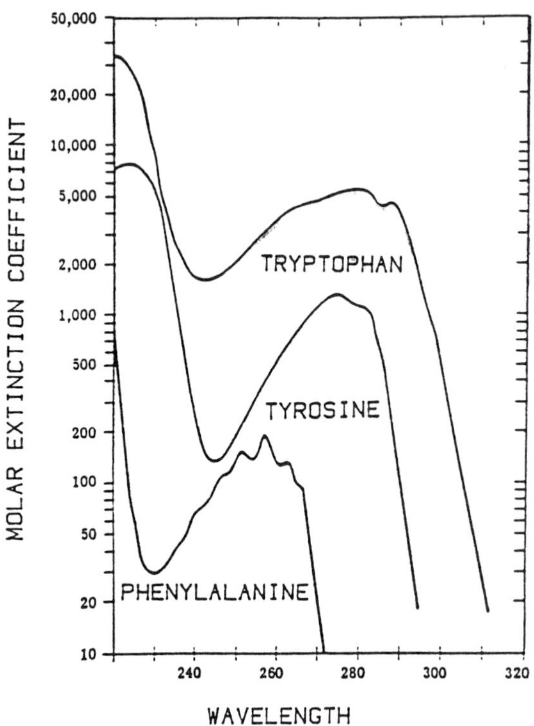

Figure 1. Absorption spectra of phenylalanine, tyrosine, and tryptophan in aqueous solution at pH 7.

Table 1. Molar extinction coefficients and absorption maxima and fluorescence quantum yields of phenylalanine, tyrosine, and tryptophan, pH 7, 20°C.

Amino Acid	λ_{max} (nm)	ϵ $(cm^{-1}M^{-1})$	Φ_F
phenylalanine	258	193	0.06
tyrosine	275	1405	0.14
tryptophan	279 290	5580 3935	0.14

indole amino acid, overlaps that of TYR so that in a protein containing both amino acids it is not possible to selectively excite the TYR residues. In the case of TRP, however, the absorption spectrum shows that it should be possible to selectively excite TRP residues using wavelengths above 290 nm where normally TYR has negligible absorbance at neutral pH. Its fluorescence spectrum is also considerably shifted to higher wavelengths, in most proteins, so that even with excitation at 280 nm of proteins containing TYR and TRP, the TRP fluorescence may be selectively monitored. Often it is the case that the number of TRP residues are limited in enzymes compared to the other aromatic amino acids. More details regarding the excited state properties and fluorescent behaviour of TRY and TRP are discussed below.

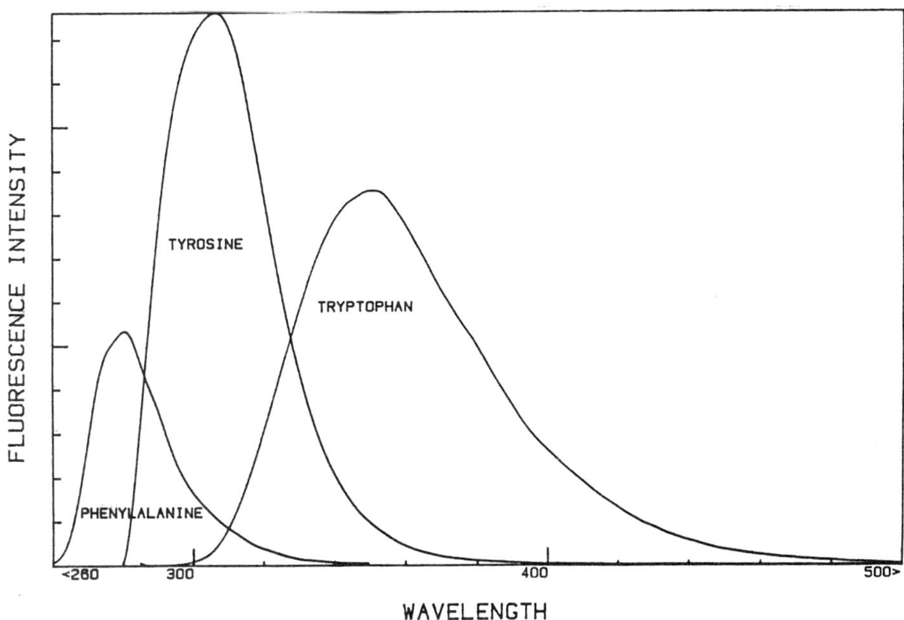

Figure 2. Corrected fluorescence spectra of phenylalanine, tyrosine, and
tryptophan in aqueous solution, pH 7, 20°C. All absorbancies were
identical at the excitation wavelengths. For phenylalanine, λ_{ex} = 250 nm,
for tyrosine and tryptophan λ_{ex} = 280 nm. Excitation and emission
bandpass, 4 nm.

MEASUREMENT OF FLUORESCENCE PARAMETERS

STEADY STATE MEASUREMENTS

Using a fluorescence spectrophotofluorimeter one measures the
variation of fluorescence intensity with the wavelength of emitted light
and obtains a fluorescence spectrum. One can readily see that if any of
the several deactivation processes of TYR and TRP are affected through
interactions or other factors then this will be reflected in the
fluorescence intensity and the value of the fluorescence quantum yield.
Integration of the spectrum of the protein under investigation and that of
an appropriate standard such as N-acetyltryptophanamide or N-acetyltyro-
sinamide, under identical optical arrangements allows one to determine Φ_F
of the protein[8]. The spectral shape and the position of the
fluorescence maximum of the chromophore can provide qualitative
information on the environment of TYR or TRP of the protein. The spectral
position of TYR, however, does not vary significantly with the polarity of
the environment. It is well known that the fluorescence of TRP is very
dependent on the polar nature of its surroundings. The large shift to
longer wavelength of the TRP fluorescence (Stokes' shift) suggests that
excited state processes such as dielectric relaxation may be occuring.

The addition of external quenchers such as iodide ion, cesium ion, or
acrylamide has been shown to give both structural and dynamic
information[9]. The fluorescence intensity of TYR or TRP will be reduced
on the addition of external quenchers since the bimolecular quenching
rate, will become more important as the quencher concentration is
increased. A linear dependence of fluorescence intensity vs. quencher
concentration, at relatively low quencher concentrations is usually
observed. From the slope and the singlet lifetime one can determine the

quenching rate constant, k_Q. This parameter will depend on the degree of exposure to the solvent of the TYR or TRP residues. If the chromophore is located in the interior of the protein then k_Q will be reduced in value.

One further parameter which should be outlined is the phenomenon of resonance energy transfer[2]. In many proteins containing both TYR and TRP residues, TYR fluorescence may not be observed. This is because the TYR emission oscillator interacts with the TRP absorption oscillator resulting in a transfer of electronic energy from the excited state of TYR to excite TRP to its singlet excited state. This process of resonance energy transfer depends on the overlap integral of the TYR fluorescence and the TRP absorption spectra, as well as the distance and angular orientation of the two residues concerned. The dependence on distance, R^{-6} of k_{ER}, allows one to obtain distance measurements between such chromophores in proteins.

TIME RESOLVED MEASUREMENTS

The relationship between the quantum yield and the kinetic rate constants indicates that it is possible to obtain information on the value of these rate constants by measuring the lifetime of the excited singlet state. These lifetimes are usually of the order of a few nanoseconds (10^{-9} s). More importantly if there is more than one fluorescence component then the fluorescence decay will be described by a sum of exponential decay components, the decay constant, τ_S, representing the decay time of each component:

$$F(t) = \sum_i \alpha_i \exp(-t/\tau_i)$$

In the case of a protein for example, multiexponential decay may result if a single chromophore exists in different conformational states with different local interactions, resulting in different non radiative deactivation processes and hence different decay times. Alternately, there may be more than one aromatic amino acid each with its own characteristic singlet lifetime.

In the cases where the initially excited chromophore undergoes a reaction in the excited state forming an excited chromophoric species with a different structure, such as would occur in an excited state protonation or deprotonation process, then the fluorescence decay parameters would describe this excited state process[10,11]. The decay kinetics associated with excited state reactions would have certain characteristic features such as negative pre-exponential terms implying that such a process was occurring.

When the fluorescence decay of a sample is described by a sum of exponentials which are associated with different chromophores or conformers, it is possible to obtain the fluorescence spectrum associated with each decay component. These decay associated spectra (DAS) are determined by measuring the fluorescence decay at several emission wavelengths. The product of the pre-exponential term, α_i, and the decay time, τ_i, of the i th component is proportional to the fractional fluorescence which that decay component makes to the total fluorescence at that particular wavelength[12]. Hence from the steady state fluorescence intensity at any wavelength λ, $F(\lambda)$, it is possible to calculate the fluorescence intensity of the i th component at the same wavelength, $F_i(\lambda)$, according to

$$F_i(\lambda) = \alpha_i \tau_i F(\lambda) / \sum \alpha_i \tau_i$$

This treatment may only be applied when the various τ_i are constant at
each wavelength, λ, and the physical model describing the fluorescence
decay behaviour indicates conformational or chromophoric heterogeneity.
In the case where the physical model describes an excited state reaction
or interconversion then a variation of the above treatment applies[13].
Recently Brand, Beechem and their coworkers have shown that it is possible
to simultaneously analyse several decay datasets measured at different
emission wavelengths[14]. This global analysis is particularly powerful
in providing essentially an overdetermination of the decay parameters and
hence better estimates, both of the decay times and the pre-exponential
terms at each emission wavelength[15]. An example of this type of
analysis is shown in Figure 3 for the fluorescence decay of Subtilisin
BPN' and this is discussed below.

There are two different instrumental techniques for measuring the
excited singlet state decay times. One is known as the Phase and
Modulation method[16-19] and the other is designated as the Time Correlated
Single Photon Counting method (TCSPC)[20-23]. It is beyond the scope of
this article to discuss the relative merits of either method. Both
methods have made significant improvements in measurement instrumentation
and data analysis procedures during the past several years. In our
laboratories the TCSPC method is used exclusively and data presented
herein has been determined by this method.

The essential features of the TCSPC instrumentation in our laboratory
include a sync pumped cavity dumped dye laser as a pulsed excitation
source. The repetition rate is 825 KHz and pulse width is 15 ps.
Convenient intensities of ultraviolet light for excitation of TYR or TRP
or other chromophores can be obtained from 275-350 nm. The detection
system includes a Microchannel Plate Photomultiplier, with a instrument
response function having a FWHM of 80 ps. A single fluorescent photon is
detected for every 100 laser pulses. The time difference between sample

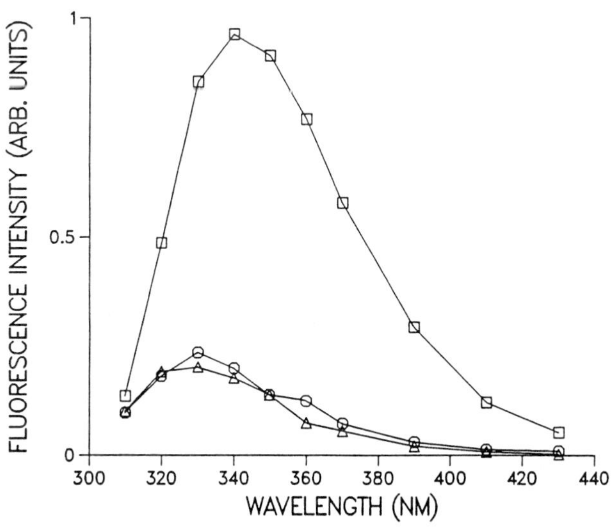

Figure 3. Decay associated spectrum of subtilisin BPN', pH 6.2, 15°C.
□, 7.90 ns component, ○ 1.22 ns component, △, 0.13 ns component,
λ_{ex} = 300 nm. The latter two spectra are x 10.

excitation and the detection of the fluorescent photon is electronically
determined. In this way a fluorescence decay histogram is accumulated
into the channels of a multichannel analyser. A typical fluorescence
decay profile of a protein (apoazurin) is shown in Figure 4, together with
the instrument response function. Commercially available instruments
using gas discharge lamps[24] having essentially the same operating
methods can provide satisfactory results for many applications.

   The interpretation of data from each method has also been the subject
of very current discussion[25-28]. Again it is beyond the scope of this
presentation to elaborate on the interpretation of data using discrete
exponential components compared to distribution functions.

ANISOTROPY MEASUREMENTS

   Quantitative information on the flexibility of enzyme segments is
available from fluorescence anisotropy measurements[1, 29-31]. The
electromagnetic nature of light allows one to selectively excite a sample
with light whose electronic vector component is oriented in only one
direction perpendicular to the direction of propagation of the light beam.
The absorption of plane polarized light will be proportional to the
resolved component of the absorption dipole oscillator on the electric
field vector of the exciting light. This absorption process results in
the formation of an anisotropic population of excited molecules. If the
excited molecules undergo motion prior to fluorescence then a fluorescent
light component will be detected at an angle of 90° from the direction of
the excitation electric field vector. By measuring the fluorescence
intensity parallel ($F_{\parallel}$) and perpendicular ($F_{\perp}$) to the plane of
polarization of the excitation source one can calculate a value of the
anisotropy which is defined as:

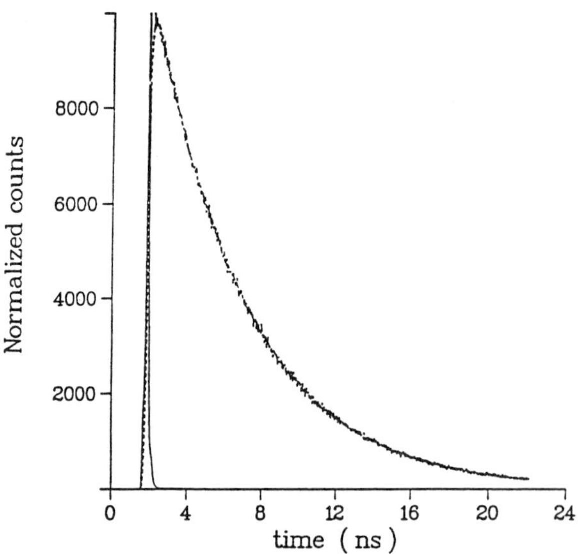

Figure 4. Instrument response function (solid curve) and fluorescence
intensity decay profile (dotted curve) for apoazurin at pH 5, λ_{ex} = 292
nm, λ_{em} = 310 nm.

$$R = F_{||} - F_{\perp} / F_{||} + 2F_{\perp}$$

The maximum value of R in a solution is 0.4. A value of R = 0 indicates
that the chromophore has become isotropic owing to rapid molecular motion.
It is possible using polarized light sources to measure the rate of
anisotropy decay just as fluorescence decay times are measured. These
measurements give values of rotational relaxation times or correlation
times, θ, according to the expresson:

$$R(t) = \sum_i R_0 (\exp\text{-}t/\theta_i)$$

The correlation time θ, is related to the global rotational diffusion time
of the whole protein, and/or the segmental motion of the chromophore in
the protein. In the case of very large proteins the rotational diffusion
time of the whole protein would be quite long. Shorter correlation times
would be associated with rapid segmental motion of the protein.

The foregoing is a cursory outline of the fluorescence spectroscopic
technique and some of the parameters describing fluorescence and how they
can be used in protein studies. The balance of this article will discuss
in greater detail some of the excited state properties of TYR and TRP and
demonstrate how detailed insights into enzyme structure, function and
dynamics can be obtained.

TYROSINE

It has been well established that tyrosine zwitterion in aqueous
solution at pH 6 has a single decay time with a value fo 3.76 ns[32].
Using a value of 0.14 for θ_F[33] then one can calculate a value of the
radiative lifetime, τ_R of 26.8 ns. There does not appear to be much
affect on the fluorescence spectral maximum by the polarity of the local
environment. A most interesting property of tyrosine is its excited state
pKa. In the ground state the pKa of dissociation of the phenolic hydroxyl
group is 10.1. Because of the redistribution of electrons occurring on
excitation the hydroxyl group becomes more acidic with an excited state
pKa* = 4.2[34]. The fluorescence of the tyrosinate anion produced on
proton dissociation has a fluorescence maximum near 340 nm. This excited
state pKa behaviour can provide useful information on the structure of
proteins. If an acidic group such as glutamic acid (GLU) or aspartic acid
(ASP) is spatially close to the tyrosine residue or there is a hydrogen
bond formed between the tyrosine hydroxyl and the acid carboxylate group
then one might observe a contribution to the fluorescence spectrum of
tyrosinate emission. There is evidence that such a situation obtains in
fluorescence studies of calcium binding proteins such as calmodulin[35]
and oncomodulin[36]. These are very acidic proteins and have no TRP
residues. On the long wavelength side of the spectrum an extended or
broader spectrum was observed than is normal for tyrosine fluorescence.
The intensity of this "red edge" fluorescence decreased on binding calcium
and the difference spectrum was characteristic of tyrosinate fluorescence.
The rationalization proposed was that calcium bound to the carboxylate
residues of the calcium binding loops, removing the availability of these
residues to act as proton acceptors for the hydroxyl proton of TYR.

The number of proteins in which this phenomenon has been clearly
demonstrated is rather limited[36-38]. Indeed the observation of
tyrosinate fluorescence in proteins is a matter of some discussion.
Clearly one has to be extremely careful in such an interpretation since
the tyrosinate fluorescence spectrum overlaps considerably with that of
TRP. Hence a small amount of TRP contaminant could be misinterpreted as
originating from tyrosinate fluorescence. Indeed such was the case in

work presented by this author[39] on small cytotoxins where the suggested tyrosinate fluorescence was subsequently shown to be due to a TRP impurity. Where there is evidence from absorption spectra that there is possible hydrogen bonding from an acid carboxylate to the TYR hydroxyl then the potential to observe tyrosinate fluorescence should be anticipated. One experimental manipulation which allows verification of such an observation is the measurement of the fluorescence of the protein under denaturing conditions such as 6M guanidinium HCl. Then the local interactions should be destroyed and only normal TYR fluorescence observed.

It has been demonstrated that time resolved fluorescence studies of proteins containing only a single TYR residue and no TRP will exhibit multicomponent fluorescence decay behaviour[40]. These results together with NMR spectroscopic studies are most consistent with a rationalization that the different decay time components correspond to different conformational states of TYR side chain or the segment of the protein containing that residue.

TRYPTOPHAN

Of all the "biological" chromophores more research effort has probably been expended on studies of the photochemical and photophysical properties of TRP than any other single biomolecular subunit[41,42]. Its photochemistry has been implicated in the photodecomposition of proteins. The relevancy of ultraviolet photochemistry to radiation damage of cells is a logical extension of these studies. Because it can be selectively excited its use as an intrinsic spectroscopic probe of protein structure, function and interactions with other biomolecules has been extensive. Concomitantly its spectroscopic properties are complex and stimulate considerable interest on their own merits.

The photochemical properties of TRP could easily take up a full article. Essentially, however, the main photochemical process occurring is electron ejection either from an upper excited singlet state or two photon absorption through the triplet state. There is evidence of an inefficient photoionization process occurring on excitation with low intensity light into the red edge of the TRP absorption band[43]. Another photochemical reaction involves photooxidation to form kynurenine derivatives.

In proteins the fluorescence spectral maximum of TRP may vary from 308 nm in azurin to 350 nm in small peptides or large globular proteins. Its fluorescence is very sensitive to the polarity of its environment; pH; temperature; and substitution on the aromatic ring. There is general agreement that TRP has two low lying excited singlet states, designated ^1L_a and ^1L_b. These states are nearly isoenergetic and their relative energies depend on substitution of the aromatic ring. It has been shown that excitation of the ^1L_a transition results in a very large change in dipole moment of ca. 3.5 Debye[44]. The transition moment vector of the ^1L_a state falls on a line between the indole nitrogen (δ^+) and carbon 4 of the aromatic ring (δ^-). There is a large charge build up on carbon 4[45]. The ^1L_b transition moment vector falls on a line nearly parallel to the long axis of the indole ring. Excitation of this transition does not result in a large change in indole moment (ca. 0.2 Debye). Many of the excited state properties may be rationalized in terms of these two transitions. Because of the large change in dipole moment on excitation into the ^1L_a state the absorption is considered to have a high degree of charge transfer character[46]. Valeur and Weber[47] using polarized fluorescence excitation spectroscopy resolved the excitation spectra of these ^1L_a and ^1L_b transitions.

One of the most interesting spectroscopic properties of TRP is the variation of the position of the fluorescence maximum and the large Stokes' shift observed in many proteins. A number of studies have been carried out using indole derivatives and TRP in solvents of different dielectric constant and refractive index. There is a good linear relationship between the spectral position and these combined properties[48] indicating that the observed Stokes' shift has its origin in a solvent dielectric relaxation mechanism. An alternate rationalization was suggested[49] that the Stokes' shift in polar solvents was due to the formation of a complex (exciplex) between the excited indole ring and the polar solvent molecule. Recently, it has been shown that polar solvents form weak ground state complexes with indoles and that the shift to longer wavelength results from a reorganization of the complexed solvent molecules reacting to the large dipole moment change on excitation[50].

This dielectric relaxation of the dipoles surrounding the TRP residue has relevancy to protein studies. If the excited TRP residue lies close to the permanent dipole of an α-helical segment for example, then the excited TRP residue may react with this dipole and undergo a restricted motion in order to achieve a dipolar orientation of lower energy. This may manifest itself in the observation of a variation of fluorescence decay times of the TRP across the fluorescence spectrum. Obviously if the protein or at least the environment of the TRP is rigid such dielectric relaxation would be restricted and the fluorescence spectral maximum would be observed at shorter wavelengths (higher energy).

The spectral position is also indicative of the degree of solvent exposure of the TRP residue. If it is protruding from the protein structure or lying on the surface of the protein then a large red shift will be observed. Interactions or structural alterations reducing the solvent accessibility will be manifested by a shift of the fluorescence maximum to lower wavelength. In addition an increased fluorescence quantum yield may result if the TRP residue is in the interior of the protein. The non-radiative deactivation occurring through collision with solvent would be rendered less efficient. These comments give an indication of some qualitative information on protein dynamic fluctuations available from steady state fluorescence studies of TRP in proteins. The time resolved fluorescence spectroscopic studies of TRP have been of particular interest in recent years. Obviously a more quantitative description of dynamic processes are available from such studies. But also fluorescence decay measurements of TRP derivatives and TRP in proteins has shown that information on the conformational heterogeneity of proteins can also be obtained. This concept of conformational heterogeneity from fluorescence studies of TRP in proteins was stimulated by the finding that the fluorescence decay of the TRP zwitterion in aqueous buffer, pH 7, did not obey single exponential decay kinetics as might have been expected for a small molecule in dilute solution. It was reported[42] that the fluorescence decay of TRP zwitterion had two exponential decay components with decay times of 3.15 ns and 0.5 ns. Furthermore the relative fractional fluorescence of each decay component varied with emission wavelength, with the short decay component becoming negligible at wavelengths > 380 nm. These results were best rationalized in terms of originating from rotamers around either the α-β bond or the β-γ bond of the alanyl side chain. Integration of the decay associated spectra of the two components provides a relative concentration of the two components if their radiative lifetimes can be assumed to be equal. The relative concentration determined in this way was similar to a value determined by NMR where hindred rotation in the alanyl side chain of TRP was suggested. The observation of dual exponential decay behaviour has been confirmed by other laboratories[51,52]. The exact assignment as to which rotamers are responsible for the heterogeneity is a matter of some debate but the concept of their origin being due to conformational heterogeneity is accepted. This obviously has led to the concept that multiexponential decay behaviour of TRP in proteins, especially proteins containing a single TRP residue originates from different conformational states of the TRP or the segment of the protein containing that residue.

The rationalization of kinetic processes leading to the short decay component of 0.5 ns has been subject to some disagreement. It was argued by Szabo and Rayner[42] that this was due to a proton transfer from a proximate ammonium residue on the alanyl side chain to the indole ring of TRP. It can be shown from model studies that this ammonium group approaches carbon 4 where there is a build up of negative charge in the excited state. Subsequently Saito and co-workers[53] performed photochemical experiments with TRP in D_2O and showed by NMR spectroscopy of the products that the hydrogen on carbon 4 was replaced by deuterium. On the other hand Fleming and co-workers[52] prefer an electron transfer mechanism to rationalize the short decay time. Fluorescence decay studies in D_2O solution at pD 7 of TRP and indole derivatives suggest that the proton transfer process dominates. The fluorescence decay behaviour of TRP in D_2O obeys double exponential decay kinetics with decay times of 6.28 ns and 0.85 ns. The Φ_F also increases from 0.14 ns in aqueous solution to 0.28 in D_2O solution. When the ammonium group is neutralized on elevating the pH then the decay time of one component lengthens to a value close to 10 ns[54]. This is consistent with the proton transfer mechanism of quenching of the TRP fluorescence at pH 7. This would require that the excited state pK_a for protonation of the indole ring increase from a value < 0 to a value between 5 and 6. Vanderdonckt[55] has suggested that such a large pK_a change for excited state protonation of indole rings may be possible. Other observations also suggest that the pK_a* for TRP ring protonation may have a value near 5. Finally fluorescence decay results on an indole derivative DAPI[56] indicate that intramolecular excited state protonation of the indole ring occurs.

The relevancy of these arguments to protein fluorescence studies becomes important if interactions between protonated HIS residues and TRP are proposed. The HIS residue would be an excellent proton source for protonation of the indole ring. In the event when the pH of a protein or enzyme is lowered so that the carboxylate group of an acid residue is neutralized then the protonated carboxylate group may also act as a proton source. Examples of these types of rationalizations have not often been invoked in protein fluorescence studies. This may be because such a rationalization has not been considered in published reports and this concept has only recently been realized. In studies of the tripeptide lysine-tryptophan-lysine a non-fluorescent component resulting from the efficient proton transfer from the n-terminal ammonium group to the excited TRP residue has been implicated[56]. Further in the hormonal peptide bombesin the variation of fluorescence decay behaviour with pH of the single TRP residue is best rationalized in terms of a close interaction with the single HIS four residues away[57].

As discussed above, the fluorescence decay behaviour of TRP zwitterion was thought to be represented by two exponential decay components. In earlier work kinetics suggesting an excited state reaction were not observed. In the case of an excited state reaction one might expect to observe in the red edge of the emission band, a decay component with a negative pre-exponential term. Hence the lack of such an observation favoured the rationalization of the dual exponential decay behaviour of the TRP zwitterion as originating from the rotamer confromational heterogeneity model. Very recently we have reexamined the fluorescence decay behaviour of TRP in D_2O and aqueous solutions, especially in the red edge of the fluorescence spectrum. Owing to the improvement of time resolved TCSPC instrumentation and data analysis methods we now report new observations on TRP fluorescence. In D_2O solution the fluorescence decay of TRP is characterized by double exponential decay kinetics with positive pre-exponential terms at wavelengths below 390 nm. At 390 nm single exponential decay kinetics are observed. Then above 400 nm double exponential decay behaviour is again observed except that the 0.85 ns component now has a negative pre-exponential term (Table 2). In the case of TRP in D_2O these decay parameters were observed with both 290 nm and 298 nm excitation. In the case of TRP in H_2O similar observations were evident except now the wavelength where single exponential decay kinetics was observed was 410nm when the excitation wavelength was 290 nm. Contrary to earlier reports[42,58] variation of the excitation wavelength did result in a change of the fluorescence decay behaviour of TRP in H_2O. A new long decay component with a decay time of ca 6.5 ns was observed at wavelengths > 410 nm when TRP was excited at 298 nm, in the red edge of the absorption band. At the present time it is considered that this data may still be consistent with the rotamer model but careful consideration has to be given to other models in order to explain these results. The negative pre-exponential term may be due to an interconversion of rotamers occurring in the excited state. The new long decay component observed with 298 nm excitation at emission wavelengths > 410 nm may have the following speculative rationalization.

If TRP exists in different conformational states in the ground electronic state, excitation at 298 nm may be due to a greater fractional absorbance of one conformer at 298 nm, while at 290 nm the fractional absorbance of this conformer may be too low to be significant. It is further speculated that this conformer may be one in which the carboxylate

Table 2. Fluorescence decay parameters of tryptophan in H_2O and D_2O solution; pH 7, 20°C. Effect of excitation wavelength.

Solvent	λ_{ex} (nm)	λ_{em} (nm)	τ_1 (ns)	τ_2 (ns)	α_1	α_2
H_2O	290	350	3.27	0.56	0.77	0.23
		400	3.28	-	1.0	-
		440	3.37	0.81	1.04	-0.04
	298	350	3.29	0.62	0.79	0.21
		400	3.78	2.69	0.54	0.46
		430	6.63	3.25	0.03	0.97
D_2O	290	350	6.10	0.84		
		390	6.28	-	1.0	-
		420	6.22	0.75	1.13	-0.13
	298	390	6.23	-	1.0	-
		420	6.29	0.92	1.09	-0.09

group is hydrogen bonded to the indole NH. Excitation of this conformer may result in a proton transfer from the indole NH to the carboxylate group. Again this excited state proton transfer mechanism requires a large change in the pK_a of the indole NH going from the ground state to the excited singlet state. In the ground electronic state the pK_a of the indole NH is near 16. It was shown that indolate anion fluorescence could be observed in a solution or TRP at pH 8.5 in an excess concentration of carbonate anion, which acted as a proton acceptor[59]. Studies with 1-methyl TRP support this rationalization.

The relevancy of this discussion to protein studies is obvious. If there is a close interaction of a carboxylate group with the indole NH one might expect to observe parameters which are consistent with a proton abstraction by the carboxylate group.

The above discussion indicates that any study of TYR or TRP fluorescence in proteins requires careful consideration and experimentation to draw structural inferences from the obtained parameters. In summary one might observe evidence of:

1. resonance energy transfer from TYR to TRP
2. solvent relaxation or interaction processes
3. dielectric relaxation owing to interactions of permanent dipole moments in the protein structure with that of the excited state of TRP
4. proton transfer from HIS or protonated carboxylic acid groups to TRP
5. proton abstraction by a carboxylate group of the TYR hydroxyl hydrogen or the indole NH.

Multiexponential fluorescence decay kinetics of TYR or TRP in proteins then strongly suggests that the different decay components may be assigned to different conformational states of the protein and the fluorescent chromophore is acting as a probe of these conformational states. Different conformers would be expected to have different local interactions which would affect the non radiative deactivation processes and hence the observed decay time.

SELECTED EXAMPLES OF TRP HETEROGENEITY

In order to demonstrate the utility of fluorescence studies in proteins and enzymes two examples have been selected from our work. The first is the fluorescence decay behaviour of the single TRP residue of the blue copper protein Azurin[60]. The fluorescence decay parameters from two different azurins are shown in Table 3. Apoazurin is the only protein, to the best of our knowledge, in which the TRP obeys single exponential decay kinetics. First of all the fluorescence maximum of the TRP in azurin occurs at 308 nm. All evidence points to this TRP being buried in the hydrophobic interior of the protein. The three decay components seen in the holoazurins are assigned to three different conformational states of the protein. The 4.9 ns component is assigned to a conformer in which the TRP is distant from the copper ligand center. The other two components are assigned to conformers in which the TRP is located in closer proximity to the copper ligand complex. Note that the fractional fluorescence values are different for the two proteins indicating that the conformational distribution of these proteins is different. It has been possible to calculate the relative concentrations of these components for the two azurin molecules. Of vital importance is the conclusive demonstration that the 4.9 ns component in the holoazurin fluorescence is not due to an apo contaminant. It is also worth pointing out that the data was fully satisfactorily fit with three distinct

Table 3. Fluorescence decay parameters of azurin and apoazurin from Pseudomonas Aeruginosa (Pae) and Pseudomonas Fluorescens (Pfl); λ_{ex} = 292 nm, λ_{em} = 310 nm. All solutions in cacodylate buffer, 0.01 M, pH 5, 20°C unless indicated otherwise.

Protein	τ_1 (ns)	τ_2 (ns)	τ_3 (ns)	F_1	F_2	F_3
Pae holo	4.89	0.36	0.098	0.49	0.08	0.44
Pae apo	5.11	-	-	1.0	-	-
Pfl holo	4.91	0.52	0.105	0.78	0.06	0.16
Pfl apo	5.10	-	-	1.0	-	-
80% glycerol/H_2O						
Pae holo 20°C	4.24	0.48	0.085	0.61	0.10	0.29
0°C	4.79	0.53	0.101	0.63	0.06	0.31
Pfl holo 20°C	4.30	0.91	0.098	0.81	0.10	0.09
0°C	4.68	1.01	0.098	0.86	0.07	0.08

exponential decay components even in glycerol water mixtures. The decay times of these components are well separated and easily resolvable under the experimental conditions used.

Finally some recent work on Subtilisin BPN' is presented. This enzyme has three TRP residues. The decay associated spectrum (λ_{ex} = 300 nm) shown in Figure 3 are associated with decay time components which have decay times of 7.90 ns, 1.22 ns, 0.13 ns. This is a case where it is difficult to resolve whether each TRP has a characteristic decay time or whether the decay time represent 3 average conformations of the most fluorescent TRP. The different spectral positions of each component suggests that the decay times may best be assigned to different TRP residues.

The potential of combining site directed mutagenesis with fluorescence studies of proteins and enzymes is fascinating and challenging. Obviously mutant Subtilisin BPN' with two of the three TRP residues changes to other amino acids would be most interesting. It would not only allow the rationalization of the fluorescence behaviour of the native enzyme but new insights into the interrelationship between the structure, function, and dynamics of the enzyme should be obtained.

CONCLUDING REMARKS

It has been shown that fluorescence studies of TYR and TRP in enzymes may provide useful information on the structure and dynamics of fluctuations of enzymes of interest. The ability to estimate conformational heterogeneity is particularly interesting. Careful experimentation can reveal specific samples of local interactions between TYR and TRP and other acidic or basic residues. Most importantly these comments show that new information requiring new interpretations of protein fluorescence are still being obtained.

ACKNOWLEDGEMENTS

The expert technical expertise of D.T. Krajcarski is gratefully acknowledged. The work on azurin was performed by C.M. Hutnik and that on Subtilisin BPN' by K.J. Willis.

REFERENCES

1.   A.J. Pesce, C.G. Rosen, and T.L. Pasby, "Fluorescence Spectroscopy. An Introduction for Biology and Medicine", Marcel-Dekker, New York (1971).
2.   R.F. Chen, H. Edelhock eds., Biochemical Fluorescence Concepts", Vol. 1 and 2, Marcel-Dekker, New Yor (1976).
3.   R.F. Steiner, ed., "Excited States of Biopolymers", Plenum, New York (1983).
4.   L. Brand, J.R. Knutson, L. Davenport, J.M. Beechem, R.E. Dale, D.G. Walbridge, and A.A. Kowalczyk, "Spectroscopy and the Dynamics of Molecular Biological Systems", Academic Press, London (1985).
5.   D.L. Taylor, A.S. Waggoner, R.F. Murphy, F. Lanni, R.R. Birge, eds. "Applications of Fluorescence in the Biomedical Sciences", A.R. Liss, New York (1986).
6.   H.H. Jaffe and M. Orchin, "Theory and Applications of Ultraviolet Spectroscopy", Wiley, New York (1962).
7.   J.B. Birks, "Photophysics of Aromatic Molecules", Wiley, London (1970).
8.   J.N. Miller, "Standards in Fluorescence Spectroscopy", Chapman and Hall, London (1981).
9.   M. Eftink and C.A. Ghiron, Anal. Biochem. 114: 119-227 (1981).
10. W.R. Laws and L. Brand, J. Phys. Chem. 83: 795-802 (1979).
11. C.M. Harris and B.K. Selinger, J. Phys. Chem. 84: 1366-1371 (1980).
12. B. Donzel, P. Gauduchon, and P. Wahl, J. Am. Chem. Soc. 96: 801-808 (1974).
13. J.M. Beechem, M. Amellot, and L. Brand, Anal. Intrum. 14: 379-402 (1985).
14. J.R. Knutson, J.M. Beechem, and L. Brand, Chem. Phys. Lett. 102: 501-507 (1983).
15. J.M. Beechem, J.R. Knutson, J.B.A. Ross, B.W. Turner, and L. Brand, Biochemistry, 22: 6054-6058 (1983).
16. J.R. Lakowicz, and B.P. Maliwal, Biophys. Chem. 21: 61-78 (1985).
17. D.M. Jameson, E. Gratton, and R.D. Hall, Appl. Spectros. Rev., 20: 55-106 (1984).
18. G. Ide, Y. Engelborghs, and A. Persoons, Rev. Sci. Instrum. 54: 841-844 (1983).
19. E. Gratton, and M. Limkeman, Biophys. J. 44: 315-324 (1983).
20. A.J.W.G. Visser, Ed. "Time Resolved Fluorescence Spectroscopy", Anal. Instrum. 14: 193-566 (1985).
21. D.V. O'Connor and D. Phillips, "Time-correlated Single Photon Counting", Academic Press, New York (1984).
22. M. Zuker, A.G. Szabo, L. Bramall, D.T. Krajcarski, Rev. Sci. Instrum. 56: 14-22 (1985).
23. R.B. Cundall and R.E. Dale, eds., "Time-resolved Fluorescence Spectroscopy in Biochemistry and Biology", Plenum, New York (1983).
24. D.J.S. Birch, R.E. Imhof, and A. Dutch, J. Phys. E. Sci. Instrum. 17: 417-418 (1984).
25. J.R. Alcala, E. Gratton, and F.G. Prendergast, Biophys. J. 51: 597-604 (1987).
26. D.R. James and W.R. Ware, Chem. Phys. Lett. 126: 7-11 (1986).
27. P. Bayley and S. Martin, in "Fluorescent Biomolecules", E. Gratton and D. Jameson, eds. in press.
28. A.K. Livesey and J.C. Brochon, Biophys. J. 52: 693-706 (1988).

29. G. Weber, Biochem. J. 51: 155-167 (1952).
30. J.R. Lakowicz, H. Cherek, I. Gryczynski, N. Joshi, and M.L. Johnson, Biophys. J. 51: 755-768 (1987).
31. J.M. Beechem, J.R. Knutson, and L. Brand, Biochem. Soc. Trans. 832-835 (1986).
32. W.R. Laws, J.B.A. Ross, H.R. Wyssbrod, J.M. Beechem, L. Brand, and J.C. Sutherland, Biochemistry, 25: 599-607 (1986).
33. R.F. Chen, Anal. Lett. 1: 35 (1967).
34. D.M. Rayner, D.T. Krajcarski, and A.G. Szabo, Can. J. Chem. 56: 1238-1245 (1978).
35. S. Pundak and R.S. Roche, Biochemistry 23: 1549-1555 (1984).
36. J.P. MacManus, A.G. Szabo and R.E. Williams, Biochem. J. 220: 261-268 (1984).
37. T. Kimura, and J. Ting, Biochem. Biophys. Res. Commun. 45: 1227-1231 (1971).
38. B. Lux, J. Baudier, and D. Gerard, Photochem. Photobiol. 42: 245-251 (1985).
39. A.G. Szabo, K.R. Lynn, D.T. Krajcarski, and D. Rayner, J. Lumin. 18: 585-585 (1979).
40. J.B.A. Ross, W.R. Laws, A. Buku, J.C. Sutherland, and H.R. Wyssbrod, Biochemistry 25: 607-612 (1986).
41. D. Creed, Photochem. Photobiol. 39: 537-562 (1984).
42. A.G. Szabo and D.M. Rayner, J. Am. Chem. Soc. 102: 554-563 (1980).
43. M. Bazin, K.L. Patterson, and R. Santus, J. Phys. Chem. 87: 189-190 (1983).
44. A. Kawski and I. Gryczynski, Bull. Acad. Polon. Sci. 21: 1061-1066 (1973).
45. E.M. Evleth, O. Chabut, and P. Barriere, J. Phys. Chem. 81: 1913 (1977).
46. S.R. Meech, D. Phillips and A.G. Lee, Chem. Phys. 80: 317-328 (1983).
47. B. Valeur and G. Weber, Photochem. Photobiol. 25: 441-444 (1977).
48. P.-S. Song and W.E. Kurtin, J. Am. Chem. Soc. 91: 4892-4906 (1969).
49. M.S. Walker, T.W. Bednar, and R. Lumry, J. Chem. Phys. 47: 1020-1028 (1967).
50. B. Skalski, D.M. Rayner, and A.G. Szabo, Chem. Phys. Lett. 70: 587-590 (1980).
51. D. Jameson and G. Weber, J. Phys. Chem. 85: 953-958 (1981).
52. R.J. Robbins, G.R. Fleming, G.S. Beddard, G.W. Robinson, P.J. Thistlewaite, and G.J. Woolfe, J. Am. Chem. Soc. 102: 6271-6279 (1980).
53. I. Saito, H. Sugiyama, A. Yamamoto, S. Muramatsu, and T. Matsuura, J. Am. Chem. Soc. 106: 4286-4287 (1984).
54. E. Gudgin, R.L. Delgado, and W.R. Ware, Can. J. Chem. 59: 1037-1044 (1981).
55. E. VanderDonckt, Bull. Soc. Chim. Belg. 78: 69-75 (1969).
56. A.G. Szabo, D.T. Krajcarski, P. Cavatorta, L.Masotti, and M.L. Barcellona, Photochem. Photobiol. 44: 143-150 (1986).
57. L. Masotti, P. Cavatorta, A.G. Szabo, G. Farruggia, and G. Sartor, "Fluorescent Biomolecules", E. Gratton and D. Jameson eds, in press (1988).
58. B. Alpert, D.M. Jameson, R. Lopez-Delgado, and R. Schooley, Photochem. Photobiol. 30: 479-481 (1979).
59. A.G. Szabo in "Time Resolved Fluorescence Spectroscopy in Biochemistry and Biology", R.B. Cundall and R.E. Dale, Plenum, New York (1983).
60. C.M. Hutnik and A.G. Szabo. submitted (1988).

THE APPLICATION OF ^{1}H NUCLEAR MAGNETIC RESONANCE SPECTROSCOPY TO THE STUDY OF ENZYMES

Christina Redfield

Inorganic Chemistry Laboratory
University of Oxford
Oxford, England

INTRODUCTION

Nuclear magnetic resonance spectroscopy has been used to study enzymes for more than 30 years. The first ^{1}H NMR spectrum of a protein was published by Saunders, Wishnia and Kirkwood in 1957 [1]. The 40 MHz spectrum of ribonuclease consisted of 4 broad peaks. The authors assigned the most downfield peak in the spectrum to the aromatic protons and the most upfield peak to hydrogens bonded to aliphatic carbon atoms attached only to other aliphatic carbon. They reported that the intensities of these two peaks were consistent with the amino acid composition of ribonuclease. Later in 1957 Jardetsky and Jardetsky used the chemical shifts of the amino acids to predict a complete NMR spectrum for ribonuclease [2]. Their predicted intensities were in good agreement with the intensities of the four peaks in the ribonuclease spectrum, and they concluded that the NMR spectra of amino acids provide a rational basis for the interpretation of the NMR spectra of proteins in solution. Today the complete interpretation of the NMR spectra of proteins is still the goal of many NMR spectroscopists; only the level of detail of this interpretation has changed. The NMR spectra of the amino acid building blocks are still used to interpret complex protein spectra.

BACKGROUND

As the magnetic field strength of available NMR spectrometers increased during the 1960's the quality of published NMR spectra of proteins improved [3,4]. By the mid 1960's resolved resonances for individual protons were observed in spectra of proteins. The downfield shifted C-2 resonances of histidine were observed in the spectrum of proteins including ribonuclease, staphylococcal nuclease and lysozyme [5].

Several resonances shifted out from the main aliphatic envelope were observed in the spectrum of hen lysozyme. These upfield shifts were thought to arise from the close juxtaposition of aliphatic protons with aromatic rings. Attempts were made to assign these resonances to specific residues using information derived from the X-ray structure of lysozyme [6,7].

Developments in the design of NMR spectrometers and experimental techniques have had a dramatic impact on the study of enzymes by NMR over the last 30 years. In 1965 NMR spectra of enzymes collected on 100 MHz spectrometers were reported in the literature [4]. In 1987, the first commercial 600 MHz spectrometer was installed at the University of Oxford. The improvement in sensitivity and resolution resulting from this six-fold increase in magnetic field strength has been dramatic; the 100 MHz and 600 MHz spectra of hen lysozyme are compared in Figure 1. Protein spectra reported in the 1960's were collected in continuous wave (CW) mode. The first Fourier transform (FT) mode NMR spectrometers were commercially available in $\sim$ 1970. The change from CW to FT NMR resulted in a great improvement in sensitivity [8]. During the 1970's a variety of techniques were used to extract information from the mass of overlapping resonances making up a protein spectrum [9,10]. Double-resonance techniques such as spin-decoupling difference spectroscopy were used to identify coupled resonances and to assign these resonances to particular types of amino acids [11,12]. Paramagnetic probes and nuclear Overhauser enhancement (NOE) effects were used to obtain the spatial information needed in order to assign resonances to specific residues in the protein [13-15]. The interpretation of this information relied on some knowledge of the X-ray structure of the protein and on the assumption that the structure of the protein in solution was similar to that in the crystalline state.

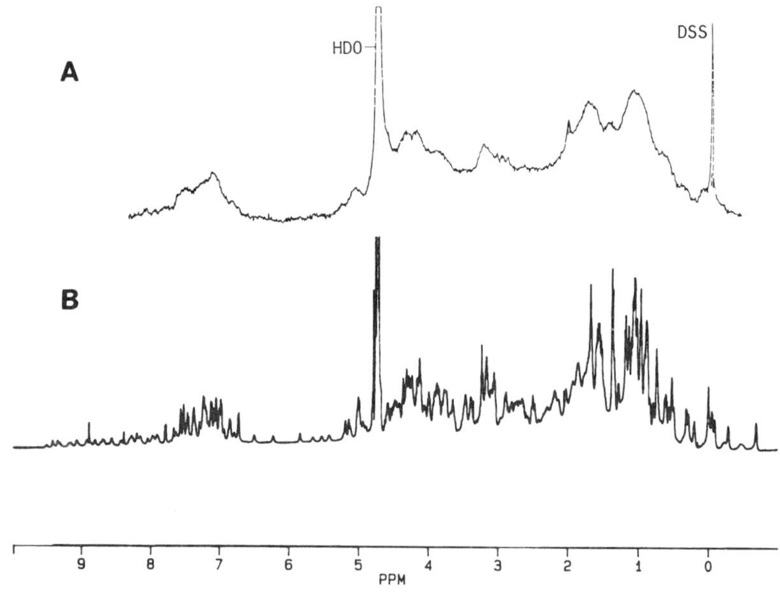

Fig. 1.　^{1}H NMR spectra of hen lysozyme collected using (a) 100 MHz [4] and (b) 600 MHz NMR spectrometers at 30°C.

Significant numbers of resonance assignments, particularly for aromatic and methyl resonances, were reported during the 1970's [15-18].

The introduction of the concept of two-dimensional NMR by Jeener in 1971 has led to a revolution in the field of protein NMR [19]. In the last ten years several hundred different 2-D NMR experiments have been reported in the literature. The first 2-D NMR spectrum of a protein, a 2-D J-resolved spectrum of bovine pancreatic trypsin inhibitor (BPTI), was published in 1977 [20]. The now widely used 2-D techniques of nuclear Overhauser enhancement effect spectroscopy (NOESY), which gives through-space distance information, and correlated spectroscopy (COSY), which gives through-bond scalar coupling information, were first applied to BPTI in 1980 and 1982 respectively [21,22]. Using these 2-D NMR techniques almost complete ^{1}H NMR spectral assignments have been reported for proteins of molecular weight up to 15,000 [23-26]. The sequential assignment procedure developed by Wüthrich and coworkers means that these assignments can now be obtained without reference to X-ray crystallographic data; only information about the amino acid sequence is required [27,28]. In fact, 2-D NMR has been used to correct errors in amino acid sequences [29].

When complete assignments are available for a protein, the through-space distance information available from NOESY spectra can be used to define the three-dimensional structure of the protein. Structures for proteins including BUSI, lac repressor headpiece, epidermal growth factor and the α-amylase inhibitor Tendamistat have been published [30-33]. Recently the preliminary structure of an enzyme, acylphosphatase, has been reported [34]; X-ray crystallography is no longer the only technique available for the determination of the 3-D structures of enzymes.

The detailed level of assignment now possible means that other types of information, in addition to structure, can be obtained. The region of the COSY spectrum containing cross peaks from NH-αCH resonances, commonly known as the fingerprint region, contains a cross peak from each amino acid residue in the protein. Changes in the position, intensity and shape of these cross peaks can be monitored as a function of pH, or the concentration of a probe such as an enzyme inhibitor or a metal ion. Thus, information about pK's of catalytically important groups, dissociation constants, and conformational changes due to interactions with substrates and inhibitors can be obtained at a level of detail never before possible. In this article the application of modern 2-D techniques to the study of enzymes will be illustrated using hen egg-white lysozyme.

HEN EGG-WHITE LYSOZYME

Hen egg-white lysozyme is an enzyme of molecular weight 14,500 which catalyzes the hydrolysis of bacterial cell wall polysaccharides [35]. The NMR spectrum of lysozyme was first reported in 1965 [4]. Since then lysozyme has been the subject of numerous NMR studies. The number of assignments in the spectrum of lysozyme increased steadily as developments in NMR spectrometers and NMR techniques were made. Lysozyme has been used as a model protein system for the development of many NMR techniques [10]. The availability of assignments for some 50 of the 129

residues has permitted detailed studies of the structure and dynamics of lysozyme to
be carried out [15,36,37]. Recently, the almost-complete assignment of the backbone resonances of hen lysozyme has been achieved using the sequential assignment method;
lysozyme is the largest protein for which such complete assignments have been reported [26]. This detailed level of assignment allows information about the structure
of the enzyme in solution, the pK's of acidic groups and the interaction with enzyme
inhibitors to be obtained.

Sequential Resonance Assignments

The sequential assignment of the backbone resonances of hen lysozyme was
carried out using the standard two-stage procedure. The first stage involved the
assignment of NH-αCH cross peaks in the fingerprint region of the COSY spectrum to
a type of amino acid. Information about spin systems was obtained from experiments
such as COSY and RELAY, which exploit through-bond scalar coupling [27]. For
residues such as alanine, glycine, threonine, valine and isoleucine, which give unique
patterns in COSY and RELAY spectra, cross peaks could be assigned to a specific
amino acid type. For other residues cross peaks could only be assigned to more

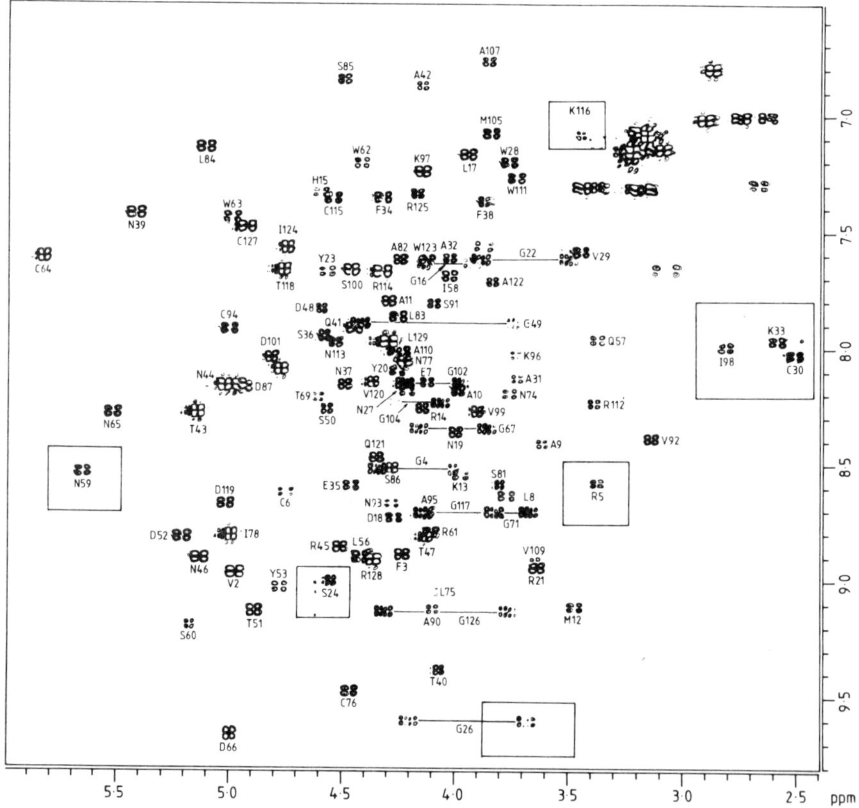

Fig. 2. Fingerprint region of the phase-sensitive 500 MHz COSY spectrum
of hen lysozyme.

general classes. For example the backbone spin systems of Tyr, Trp, Phe, His, Asn, Asp, Ser and Cys could not be easily distinguished so the fingerprint region cross peaks of these residues were assigned as Type-J. Similarly, the spin systems of Gln, Glu, Arg, Lys and Met were assigned as Type-U. The second stage involved the assignment of the NH-αCH cross peaks to specific amino acid residues in the protein sequence. It has been demonstrated by Wüthrich and coworkers that, for all sterically allowed backbone conformations, the distance between either the NH protons on adjacent residues i and $i + 1$ (d_{NN}) or the alpha proton on residue i and the NH proton on residue $i + 1$ ($d_{\alpha N}$) is short enough ($< \sim3.5$Å) to give rise to an NOE effect [27]. A short value of d_{NN} (~2.8Å) is characteristic of helices and turns whereas a short value of $d_{\alpha N}$ (~2.2Å) is characteristic of extended conformations such as those found in β-sheets [27]. Sequential NOE effects identified in the NOESY spectrum were used to build up short peptide segments. The sequences of these segments were compared with the known amino acid sequence of lysozyme. Sequence specific assignments were obtained when there was a unique occurrence of the peptide segment in the protein sequence.

The sequential assignment procedure is now fairly straightforward for small proteins containing less than ~75 amino acid residues. In general there is not much overlap in the NH and αCH regions of the spectra of these proteins so NOE peaks can be interpreted unambiguously. As the molecular weight of a protein increases there are proportionately more cross peaks in the fingerprint region of the COSY spectrum and resonance overlap becomes a severe problem. The fingerprint region of the COSY spectrum of lysozyme is illustrated in Figure 2. It can be seen from inspection of the spectrum that many cross peaks share the same NH or αCH chemical shift. For example, the cross peaks of Gly-71, Ala-95 and Gly-117 share the same NH chemical shift of 8.70 ppm. Thus, NOE effects involving an NH at 8.70 ppm could arise from any one of these three residues. There are several procedures for overcoming the problems of overlap. COSY and NOESY spectra can be collected at different pH or temperature values where chemical shifts will vary somewhat. Spectral simplification can be achieved in COSY spectra by using appropriate multiple-quantum filters. For example, the fingerprint region of the triple-quantum filtered COSY spectrum of lysozyme contains cross peaks from glycine residues only [38]. This simplification occurs because glycine is the only amino acid residue with two αCH protons. In the study of lysozyme problems of overlap were overcome, to some extent, by exploiting the range of hydrogen exchange rates observed for the amide hydrogens of lysozyme. The COSY spectrum shown in Figure 2 was collected using a sample of lysozyme dissolved in H_2O; this spectrum contains resonances from all amide hydrogens. A spectrum of lysozyme containing resonances from the 58 slowly exchanging amide hydrogens is obtained when lysozyme is dissolved in D_2O. A spectrum containing resonances from the 57 rapidly exchanging amides is obtained when lyophilized fully-exchanged lysozyme is dissolved in H_2O. Part of the fingerprint regions of COSY spectra collected with lysozyme samples prepared in these three ways are shown in Figure 3. The amide of Ala-95 is slowly exchanging and gives rise to a cross peak in the spectrum of lysozyme dissolved in D_2O (Figure 3b) whereas the amides of Gly-71 and Gly-117 are rapidly exchanging and give peaks in the spectrum of fully-exchanged lysozyme dissolved in H_2O (Figure 3c). Similar spectral simplification occurs in the

NH-αCH region of the NOESY spectra of these samples. Additional simplification
occurs in the NH-NH region of the NOESY spectrum because a cross peak between
a pair of amide hydrogens is observed only if the two amides have similar hydrogen
exchange properties.

Detailed analysis of COSY and NOESY spectra using the methods described
above for spectral simplification resulted in the sequential assignment of 121 of the
129 residues of hen lysozyme. The $d_{\alpha N}$ and d_{NN} NOE connectivities identified and
the hydrogen exchange classification of residues are presented in Figure 4. The complete assignment was based on the identification of 21 sequential peptide segments.
Breaks between these segments occured because some of the d_{NN} or $d_{\alpha N}$ connectivities were absent from the spectrum as a result of degeneracy of NH chemical shifts
or the overlap of αCH resonances with the H_2O solvent resonance. The probability
of identifying a unique peptide segment within the lysozyme sequence generally increases as the length of the sequentially assigned peptide increases. In this study it
was found to be particularly important to identify, initially, sequences which contain

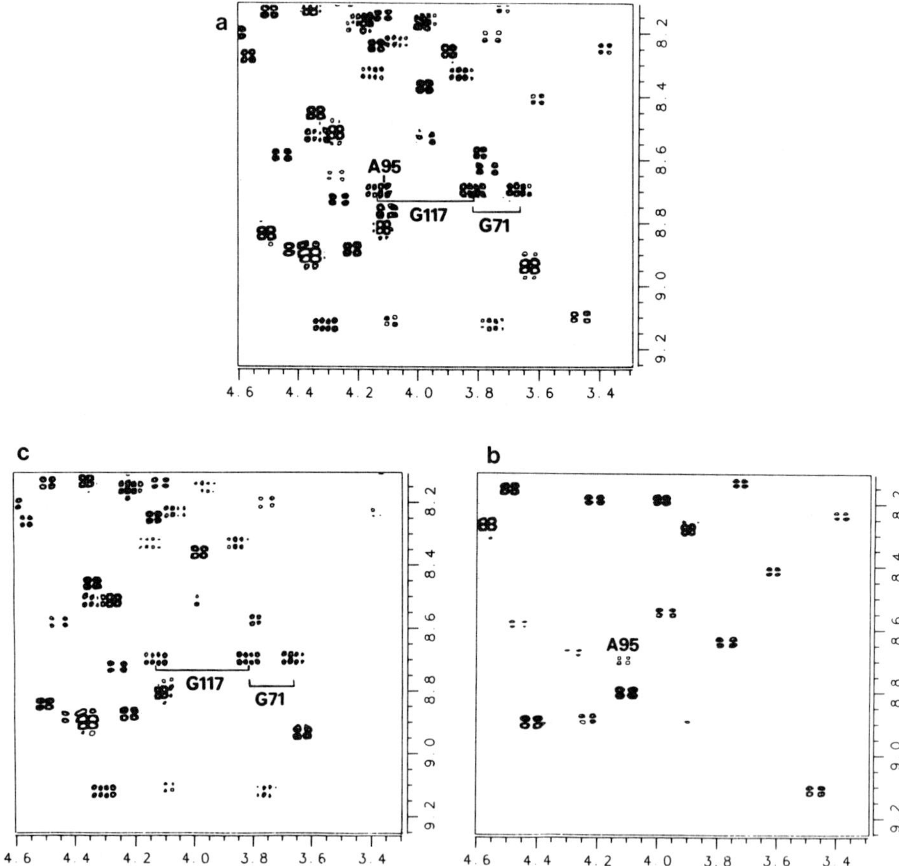

Fig. 3. Phase-sensitive 500 MHz COSY spectra of (a) lysozyme dissolved
in H_2O, (b) lysozyme dissolved in D_2O, and (c) fully-exchanged
lysozyme dissolved in H_2O. The cross peaks of Gly-71, Ala-95 and
Gly-117 are labelled.

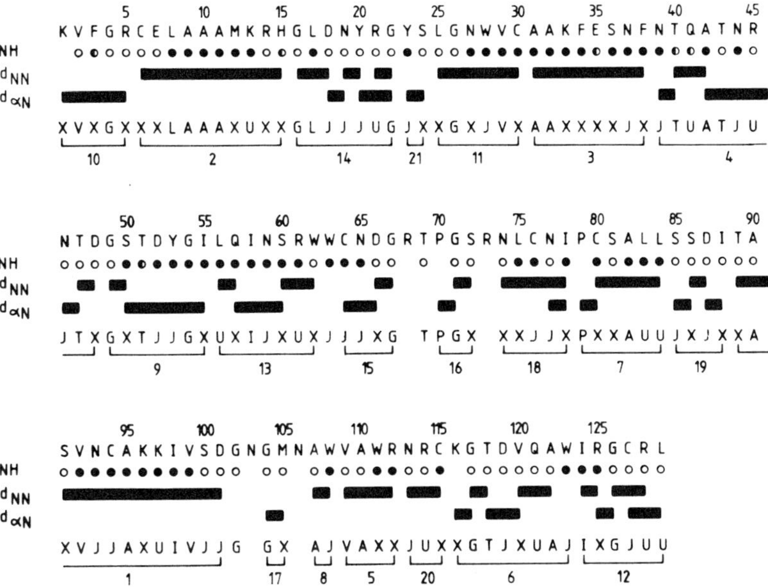

Fig. 4. Amino acid sequence of hen lysozyme, summary of hydrogen exchange rates and summary of the NOE connectivities used in the sequential assignment procedure. Filled and empty circles indicate amides with slow and rapid exchange rates, respectively. Peptide segments identified in the assignment procedure and the amino acid types of the residues in these segments are shown. The numbers beneath these segments indicate the order in which they were assigned.

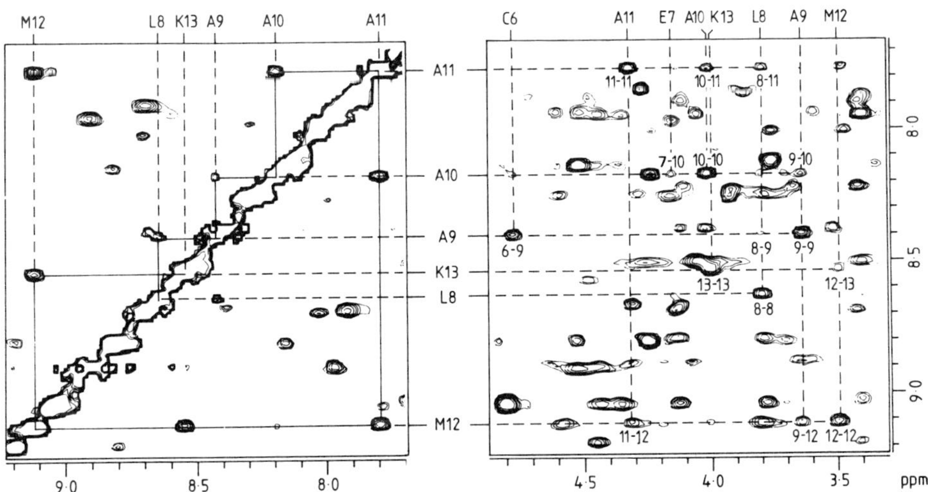

Fig. 5. Phase-sensitive 500 MHz NOESY spectrum of hen lysozyme dissolved in D_2O. Sequential d_{NN} connectivities for residues 8 through 13 are shown on the left; $d_{\alpha N}(i,i)$, $d_{\alpha N}(i,i+1)$ and $d_{\alpha N}(i,i+3)$ connectivities for residues 6 through 13 are shown on the right.

several residues, such as glycine and alanine, which had been assigned uniquely in the first stage of assignment. Lysozyme contains 12 alanine residues which are scattered throughout the sequence. These residues were chosen as the starting point for the assignment task; the first 8 peptide segments assigned, which account for 61 of the 129 residues, all contain alanine residues. An example of the sequential assignment is shown in Figure 5. Peptide segment 2 was assigned to residues Cys-6 through His-15 on the basis of the three consecutive alanine residues, Ala-9, Ala-10 and Ala-11. Residues 8-13, in this segment, have slowly exchanging amide hydrogens. The d_{NN} connectivities used to assign these resonances are shown in the NOESY spectrum of lysozyme collected in D_2O shown in Figure 5.

A further nine segments, which account for another 46 residues, were assigned on the basis of the Thr, Val, Ile, Leu and Gly residues they contain. The restrictions on the length and sequence of the remaining unassigned groups of residues made assignment on the basis of the broader Type-J and Type-U classifications feasible at this stage. This assignment strategy, which relies initially on the identification of some key amino acid spin systems such as alanine and glycine, is likely to be important in studies of larger proteins where complete identification of side-chain resonances in COSY and RELAY spectra is impaired due to spectral overlap or broad lines. If residues such as alanine or glycine, whose spin systems are easily identified by 2-D ^{1}H NMR techniques, are not present in sufficient numbers or at strategic positions in the sequence then any amino acid type could be selected as a key amino acid by incorporating appropriate ^{15}N or ^{13}C labels and using heteronuclear 2-D techniques. This approach requires that the protein be expressed in an appropriate organism for label incorporation. Resonances in the spectrum of thioredoxin and T4 lysozyme have been assigned using labelling techniques and heteronuclear 2-D NMR [39,40].

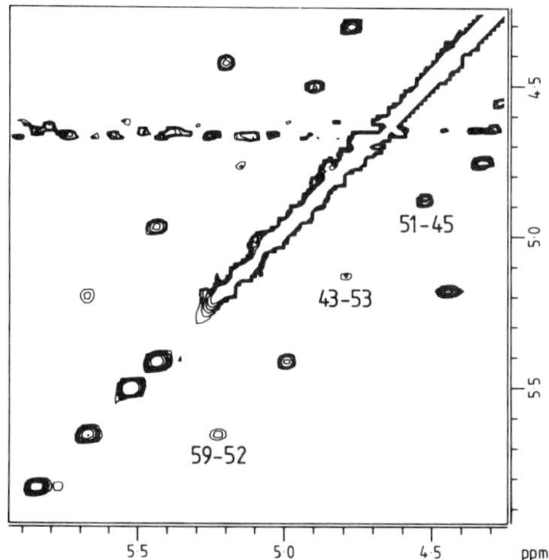

Fig. 6. Phase-sensitive 500 MHz NOESY spectrum of fully-exchanged hen lysozyme. Long-range $d_{\alpha\alpha}$ connectivities characteristic of an antiparallel β-sheet are labelled.

In addition to sequential assignment information, NOE effects involving NH and αCH protons also provide information about the secondary structure within the protein. Helical regions of structure are characterized by a short distance (2.8Å) between neighboring NH protons, d_{NN}, and by a short distance (3.4Å) between the αCH proton of residue i and the NH proton of residue $i+3$, $d_{\alpha N}(i,i+3)$ [27]. Extended regions of d_{NN} sequential NOE connectivities were observed for residues 6-15, 25-38 and 89-101 as shown in Figure 4. Helical conformation was confirmed for these regions by the identification of many $d_{\alpha N}(i,i+3)$ NOE connectivities. These NOE effects are shown in Figure 5 for the helical residues 6-15. Regions of β-strand are characterized by a short distance (2.2Å) between the αCH proton and the NH proton on the following residue, $d_{\alpha N}$. The tight turns between strands in the sheet are characterized by short d_{NN} distances. The interaction of β-strands in β-sheets can be extracted from long-range NOE effects between NH and αCH protons on the individual strands [27]. The pattern of NOE effects observed for residues 42-60 of hen lysozyme is characteristic of a triple-stranded antiparallel β-sheet. Three segments of $d_{\alpha N}$ NOE effects are connected by two short segments of d_{NN} effects. Several long-range inter-strand NOE effects, involving NH and αCH resonances, confirm this secondary structure. The NOE effects between pairs of αCH protons shown in Figure 6 are characteristic of antiparallel β-sheet [27].

The secondary structure determined from NOE connectivities can be compared with the secondary structure determined by X-ray diffraction studies. The backbone structure of hen lysozyme in the crystalline state is illustrated schematically in Figure 7 [41]. Regions of the sequence which give d_{NN} NOE connectivities are shaded in Figure 7a; regions which give $d_{\alpha N}$ connectivities are shaded in Figure 7b. Regions which were identified as helical from NMR data correspond to the major helices in the X-ray structure. The triple-stranded β-sheet identified from NOE effects agrees with the β-sheet identified in the X-ray structure. Further analysis of the NOESY spectrum

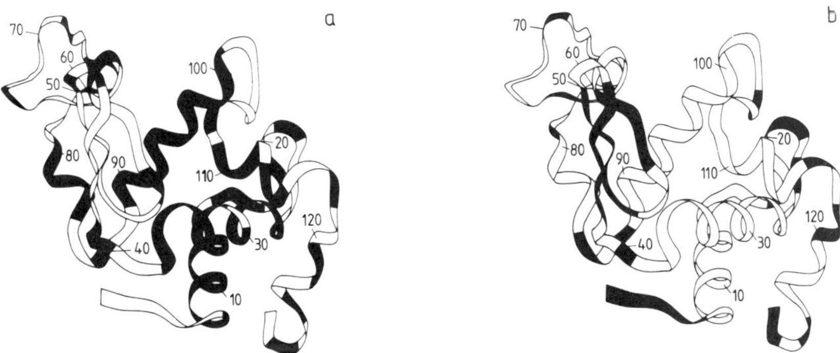

Fig. 7. Schematic ribbon representation of the backbone of hen lysozyme derived from the X-ray structure. Regions characterized by d_{NN} NOE connectivities are shaded in (a) whereas regions characterized by $d_{\alpha N}$ connectivities are shaded in (b).

of hen lysozyme will yield additional distance constraints which can be used as input
for 3-D structure determination algorithms such as distance geometry and restrained
molecular dynamics [42,43].

Information about protein structure can also be obtained from measurements
of vicinal ^{1}H-^{1}H coupling constants. The value of ^{3}J$_{N\alpha}$ is related to the torsion angle
ϕ by the Karplus equation [44]

$$^3\mathrm{J}_{N\alpha} = A\cos^2\theta - B\cos\theta + C \qquad (\theta = |\,\phi - 60°\,|)$$

A, B and C are coefficients which can be obtained from measurements on model
compounds [45]. NH-αCH coupling constants can be extracted from fingerprint region
cross peaks of COSY spectra collected with high digital resolution in the F2 dimension. These experimental coupling constants provide angular constraints which
supplement the NOE distance information used in distance geometry and molecular
dynamics calculations [27].

NH-αCH coupling constants have been measured for $\sim$100 of the 129 residues of
lysozyme. These values are plotted in Figure 8 against the torsion angles determined
from the X-ray structure of lysozyme [41]. For the majority of residues the experimental coupling constants fall within 1.0 Hz of the theoretical curve. The good agreement
between experimental coupling constants and crystallographic torsion angles is further evidence for the close similarity between the average structure of lysozyme in
solution and the structure in the crystalline state.

The results presented here indicate that complete resonance assignments and
information about structure in solution can be obtained for a protein which is

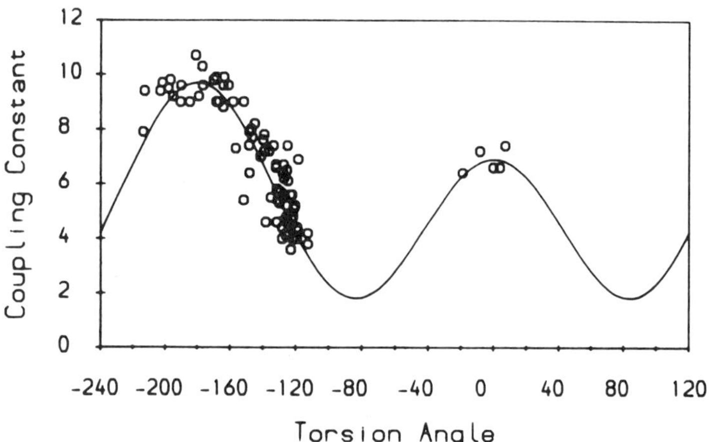

Fig. 8. Plot of experimental coupling constants versus torsion angles determined from the X-ray structure. The solid line shows the theoretical
Karplus equation using 6.4, 1.4 and 1.9 for A, B and C [45].

150

relatively large in NMR terms without using information obtained from the X-ray structure. Hen lysozyme is the first enzyme for which such a detailed study has been reported. These resonance assignments provide the starting point for a variety of important studies.

Measurement of pK Values

Acidic and basic amino acids including aspartic acid, glutamic acid and histidine are found in the active site of many enzymes. Enzyme mechanisms often involve these amino acids acting in key roles as acid or base catalysts [46]. A knowledge of accurate pK values for these groups can be important evidence in the formulation of a particular enzyme mechanism. Information about pK values for residues not in the active site is useful for understanding protein stability and interactions with other proteins including antibodies. NMR has been used extensively in the past to measure pK values for histidine residues in proteins [5,47]. Individual pK values for the four histidine residues in ribonuclease were measured in 1967 by following the chemical shifts of the resolved C-2 resonances [5]. Reports of pK values for other titratable residues have been less common owing to the absence of assignments and resolved resonances for these residues in the NMR spectrum. The level of assignment which can now be achieved and the improved resolution available in 2-D NMR spectra, allow measurements of pK values of Asp and Glu residues in lysozyme to be made.

Chemical shift values are sensitive to the electronic environment around the nucleus. Changes in this environment resulting, for example, from a change in ionization state lead to a change in chemical shift. These changes in electron density are transmitted, through-bond, to other nuclei of the ionizable residue [10]. For example, the ionization of the γ-carboxyl group of aspartic acid will lead to a large change in chemical shift for the βCH resonances and smaller changes for the αCH and NH resonances. At any pH value the protonated and deprotonated species are usually

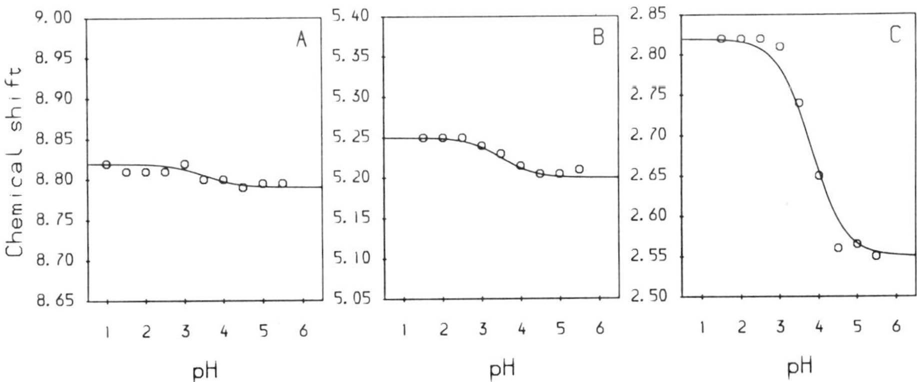

Fig. 9. Plots of chemical shift versus pH for the (a) NH, (b) αCH, and (c) βCH resonances of Asp-52. Experimental points are indicated with open circles. The solid curves represent the theoretical behaviour based on the measured values of pK and chemical shift extremes.

in fast chemical exchange and the observed resonance chemical shift is a weighted average of the chemical shifts of the two species. A plot of chemical shift versus pH gives the expected Henderson-Hasselbach curve from which the pK value can be measured.

COSY spectra were collected for samples of lysozyme ranging from pH 1.0 to pH 6.5 in 0.5 pH unit steps. The excellent sensitivity of the NMR spectrometer made it possible to collect each 2-D COSY spectrum in approximately 1 hour; the whole series of COSY spectra was collected in less than 24 hours of NMR time. The NH, αCH and βCH chemical shift values of resonances of the acidic residues were followed. The NH, αCH and βCH chemical shifts for one of the aspartic acid residues, Asp-52, are plotted as a function of pH in Figure 9. As expected the change in chemical shift is largest for the βCH resonance and smallest for the NH. From these plots an average pK value of 3.6 is obtained for Asp-52. PK values have been obtained for all the acidic residues and for the C-terminal residue of hen lysozyme using this procedure.

The chemical shifts of protons of non-acidic amino acid residues which are near to a carboxyl group are also sensitive to the ionization state of this group. Ionization of the carboxyl group causes a change in the electrostatic interactions in the protein and may cause slight conformational changes [10]. The NH and αCH chemical shifts of all cross peaks in the fingerprint region have been monitored as a function of pH. Small sections of the fingerprint region of the COSY spectra collected at pH 1.5, 3.5 and 5.5 are shown in Figure 10. For many residues in lysozyme, such as Ile-78, the NH and αCH chemical shifts are not found to vary with pH. This is because these residues are distant from any acidic side chains. For other residues, such as Asn-46, the chemical shifts vary substantially with pH. The complex pH variation of the NH chemical shift of Asn-46 results from its proximity to the carboxyl groups of Asp-48 (pK < 2) and Asp-52 (pK = 3.6). These measurements provide information about the local electrostatic environment around individual amino acid residues in lysozyme.

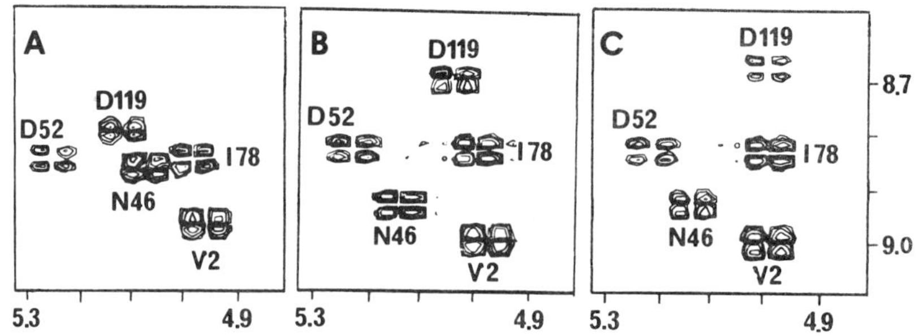

Fig. 10.   Small regions of the phase-sensitive 500 MHz COSY spectra of lysozyme at (a) pH 1.5, (b) pH 3.5, and (c) pH 5.5.

Some of the first studies of proteins by NMR involved the investigation of the interaction of enzymes with their inhibitors [47,48]. In 1967 Meadows et al. reported the results of NMR studies of ribonuclease and hen lysozyme and their respective inhibitors cytidine 3'-phosphate (3'-CMP) and N-acetylglucosamine (NAG) [48]. These authors observed that the addition of 3'-CMP to a solution of ribonuclease led to the perturbation of two histidine C-2 resonances. They indicated that these were likely to arise from His-12 and His-119, two residues known to be involved in enzymatic activity. The addition of NAG to a solution of lysozyme had no effect on the single histidine residue and the authors concluded that this histidine was not involved in interactions with NAG. This result was consistent with the results of X-ray crystallography [48].

Since 1967 more detailed studies of the interaction of lysozyme with mono-, di- and tri- saccharide inhibitors have been reported [49]. The resolved resonances of residues including Leu-17, Trp-28, Ile-98 and Trp-108 have been followed in titration experiments. Until very recently assignments for active site residues have not been available. The few active site resonances which had been assigned, for example Asp-52, were not resolved in the 1-D spectrum and consequently were very difficult to follow. The complete assignment of the lysozyme spectrum and the advent of 2-D techniques facilitate the study of lysozyme-inhibitor interactions.

The specific techniques used to study the interaction of enzymes and their inhibitors depend on the relative values of the chemical shift difference between protein resonances in the free and bound forms and the rate of chemical exchange between the free and bound species [10]. When the chemical shift difference is small compared with the exchange rate the enzyme is in fast chemical exchange. A single resonance with a chemical shift defined as the weighted average of the free and bound chemical shifts is observed. As inhibitor is titrated into the protein sample the protein resonances are observed to shift. A value for the enzyme-inhibitor dissociation constant can be obtained from a plot of chemical shift versus inhibitor concentration. N-acetylglucosamine (NAG) and di-N-acetylglucosamine (diNAG) are examples of inhibitors which are in fast exchange with lysozyme at 35°C and 500 MHz. The effects of added NAG or diNAG on the chemical shifts of resonances arising from all 129 residues of lysozyme can be monitored by collecting 2-D COSY spectra for samples containing varying concentrations of inhibitor. Such studies are currently in progress.

When the chemical shift difference between free and bound forms is large compared with the exchange rate the enzyme is in slow chemical exchange. As inhibitor is titrated into the enzyme sample the resonances from the free enzyme decrease in intensity and new resonances appear for the bound enzyme. The ratio of intensities gives a measure of the relative amounts of free and inhibitor-bound enzyme. Pairs of resonances corresponding to a specific proton in the free and bound forms can be identified using 1-D saturation-transfer experiments or using 2-D NOESY experiments [10,50]. NOESY spectra of systems in chemical exchange contain cross peaks which arise from the chemical exchange process as well as the through-space dipolar NOE interaction [50]. Comparison of NOESY spectra of free enzyme, fully inhibitor-bound enzyme and a 50:50 mixture of free and bound enzyme allows the chemical exchange cross peaks in the spectrum of the mixture to be distinguished from the NOE cross peaks. Tri-N-acetylglucosamine (triNAG) is an example of an inhibitor which is in slow chemical exchange with lysozyme at 35°C and 500 MHz. Small regions of the 1-D spectra of free lysozyme, fully triNAG-bound lysozyme and a 50:50 mixture of free and bound lysozyme are shown in Figure 11a-c. The two resonances at 6.20 and 6.05 ppm can be shown to arise from the same proton in the free and bound enzymes using a 2-D NOESY experiment. The presence of off-diagonal cross peaks in the NOESY spectrum of the 50:50 mixture of free and triNAG-bound lysozyme indicates that the resonances at 6.20 and 6.05 ppm are in chemical exchange. Using this technique it will be possible to assign resonances in the NMR spectrum of enzyme-inhibitor complexes on the basis of the already known assignments of the free enzyme spectrum. When the spectrum of the inhibitor-bound enzyme has been assigned, information about the structure of the complex can be obtained from NOE effects between enzyme and inhibitor resonances.

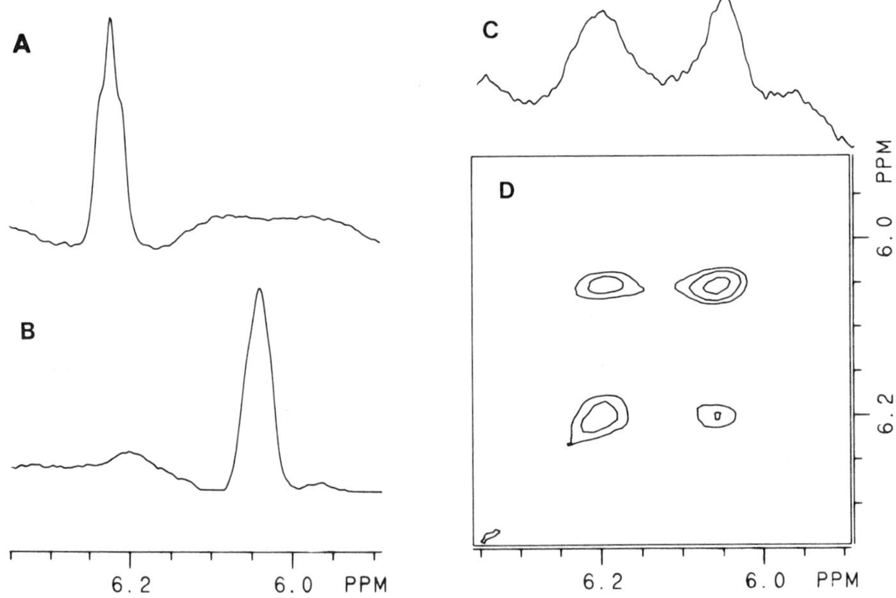

Fig. 11. A small region of the 1-D 500 MHz spectrum of (a) lysozyme, (b) lysozyme fully bound with triNAG, and (c) a 50:50 mixture of free and triNAG-bound lysozyme. (d) A small region of the phase-sensitive 500 MHz NOESY spectrum of a 50:50 mixture of free and bound lysozyme. These resonances have been assigned to Trp-28 $H^{\zeta 3}$.

CONCLUSION

The development of high-field NMR spectrometers and the introduction of 2-D NMR techniques have had a dramatic impact on NMR studies of enzymes during the last ten years. NMR is now an accepted method for the determination of the three-dimensional structure of proteins. The availability of complete resonance assignments now permits information about the behaviour of residues in the active site to be obtained by NMR. The plans for even higher field spectrometers, the recently reported 3-D NMR techniques [51], and heteronuclear NMR methods indicate that future prospects for NMR studies of enzymes are excellent.

ACKNOWLEDGEMENTS

The work on lysozyme described in this article was carried out at the University of Oxford with Dr. C. M. Dobson, J. Blazer and K. Lumb. The work was supported by the U. K. Science and Engineering Research Council.

REFERENCES

1. M. Saunders, A. Wishnia, and J. G. Kirkwood, The NMR Spectrum of Ribonuclease, *J. Am. Chem. Soc.* 79:3289 (1957).
2. O. Jardetsky and C. D. Jardetsky, An Interpretation of the Proton Magnetic Resonance Spectrum of Ribonuclease, *J. Am. Chem. Soc.* 79:5322 (1957).
3. A. Kowalsky, NMR Studies of Proteins, *J. Biol. Chem.* 237:1807 (1962).
4. M. Mandel, Proton Magnetic Resonance Spectra of Some Proteins, *J. Biol. Chem.* 240:1586 (1965).
5. D. H. Meadows, J. L. Markley, J. S. Cohen, and O. Jardetsky, NMR Studies of the Structure and Binding Sites of Enzymes, I. Histidine Residues, *Proc. Natl. Acad. Sci. U.S.A.* 58:1307 (1967).
6. C. C. MacDonald and W. D. Phillips, Manifestations of the Tertiary Structures of Proteins in High-Frequency NMR, *J. Am. Chem. Soc.* 89:6332 (1967).
7. H. Sternlicht and D. Wilson, Magnetic Resonance Studies of Macromolecules, I. Aromatic-Methyl Interactions and Helical Structure Effects in Lysozyme, *Biochemistry* 6:2881 (1967).
8. R. R. Ernst and W. A. Anderson, Application of Fourier Transform Spectroscopy to Magnetic Resonance, *Rev. Sci. Instr.* 37:93 (1965).
9. K. Wüthrich, "NMR in Biological Research: Peptides and Proteins", North Holland, Amsterdam (1976).
10. I. D. Campbell and C. M. Dobson, The Application of High Resolution NMR to Biological Systems, in: "Methods of Biochemical Analysis Vol. 25", D. Glick, ed., Wiley, New York (1979).
11. I. D. Campbell, C. M. Dobson, and R. J. P. Williams, Proton Magnetic Resonance Studies of the Tyrosine Residues in Hen Lysozyme- Assignment and Detection of Conformational Mobility, *Proc. R. Soc. London B* 189:503 (1975).
12. I. D. Campbell and C. M. Dobson, Spin Echo Double Resonance: A Novel Method for Detecting Decoupling in Fourier Transform NMR, *J. Chem. Soc., Chem. Commun.* 1975:750 (1975).

13. I. D. Campbell, C. M. Dobson, R. J. P. Williams, and A. V. Xavier, Resolution Enhancement of Protein PMR Spectra Using the Difference Between a Broadened and a Normal Spectrum, *J. Mag. Res.* 11:172 (1973).

14. G. E. Chapman, B. D. Abercrombie, P. D. Cary, and E. M. Bradbury, The Measurement of Small Nuclear Overhauser Effects in the [1]H Spectra of Proteins, and Their Application to Lysozyme, *J. Mag. Res.* 31:459 (1978).

15. F. M. Poulsen, J. C. Hoch, and C. M. Dobson, A Structural Study of the Hydrophobic Box Region of Lysozyme in Solution Using Nuclear Overhauser Effects, *Biochemistry* 19:2597 (1980).

16. K. Wüthrich and G. Wagner, NMR of Labile Protons in Basic Pancreatic Trypsin Inhibitor, *J. Mol. Biol.* 130:1 (1979).

17. J. A. Lenstra, B. F. J. M. Bolscher, S. Stob, J. J. Beintema, and R. Kaptein, The Aromatic Residues of Bovine Pancreatic Ribonuclease Studied by ^{1}H NMR, *Eur. J. Biochem.* 98:385 (1979).

18. G. R. Moore and R. J. P. Williams, NMR Studies of Eukaryotic Cytochrome c: Assignment of Resonances of Aliphatic Amino Acids, *Eur. J. Biochem.* 103:503 (1980).

19. J. Jeener, Ampere International Summer School, Basko Polje, Yugoslavia (1971).

20. K. Nagayama, K. Wüthrich, P. Bachmann, and R. R. Ernst, Two-Dimensional J-resolved ^{1}H NMR Spectroscopy for Studies of Biological Macromolecules, *Biochem. Biophys. Res. Commun.* 78:99 (1977).

21. G. Wagner and K. Wüthrich, Sequential Resonance Assignments in Protein ^{1}H NMR Spectra: Basic Pancreatic Trypsin Inhibitor, *J. Mol. Biol.* 155:347 (1982).

22. Anil Kumar, R. R. Ernst, and K. Wüthrich, A Two-Dimensional Nuclear Overhauser Enhancement (2D NOE) experiment for the Elucidation of Complete Proton-Proton Cross Relaxation Networks in Biological Macromolecules, *Biochem. Biophys. Res. Commun.* 95:1 (1980).

23. T. A. Holak and J. H. Prestegard, Secondary Structure of Acyl Carrier Protein as Derived from Two-dimensional ^{1}H NMR Spectroscopy, *Biochemistry* 25:5766 (1986).

24. A. D. Kline and K. Wüthrich, Complete Sequence-specific ^{1}H NMR Assignments for the α-amylase Polypeptide Inhibitor Tendamistat from *Streptomyces tendae*, *J. Mol. Biol.* 192:869 (1986).

25. P. C. Driscoll, H. A. O. Hill and C. Redfield, ^{1}H NMR Assignments and Cation Binding Studies of Spinach Plastocyanin, *Eur. J. Biochem.* 170:279 (1987).

26. C. Redfield and C. M. Dobson, Sequential ^{1}H NMR Assignment and Secondary Structure of Hen Egg-White Lysozyme in Solution, *Biochemistry* 27:122 (1988).

27. K. Wüthrich, "NMR of Proteins and Nucleic Acids", Wiley, New York (1986).

28. M. Billeter, W. Braun, and K. Wüthrich, Sequential Resonance Assignments in Protein ^{1}H NMR Spectra: Computation of Sterically Allowed Proton-Proton Distances and Statistical Analysis of Proton-Proton Distances in Single Crystal Protein Conformations, *J. Mol. Biol.* 155:321 (1982).

29. P. Štrop, G. Wider, and K. Wüthrich, Assignment of the ^{1}H NMR Spectrum of the Proteinase Inhibitor IIA from Bull Seminal Plasma by Two-dimensional NMR at 500 MHz, *J. Mol. Biol.* 166:641 (1983).

30. M. P. Williamson, T. F. Havel, and K. Wüthrich, Solution Conformation of Pro-

teinase Inhibitor IIA from Bull Seminal Plasma by ^{1}H NMR and Distance Geometry, *J. Mol. Biol.* 182:295 (1985).

31. R. Kaptein, E. R. P. Zuiderweg, R. M. Scheek, R. M. Boelens and W. F. van Gunsteren, A Protein Structure from NMR Data: lac Repressor Headpiece, *J. Mol. Biol.* 182:179 (1985).

32. R. M. Cooke, A. J. Wilkinson, M. Baron, A. Pastore, M. J. Tappin, I. D. Campbell, H. Gregory, and B. Sheard, The Solution Structure of Human Epidermal Growth Factor, *Nature* 327:339 (1987).

33. A. D. Kline, W. Braun, and K. Wüthrich, Studies by ^{1}H NMR and Distance Geometry of the Solution Conformation of the α-Amylase Inhibitor Tendamistat, *J. Mol. Biol.* 189:377 (1986).

34. V. Saudek, and R. J. P. Williams, Secondary Structure of Acylphosphatase from Rabbit Skeletal Muscle: A NMR Study, *J. Mol. Biol.* 199:233 (1988).

35. T. Imoto, L. N. Johnson, A. C. T. North, D. C. Phillips, and J. A. Rupley, Vertebrate Lysozymes, in: "The Enzymes", P. D. Boyer, ed., Academic Press, New York (1972).

36. C. M. Dobson, Studies of Protein Structure and Dynamics by Proton NMR, in: "Structure and Dynamics: Nucleic Acids and Proteins", E. Clementi and R. H. Sarma, ed., Adenine Press, New York (1983).

37. C. M. Dobson and P. A. Evans, Protein Folding Kinetics from Magnetization Transfer NMR, *Biochemistry* 23:4267 (1984).

38. J. Boyd, C. M. Dobson, and C. Redfield, Identification of Glycine Spin Systems in ^{1}H NMR Spectra of Proteins Using Multiple-Quantum Coherences, *FEBS. Lett.* 186:35 (1985).

39. D. M. LeMaster and F. M. Richards, NMR Sequential Assignment of *Escherichia coli* Thioredoxin Utilizing Random Fractional Deuteration, *Biochemistry* 27:142 (1988).

40. L. P. McIntosh, F. W. Dahlquist, and A. G. Redfield, Proton NMR and NOE Structural and Dynamic Studies of Larger Proteins and Nucleic Acids Aided by Isotope Labels: T4 Lysozyme, *J. Biomol. Struc. Dyn.* 5:21 (1987).

41. C. C. F. Blake, G. A. Mair, A. C. T. North, D. C. Phillips, and V. R. Sarma, On the Conformation of the Hen Egg-White Lysozyme Molecule, *Proc. R. Soc. London B* 167:365 (1967).

42. W. Braun, G. Wider, K. H. Lee, and K. Wüthrich, Conformation of Glucagon in a Lipid-Water Interphase by ^{1}H NMR, *J. Mol. Biol.* 169:921 (1983).

43. A. T. Brunger, G. M. Clore, A. M. Gronenborn, and M. Karplus, Three-dimensional Structure of Proteins Determined By Molecular Dynamics with Interproton Distance Restraints: Application to Crambin, *Proc. Natl. Acad. Sci. U.S.A.* 83:3801 (1986).

44. M. Karplus, Contact Electron-Spin Coupling of Nuclear Magnetic Moments, *J. Chem. Phys.* 30:11 (1959).

45. A. Pardi, M. Billeter, and K. Wüthrich, Calibration of the Angular Dependence of the Amide Proton-Cα Proton Coupling Constants, $^3J_{HN\alpha}$, in a Globular Protein, *J. Mol. Biol.* 180:741 (1984).

46. C. Walsh, "Enzymatic Reaction Mechanisms", W. H. Freeman, San Francisco (1979).

47. J. S. Cohen, Proton Magnetic Resonance Studies of Human Lysozyme, *Nature* 223:43 (1969).

48. J. S. Cohen and O. Jardetsky, NMR Studies of the Structure and Binding Sites of Enzymes, II. Spectral Assignments and Inhibitor Binding in Hen Egg-White Lysozyme, *Proc. Natl. Acad. Sci. U.S.A.* 60:92 (1968).

49. C. C. F. Blake, R. Cassels, C. M. Dobson, F. M. Poulsen, R. J. P. Williams, and K. S. Wilson, Structure and Binding Properties of Hen Lysozyme Modified at Tryptophan 62, *J. Mol. Biol.* 147:73 (1981).

50. J. Jeener, B. H. Meier, P. Bachmann, and R. R. Ernst, Investigation of Exchange Processes by Two-dimensional NMR Spectroscopy, *J. Chem. Phys.* 71:4546 (1979).

51. H. Oschkinat, C. Griesinger, P. J. Kraulis, O. W. Sorensen, R. R. Ernst, A. M. Gronenborn, and G. M. Clore, Three-dimensional NMR Spectroscopy of a Protein in Solution, *Nature* 332:371 (1988).

THERMODYNAMIC FLUCTUATIONS AND FUNCTION IN PROTEINS

Alan Cooper and David T.F. Dryden[1]

Department of Chemistry
Glasgow University
Glasgow G12 8QQ
Scotland, U.K.

INTRODUCTION

Although protein molecules have reasonably well defined structures
under   physiological   conditions,   with the elegantly complex   and
functionally refined architectures revealed by X-ray crystallography,
there is an additional level of complexity which we must now consider.
For proteins and other biological macromolecules are inherently dynamic
objects which are in a perpetual state of conformational fluctuation.
Superimposed upon the average conformation of any molecule is a spectrum
of atomic motions covering a range of frequencies and amplitudes, and
involving both the uncorrelated movement of single atoms or groups and
the more orchestrated motions of larger structural domains. With a few
notable exceptions it was not until the 1970's that most of us began
(reluctantly in many cases!) to accept this point of view, despite a
significant   amount of experimental evidence that could   not   be
interpreted in terms of the hitherto conventional static picture of
protein structures (Cooper,1976). It is not hard to understand the basis
for this reluctance: protein structures are difficult enough to
comprehend as it is without the additional complications of visualising
them in perpetual motion as well; and why bother if we can understand
protein function in terms of their static structures anyway? But, of
course, the answer is that we can't. Many aspects of protein function,
not least the various aspects of enzyme catalysis which concern us here,
are inherently dynamic processes which would not, and could not take
place in the absence of fluctuations. There has been considerable
theoretical progress in recent years, notably in the area of computer
simulation of macromolecular dynamics (McCammon & Harvey, 1987) which,
together with some of the experimental advances, have been discussed
elsewhere here. Our role here is to review the subject from a more
general thermodynamic viewpoint and to consider some of the possible
implications to our understanding of protein function.

____

[1]Present address: Department of Biochemistry, University of Newcastle,
Newcastle-upon-Tyne NE1 7RU, U.K.

THE THERMODYNAMIC UNCERTAINTY PRINCIPLE

The chaotic nature of the molecular world was first observed not by
a physicist or chemist, but by a biologist - the Scottish botanist
Robert Brown - who showed in 1828 that the incessant, irregular motion
of microscopic particles is an innate property of small objects.
Einstein much later described the phenomenon in terms of random
collisions of solvent molecules and laid the foundations for what has
become a general theory of thermodynamic fluctuations. What we term
"heat" is the subjective manifestation of the internal energy and
incessant motion of all atoms and molecules at any temperature other
than absolute zero. As a consequence any system that we care to imagine,
be it a molecule, macromolecule or macroscopic object, provided that it
is not totally isolated from the rest of the world, will be subject to
perpetual collisions from the molecules around it. Any one collision
will result in an exchange of energy and momentum which may alter the
trajectory of our particle (as Robert Brown observed) or change the
internal energy and motions of the system. The frequency and number of
such collisions is such that we find it impossible to describe or
predict the behaviour of any one system and, moreover, otherwise
identical objects will have entirely different dynamic trajectories
because they are subject to a different set of molecular collisions at
any one time. It is perhaps surprising, therefore, that in our
macroscopic world, at least, things are so relatively ordered and
predictable. But this is because the things we normally observe are
usually so large, or there are so many of them, that the microscopic
differences due to these fluctuations tend to average out.

So, let us imagine some small object - a single protein molecule
for instance, though it does not need to be - surrounded by an
incessantly writhing sea of solvent molecules. And let's concentrate
here on the internal energy of our object since this will give some
measure of its internal motions. [Note that the theoretical concepts
which follow can be generalised and applied to many other extensive
thermodynamic properties of the system, such as its volume for example,
though we usually lack sufficient experimental information to apply this
usefully in the case of proteins (Cooper,1976,1984).] As outlined above,
we have no way of predicting the energy of our system at any one time,
and nor would we really want to because it is unlikely to be the same as
any other similar system in similar surroundings and, in any case, it
will be continually changing with time. The best we can do is to write
down the probability, p_i, that our system is in a particular state i
with energy E_i at any one time, and that is given by the Boltzmann
expression:

$$p_i = \exp(-E_i/kT)/Z \quad , \quad \text{with} \quad Z = \sum_i \exp(-E_i/kT)$$

where k is Boltzmann's constant and T is the absolute temperature. Note
that the quantity Z, although it is simply a normalisation constant that
guarantees that the sum of all probabilities is one (i.e. the system has
to be in some state or other), plays a central role in the basic theory
of the statistical thermodynamics of matter (see McQuarrie,1976, for
example). It is given the special name of "partition function" and, if
it can be evaluated (not a trivial procedure in most realistic
situations!), then all thermodynamic properties of the system may be
calculated. For example, a relation that we will use later is that the
Free Energy of a system is given simply by $G = -kT.\ln Z$. (Though,

strictly, this is the Helmholtz free energy for systems at constant volume, it is the same as the Gibbs free energy to a good approximation for most systems under normal constant pressure conditions.)

You may already be puzzled, since the exponential nature of the Boltzmann expression suggests that the most probable state of the system is at zero energy (i.e. no molecular motion), and we know that is not the case except for systems at absolute zero. But what we have written down so far is the probability of finding our system in a particular state, which just happens to have a particular energy, and there may well be many different states having this same energy. In other words, for a given energy E, the system may be found in various different states in which, at one extreme maybe, all the internal energy is in the form of kinetic motion (molecular translation, group rotations and vibrations, for instance), whilst at another extreme the energy might all be potential (stored in bond strains, torsions, and so forth). This availability of a multitude of possible states means that if we are interested in just the energy of the system, regardless of what form that energy takes, then we can write the energy probability as:

$$p(E) = g(E).p_i = g(E).\exp(-E/kT)/Z$$

where $g(E)$ is the degeneracy of the energy level E. [This degeneracy can be related rather loosely to the concept of the "entropy" of a particular energy level ($S = k.\ln g$), i.e. the higher the degeneracy then the higher the entropy. But this is dangerous ground and subject of much philosophical discussion, so we will not carry it any further here.] It follows that the higher the energy then the more ways there are of distributing this energy within the system, that is $g(E)$ increases with energy. This combination of factors, the decrease in the Boltzmann exponential factor with energy, coupled with the simultaneous increase in internal degeneracy, means that the most probable energy is now not at zero but at some higher mean value where the two trends just balance. We can calculate this mean energy, $\langle E \rangle$, from the probabilities:

$$\langle E \rangle = \Sigma E.p(E) = kT^2.\frac{d \ln Z}{dT}$$

where the summation is now taken over all possible energies, and the second relationship is merely an algebraic consequence of the definition of Z.

But this just tells us what the average energy of the system will be over a period of time or, if we have a large number of identical but uncorrelated systems, what the average of the whole ensemble will be. Because of the statistical uncertainties in the molecular collisions with the surrounding environment our system will be fluctuating about this mean position, and we need to get some idea of the magnitude of these fluctuations and the shape of the probability distribution function, $p(E)$. As a first step we can calculate the approximate width of the distribution from the mean-square energy fluctuations, $\langle \delta E^2 \rangle$, as first done by Einstein from simple probability theory:

$$\langle \delta E^2 \rangle = \Sigma (E - \langle E \rangle)^2 p(E) = kT^2.\frac{d\langle E \rangle}{dT}$$

where, again, the second equality arises from algebraic manipulation. This is very useful because $d\langle E \rangle/dT$, the rate of change of average energy of the system with temperature, is what we know as the heat capacity of the system, C_V (strictly at constant volume in this case, but the details are not numerically important), and is experimentally measurable. So, if we know this, we may estimate the size of the energy fluctuations in our system using:

$$\langle\delta E^2\rangle \;=\; kT^2 C_V$$

But we can go further using probability theory (Cooper,1984) and estimate the higher moments of the distribution, $\langle\delta E^3\rangle$, $\langle\delta E^4\rangle$, and so on, again in terms of the heat capacity of the system and, in this case, its temperature dependence. This in principle gives us sufficient information to calculate the actual shape of the energy probability distribution function for any system.

There are several points to note before we proceed:

(i) The theory we have outlined above is standard textbook material relating to systems at equilibrium. In other words, any object at thermodynamic equilibrium, even though it may give the appearance of stasis and stability, is at the molecular level in a perpetual state of flux and motion as it exchanges energy to and fro with its environment.

(ii) Nothing we have derived so far is specific to protein molecules, rather they are general and immutable properties of all matter.

(iii) There are in fact two kinds of fluctuation inherent in our system, external and internal. The external ones are those due to energy exchange with the surroundings that we have derived above. But we have already emphasized the degeneracy of energy levels in a system arising from the many different dynamic states of the system which share a common energy. Thus, even a system at constant internal energy, kept totally isolated from its environment and unable, therefore, to undergo external fluctuations, will still demonstrate dynamics and uncertainty as the system fluctuates between the various iso-energetic states available, converting energy from one form to another throughout the system. In the case of proteins and other biological macromolecules it is, in the main, these latter internal fluctuations that are studied by molecular dynamics simulations since, with a few exceptions, solvent has so far proved too complex and expensive to incorporate into the computer models. These simulations are therefore of molecules in a vacuum, with no surroundings to interact with, and will consequently underestimate the magnitudes of the dynamic processes taking place in real systems.

(iv) Although we have derived expressions for thermodynamic fluctuations using abstract and, usually, uncalculable concepts such as probability, partition functions and degeneracy, the end results are expressed entirely in measurable quantities, for instance the heat capacity, and are not dependent on any specific model of the system.

Having now established the formal basis of fluctuation theory we can do some simple and instructive sums to estimate the magnitudes of energy fluctuations in various systems to show how their relative importance depends on the size of the system under consideration.

Consider first a typical macroscopic system consisting of 1ml of water at room temperature. We don't normally think of this as a fluctuating system – it usually just sits there looking pretty static and uninteresting – but it is a thermodynamic system like any other, and will be subject to fluctuations due to molecular collisions from the surrounding air, the container, and so forth. Its heat capacity is about 1 cal K^{-1} (4.184 JK^{-1}) so that at 300K, and with k = 1.38×10^{-23} JK^{-1}, $\langle \delta E^2 \rangle$ is about 5×10^{-18} J^2, or taking the square root, $\delta E_{RMS} \approx 5 \times 10^{-10}$cal. This is tiny compared to the total heat content, or internal energy of our 1ml sample and corresponds to temperature fluctuations of only about $\pm 5 \times 10^{-10}$C. So, although fluctuations are occurring, the distribution function is very sharp on our macroscopic observable scale and it is not too surprising that we are not normally conscious of them.

Going to the opposite and almost absurd microscopic extreme, we might estimate the similar fluctuations in just one of the water molecules in our sample as it gets kicked around by its neighbours. It might be argued, with some justification, that a single water molecule is too small an object to be considered as an independent thermodynamic system, but let's do the sum anyway. Its heat capacity will be about $1/6 \times 10^{23}$ cal K^{-1}, giving $\delta E_{RMS} \approx \pm 10^{-21}$cal per molecule or 0.6 kcal per mole at room temperature, which is comparable to kT. Bearing in mind that the energy of any individual molecule is also of the order of kT (≈ 0.6 kcal mol^{-1}) this is much more impressive and confirms that the behaviour of individual molecules in liquid water is totally dominated by the dynamics of interaction with their neighbours. [Water is, of course, a slightly special case in that its relatively high heat capacity (for a liquid) arises from the strong and highly-structured interactions between adjacent molecules. Hence our reservations about taking too seriously the concept of a single molecule as an independent thermodynamic system. Interestingly, the heat capacities of proteins are not markedly different in the dry or fully hydrated states, implying that our assumption is not quite so outrageous when we come to apply it below to individual protein molecules.]

So, for macroscopic objects the fluctuations are, except in special circumstances, of little consequence; whereas for the microscopic regime they assume overwhelming importance. Midway between these extremes there lies the range of "mesoscopic" systems, neither very big nor very small, into which proteins and other biological macromolecules fall. The specific heats of proteins under a range of conditions are remarkably uniform at about 0.32 cal $K^{-1}g^{-1}$ (Suurkuusk,1974; Privalov,1979), and are typical of organic solids in general, though they increase during unfolding transitions and are usually slightly higher for the unfolded polypeptide in water. But a typical protein molecule of 25,000 molecular weight will have a heat capacity of about 1.3×10^{-20}cal K^{-1} with consequential δE_{RMS} energy fluctuations, using the same formula as above, of about 3×10^{-19}J, corresponding to ± 40 kcal mol^{-1} or ± 5°C in temperature. With more detailed calculations (Cooper,1984), taking into account the empirical temperature variation of the heat capacity for real proteins under physiological conditions (Privalov,1979; Sturtevant,1987), we can obtain the complete energy probability distribution, p(E). The general shape of this shows nothing particularly surprising. At temperatures well above or below the thermal unfolding transition of the protein the distributions are very close to single symmetric Gaussians in form, and only during the unfolding transition itself does the distribution break up into a non-symmetric bi-modal form showing peaks corresponding to the folded and unfolded states of the polypeptide, though even here the widths of the distributions are such that there is considerable overlap between the two. It is this which is the most intriguing aspect of these numerical estimates. The magnitudes of the mean square fluctuations of proteins are not in themselves very remarkable, with hindsight, since they are typical of any piece of organic matter in this mesoscopic size range. But because the folded conformations of proteins, dependent as they are on weak non-covalent interactions, are only marginally stable with folding free energies in the region of just a few kcal mol^{-1} and enthalpies in the range 40-200 kcal mol^{-1} which are comparable to our estimated energy fluctuations, we must accept that our view of protein structure and function must incorporate a considerable dynamic component.

SO WHAT ?

Does all this have anything to do with the function of proteins in general, or the catalytic efficiency of enzymes in particular ? Well, in one sense it does. If we think about each of the stages in an enzymic reaction, insofar as they can be resolved into separate steps, then it is clear that each one is likely to involve fluctuations in the protein. The binding and release of substrates at the active site frequently requires transient conformational changes - the opening of gates or channels - to give access to the site. Haemoglobin, though not an enzyme, is the classic example here with its totally enclosed oxygen binding site (Perutz,1978). After substrate binding many enzymes require another change in conformation before actual catalysis can take place. This may be necessary to bring substrates or functional groups into close proximity or, particularly important in the kinases, to exclude water from the site during the reaction. In phosphoglycerate kinase, for example, which is responsible for ATP production by catalytic transfer of the phosphate group from 1,3-diphosphoglycerate to ADP, the two substrate binding sites are on separate domains of the protein which must move together during catalysis; firstly to juxtapose the appropriate substrate groups, but secondly, and probably even more important in this case, to evict solvent water from the site and prevent it competing for the phosphate group to give hydrolysis rather than transfer (Watson et al,1982). Synthetic chemists would never dream of using water as solvent for such reactions, but are forced to use organic solvents instead. Enzymes cannot avoid the water but can create their own transient mesoscopic organic solvent/reaction vessel by conformational fluctuation at the appropriate moment.

The catalytic act itself is clearly a dynamic process involving transient fluctuation. We are all familiar with the classical Arrhenius expression describing the empirical temperature dependence of the rates of chemical reactions:

$$k = A.exp(-E_A/RT)$$

where E_A is the activation energy or height of some rate-determining barrier on the reaction pathway, and A is some sort of frequency factor. (More sophisticated absolute rate theories have similar expressions.) The exponential energy term is immediately reminiscent of the Boltzmann probability factor, which shows that all chemical processes require transient fluctuations away from the thermodynamic mean in order that they might proceed. Enzymes are no exception here and, at least over the temperature range in which they are stable, show exponential temperature dependence in their k_{CAT}.

But the astute reader will have noticed that there are various problems relating to our discussion so far in this section. Firstly, as we have emphasized in previous sections, there is nothing remarkable about the magnitudes of the thermodynamic fluctuations that occur in a protein molecule - they are simply a consequence of the finite size of the molecule and the unexceptional nature of the thermal properties of the material from which it is made. So we cannot invoke the size of the fluctuations as an explanation of catalytic rate enhancements. Indeed, one feature of enzyme catalysed reactions is that they usually appear to have a lower Arrhenius activation energy than the un-catalysed reaction so that, instead of actually utilising the large fluctuations that are available in the whole molecule, enzymes seem to have gone to some pains to reduce the necessary barrier. (Possible mechanisms for this are discussed elsewhere in this volume.) Others have suggested that enzymes may have evolved the ability to somehow "focus" their fluctuations onto the catalytic site (Welch et al,1982) in such a way as to enhance the effective local temperature and speed up the reaction. Again, some account of this is given elsewhere here and we will not discuss it further - mainly because our own opinions of these un-thermodynamic theories are, frankly, sceptical.

A second problem we face is that all the enzyme-related fluctuation steps mentioned so far - opening and closing of gates or channels, domain movements, barrier crossings, etc. - are dynamic, kinetic rate processes. Yet our formal thermodynamic theory is strictly an equilibrium theory in which rates do not appear. We will have a little to say about the rates of fluctuations in a general sense later. But first we will describe one area, unique to proteins and enzymes, in which fluctuations seem likely to contribute in a positive way.

ALLOSTERY WITHOUT CONFORMATIONAL CHANGE ?

The complex feedback control systems upon which living organisms must rely for a stable existence result, in part, from the ability to modulate the catalytic activities of certain key enzymes by molecules produced at other stages of the biochemical process. In such allosteric enzymes the binding of one molecule at a control site on the protein can affect the binding of substrates or rate of the reaction at the catalytic site in ways that may either enhance or inhibit the overall reaction rate at this stage. The remarkable thing is that these control and catalytic sites are frequently found to be at quite different and distant locations on the protein, or even on different subunits in some cases. This ability to communicate across relatively large molecular distances, unique to biological macromolecules as far as we know, was first extensively described by Monod, Wyman & Changeux(1965) and has since been popularly explained in terms of ligand-induced global conformational changes in the protein. That is, binding of a molecule ("ligand") at one site induces a conformational change in the protein that extends, possibly across subunit boundaries, to a second site and thereby affects binding or catalysis at this site. But even in their early publications Monod et al. were at pains to point out that the terms "conformational transition" or change were to be understood in their widest possible connotations, and not necessarily in the strict stereochemical sense that has become commonplace. It occurred to us some time ago (Cooper,1980,1984;Cooper & Dryden,1984) that there might be an alternative explanation based on protein dynamics that was worth exploring.

The basis of the idea is that, given that protein structures are dynamic, it should be possible to transmit information across a molecule not only by changes in conformation but also by ligand-induced changes in the frequency and amplitude of dynamic motion within this conformation. But before developing the theory of this, we should explain what we mean by "conformation" and "conformational change" in this context, since we now know that conformational fluctuations are occurring all the time and are unstoppable. We take conformation to mean the average coordinates of atoms and groups within the macromolecule that would be determined by any experimental technique, such as X-ray crystallography, that gives a time-averaged and/or number-averaged picture of the molecule. Conformational change here means any change in these averaged coordinates. In the same way that the mid-point of a swinging pendulum is independent of the amplitude or frequency of swing, we can at least imagine changes in frequencies and amplitudes of motion of protein atoms that do not also entail changes in mean position, that is without conformational change. Leading on from this, we know from experimental considerations that binding of ligands to often seems to "stiffen" the protein, that is to increase the frequency or decrease the amplitude of its dynamic motion. It is not inconceivable that such changes might extend across the molecule and be sensed at another distant binding site.

To develop these ideas quantitatively we will consider the simpler case of cooperativity, rather than general allostery, in the sequential binding of two identical ligands (L) at formally identical sites on a protein (E). [Note that our original publications on this subject (Cooper, 1984; Cooper & Dryden,1984) contained some trivial algebraic errors which are corrected here.] We wish to compare the binding free energies of the following two equilibrium processes:

$$E \; + \; L \; \rightleftharpoons \; EL \qquad ; \quad \Delta G_1 \; = \; G_{EL} \; - \; G_E \; - \; G_L$$

$$\text{and} \qquad EL \; + \; L \; \rightleftharpoons \; ELL \qquad ; \quad \Delta G_2 \; = \; G_{ELL} \; - \; G_{EL} \; - \; G_L$$

$$\text{with} \qquad \Delta\Delta G \; = \; \Delta G_2 \; - \; \Delta G_1 \; = \; G_{ELL} \; - \; 2G_{EL} \; + \; G_E$$

In the absence of cooperativity the difference in binding free energy between the first and second ligand ($\Delta\Delta G$) is zero. For positive cooperativity, with the second ligand binding more strongly, $\Delta\Delta G$ is negative; and vice versa for negative cooperativity effects.

We saw earlier how the free energy of any system is related to its partition function, $G_{EL} = -kT.\ln Z_{EL}$, and so forth, so we may write:

$$\Delta\Delta G \; = \; -kT.\ln\{Z_E Z_{ELL}/Z_{EL}^2\}$$

and we merely have to evaluate the partition functions for each liganded state of the protein. But this is easier said than done and we must adopt some simple models. Fortunately it is a property of partition functions that different kinds of contribution to Z (translational, rotational, vibrational, conformational energies) may, to a first approximation, be treated separately and any contributions that do not change due to ligand binding, or change to the same extent, will cancel in $\Delta\Delta G$. Consequently we will consider two rather simple pictures of the dynamical contributions: firstly in terms of low-frequency normal mode vibrations that might span the entire molecule, and secondly, at the opposite extreme, in terms of non-periodic, haphazard, uncorrelated motions of individual groups. But please remember that these are just models for the purposes of illustration. We certainly do not believe, for example, that a protein molecule rings like a bell ! Even though formal normal mode calculations can be done (Go et al,1983; Brooks & Karplus,1983) which do show global low-frequency modes in these models, in any real protein the solvent damping and anharmonicity effects will be sufficient to quell any such motion. Indeed, in recent experiments in Glasgow using low-frequency depolarized Rayleigh scattering (Cooper & Sanders, work in progress) we have shown that the motion in some proteins, though it can be affected by ligand binding, shows no resemblance to the calculate harmonic vibration spectrum.

But, suspend disbelief for a moment, and consider the possible dynamic contribution from a low-frequency protein mode that might be affected by ligand binding, and remember that low-frequency here means those modes of frequency f for which $hf \approx kT$, or less, which might be excited at normal temperatures (i.e. $f \leq 200cm^{-1}$). Vibrational partition functions are quite easy to calculate (e.g. McQuarrie,1976) but we will omit the full algebraic details here (see Cooper & Dryden,1984) and simply write down the high and low frequency limit results:

$$\text{For} \quad hf \gg kT \quad , \quad Z(f) \approx 1 \qquad \text{(quantum limit, mode "frozen")}$$

$$\text{For} \quad hf \ll kT \quad , \quad Z(f) \approx kT/hf \qquad \text{(classical equipartition limit)}$$

We now have to imagine what might happen to the normal mode frequencies during ligand binding and there are obviously many possibilities, but a simple example will serve to illustrate the point. Imagine that during the sequential binding process E --> EL --> ELL, the normal mode frequency shifts are f_0 --> f_1 --> f_2 . If the frequency shifts are small and in the classical limit then, using the general expression for $\Delta\Delta G$ above:

$$\Delta\Delta G_{vib} \approx -kT.\ln(f_1{}^2/f_0 f_2)$$

If, as we might imagine for a ligand-induced stiffening of the protein, $f_1 \approx f_2 > f_0$ then we get a small negative $\Delta\Delta G_{vib}$ contribution, i.e. positive cooperativity, from this mode. Many such modes, similarly affected, would reinforce this effect. An extreme limit might be one in which a low-frequency mode (say about $50cm^{-1}$) in the unliganded protein is completely frozen out by binding of the first ligand and shifted to even higher frequency on binding the second. Here we would obtain a positive cooperativity effect $\Delta\Delta G_{vib} \approx -kT.\ln(kT/hf_0)$, of the order of kT per affected mode in this case, and quite sufficient to account for the magnitude of observed allosteric effects. In words, what is happening here is as follows: in the absence of any ligand the protein is loose and floppy, with a lot of energy stored in its thermal motions. Binding of the first ligand stiffens the protein, releasing the thermal energy as heat. Although this is an exothermic contribution to binding, the loss of entropy due to inhibition of the conformational motion more than offsets this effect, and the nett result is a weakening of the binding free energy. But when the second ligand comes along there is no more conformational entropy or thermal energy to lose - it has already been frozen out in the first step - so binding, though less exothermic, is more favourable. Incidentally, these additional predictions about entropic and enthalpic differences in cooperative binding processes seem to be borne out in the rather few appropriate systems that have been studied calorimetrically.

We have already explained why it is really unreasonable to picture the dynamic motion of proteins as harmonic oscillations but, nevertheless, the overall conclusions are, at least qualitatively, valid in general. To emphasize this point let's look at another, probably more realistic model in which we view the dynamics as random motions of all the individual atoms or groups in the protein which, though uncorrelated, might still be affected by ligand binding. The motion of an individual group, i, would be described by a mean-square amplitude of displacement, σ_i, from its mean position and a random-noise-like frequency distribution. Evaluation of the exact partition functions in this case is not feasible, but we can show, with a few simplifying assumptions, that Z for this kind of motion is at least proportional to σ, so that:

$$\Delta\Delta G_i \approx -2kT.\ln(\sigma_0\sigma_2/\sigma_1{}^2) \qquad \text{per group}$$

where, as before, the subscripts 0,1,2 refer to the free, mono- and bi-liganded enzymes, respectively. This, at least in principle, is amenable to experimental test since the mean-square displacements, σ_i, can be obtained by crystallography - though we would need a case in which each state of ligation of the protein is studied independently. Meanwhile let's just imagine a situation in which the mean-square displacement σ_0 of a group in the free protein is reduced by a small factor δ by ligand binding at either site, i.e. a slight ligand-induced stiffening of the protein with $\sigma_1 \approx \sigma_2 \approx \sigma_0(1-\delta)$, and note that this might take place without any actual conformational change in the sense noted above. For small shifts, δ, we get $\Delta\Delta G_i \approx -2kT\delta$ per group. In other words, a cooperative contribution to the binding free energy which, summed over all participating groups in the protein, would be quantitatively significant even if the individual shifts were small.

We do not imagine, of course, that all cooperative or allosteric effects in proteins are explicable in terms of dynamic changes alone, nor that there is no role for the conventional ligand-induced conformational change interpretation. Rather, we would argue that both are feasible and equally plausible, and that both might be present to varying extents in particular cases. But one advantage of the dynamic view is that it is possible to make quantitative and verifiable predictions regarding the thermodynamics of the process, which is not yet feasible with the conformational change hypothesis.

FLUCTUATION KINETICS

Central to our understanding of the source of rate enhancement in enzyme catalysed reactions is the rate of occurrence of the transient fluctuations responsible for various stages of the process. But this whole area of the kinetics of chemical reaction in condensed media, not just in enzymes, is fraught with difficulties, many of which have been discussed by others in this volume. We still lack a tractable theory of non-equilibrium rate processes which has the power, rigour and general applicability of equilibrium thermodynamics. But there is a somewhat general, phenomenological approach to fluctuation rates based on a generalized Langevin equation approach which can give some insights into these matters (Cooper,1984,1988). To paraphrase what has gone before: a protein molecule has many available states consisting of different conformational arrangements and states of motion of the individual atoms or groups of which it is made up. In the course of time it will evolve from state to state under its thermal motion and in interaction with its surroundings. This complex trajectory will tend to cluster around an appropriate mean point, but there will be excursions or fluctuations from the mean whose amplitudes may be estimated quite accurately from equilibrium theory and which can be broadly classified into two types: (a) the adiabatic or internal transitions involving redistribution of energy within the macromolecule, and which may be studied over a limited time range, at least, by molecular dynamics simulation, and (b) non-adiabatic or external fluctuations involving exchange of energy with surrounding solvent due to molecular collisions, frictional and viscous damping, and so on. It is the latter, external changes which are so difficult to simulate. One way to approach the problem is to consider the energy trajectory of the protein molecule as a sort of Brownian motion in energy space. This lacks, of course, the fine structural detail available from molecular dynamics models, but does point out some of the general features that might be expected.

Imagine, yet again, our dynamic protein molecule writhing in a bath of solvent molecules, and suppose for a moment that it is a macroscopic object. At any instant in time it might, due to fluctuations, have slightly more energy than average. Being, therefore, slightly "hotter" than the bath it will begin to "cool". Similarly, low energy states will tend to "warm" towards the mean. But while all this is going on, and usually before it has had chance to get very far, the protein will be kicked into some other state by random collisions with solvent molecules. You will see at once the analogy with Brownian motion, and the rate of change of protein internal energy can be formally described by a stochastic Langevin equation appropriate for such random walks:

$$\frac{dE}{dt} = -\alpha(E - \langle E \rangle) + R(t)$$

where α is a phenomenological linear relaxation coefficient for energy dissipation to the bath, $\langle E \rangle$ is the thermodynamic mean energy, and $R(t)$ is a random impulse term representing energy changes due to spontaneous collisions with solvent molecules. This equation can be solved by standard methods (McQuarrie,1976) to show that the distribution of energy states generated by this process is Gaussian with a width given by:

$$\langle \delta E^2 \rangle = \langle R^2 \rangle \tau/2\alpha = kT^2 C_V$$

where the second equality arises because this quantity must be the same as that derived earlier by equilibrium techniques. $\langle R^2 \rangle$ is the mean square impulse and τ is an arbitrary small period of time over which any collision might be thought to act. This result, incidentally, shows that the dissipation and fluctuation (impulse) parameters, α and R, are not independent but are related by equilibrium properties of the system. This makes sense because they are really different aspects of the same microscopic phenomenon, that is molecular collisions.

But there are at least two problems with this so far: (i) it is altogether too simple minded, and (ii) it doesn't work, in the sense that we know from earlier sections that the energy distribution function of a protein molecule is never exactly Gaussian, and can deviate quite markedly from this in the region of conformational transitions. The mistake we have made is to assume that the parameters α and R are single valued quantities having the same magnitude whatever the dynamic state of the protein. This cannot be true. For instance, some dynamic states will consist of short-range, high frequency motions of individual atoms and groups (rotation, vibration, libration) whose dissipation might resemble thermal conductivity in a macroscopic object. Other states might involve large scale, low frequency cooperative motions of structural domains, hinge bending, partial unfolding or breathing modes, for example, whose relaxation would be better described by frictional or viscous damping parameters, maybe. Impulse parameters for collisions with solvent molecules will similarly depend on the instantaneous state of the protein.

We can formally extend the Langevin equation to cover this by assuming that there exist different classes of dynamic states with different relaxation and impulse properties (α_i, R_i), so that:

$$\frac{dE}{dt} = \sum_i \Phi_i(E,t).\{-\alpha_i(E-\langle E_i \rangle) + R_i(t)\}$$

where $\Phi_i(E,t)$ is a stochastic function giving the probability that the protein is in a particular dynamic class at any one time, and $\langle E_i \rangle$ is the mean energy of that particular class, in other words the mean energy that would be observed for this class on its own with all others inaccessible.

To get a mental picture of what this equation describes, we can imagine the dynamic conformational motion of the protein as a sort of weighted random walk across a multi-dimensional energy surface with various depressions or valleys corresponding to the different available dynamic classes. Analytical solution of this equation is difficult in most cases, but it can form the basis of numerical simulations which can illustrate some of the properties that may emerge (Cooper,1988). Take a simple two class model, for example, in which we imagine that the protein molecule may exist in one of just two different dynamic classes with different mean energies and thermal relaxation parameters. (This might represent folding/unfolding equilibrium dynamics, for instance.) Numerical integration of the generalized Langevin equation in this case shows that the equilibrium probability distribution that one gets depends critically on the relative rates of internal versus external fluctuations; that is, the rate of transitions from one dynamic class to the other without change in internal energy, compared to the rates of energy exchange with the surroundings. In the case that internal fluctuations are slow compared to interactions with the solvent, we obtain a bimodal distribution comparable to that seen in real proteins during the unfolding transition (Cooper,1984). At the opposite extreme, if internal interconversions are much faster than external energy exchange, then we see just a single broad Gaussian distribution regardless of however many dynamic classes may exist. Unfortunately, time does not allow us to go into some of the other interesting features that emerge from these simulations, including explanations observations of "conformational drift" (Xu & Weber,1982; King & Weber,1986; Weber,1986) and hysteresis effects in enzymes, but detailed explanations of these may be found elsewhere (Cooper,1988).

We have looked here at just some of the possible consequences of equilibrium thermodynamic fluctuations and their kinetics for our understanding of macromolecular properties. We could have chosen others, such as their role in energy transduction processes and muscular contraction for example (Cooper,1984), or the way in which dynamic effects can dramatically alter the way in which we interpret experimental observations (Cooper,1981). But we have seen enough to say that even if fluctuations as described here have no special or unusual part to play in the enzyme catalytic process, that is, any way different from normal chemical reaction processes, there are other areas such as allostery and cooperativity (amongst others) where biological macromolecules may have developed in the course of evolution to take advantage in special ways of the inevitable dynamic properties of molecular systems of this size.

REFERENCES

Brooks,B. & Karplus,M.(1983) Proc.Nat.Acad.Sci., 80:6571.
Cooper,A.(1976) Proc.Nat.Acad.Sci., 73:2740.

Cooper,A.(1980) <u>Sci.Prog.(Oxford)</u>, 66:473.
Cooper,A.(1981) <u>Proc.Nat.Acad.Sci.</u>, 78:3551.
Cooper,A.(1984) <u>Prog.Biophys.molec.Biol.</u>, 44:181.
Cooper,A.(1988) <u>J.Mol.Liqu.</u>, (in press).
Cooper,A. & Dryden,D.T.F.(1984) <u>Eur.Biophys.J.</u>, 11:103.
Go,N.,Noguti,T. & Nishikawa,T.(1983) <u>Proc.Nat.Acad.Sci.</u>, 80:3696.
King,L. & Weber,G.(1986) <u>Biophys.J.</u>, 49:72.
McCammon,J.A. & Harvey,S.C.(1987) "Dynamics of Proteins and Nucleic Acids", Cambridge University Press, Cambridge, U.K.
McQuarrie,D.A.(1976) "Statistical Mechanics", Harper & Row, New York.
Monod,J.,Wyman,J. & Changeux,J-P.(1965) <u>J.Mol.Biol.</u>, 12:88.
Perutz,M.F.(1978) <u>Scientific American</u>, 239:68.
Privalov,P.L.(1979) <u>Adv.Protein Chem.</u>, 33:167.
Sturtevant,J.M.(1987) <u>Ann.Rev.Phys.Chem.</u>, 38:463.
Suurkuusk,J.(1974) <u>Acta Chem.Scand.</u>, B28:409.
Watson,H.C., Walker,N.P.C., Shaw,P.J., Bryant,T.N., Wendell,P.L., Fothergill,L.A., Perkins,R.E., Conroy,S.C., Dobson,M.J., Tuite,M.F, Kingsman,A.J. & Kingsman,S.M.(1982) <u>EMBO J.</u>, 1:1635.
Weber,G.(1986) <u>Biochemistry</u>, 25:3626.
Welch,G.R.,Somogyi,B. & Damjanovich,S.(1982) <u>Prog.Biophys.molec.Biol.</u>, 39:109.
Xu,G-J. & Weber,G.(1982) <u>Proc.Nat.Acad.Sci.</u>, 79:5268.

USING RESONANCE RAMAN SPECTROSCOPY TO STUDY THE STRUCTURE AND DYNAMICS OF

ENZYME-BOUND SUBSTRATES[1]

P.R. Carey

Division of Biological Sciences
National Research Council
Ottawa, Canada K1A OR6

ABSTRACT

The chemical nature of the enzyme-substrate intermediates studied by
resonance Raman (RR) spectroscopy is defined. The information available on
these intermediates from RR spectroscopy is compared to that obtained from
absorption and fluorescence techniques. Two major classes of substrates
useful for RR studies are discussed. The first class involves the use of
substrates which are chromophoric by virtue of the fact that they have acyl
groups, e.g. cinnamoyl, based on delocalised π-systems. Their use to probe
the effect on the substrate's C=O group of going to active-pH in serine
proteases, and to probe the strong electron polarisation forces available
in cysteine proteases, is detailed. The second class of substrate involves
thionoesters $RC(=S)OCH_3$ which form chromophoric dithioester intermediates,
$RC(=S)S$-enzyme, with cysteine proteases, HS-enzyme. Because the
dithioester RR spectrum is sensitive to conformational changes in the
R-C(=S)-S-C bonds it is used to characterise conformational activation in
the scissile linkages which occur for substrates specific for the cysteine
protease papain. In addition RR spectra of these dithio-intermediates have
been obtained under cryoenzymological conditions at 77°K and offer the
opportunity to characterise such dynamical events as the thermal activation
of the substrate in the active-site.

REACTION SCHEME FOR SERINE AND CYSTEINE PROTEASES

Serine and cysteine proteases are two classes of enzymes found in
nature which break down proteins by hydrolysing their peptide linkages:

$$\text{polypeptide-}\overset{\displaystyle O}{\overset{\displaystyle \|}{C}}\text{-}\underset{\displaystyle H}{N}\text{-polypeptide'} + H_2O \xrightarrow{\text{enzyme}}$$

$$\text{polypeptide-}\overset{\displaystyle O}{\overset{\displaystyle \|}{C}}\text{-O}^- + \overset{+}{H_3}N\text{-polypeptide'}$$

---

[1] Published as NRCC No. 29252

The serine and cysteine proteases derive their names from the fact
that they have an essential serine ($-CH_2OH$ side chain) and cysteine
($-CH_2SH$ side chain) amino acid in their active sites, respectively (1,2).
The best studied members of both families are globular proteins in the
25,000D MW range. A generalised reaction scheme for hydrolysis is:

$$En\text{-}XH + R\text{-}\overset{\displaystyle O}{\overset{\|}{C}}\text{-}Y \rightarrow [En\text{-}XH.RCOY] \rightarrow En\text{-}X\text{-}\overset{\displaystyle O^-}{\underset{\underset{\textstyle R}{|}}{\overset{|}{C}}}\text{-}Y$$

Michaelis complex   T.I.

$$\xrightarrow[\displaystyle -HY]{} En\text{-}X\text{-}\overset{\displaystyle O}{\overset{\|}{C}}\text{-}R \xrightarrow{H_2O} En\text{-}X\text{-}\overset{\displaystyle O^-}{\underset{\underset{\textstyle OH}{|}}{\overset{|}{C}}}\text{-}R \rightarrow En\text{-}XH + \overset{\displaystyle O}{\underset{\underset{\textstyle HO}{|}}{\overset{\|}{C}}}\text{-}R$$

acyl-enzyme   T.I.

Where X is a serine oxygen or a cysteine sulfur atom. As can be seen
both classes of enzyme generate intermediates known as acyl enzymes wherein
the substrate is linked to the enzyme by a transient covalent bond. The
acyl enzyme is preceded and followed by a tetrahedral intermediate (T.I.)
for acylation and deacylation, respectively. The tetrahedral intermediates
are very short lived and are not candidates for direct structural or
spectroscopic examination. However, acyl enzyme intermediates can have
life-times of milliseconds or longer and are prime targets for
spectroscopic analysis. The strategy is to use ester (Y=OR') rather than
peptide (Y=NHR'') substrates since when the former are hydrolysed by serine
or cysteine proteases acylation is faster than deacylation. Then, by using
an excess of substrate it is possible to form a quasi steady-state
population of acyl enzyme for spectroscopic characterisation. The chemical
bonding between the acyl group and the enzyme is the same regardless of
whether the acyl moiety originated from a peptide or ester substrate.

GENERATING A RESONANCE RAMAN (RR) SPECTRUM FROM AN ACYL ENZYME

Resonance Raman (RR) spectroscopy can be looked upon as the
vibrational equivalent of absorption spectroscopy (3). Just as the latter
enables us to focus on the electronic transition associated with a 'small'
chromophore in a complex biological system so the RR effect focusses on the
vibrational spectrum of the chromophore. Thus to use the RR effect to
study enzyme-bound substrates there must be a chromophore in the
active-site; this is achieved either by using a substrate which is itself
chromophoric or by generating the chromophore during the catalytic act (see
dithioesters, below). An example of the former situation is the hydrolysis
of the imidazole ester of 5-methylthienylacrylic acid by chymotrypsin (4).
The acyl enzyme 5-methylthienylacryloyl-chymotrypsin can be purified by
manipulating pH and by column chromatography and has the absorption
spectrum shown in Fig. 1. The bound acyl group gives rise to the
absorbance near 340 nm and by focussing a laser of wavelength 350.6 nm into
a solution of the acyl-enzyme (Fig. 2) the RR spectrum shown in Fig. 3 is
generated. Fig. 3 demonstrates that the carbonyl of the acyl group; the
ethylenic linkage and the C-O bond (from the linkage being cleaved) can be
monitored via the features near 1700, 1612.5 and 1263 cm^{-1}, respectively.

Since the positions of these spectral features can be related to chemical
effects, for example how the strength of hydrogen bonding about the
carbonyl affects $\nu_{C=O}$, the carbonyl stretching mode, the RR spectrum is a
rich source of chemical information on the enzyme-bound substrate (5).

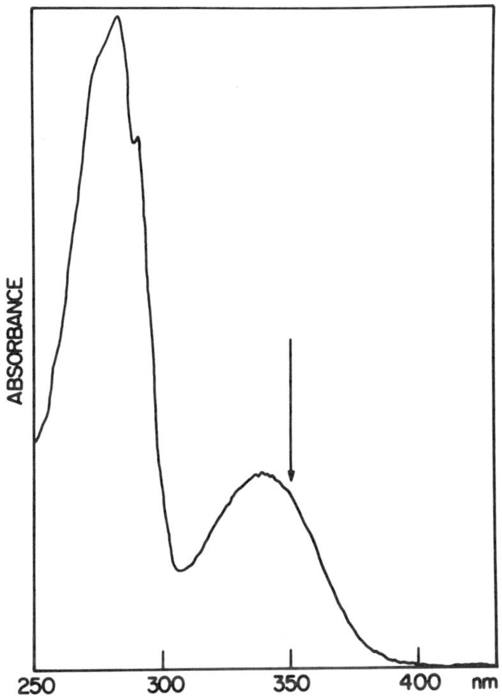

Figure 1. The absorption spectrum of 5-methylthienylacryloyl-chymo-
trypsin. Laser excitation wavelength for RR spectroscopy is indicated by
arrow.

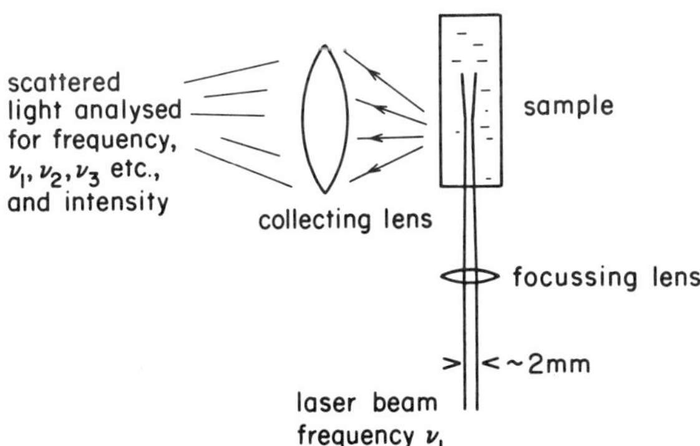

Figure 2. Schematic of a Raman experiment. The scattered light at
frequencies different from the incident beam, viz. ν_2, ν_3 etc, constitutes
the Raman spectrum. A large number of different experimental
configurations are possible e.g. flow cells, cryostats. Sophisticated
optical and photon detection equipment is used to detect the feeble Raman
or resonance Raman spectrum.

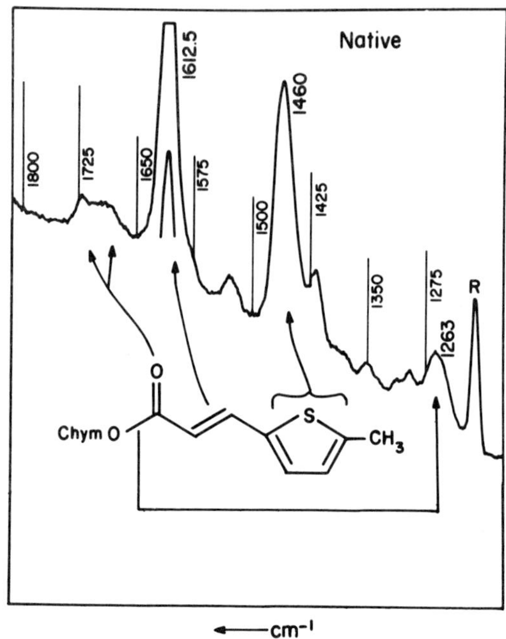

Figure 3. RR spectrum of 5-methylthienylacryloyl-chymotrypsin.

COMPARISON OF RESONANCE RAMAN, ABSORPTION AND FLUORESCENCE SPECTROSCOPIES
FOR CHARACTERISING ENZYME-SUBSTRATE COMPLEXES

Each of these techniques makes a substantial contribution to our
understanding of enzyme structure and dynamics and a comparison of the
three methods is of interest since the physical processes underlying each
are related. The model used to depict absorption, fluorescence and Raman
processes is shown in Figure 4. It is based on a quantised description of
molecular energy where the molecular energy E_{mol} can be written as a sum of
terms

$$E_{mol} = E_{elec} + E_{vib}$$

where the subscripts refer to electronic and vibrational components,
respectively of the total energy. For present purposes the contributions
due to molecular rotation and translation can be ignored. Electronic
energy transitions involve much large quantities of energy than vibrational
transitions do, with values of 10,000-50,000 cm⁻¹ for the former and 10-
4000 cm⁻¹ for the latter and this situation is reflected in the energy
level spacings in Figure 4.

The energy level situation seen in Figure 4 may obtain for a small
molecule in the gas phase or, to a good approximation, for a chromophore in
a complex biological environment e.g. for a porphyrin ring in a heme
protein. The population of molecules within the energy levels is governed
by a Boltzmann distribution. Therefore, under normal conditions
essentially all molecules are in the ground electronic state and most are
in the ground vibrational state of the ground electronic state.
Absorption, fluorescence and resonance Raman processes are understood in

176

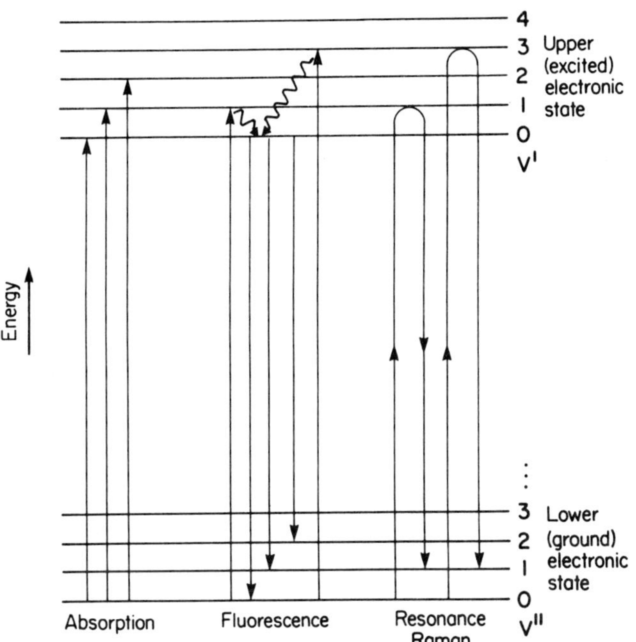

Figure 4. Absorption, fluorescence and the resonance Raman effects. Some
of the possible consequences of a photon-molecule interaction.

terms of interactions of photons ('particles' or quanta of light) with the
energy levels schematised in Fig. 4.

In the absorption process a photon is absorbed by the molecule,
raising the molecule to a vibrational level in the upper electronic state.
This process occurs on the ~10^{-15}s time scale and is depicted by the single
upward arrow in Fig. 4.

In the fluorescence process a photon is absorbed by the molecule,
raising the molecule to a vibrational level in the upper electronic state.
Relaxation processes occur within the excited state and the molecule
remains in the lowest vibrational state of the upper electronic state for
~10^{-9}s (Fig. 4). The molecule returns to the ground state by emitting a
photon-which, of course, is of lower energy (red-shifted) vis à vis the
absorbed photon (Fig. 4).

The RR process is a scattering event whose time scale is <10^{-11} s (3).
It involves the interaction between a photon, with energy corresponding to
the gap between the excited and ground electronic states, and a molecule
such that 'absorption' is followed by prompt re-emission of the photon.
The selection rules for the process are that the molecule changes in energy
by a single vibrational quantum number in the ground electronic state. The
complementarity between photon energy and electronic energy gap ensures
that the scattering probability is greatly increased and that an intense RR
spectrum associated with the chromophore dominates the scattering
spectrum.

The key question now is "What do we measure when we use one of these
three optical spectroscopies to probe an enzyme system"?

<u>Absorption</u>

The measurement (the absorbed photon) represents the difference between ground and excited electronic states, and therefore the measured quantity is a property of both states.

<u>Fluorescence</u>

The measurement (the emitted photon) also reflects the difference between electronic ground and excited state properties and the wavelength of emission is a property of both states. Importantly emission occurs on the nanosecond (10^{-9}s) time-scale which provides access to many important dynamical protein events occurring in this time frame (6).

<u>Resonance Raman</u>

The measurement consists of a number of features in the resonance Raman spectrum whose position is a vibrational transition in the ground electronic state. The intensities of RR peaks are a property of both ground and excited electronic states. Hence a major advantage of the RR approach is via <u>peak positions</u> it provides a direct probe of the electronic ground state, unencumbered by excited-state considerations (3,5). This assumes special importance when we are using chromophoric substrates to study enzyme catalysis since in the vast majority of reactions the latter occurs solely in the ground electronic state.

<u>Obtaining structural information by absorption and resonance Raman</u>

A number of absorption studies have been carried out by Bernhard and co-workers (7) which attempt to derive molecular information on scissile linkages in chromophoric enzyme-substrate complexes using substrates based on cinnamoyl- or indoleacryloyl-(IA) groups. A comparison of the absorption and RR properties of IA-chymotrypsin illustrates the strengths and weaknesses of both these approaches. Fig. 5 demonstrates a common finding - that the absorption maximum of IA-chymotrypsin is red-shifted compared to IA-methyl ester in water. Several molecular explanations have been put forward to explain the red-shift:

(i) structural deformation of the chromophore which converts the acylenzyme ester linkage into the electronic equivalents of an aldehyde or ketone

(ii) protonation of, or strong hydrogen binding to, the carbonyl oxygen of the acyl derivative

(iii) excited state stabilisation by the active site's electrostatic field (8)

The strength of the absorption spectroscopic approach is that it is relatively easy and inexpensive to obtain data and the method provides a facile means of monitoring the formation of an acyl enzyme. The weakness is that from a single piece of datum, the red-shift, it is difficult to extract molecular information.

The situation is reversed for the RR method, which requires sophisticated instrumentation and technical skill in data acquisition but which can yield quite detailed molecular information. In the RR spectrum of IA-chymotrypsin (Fig. 6) $\nu_{C=O}$ in the enzyme is observed as a single band at 1704 cm^{-1}. This lies between the $\nu_{C=O}$ values for IA methyl ester in H$_2$O

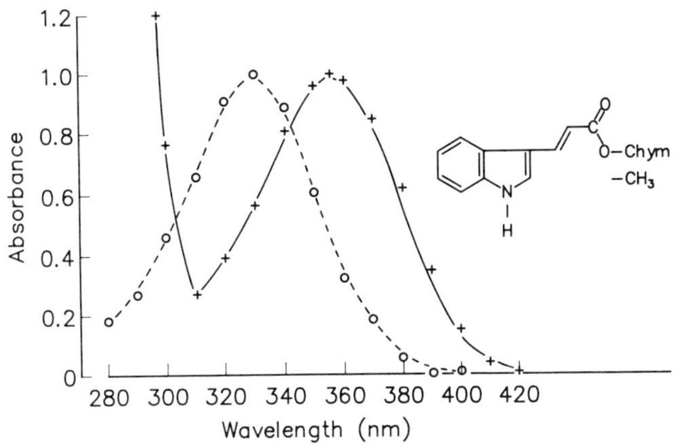

Figure 5. UV-red shift observed on binding indoleacryloyl chromophore to
chymotrypsin. Methyl ester in water (- - - -); acyl chymotrypsin (———).

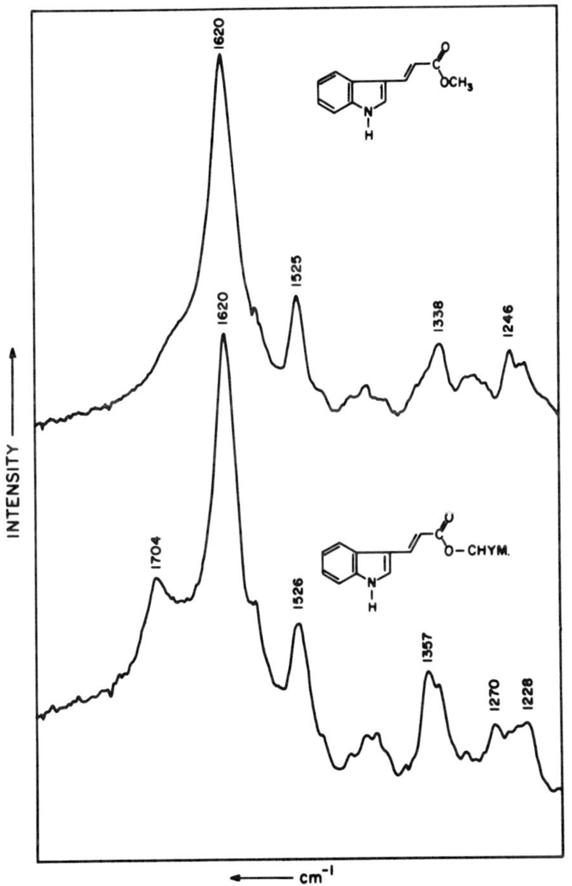

Figure 6. RR spectra of indoleacyloyl-methyl ester in H_2O (top) and
-chymotrypsin (bottom). λ_{ex} = 324 nm.

(1685 cm^{-1}, by FTIR) and in CH_3CN (1709 cm^{-1}). As $\nu_{C=O}$ for an aldehyde-like carbonyl is expected below 1685 cm^{-1}, the RR data strongly suggests that the acyl-group's carbonyl is present in an ester (not aldehyde- like) grouping, ruling out possibility (i) above.

The ethylenic stretching vibration $\nu_{C=C}$ is at 1620 cm^{-1} for both IA-chymotrypsin and IA methyl ester in H_2O (Fig. 6). This suggests that the IA moiety is in the same conformation in the C-O-C linkages in H_2O as in the enzyme (s-<u>trans</u>). Comparison of the $\nu_{C=O}$ values indicates therefore that the IA carbonyl group is less strongly hydrogen bonded in the enzyme (1704 cm^{-1}) than it is in H_2O (1685 cm^{-1}).

Thus, there is no evidence for an aldehyde-like or strongly hydrogen bonded carbonyl group in IA chymotrypsin. The observed red shift in λ_{max} must be due to excited state stabilisation. This is consistent with Warshel's proposal that the protein's electrostatic field stabilises the build-up of negative charge on the carbonyl oxygen in the excited state of the acyl group (8).

Red shifts are a very common phenomenon for protein-bound chromophores (9) and since excited states usually involve greater charge separation than ground states, excited-state stabilisation by the protein is a prime candidate to explain all, or most, observed red-shifts. Why, however, the chromophore orientation in the protein is always correctly configured for stabilisation is not obvious at this time. It might be thought that the protein's electrostatic field in some binding configurations would oppose charge-separation in the excited state and, of course, on the time scale of the absorption experiment, 10^{-15}s, solvent (i.e. protein) reorganisation about the excited state cannot occur.

<u>Dynamical information from fluorescence spectroscopy</u>

While the major emphasis from RR studies is on structural conclusions, fluorescence spectroscopic analysis is mainly concerned with dynamical information (6). A pertinent example concerns mapping the mobility of a portion of an inhibitor bound to the active site of the cysteine protease papain. Papain is known to have an extended active site accomodating up to four residues on the acyl side (ref. 2 p. 29). With this is mind, inhibitors were synthesised which placed a fluorescent probe progressively further from the active site residue Cys 25 with the strategy of monitoring the dynamics of the probe at varied distances. Fluorescence studies were performed on papain bound to active-site-directed inhibitors of the type mansyl-(Gly)$_n$-Phe-glycinal, where n=0,1,2 (10). Mansyl [6-(N-methylanilino)-2-napthalene sulfonyl] was used as the fluorescence probe. It was found that the fluorescence of mansyl-Phe-glycinal was greatly enhanced upon binding to the enzyme. Moreover measurement of fluorescence polarisation and rotational relaxation time (ρ= 29 nsec) suggested that the mansyl probe does not have appreciable rotation independent of that of the papain-ligand complex. Mansyl-Phe-glycinal meets papain's known propensity for Phe in the P_2 position (ref. 2 p. 29) and it is very likely that the Phe-ring - active site contacts taken together with inhibitor-enzyme -NH...O=C H-bonding contacts essentially anchor the mansyl to the enzyme. The situation for the other two inhibitors is quite different, however. The fluorescence of mansyl-Gly-Phe-glycinal (n=1 or 2) is not enhanced upon binding although both are equally as good inhibitors as the n=o compound. Moreover the longer inhibitors exhibits less fluorescence polarisation and shorter rotational relaxation times (6 and 3 nsec for n=1 and 2, respectively) compared to mansyl-Phe-glycinal. These results suggest that the mansyl groups on the larger inhibitors are not effectively anchored to the enzyme and, instead, protrude into solution. Obviously any contacts made by the peptide

linkages associated with additional (n=1 or 2) glycines do not markedly
limit the mobility of the mansyl group. This result makes an interesting
comparison with some RR data where it is shown that structural changes
occur in the scissile cysteine-to-acyl linkages of acyl enzymes formed from
PheGly compared to Gly-based substrates but that the results from Gly
PheGly and GlyGlyPheGly are identical to PheGly - again suggesting that the
additional glycines in the P_3 and P_4 sites are not causing observable
perturbations to the system (see Fig. 12).

RR STUDIES USING SUBSTRATES BASED ON DELOCALISED Π-ELECTRON SYSTEMS

The acyl groups of these substrates are chromophoric by virtue of the
fact that a delocalised π-electron system is formed by a ring system
conjugated to an acryloyl group (Fig. 7).

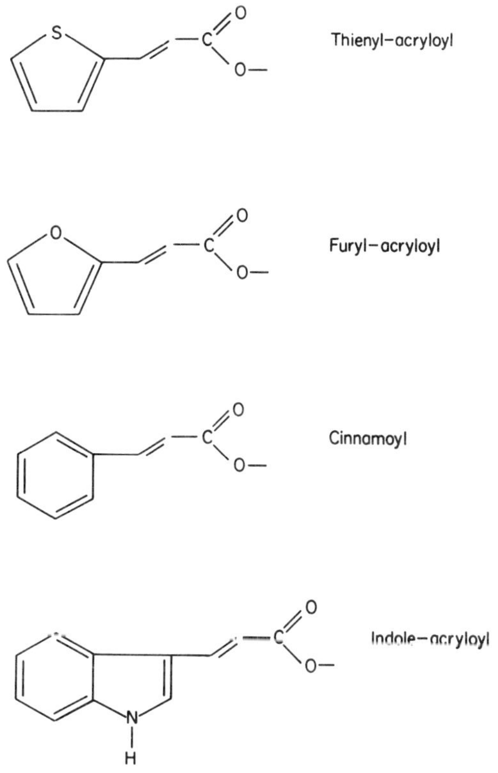

Thienyl-acryloyl

Furyl-acryloyl

Cinnamoyl

Indole-acryloyl

Figure 7. Chromophoric acyl groups.

Examples of thienyl-acryloyl and indole-acryloyl chymotrypsins have
already been given (Fig. 3 and 6, respectively). From the above discussion
centering on these acyl-enzymes is apparent that the RR data provides
detailed molecular information on the structure of the bound acyl group
(3,5,9). The data can also provide considerable insight into active site
forces and we now consider two examples, (i) the activation of the serine
protease catalytic mechanism and (ii) the strong electron polarisation
forces in cysteine protease active sites, which illustrate how active site
effects can be probed. Of course, the RR approach has considerable
potential for quantitating the changes in an active site, at the structural
and electrical levels, brought about by site selected mutations.

(i) <u>Probing the activation of serine protease catalytic mechanism</u>

The hydrolysis, or deacylation, of acyl-enzymes based on serine proteases such as chmyotrypsin is facilitated by residues aspartic acid-102 and histidine-57. At low pH His-57 is protonated and deacylation does not occur - this can be used to prepare a stable population of acyl enzymes by running the enzyme substrate reaction near neutral pH and subsequently dropping the pH to 3.0. Fig. 8 demonstrates that the effect on the key bonds of deprotonating His-57 can be monitored in the RR spectrum. The RR data is obtained by mixing stable acyl enzyme (pH 3.0) with base in a flow system and monitoring the unstable acyl enzyme in the flow cell before deacylation occurs. For the stable acyl enzyme, $4NH_2 3NO_2$ cinnamoyl-chymotrypsin at pH 3.0 the carbonyl stretching frequency $\nu_{C=O}$ is

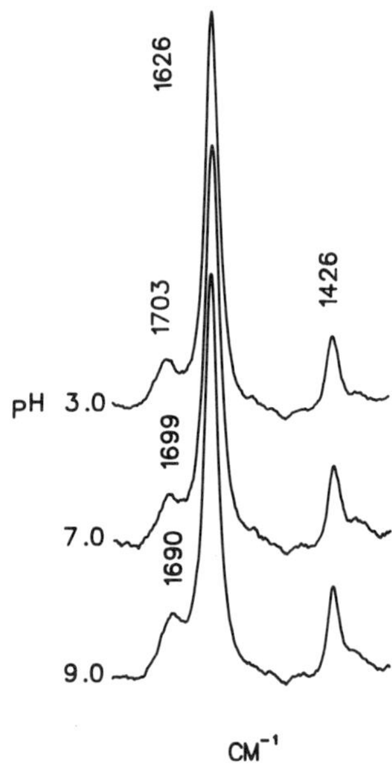

Figure 8. Changes in the carbonyl stretching region (near 1700 cm^{-1}) upon activating 4-NH$_2$,3-NO$_2$cinnamoyl-chymotrypsin. The intense feature at 1626 cm^{-1} is due to $\nu_{C=C}$.

at 1703 cm^{-1}. Upon activation this shifts to 1690 cm^{-1} (Fig. 8) with a pKa of 7.2, - which is identical to the deacylation pK$_a$ determined by enzyme kinetics. The shift in $\nu_{C=O}$ indicates that the carbonyl is more polarised at active pH and this could come from one of, or a combination of, two causes.

1. The deprotonation of the imidazole ring of His-57 leads to binding and correct orientation of a water molecule as a prelude to deacylation:

$$\text{Im : H-O} \overset{\displaystyle \text{acyl} \atop \displaystyle |}{} \;\; \overset{\displaystyle |}{\underset{\displaystyle \text{O-Enz}}{\underset{\displaystyle |}{\text{C=O}^{\delta-}}}}$$

2. Stronger hydrogen bonding, e.g. in the oxyanion hole, brought about by
a minor conformational change:

In any event polarisation of the C=O group is catalytically
advantageous since it renders the C=O more prone to nucleophilic attack and
makes the C=O more C^+-O^--like - which is closer to the transition state
(and the tetrahedral intermediate) for deacylation.

ii) <u>Strong electron polarisation in cysteine protease active sites</u>

In the examples discussed above involving serine proteases the RR
spectrum of the acyl-enzyme bears a fairly close resemblance to the RR
spectrum of a model compound such as the corresponding methyl ester e.g. in
Fig. 6 the RR specra of indoleacrylolyl-chymotrypsin and -methyl ester are
quite similar. However, early studies on acyl papains revealed remarkable
differences between enzyme and model spectra (11). The RR data from
4-dimethylamino, 3-nitrocinnamoyl-papain are shown in Figure 9. Figure 9
demonstrates that the RR spectrum of the acyl-papain is distinct from that
of the product. The active site obviously produces drastic changes in the
properties of the acyl residue. As can be seen in Figure 9, the spectrum

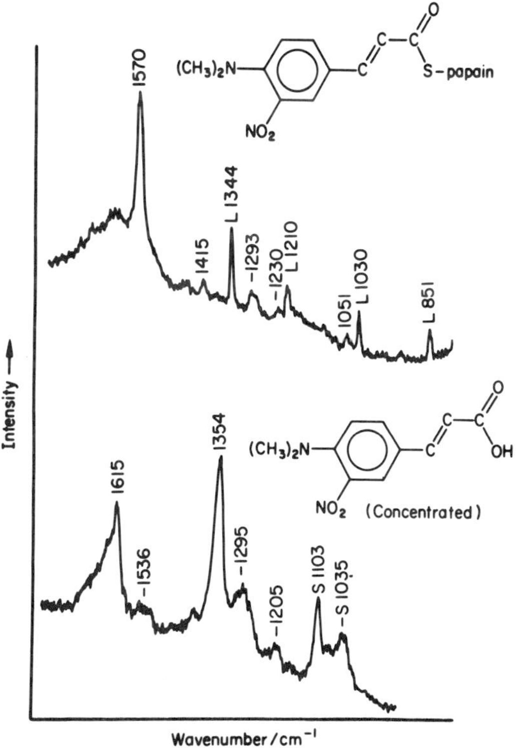

Figure 9. Comparison of the RR spectra of the papain-substrate
intermediate and product based on the 4-dimethylamino-3-nitrocinnamoyl
chromophore. L and S denote laser lines and solvent peaks, respectively.
441.6 nm excitation.

of the product is dominated by bands due to ethylenic and ring modes in the 1600-cm^{-1} region and a nitro feature near $1350\ \text{cm}^{-1}$. In contrast, the spectrum of the acyl papain shows a very intense peak at $1570\ \text{cm}^{-1}$. There is little evidence for peaks from the product in the spectrum of the intermediate or vice versa.

The major differences in the RR spectra of the cinnamoyl chromophore shown in Figure 9 are interpreted in terms of the polarization of π electrons in the bound cinnamoyl group (3,11). A series of model compounds, based on the imidazole esters of cinnamic acid, mimic the absorption and RR properties of the acyl papain studied. The crucial property of the imidazole esters is that they have a very strong electron-attracting group (imidazole) attached to the carbonyl and a strong electron-donating group (e.g., para-dimethylamino) at the other extremity of the cinnamoyl skeleton. Thus, acting in concert through the chemical bonds, these groups set up a highly polarized π-electron system. It is proposed that essentially the same sort of electron polarization occurs in the acyl group bound at the active site. However, in the active site the polarization probably occurs indirectly, such as by interaction of the acyl residue with charged amino acid side chains or a dipole due to the α-helix terminating at cysteine 25 of papain.

For both serine and cysteine proteases it will be of high interest to determine how site selected changes affect the RR results obtained for the wild-type enzymes.

SUBSTRATES WHICH GENERATE DITHIOESTER CHROMOPHORES

An important approach to the study of enzyme action by the RR technique is to create a dithioester chromophore at the time and place of catalysis - as the enzyme and substrate come together in the active site (12). This allows us to focus on the critical region of a large molecular complex at the critical time. A prototype reaction involving chromophore creation is the hydrolysis of thionoesters by the cysteine proteinase papain:

$$
\begin{array}{l}
\overset{O}{\overset{\|}{}}\ \overset{S}{\overset{\|}{}} \\
RCNHCH_2COCH_3 + \text{HS-papain}
\end{array}
$$

$k_s \updownarrow$

$$
\begin{array}{l}
\overset{O}{\overset{\|}{}}\ \overset{S}{\overset{\|}{}} \\
[RCNHCH_2COCH_3 \cdot \text{HS-papain}]
\end{array}
\qquad \text{Michaelis Complex}
$$

$k_2 \downarrow$

$$
\begin{array}{l}
\overset{O}{\overset{\|}{}}\ \overset{S}{\overset{\|}{}} \\
RCNHCH_2C\text{-S-papain} + CH_3OH
\end{array}
\qquad \text{Dithioacyl enzyme} \quad \lambda_{max}\ 315\ nm
$$

$k_3 \downarrow H_2O$

$$
\begin{array}{l}
\overset{O}{\overset{\|}{}}\ \overset{S}{\overset{\|}{}} \\
RCNHCH_2C\text{-OH} + \text{HS-papain}
\end{array}
$$

In this reaction mixture a transient dithioester is formed from the
C=S group of the substrate and the HS- group from cysteine-25 in papain's
active site. The dithioester chromophore has a λ_{max} near 315 nm and is the
only species in solution with an electronic absorption band to the red of
300 nm. Thus, it is possible to excite specifically the RR spectrum of the
dithioester group. The resulting RR spectrum is very sensitive to
conformational events in the $RC(=O)-NH-CH_2-C(=S)-S-CH_2$ bonds and, combined
with X-ray crystallographic and kinetic studies, has been used extensively
to elucidate the details of catalytic events (9,13 references therein).

As can be seen in Fig. 10 by excitation with the 324 nm Kr^+ line, the
intermediate gives rise to many bands in the 500-1200 cm^{-1} range. The
peaks in Fig. 10 contain contributions from the stretching and other
motions of the C=S and C-S-C bonds, and thus intense modes from the
catalytically crucial bonds are observed in the RR spectrum. Moreover, the
reaction scheme can be generalized to use, for example, a polypeptide
sequence as substrate and any enzyme forming a transient ester linkage
involving an active-site thiol. The "natural" reaction involves the
formation of the -C(=O)S-group, so the RR experiment involves a single atom
replacement-sulfur for oxygen. Kinetic studies (14) have shown that the
reaction mechanisms involving "natural" esters and "model" thionoesters are
the same; k_{cat}/K_m's are similar for the two classes of substrate and k_{cat}'s
are 20-30 times slower for the thionoesters. The spectroscopic data are
obtained by examining a reaction mixture (in contrast to the examples
discussed above where the acyl enzymes had to be separated from substrate
and product) that contains an excess of substrate over enzyme; since for
ester substrates acylation is faster than deacylation, there is buildup of
a quasi-steady-state population of intermediate.

Spectroscopic and crystallographic investigations of $\underline{N}$-acylglycine
dithioesters, $RC(=O)NHCH_2C(=S)SC_2H_5$, have provided the key to understanding
the RR spectra of the enzyme-substrate transients. Approximately fifteen
different dithioesters of this type, with differing R groups, have been
analysed to date. The model $\underline{N}$-acylglycine ethyl dithioesters give rise to
intense RR bands in the 500-700 and 1000-1200 cm^{-1} regions of the spectrum.
In solution, the relative intensities of the bands were found to be very
sensitive to temperature and solvent. These facts, taken with other
considerations, indicated the presence of more than one conformer. It was
found that in aqueous or acetonitrile solutions there are two major
conformational states, designated conformers A and B (shown in Fig. 11),
and that each conformer has a characteristic and separate RR spectrum in
both the 600 and 1100 cm^{-1} regions. In order to form an exact description
of conformers A and B, combined X-ray crystallographic-Raman analyses were
undertaken on single crystals of $\underline{N}$-acylglycine ethyl dithioesters. The
combined X-ray crystallographic-RR approach could, in turn, be used to
understand the solution RR spectra of the $\underline{N}$-acylglycine dithioesters and
the RR spectra of the dithioacyl papains. In keeping with the higher
thermodynamic stability of conformer B, most of the $\underline{N}$-acylglycine
dithioesters crystallize in this form.

Figure 10 compares the RR spectrum of crystalline
$PhC(=O)NHCH_2C(=S)SC_2H_5$ with the RR spectrum of this molecule in H_2O. In
solution (top spectrum), the spectral signatures of forms A and B are
present, but for the crystals only the signature of conformer B is found,
e.g. in the upper trace in Figure 10 most of the intensity of the 1165 cm^{-1}
peak is due to a conformer A mode which is absent in the spectrum of the
crystalline material. An X-ray diffraction analysis of the crystal
provides an accurate structure of conformer B, and this is shown in Fig.
11. Figure 11 also depicts the crucial fragment ($-C(=O)NHCH_2C(=S)SC_2H_5$)
from an A conformer found for N($\underline{p}$-nitrobenzoyl)glycine ethyl dithioester
crystals. The major conformational difference between conformers A and B

185

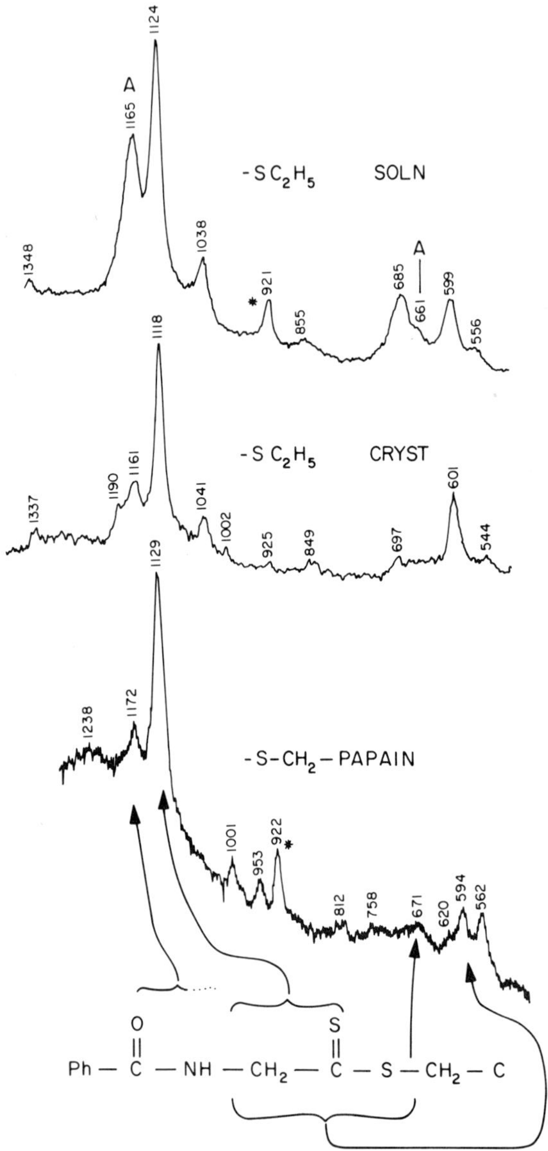

Figure 10. RR spectra (324 nm excitation) of <u>N</u>-benzoylglycine ethyl dithioester in aqueous solution and in the crystalline state and of the corresponding <u>N</u>-benzoylglycine dithioacyl papain. Conformationally sensitive peaks are indicated.

186

is a rotation of ~150° about the C(3)-C(4) bond. The rotation about the
C-C linkage changes the vibrational coupling in and about the dithioester
moiety and accounts, at least in part, for the different Raman spectral
signatures of conformers A and B. The RR spectra of N-acylglycine
dithioacyl papains, e.g. the RR spectrum of PhC(=O)NHCH$_2$C(=S)S-papain seen
in Fig. 10,demonstrate unequivocally that between pH 4 and 9 the great
majority of acyl groups assume a B-type conformation. In Fig. 10 the
intense peak at 1129 and the peak at 594 cm^{-1} are modes characteristic of
a B conformer - the corresponding peaks are seen in the RR spectrum of the
crystalline ethyl dithioester. The situation is the same for every
N-acylglycine papain studied. There is a very close correspondence between
the RR spectrum of each dithioacyl papain and that of its corresponding
ethyl ester in the B form. All major peaks and most minor peaks in the
acyl enzyme spectra are due to conformer B. Although it is possible that
some minor peaks are due to a small population of non-B conformers, there
is no convincing evidence to support this notion. However, denaturation of
the dithioacyl enzymes below pH 3.0 does result in conformer A peaks
appearing in the spectrum and, under these conditions, the conformational
population of the covalently linked acyl group reverts to that found for
the corresponding ethyl ester. Thus, the first conclusion to emerge is
that the native active site exerts conformational selection; it binds one
of the conformational states available to N-acylglycine dithioesters.

Substrates based on di- tri- and tetra-peptides: conformational activation
in the active site

For N-acylglycine dithioacyl papains several studies have been
undertaken to determine whether substrate strain occurs in the active site.
One way to approach this question is to take the most stable conformation
of the model N-acylglycine ethyl dithioesters in H$_2$O/CH$_3$CN solution,
conformer B, as the standard relaxed state and to compare its spectroscopic properties with those of the corresponding dithioacyl enzyme. This
technique was used for N-benzoylglycine and N-(β-phenylpropionyl)glycine
dithioacyl papains, their corresponding ethyl dithioesters and a number of
^{13}C, D and ^{15}N substituted analogues. Comparison of the model dithioester
RR spectra and those of the dithioacyl enzymes led to the conclusion that
major substrate strain does not occur in the acyl group of N-acylglycine
dithioacyl papains. For example, no evidence was found for pyramidalisa-
tion at the C=S carbon atom or for energetically expensive distortion about
the torsional angles in the -NH-CH$_2$-C(=S)S-CH$_2$-bonds. Recently, additional
studies have indicated that upon going to larger, more specific substrates
there is a change in the -CH$_2$-C(=S) torsional angle (15).

Figure 12 compares the RR spectra of a series of dithioacyl papains
starting with a substrate based on a single amino acid (top) and
progressing to a tetrapeptide based substrate (bottom) which has the
potential to completely occupy the S$_1$-S$_4$ sites in the active site. Phenylalanine was placed at the P$_2$ position of the substrate to meet papain's
known requirement for a hydrophobic residue at this point. The positions
of the major RR features near 1141 and 590 cm^{-1} for the bottom three
spectra are identical but these positions vary by 3-4 cm^{-1} compared to the
corresponding peaks in the top spectrum.

The significance of the modest shift in the feature near 1140 cm^{-1} is
demonstrated by X-ray crystallographic and RR analyses of the model
compounds C$_6$H$_5$(CH$_2$)$_2$C(=O)NHCH$_2$C(=S)S-C$_2$H$_5$ and CH$_3$OC(=O)PheNHCH$_2$C(=S)S-C$_2$H$_5$.
Table 1 compares the values of the '1140 cm^{-1}' peak for the models in
aqueous solution and is the crystalline phase. In addition Table 1 gives
the crystallographically derived result for Φ', the NHCH$_2$-CS(thiol)
torsional angle.

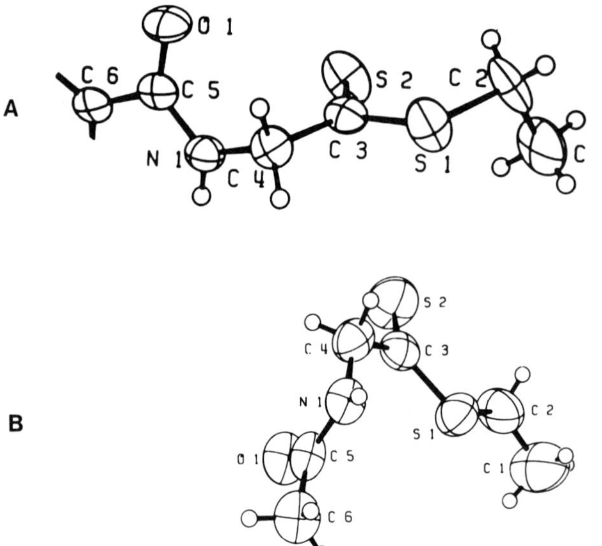

Figure 11.   Conformers B and A.

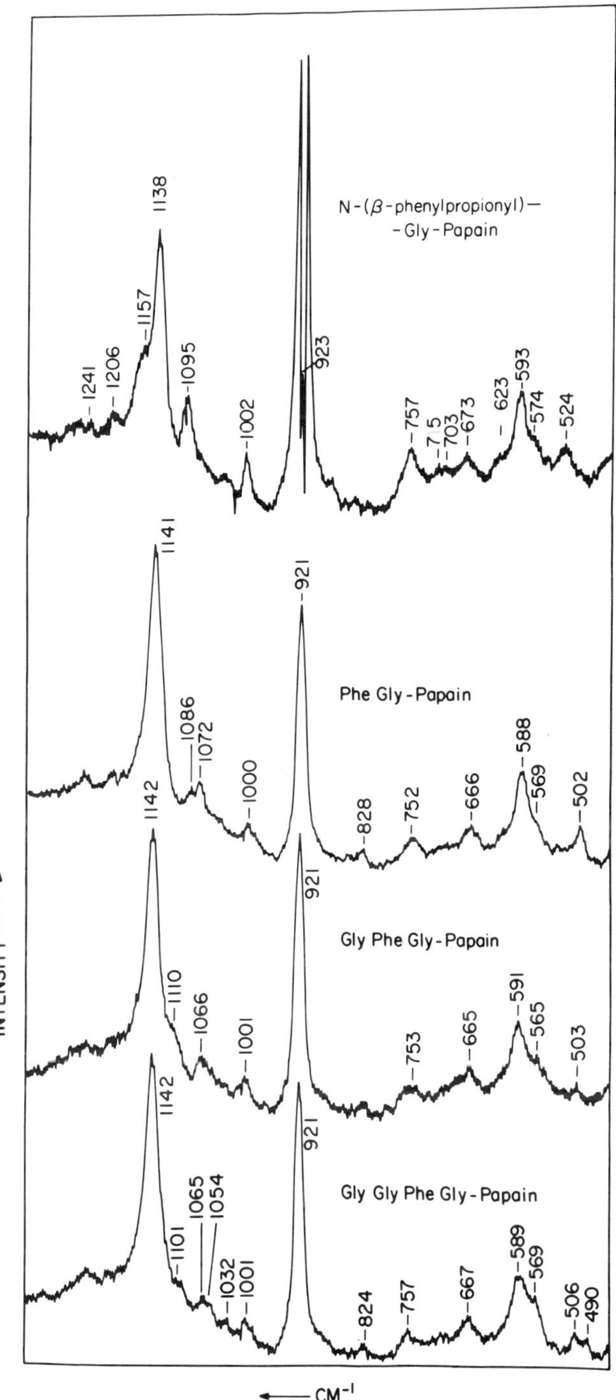

Figure 12. The 324 nm excited RR spetra of the dithioacyl papains:

$C_6H_5(CH_2)_2C(=O)NHCH_2C(=S)S$-papain (top)
$CH_3OC(=O)PheNHCH_2C(=S)S$-papain
$CH_3OC(=O)GlyPheNHCH_2C(=S)S$-papain
$CH_3OC(=O)GlyGlyPheNHCH_2C(=S)S$-papain

Table 1. Comparison of exact position of "1140 cm^{-1} band" can be used to estimate Φ' in dithioacyl-papain.

	dithioester 1140 cm^{-1} band in H$_2$O	dithioester crystal	In acyl enzyme	Φ' from crystal
$C_6H_5(CH_2)_2C(=O)NHCH_2C(=S)S$	1132 cm^{-1}	1134	1138	$-18.9°$ (16)
$CH_3OC(=O)PheNHCH_2C(=S)S$	1133	1137	1141	-29.6 (17)

The data from the crystalline compounds provide a key to the relationship between the exact position of the 1140 cm^{-1} peak and Φ' since a 10° increase (in the negative direction) brings about a 3 cm^{-1} increase in the intense 1140 cm^{-1} feature. If we now assume that the B-conformer in solution for both the PheGly and β-phenylpropionyl model compounds is a 'relaxed' standard state with Φ' in the ±10-$20°$ range we see that Φ' for the β-phenylpropionyl lies in the 20-30° region while Φ' for the larger, more specific substrate based on PheGly is 40-50°. Thus, by extending the size of the substrate and meeting papain's specificity requirement significant distortion of the Φ' torsional acyl occurs away from a characteristic B conformer value. Since the proposed mechanism for deacylation involves breaking or weakening the N...S contact in conformer B (Fig. 13) it can be seen how increasing Φ', which will 'open up' the N...S contact, will make the acyl enzyme more like the transition state for deacylation and result in conformational activation. Thus, the observed 3 fold increase in k_3 (15), the rate constant for deacylation for PheGly, GlyPheGly and GlyGlyPheGly (k_3s, 0.5-0.7 s^{-1}) compared to the N-(β-phenyl-propionyl)glycine dithioacyl enzyme (k_3=0.16 s^{-1}) may be accounted for by favourable changes in Φ'.

<u>Dynamical information from dithioacyl-papains at low temperature</u>

Since the dithioacyl group's RR spectrum is sensitive to changes in the force field in the active site, variable temperature studies provide an excellent opportunity to probe dynamical events such as the effects of stopping solvent bombardment upon freezing, phase transitions in the enzyme, and thermal activation and energy migration in the active site.

Recently we constructed a cryogenic apparatus which provides very high quality RR data from enzyme-substrate complexes in the 10°K range. The initial results have been obtained at liquid nitrogen temperatures 77°K. These studies taken with earlier observations on frozen dithioacyl papains at 230-240°K have led to the following findings.

The effects of freezing is different for poor substrates or good substrates e.g. N-benzoylglycine or PheGly (which meets papain's specificity requirements), respectively. Upon freezing at -20 to -40°C the 1128 cm^{-1} feature of N-benzoylglycine dithioacyl papain shifts to 1125 cm^{-1}. This in itself is not surprising since small shifts in vibrational frequencies often accompany phase changes. However, for PheGly dithioacyl-papain the intense 1141 cm^{-1} band remains unaffected by freezing of the solvent. Thus, the additional active-site acyl group contacts made by this substrate insulate it from the effects of solvent solidification. By the same token it is apparent that freezing the solvent does not bring about a conformational change in PheGly-papain - the latter must be highly resistant to solvent changes.

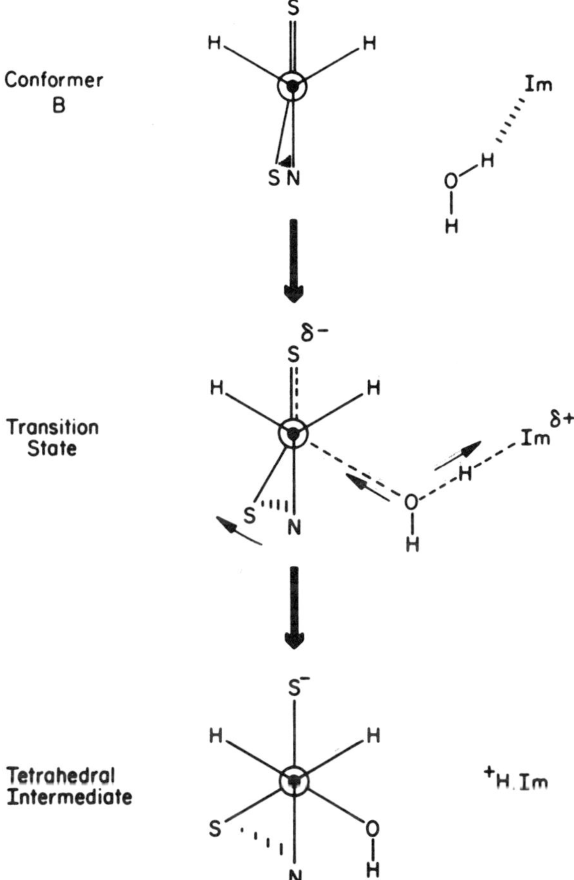

Figure 13. Changes occuring near the catalytic centre as the acyl enzyme
is converted to the tetrahedral intermediate for deacylation.

Upon taking the acyl enzyme down from c. 250°K to 77°K no further change is seen in the position of the 1125 cm^{-1} N-benzoyl glycine dithioacyl papain feature (Figure 14). However, at 77°K the corresponding Phe Gly peak shifts to 1136 cm^{-1} compared to 1140 cm^{-1} at room temperature or at 250°K (Figure 15). Thus the effect of lowering the temperature is to reduce Φ by ~20° (see previous section) along the coordinate corresponding to kinetic deactivation. Whether this reduction is accomplished smoothly from 250 to 77°K, or results from a sudden phase transition is not known at this time. However, it is apparent that thermal inactivation makes the dithioacyl enzyme less distorted about Φ' and more relaxed or model like (previous section).

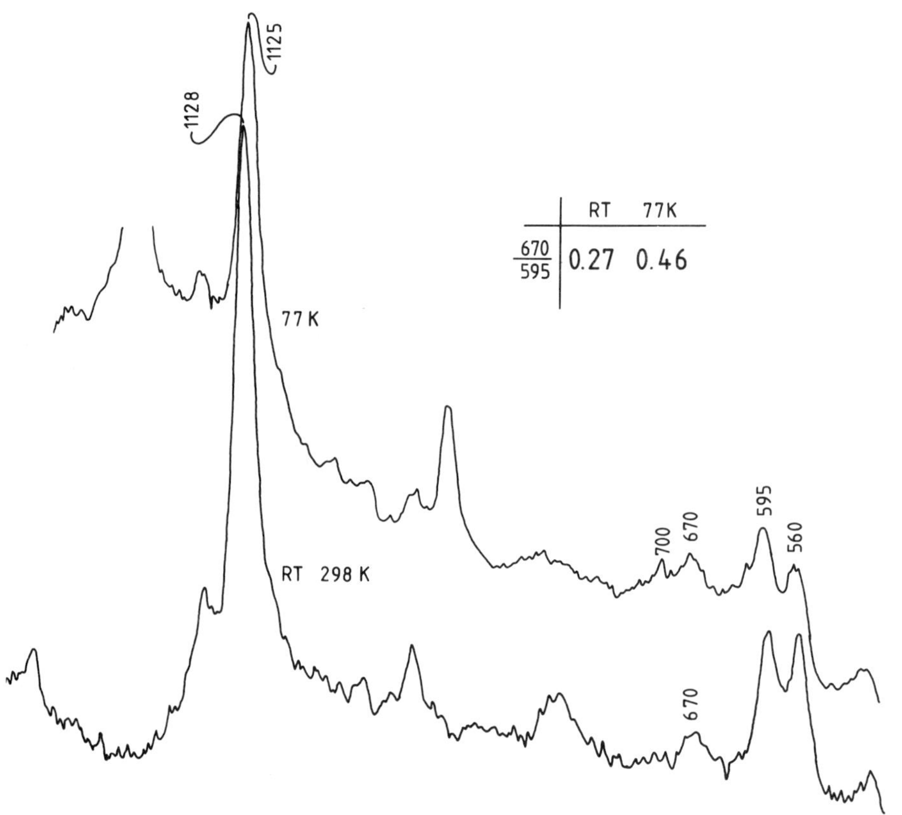

Figure 14. Comparison of the 324 nm excited RR spectra of N-benzoylglycine dithioacyl papain at 298°K (RT) and 77°K.

Reducing temperature might be expected to narrow RR line shapes due to reduced heterogenous broadening (the populations in the active site becomes more homogenous due to fewer lower energy thermal excursions). However, at 77°K, and at the present rather low spectral resolution, no evidence was found for line narrowing.

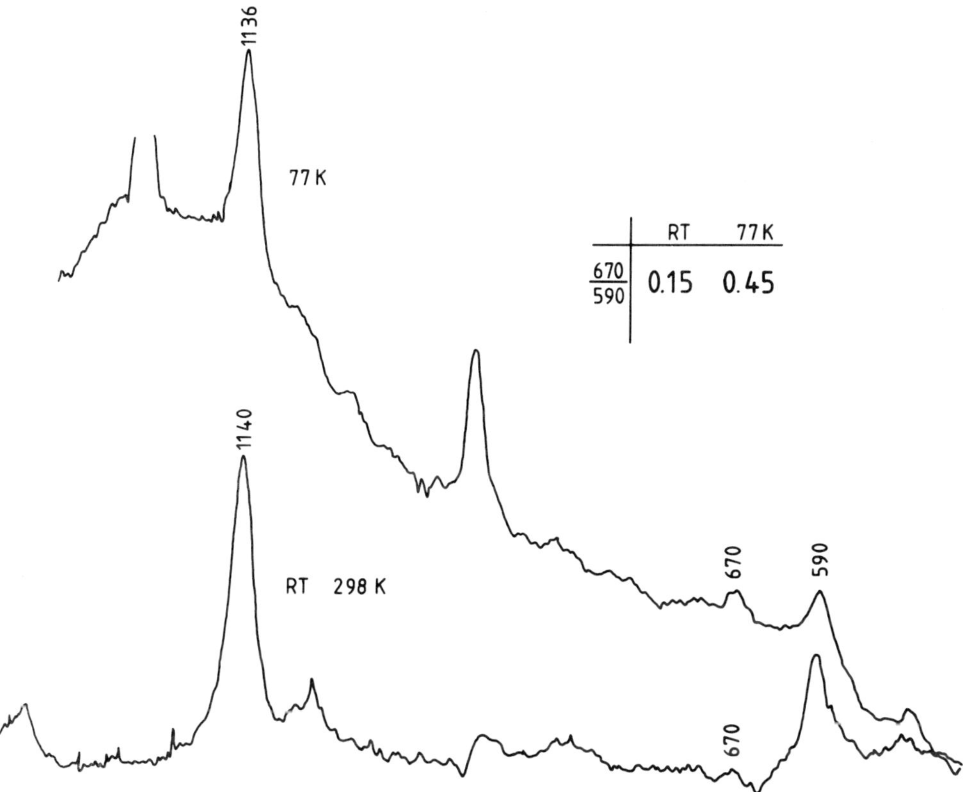

Figure 15.   Comparison of the 324 nm excited RR spectra of $CH_3OC(=O)PheGly$ dithioacyl papain at 298°K and 77°K.

A further notable change for both N-benzoylglycine and PheGly-dithio-acyl papains at 77°K concerns the spectral region near 680 cm^{-1}. At room temperature both acyl enzymes show a weak band near 670 cm^{-1}. This has been assigned to a normal model with a high degree of S-C (from cysteine 25) stretching character, that <u>it is due to a bond associated with the enzyme</u>. As has been noted a number of times the peak occurs at a position markedly lowered compared to that found for ν_{s-c} from a model compound. Upon freezing little change is observed in the 680 cm^{-1} region but on lowering the temperature to 77°K a really marked change in the 680 cm^{-1} profile occurs. For both intermediates the relative intensity of the 680 cm^{-1} band increases 2-3 fold at low temperature (Figs. 14 and 15). Additionally, for the N-benzoylglycine derivative there is clear evidence for a second feature, suggesting conformational heterogeneity in the S-C environments, at 700 cm^{-1} (Fig. 14). Two explanations are possible for the observed intensity increase. Either there is a change in the force field about the S-C linkage such that the normal mode at low temperature better maps the changes upon going to the excited electronic state (thus, increasing RR intensity enhancement), or alternatively, some thermal process involving the S-C linkage may be 'washing out' intensity in the 670 cm^{-1} region at room temperature. The latter possibility is, of course, of special interest since the S-C bond is a property of the enzyme. Thermal processes could include large anharmonic motions which are used to explore the reaction pathway or more esoteric ideas such as non-Boltzmann equilibriation. In any event it will be of high interest to extend the measurements to liquid helium temperatures.

<u>Acknowledgement</u>

It is a pleasure to thank my collaborators who appear in the list of references. I also wish to thank Dr. Peter Tonge for his help in providing material for part of this article.

REFERENCES

1. P. D. Boyer (Ed.) "The Enzymes" Vol 3 Academic Press, N.Y. (1971).
2. A. Fersht, "Enzyme Structure and Mechanism", W.H. Freeman, N.Y. (1985).
3. P. R. Carey, "Biochemical Applications of Raman and Resonance Raman Spectroscopies", Academic Press, N.Y. (1982).
4. B. A. E. MacClement, R. G. Carriere, D. J. Phelps and P. R. Carey, Biochemistry, 20, 3438-3447 (1981).
5. P. R. Carey, "Resonance Raman Labels and Enzyme-Substrate Reactions", in "Biological Applications of Raman Spectrometry", Vol. 2, (T.G. Spiro, Ed) J. Wiley, N.Y. (1987).
6. I. D. Campbell and R. A. Dweck, "Biological Spectroscopy", Chapter 5, Benjamin-Cummings, Menlo Park, Calif. (1984).
7. S. A. Bernhard and S.-J. Lau, "Cold Spring Harbor Symp. Quant. Bol. 36, 75-83 (1972).
8. A. Warshel and S. T. Russell, Q. Rev. Biophys. 17, 283-422 (1984).
9. P. R. Carey and A. C. Storer, Ann. Rev. Biophys. Bioeng., 13, 25-49 (1984).
10. J. B. Henes, J.A. Mattis and J. S. Fruton, Proc. Natl. Acad. Sci. USA, 76, 1131-1134 (1979).
11. P. R. Carey, R. G. Carriere, D. J. Phelps and H. Schneider, Biochemistry, 17, 1081-1087 (1978).
12. A. C. Storer, W. F. Murphy and P. R. Carey, J. Biol. Chem. 254, 3163-3165 (1979).
13. P. R. Carey and A. C. Storer, Pure Appl. Chem., 57, 225-246 (1985).
14. A. C. Storer and P. R. Carey, Biochemistry, 24, 6808-6818 (1985).

15. R. H. Angus, P. R. Carey, H. Lee and A. C. Storer, Biochemistry, <u>25</u>-, 3304-3310 (1986).
16. C. P. Huber, Acta Cryst. <u>C43</u>, 902-904 (1987).
17. K. I. Varughese, C. P. Huber, A. C. Storer and P.R. Carey, ms. in preparation.

STRUCTURAL DISTRIBUTIONS, FLUCTUATIONS AND CONFORMATIONAL CHANGES IN
PROTEINS INVESTIGATED BY MÖSSBAUER SPECTROSCOPY AND X-RAY STRUCTURE
ANALYSIS

Fritz Parak

Institut für Physikalische Chemie
der Universität, Schloßplatz 4-7
4400 Münster, Federal Republic of Germany

ABSTRACT

This contribution shows how X-ray structure analysis and Mössbauer
spectroscopy can be used as complementary methods in the investigation of
protein dynamics. X-ray analysis measures structural distributions via
mean square displacements without having any time resolution. Mössbauer
spectroscopy on ^{57}Fe labels dynamics on a time scale faster 100 ns. Even
in the extrapolation to $T = 0$ K myoglobin has no well defined structure
with one energy minimum. The molecules are in different conformational
substates. At physiological temperatures the structural distribution
in the liganded (or unliganded) conformation is larger than the structural differences between these conformations. Fluctuations through conformational substates and changes of the conformation are highly correlated. The structure distribution around the heme iron influences the
oxygen binding.

Fluctuations in myoglobin envolve diffusion-like segmental motions in
a restricted space. The characteristic size of the segments is about 5 A.
Drying of the sample or freezing of the hydration water reduces these
fluctuations drastically. The Brownian type motion of segments can also
be found in membranes.

INTRODUCTION

Since more than 30 years X-ray structure analysis provides us with
the three-dimensional architecture of proteins (compare[1]). Catalytic

centers of enzymes can be studied with and without substrate and the structures of different conformations of one molecule can be compared. In this way, protein structure analysis supplied substantial contributions to our understanding of working mechanisms of enzymes. The three-dimensional arrangement of a protein provides a frame for bringing interacting groups into their necessary geometric relations. It fixes prostetic groups in space and defines the complementary shapes of enzyme and substrate[2]. However, an absolutly rigid molecule could not fulfill its function. Conformational changes from an unliganded to a liganded structure can only occur if the molecule has some flexibility. Even within one conformation as obtained by X-ray structure analysis each individual molecule of the ensemble differs slightly from the average structure. There exists a structural distribution of molecules in substates[3]. Molecules in different conformational substates perform the same biological function but vary in the rate at which this function is performed. Thermal fluctuations of the structure within the structural distribution occur at physiological conditions[4]. Such fluctuations can be necessary to render possible energy dissipation.

Based on the knowledge of the structure the discussion of the enzyme catalysis process can go into details using information obtained by a large number of different spectroscopic methods. In this contribution aspects will be discussed dealing with dynamic properties and structural distributions. To elucidate principles of protein dynamics most of the experiments have been performed on a rather simple protein, namely the myoglobin. Since myoglobin contains an iron atom Mössbauer spectroscopy on the ^{57}Fe nucleus can be applied. This technique employs the iron nucleus as a label for the motion of the protein. From the Mössbauer spectrum a mean square displacement, $\langle x^2 \rangle$, of the iron can be determined. This spectroscopic method has a time resolution depending on nuclear properties. Only processes contribute to the $\langle x^2 \rangle$-value which occur in characteristic times faster 100 ns.

To a certain extent Mössbauer spectroscopy is a method complementary to high resolution X-ray structure analysis. There, mean square displacements, $\langle x^2 \rangle$, for each individual atom of the molecule can be determined. However, X-ray structure analysis has no time resolution. Therefore, these values represent a distribution around the average

structure. At one temperature $\langle x^2 \rangle$-values from dynamic processes and
from static distributions cannot be separated. A structure determination
at different temperatures overcomes this restriction. At sufficiently
low temperature fluctuations can be frozen out. A comparison between results from Mössbauer spectroscopy and X-ray structure analysis in a wide
temperature range gives valuable information on protein dynamics. Pictures obtained from such investigations on myoglobin can be used in order
to understand Mössbauer data measured on iron containing proteins which
do not crystallize or which are investigated in their physiological
environment.

THE EXPERIMENTAL TOOLS

Mössbauer spectroscopy

In principle a Mössbauer absorption spectrometer has a setup similar to an optical photometer. As shown in Fig. 1 monochromatic radiation from a light source (1) passes the sample under investigation
(2). The transmitted intensity is registered in a detector (3). A spectrum of the transmitted intensity as a function of energy can be observed. It reflects the energy (or wavelength) dependent absorption of
the sample.

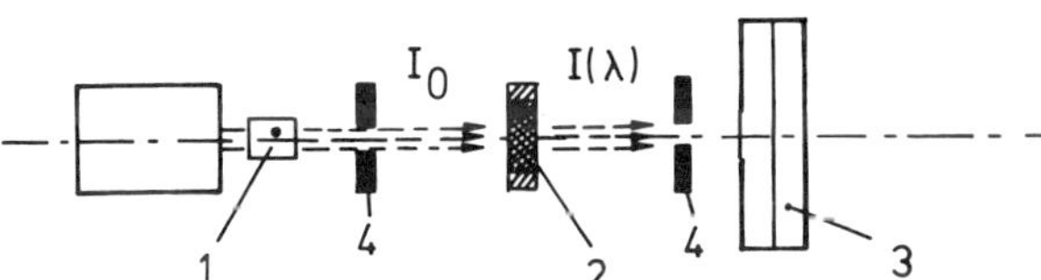

Fig. 1. Principle of spectroscopic absorption experiments. Light of the
        wavelength λ (1) is emitted with the intensity I_0 and transmitted
        through the sample (2). The transmitted intensity $I(\lambda)$ is measu-
        red in a detector (3) as a function of λ. Diaphragms (4) confine
        the light path. In Mössbauer spectroscopy of ^{57}Fe (1) is a ^{57}Co-
        source. The sample (2) contains ^{57}Fe nuclei. The transmitted ra-
        diation is detected in a proportional counter (3). The energy of
        the gamma-radiation is changed by the Doppler effect by moving
        the source with the velocity v.

For investigating biomolecules by Mössbauer spectroscopy the 14.4 keV-radiation of ^{57}Fe nuclei is used (wavelength λ = 0.86 A). This radiation can be absorbed in ^{57}Fe nuclei of the sample only under certain conditions which establish the peculiarity of this spectroscopic method.

The 14.4 keV radiation of a ^{57}Fe nucleus is extremely monochromatic (ΔE = 4.66 $\cdot 10^{-9}$ eV). The absorption in another ^{57}Fe nucleus is a resonance process which can occur only if the incoming radiation has precisely the resonance energy. Let us assume for a moment that the ^{57}Fe nucleus emitting the gamma-radiation and the ^{57}Fe nucleus in the sample are fixed absolutely rigid in space. Moreover, the chemical environment is assumed to be identical in source and absorber. The resonance condition is then fulfilled necessarily. The absorption is described by the Beer Lambert law where the cross section δ_0 equals 2.56 $\cdot 10^{-18}$ cm^2 / ^{57}Fe nucleus. The thickness of the absorber is measured by n_{Fe} which gives the number of ^{57}Fe nuclei per cm^2. Due to the energy distribution ΔE of the incoming wave and the same distribution in the acceptance energy in the absorber the observed absorption line has a Lorentzian energy distribution with a width of 2 ΔE.

In order to measure the absorption spectrum one has to vary the energy of the incoming radiation. In case of Mössbauer spectroscopy energy changes of about 100 ΔE are typical. These small changes (10^{-7} eV) can be performed by means of the Doppler effect. The ^{57}Fe-source is mounted on an electromechanical driving system and moved with the velocity v. The energy of the beam incident on the sample becomes then

$$E = E_s (1 + v/c) \qquad\qquad (1)$$

where E_s is the center of the emission energy of the ^{57}Fe nuclei in the source and c is the velocity of light. Unfortunately, each spectroscopic method defines its own energy scale. In case of Mössbauer spectroscopy the energy is measured in terms of Doppler velocity v. Thus, a Mössbauer spectrum is the gamma intensity T(v) transmitted through the sample as function of the Doppler velocity v. The conversion to the energy unit eV is performed by multiplying v(mm/s) with 4.80768 $\cdot 10^{-8}$. At the present state of discussion a Mössbauer spectrum exhibits one Lorentzian-shaped absorption line with a full width at half height of Γ_{exp} = 0.194 mm/s if 2 ΔE is converted in Mössbauer energy units.

We now come to the real situation. The ^{57}Fe nucleus never can be kept absolutely rigid in space. Then, the recoil of the emitted radiation can shift the radiation energy E by about 10^{-3} eV. The same shift can occur in the resonance absorption process but with opposite sign. The possibility of absorption is completely destroyed. However, there still exists a probability to emit and to absorb radiation without energy change by recoil effects if the iron is part of a sample in the condensed state. This probability is called Lamb Mössbauer factor and is given by

$$f = \exp(-k^2 \langle x^2 \rangle) \tag{2}$$

Here, $k = 2\pi/\lambda$. The crucial parameter is the mean square displacement, $\langle x^2 \rangle$, which measures the motion of the iron during the absorption (or emission) process of the gamma-quantum. Therefore, contributions to the $\langle x^2 \rangle$-value arise from vibrations of the iron with respect to the ligands, motions of segments containing the iron and intermolecular long range vibrations of solids which usually are called acoustic modes. Since f gives the probability of resonance absorption an experimental determination of the area of the absorption spectrum allows the determination of the mean square displacement, $\langle x^2 \rangle$, of the iron. Measuring $\langle x^2 \rangle$ as a function of temperature one can separate different contributions to the motion of the iron and this way analyze the dynamics of the sample in detail. It is this property of Mössbauer spectroscopy which we employ for the investigation of proteins in the following.

Normally, Mössbauer spectra do not consist of one absorption line only. Differences in the chemical environment of the ^{57}Fe nucleus in the source and the absorber shift and/or split the resonance energy. The environment is seen from the nucleus via hyperfine interactions with the electronic shell of the iron atom. One can distinguish between three contributions: (i) The overlap of the electron density with the nucleus determines the energetic center of the absorption lines. Per definition the energy center of the absorption is set to v = 0 for metallic iron. Deviations from this value are called isomer shift, IS, and measure the valence and spin state of the iron. (ii) Usually, the ^{57}Fe absorption line is split into two lines by quadrupole interactions. This occurs if the iron environment has a symmetry less than cubic which is always the case in proteins. The quadrupole splitting, QS, gets larger with increasing assymmetry of the neighbors of the iron. (iii) In Fe^{3+}

compounds a further splitting can be observed. A magnetic field at the
position of the iron nucleus yields a six line absorption pattern
due to the nuclear Zeeman effect. The splitting is a measure for the
strength of the magnetic field B. A magnetic field may either be applied
externally or results from the total spin of the 3d electrons of the iron.
Complex patterns are obtained if the electron spin is not completely frozen in but performs spin-lattice relaxations or spin-spin relaxations
on a time scale comparable to the reciprocal of the Larmor frequency of
the nuclear spin of the iron around the magnetic field of the electrons.   At room temperature the spin-lattice relaxation is usually so
fast that no magnetic field remains from the electrons at the nucleus
and so the magnetic splitting disappears. An example of a Mössbauer spectrum with IS = 0.85 mm/s and QS = 1.85 mm/s is   given   in Fig. 2.

Mössbauer spectroscopy has a time threshold and a time window. The
time threshold can be understood with the help of Fig. 3. The radiation
emitted by the source can be visualized by a damped wave train travelling through the absorber with the velocity of light. Absorption by the
iron nucleus can only occur during the time the wave packet overlaps with
the nucleus (situation b and c). The coherence length of the wave   train

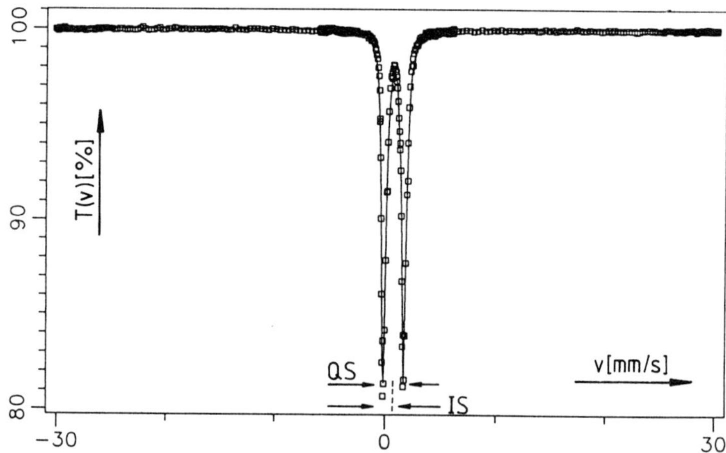

Fig. 2. Mössbauer spectrum of deoxymyoglobin crystals at T = 173 K.   The
        transmitted intensity T(v) as function of v can be fitted with a
        doublet of Lorentzians having a quadrupole splitting QS. The cen-
        ter of gravity of the spectrum is shifted by IS from v = 0.

defined as the length where the amplitude is reduced by a factor of 1/e
is about 42 m. The absorption has to occur in about 140 ns. With respect
to the Lamb Mössbauer factor this means that only motions with a characteristic time of about 100 ns or faster contribute to the $\langle x^2 \rangle$-value.

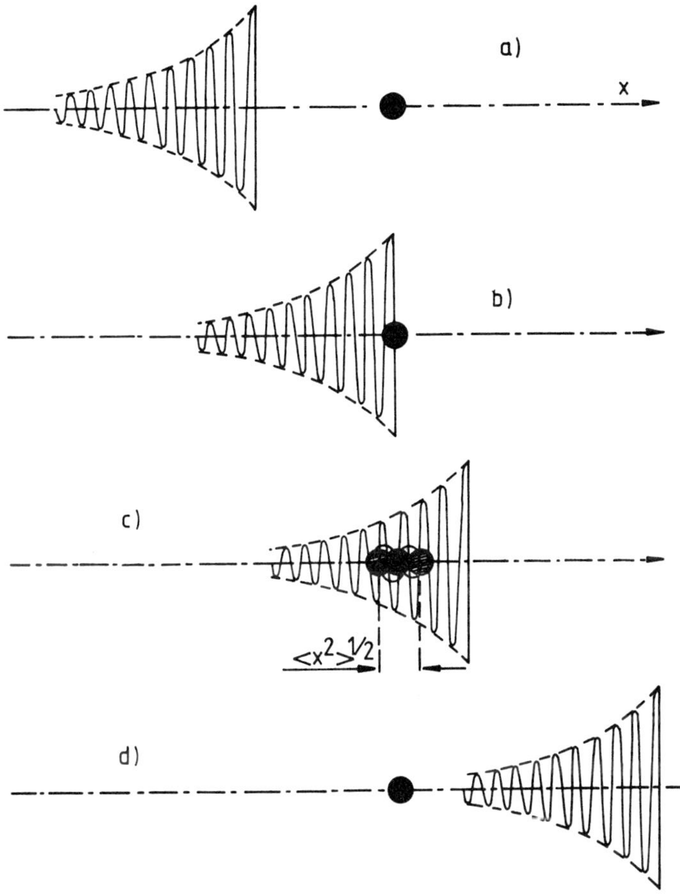

Fig. 3. Damped wave travelling across an atom (^{57}Fe absorber or X-ray
scatterer). Absorption (or scattering) occurs only during the
time of the overlap of the wavetrain and the atom (situation b
and c). Atomic motions during this time modify the absorption (or
scattering) process.

In viscous liquids the presence of diffusional motions may lead to
a very small displacement in 140 ns. If the iron takes part in such a
motion the absorption line is broadened without changing its area. The

line broadening   gives   the characteristic time τ_c of this motion   via the Heisenberg   uncertainty relation:

$$\Delta\Gamma\tau_c = \hbar \qquad\qquad (3)$$

where $\hbar$ is $0.6582 \cdot 10^{-15}$ eV s or $1.369 \cdot 10^{-8}$ mm if $\Delta\Gamma$ is measured in the Mössbauer energy units mm/s and τ_c in   seconds. Details   of Mössbauer spectroscopy and its application to   biochemical problems can   be   found in [5-8].

X-ray analysis of structural distributions

X-rays scattered by the atoms of a molecule   superpose   coherently. The   angular dependent amplitudes of the   resultant waves carry the   information   about the spatial arrangement. In principle   these   amplitudes are   modified by motions of atoms in   the   molecule during the scattering process. The time   resolution can again be visualized with the   help   of Fig. 3. The structure   amplitude of one individual molecule changes   if the atomic   positions   change during the time in which the damped   X-ray wave   train   travels across   the   molecule. One often uses CuK_α-radiation as a   X-ray source which has an energy   E = 8 keV corresponding to a wavelength of 1.54 A.   In contrast to ^{57}Fe-radiation the energy is rather poorly defined ($\Delta E \approx 1eV$) which means that the wave train is strongly damped.   The   coherence length is only about 2000 A and the main part of the wave passes the molecule in about 10^{-15} s. Only motions   during   this time interval can modify   the   structure amplitude   substantially. Merely vibrations of hydrogen   atoms occur on this time scale. Generally,   motions are slower especially if larger molecular segments of a molecule are   envolved.   For a   X-ray wave packet a molecular structure is   practically frozen in during the scattering time.   An f-factor in strict   analogy   to that of Eq.   (2) becomes 1.

However, X-ray scattering of a   single   molecule is too weak to be detected. Single   crystals are employed serving as amplifiers for special radiation scattered into   angles   given by the Bragg law.   In   a   protein crystal   typically 10^{15} molecules   contribute to the measured   intensity. At a certain   instant the molecules   of the crystal   do   not have   an identical structure because of   intramolecular motions. In addition, the molecules   oscillate   around their   mathematical position   in the   lattice because of acoustic modes involving collective motions of more   than one   molecule. Also static   disorder   makes the unit cells of the crystal

unequal. The X-rays perform a flash photograph at a certain
moment. Via the Bragg law an averaging over all molecules in the crystal
occurs. This way, the coordinates X_j, Y_j, Z_j of the atom j determined from
X-ray structure analysis become average coordinates. The distribution
around these coordinates is measured by the mean square displacement,
$\langle x^2 \rangle$. The wave scattered by a crystal into the Bragg reflexion with the
Miller indices hkl is characterized by the structure amplitude

$$\bar{F}_{hkl} = \sum_j f_j e^{-2\Pi i (hX_j + kY_j + lZ_j)} \cdot e^{-8\Pi \langle x_j^2 \rangle \sin^2 \vartheta / \lambda^2} \qquad (4)$$

Here, f_j is the atomic form factor of the atom j. The second exponential function is called Debye Waller factor. The sum j runs over all
atoms in the unit cell of the crystal which contains at least one molecule.

After solving the phase problem by the method of multiple isomor-
phous replacement a Fourier transformation of Eq. (4) gives the average coordinates X_j, Y_j, Z_j for all atoms j in the molecule or in the
unit cell, respectively. From high resolution data the individual mean
square displacements, $\langle x^2 \rangle$, can also be obtained. Besides the average
structure X-ray analysis yields also the structural distribution.
Since there exists no time resolution everything contributes to
this distribution which makes the molecules unequal. Static and dynamic
disorder can only be separated if the structure is determined at different temperatures. A reduction of the temperature allows to freeze all
modes of motions except the zero point vibrations. At low temperatures
the relative contribution due to static disorder increases.

STRUCTURAL DISTRIBUTIONS IN MYOGLOBIN

In the following the discussion is based on results from a X-ray
structure analysis of myoglobin (Mb) at 300 K, 260 K, 180 K, 165 K, 115 K,
80 K and 40 K. Apart from the measurements at 260 K and 40 K details are
found in [9]. The structural distribution is measured by the $\langle x^2 \rangle$-values. In
order to increase the accuracy for each residue values averaged over the
backbone atoms N-C-C or over the side chain atoms have been taken. Fig. 4
gives some examples. The mean square displacements decrease when lowering
the temperature indicating a freezing of dynamic processes. Within the
limits of error the temperature dependence of the $\langle x^2 \rangle$-values can be

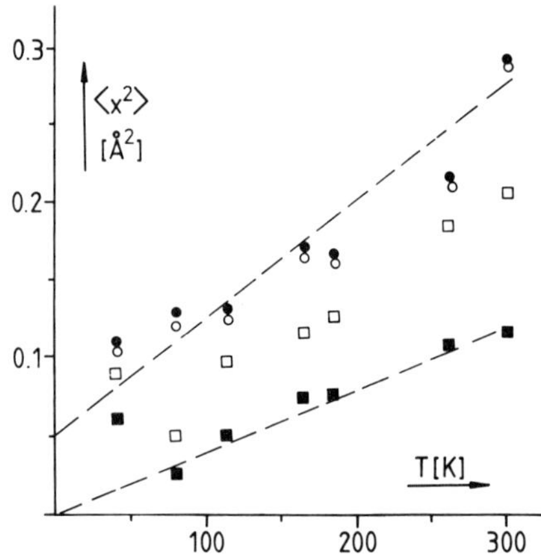

Fig. 4. Mean square displacements obtained from X-ray structure analysis. Average values of the backbone atoms (closed symbols) and the side chain atoms (open symbols) are given[9] . Circles: Pro120(GH2); squares: Lys56(D6)

treated as linear down to about 80 K. An extrapolation of this
behaviour to T = 0 yields $\langle x^2 \rangle$ = 0 for the residue Lys 56(D6). This can
be understood easily from the Debye law describing vibrations in solids.
The linear extrapolation of the $\langle x^2 \rangle$-values to T = 0 K for the
residue Pro120(GH2) contradicts this easy view. Here, a remarkable
zero point distribution exists even at T = 0 K which is much larger than
any zero point vibration. The low temperature measurement at 40 K indicates the vanishing of the slope of the function $\langle x^2 \rangle$ of T. Average
values over all backbone atoms or all side chain atoms as shown in Fig.
5 reveal the same tendency. Fig. 6 gives an extrapolation to T = 0 K
of the $\langle x^2 \rangle$-values of the backbone atoms and the side chain atoms of the
153 residues of myoglobin. A linear temperature dependence of the $\langle x^2 \rangle$-

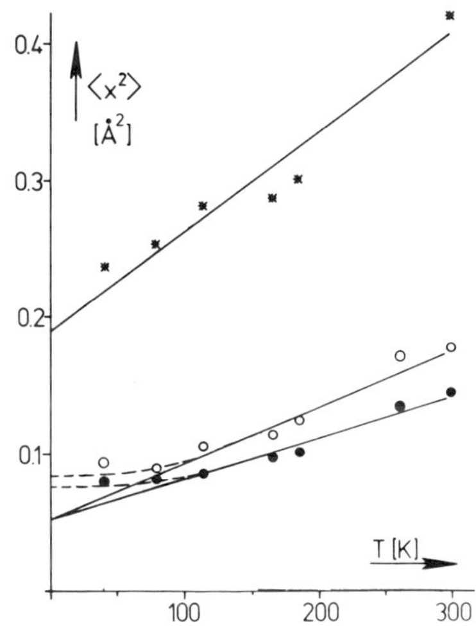

Fig. 5. Mean square displacements, $\langle x^2 \rangle$, determined by X-ray structure
analysis of metmyoglobin crystals [9]. Closed circles: average va-
lues of backbone atoms, open circles: average values of side
chain atoms; crosses: average values of 160 water molecules which
can be resolved in X-ray structure analysis. Solid lines: linear
regression not including data at 40 K and 260 K; dashed lines:
guide for the eye.

values between 300 K and 80 K has been assumed. Figs. 4 to 6 show that
structural distributions remain even at 0 K. A myoglobin molecule has
no structure with one well defined energy minimum. It is essentially
this property which often gives rise to a comparison with a glass.

At low temperatures each molecule is frozen in one conformational
substate. An increase of the temperature rises the probability to surmount

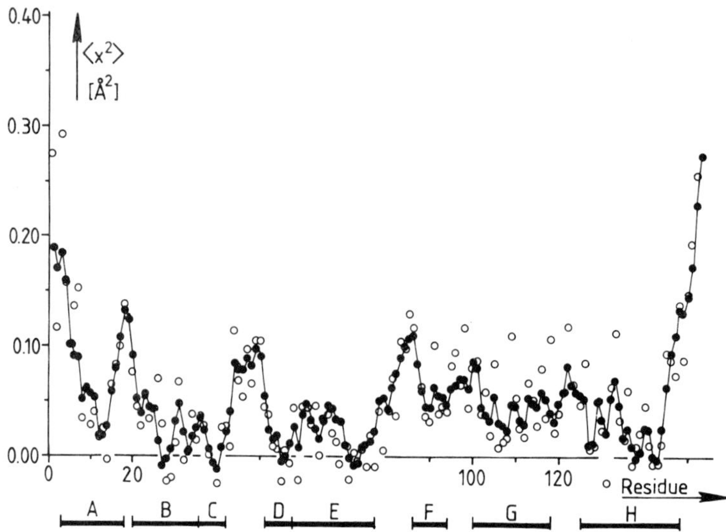

Fig. 6. Mean square displacements of the backbone atoms (full symbols) and the side chain atoms (open symbols) performing a linear regression to T = 0 K of the data from measurements at 5 temperatures between 80 K and 300 K [9].

the energy barrier which separates different conformational substates. At physiological temperatures each molecule fluctuates through the conformational substates within the structural distribution.

STRUCTURAL DISTRIBUTIONS AND FUNCTION

The hemoglobin CTT3 of the insect larve of chironomus thummi thummi consists of 136 residues. The sequence overlaps only at 28 positions with that of Mb. Nevertheless, the structures of CTT3 and Mb resemble each other [10]. Fig. 7 shows that the structural distributions along the sequence are also rather similar [11]. Tertiary structure and structural distributions appear to be highly correlated.

In two regions the structural distributions in CTT3 and myoglobin are clearly different. Parts of the F-helix show a wider distribution in Mb. Can this be correlated with functional properties? As already mentioned molecules in different conformational substates perform the same function but with different rates. If oxygen binds to the heme the iron moves into the heme plane inducing a motion of the histidin together with parts of the F-helix. Molecules in the deoxy state occupying a confor-

208

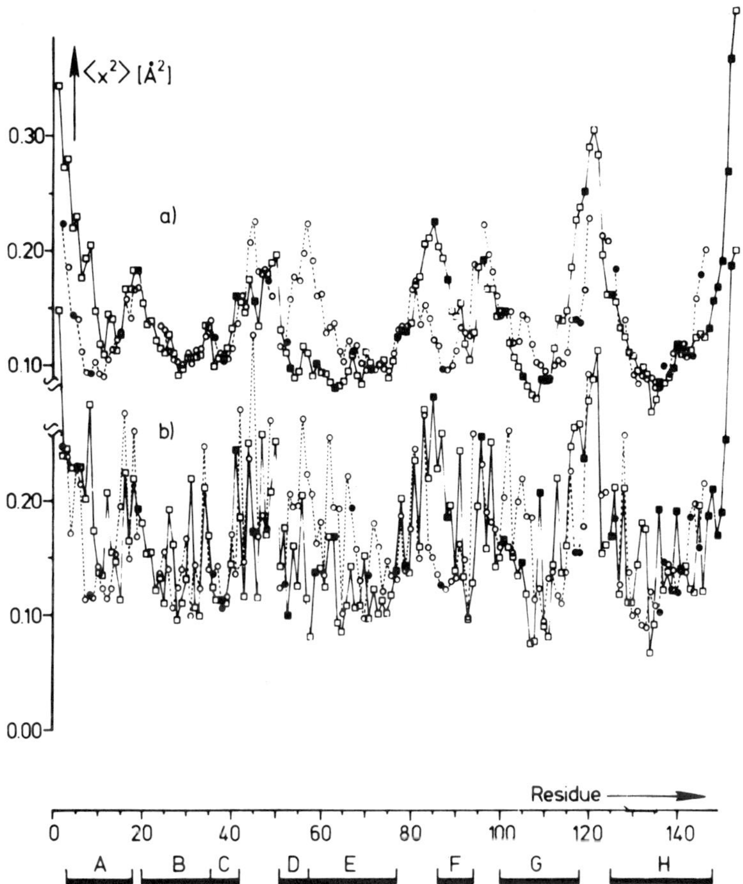

Fig. 7. Mean square displacements, $\langle x^2 \rangle$, as obtained from X-ray structure analysis. Squares: Mb circles: CTT3; a) average values of the backbone atoms, b) average values of the side chain atoms of one residue, respectively. The helical regions are marked below (taken from [11]).

mational substate with a structure closer to that of the oxygenated conformation should bind the oxygen easier. In Mb the structural distribution is larger. The probability to find molecules close to the oxygenated structure is, therefore, larger than in CTT3. This may explain in part the higher oxygen affinity of myoglobin[11].

So far we have assumed that the rate limiting step of the oxygenation is the binding of the oxygen to the iron and not the penetration of the oxygen into the heme pocket. This picture is proved by several investigations[12]. Nevertheless, a careful inspection of the X-ray coordinates of deoxymyoglobin shows that a completely rigid average deoxymyoglobin molecule offers no path wide enough to allow the penetration of an O_2-molecule into the heme pocket. A rigid molecule of the average structure would not fulfill any function. Case and Karplus have shown that there exist two major path ways where the oxygen access to the heme pocket is less hindered than for any other route[13]. In path 1 the potential barriers hindering the oxygen passage are formed by the residues His64(E7), Thr67(E10) and Val68(E11), while in the second path[14] the main obstacles are the Leu61(E4) and the Phe33(B14). Ringe et al. have shown that a ligand like phenylhydrazine can open a path into the heme pocket by forcing aside the residues His64(E7), Arg45(CD3) and Val68(E11). This marks a possibility for the entrance of the O_2-molecule.

The X-ray[9] structure determination of myoglobin at different temperatures has shown that the methyl groups of the Leu61(E4) have one well defined position at 80 K while the alternative positions obtained by rotations of about 120° and 240° around the CB-CG bond are almost equally populated at room temperature. However, all three conformations hinder the oxygen access. The door is open only during the time when the side chains jump from one position to another. Dynamic structure fluctuations allow the adjustment of an oxygen preequilibrium in the heme pocket.

We now come to the general mechanism of the change of conformations. When Mb is liganded with O_2 it goes from the deoxy conformation into the oxy conformation. Both structures are well known from X-ray analysis. Fig. 8 compares these conformations with respect to the backbone atoms. Average coordinates for the N-C-C atoms of each residue have been calculated for metmyoglobin (liganded conformation) and deoxymyoglobin (unliganded conformation). The differences in the coordinates of the two

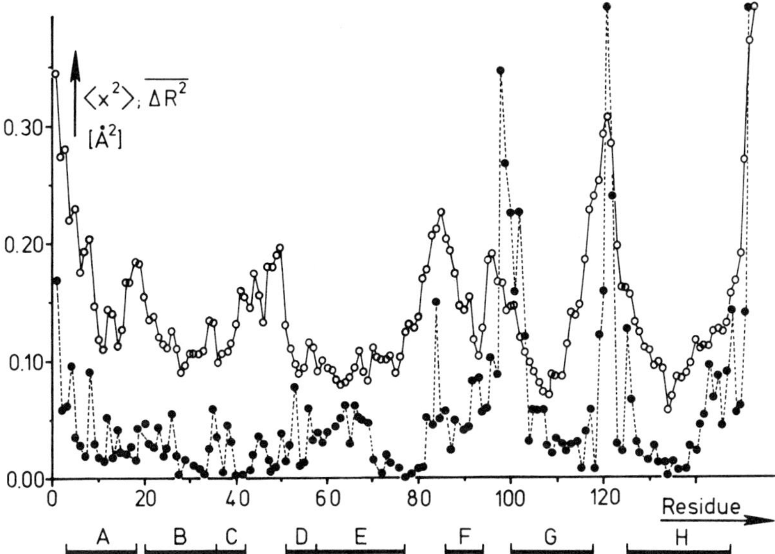

Fig. 8. Structural distribution in liganded Mb (metmyoglobin, open circles) measured by the mean square displacement, $\langle x^2 \rangle$, compared to the squared structural changes, $\overline{\Delta R^2}$ between metmyoglobin and deoxymyoglobin (full circles)^{20}. Backbone averages are shown. The helical regions are marked.

conformations have been squared. The obtained $\overline{\Delta R^2}$-values are plotted in Fig. 8 as a function of the residue number and are compared with the mean square displacements from metmyoglobin. Astonishingly, the squared structural difference between the two conformations, $\overline{\Delta R^2}$, are smaller than the structural distributions, $\langle x^2 \rangle$, within one conformation. The structural distribution around the average coordinates of the liganded conformation contains also molecules with coordinates close to the average coordinates of the unliganded conformation and vice versa. This gives a very simple picture for the oxygen binding to deoxymyoglobin. At room temperature a molecule in the deoxy conformation fluctuates through conformational substates. Some of these substates have a structure close to the oxygenated conformation. It is reasonable to assume that the oxygen binds preferably to molecules in these substates. So, step by step the number of molecules belonging to the deoxy average structure decreases and after the ligation of all molecules the ensemble average gives the coordinates of oxygenated molecules. Structural fluctuations then occur   around this conformation overlapping with substates of the deoxy conformation. Only in the average these are energetically unfavourable.

THE TIME SCALE OF STRUCTURAL FLUCTUATIONS

In a complex system like an ensemble of protein   molecules a number of different physical processes contributes to   dynamics. Charac- teristic times for protein dynamics cover the large range from 0.1 ps to several seconds. Each method of investigation sets a time window and is only sensitive to selected channels of motion. As already discussed X-ray structure analysis of single crystals has practically no time sen- sitivity. In the following the Mössbauer spectroscopy is used as a com- plementary method to X-ray structure analysis. Fig. 9 shows a Mössbauer spectrum of ^{57}Fe in deoxymyoglobin (compare [15,20]). The investigated sample consisted of a large number of single crystals similar to those used in X-ray structure analysis. The assymetric environment of the iron causes the observed splitting into two lines. As usual the spectrum is

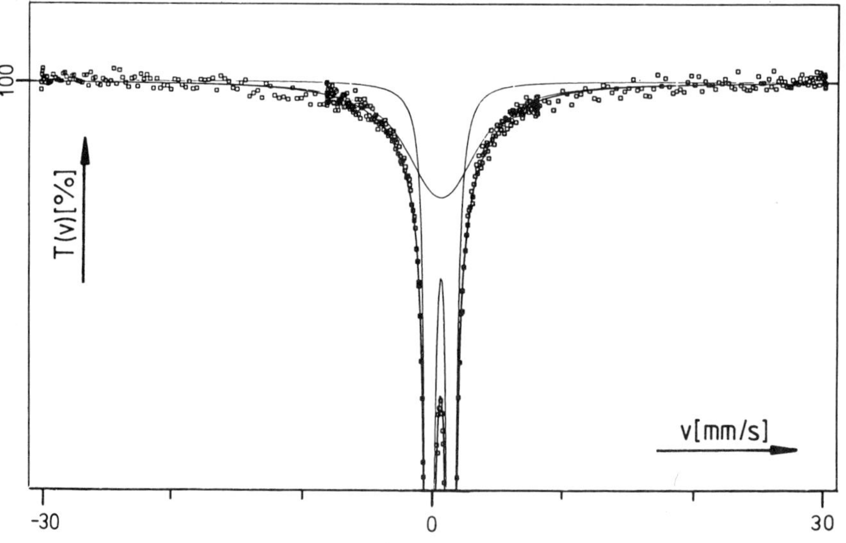

Fig. 9. Mössbauer spectrum of deoxymyoglobin   crystals at T = 245 K. It
     is obvious that a doublet of Lorentzians does not fit the expe-
     rimental data. Broad Lorentzians have to be added.

fitted by a pair of Lorentzians. However, it is obvious that the fit does not explain the experimental data. A pair of broad Lorentzians has to be added to give agreement between experiment and theory. The broad lines clearly indicate that diffusive motions are registered at the position of the iron. The width of the broad lines indicates that these motions occur on a time scale of about 1 ns at 265 K.

The area A of the narrow lines allows the determination of a mean square displacement, $\langle x^2 \rangle$, (compare Eq. (2)). All processes moving the iron on a time scale faster 100 ns contribute. The $\langle x^2 \rangle$-values determined from 4.2 K to 300 K are shown in Fig. 10. The temperature dependence up to about 180 K can be described by a Debye law with Θ_D = 195 K.

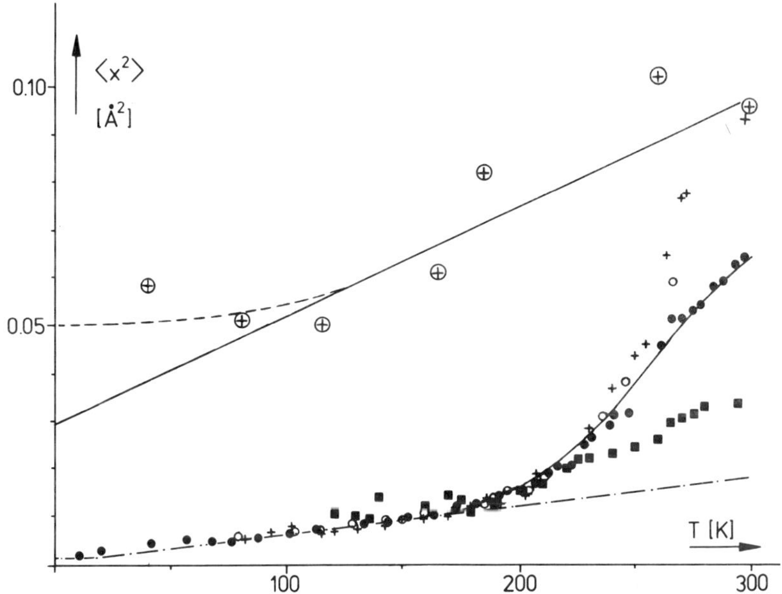

Fig. 10. Mean square displacements, $\langle x^2 \rangle$, of iron in myoglobin as a function of temperature. Open circles with crosses: X-ray data on metmyoglobin; solid line: linear regression not including the values at 40 K and 260 K; dashed line: guide for the eye. Mössbauer experiments on deoxymyoglobin crystals: closed circles: taken from [15]; open circles: experiments with extreme accuracy [20]. Dashed dotted line: Debye law with Θ_D = 195 K; solid line: fitting of the model of the extended Brownian oscillator [25] to closed circles [19]. Squares: freeze dried metmyoglobin [15]. Crosses: Fe (CN)$_6^{4-}$ diffused into metmyoglobin crystals [57].

Such a behavior reflects ordinary solid state vibrations. At about 180 K the $\langle x^2 \rangle$-values increase more than linearly with temperature. A new channel of motion is thermally opened. The already mentioned additional broad lines also appear in this temperature regime. These findings can be   correlated with the measurement of the dielectric properties of water in myoglobin crystals. Below 200 K the water mobility is that of a solid. It increases between 200 K and 300 K. At room temperature it is still below that of free bulk water[16].

The protein specific modes above 180 K can be understood as a diffusive motion of molecular segments in limited space. This picture can be used to develop models which allow a quantitative formulation of a theory that can be   compared with experimental results. A detailed discussion of different models can be found in[17]. Here, only the main features are reviewed. The models are based on the concept of conformational substates (compare[4]) but differ in the description of jump processes. In all these models two situations can be distinguished: A molecule is either frozen in a conformational substate or it is just jumping across the   barrier between substates. In the frozen state the molecule performs solid state vibrations. The fluctuations between substates yield the protein specific motions. In analogy to reaction kinetics one can say that the molecule is in a transition state if it is just at the top of the barrier between substates. One has to differentiate between   models where the transition state is populated for a practically immeasurable time only or where the molecule spends a time in the transition state which is comparable to τ_N = 141 ns.

The simplest discrete jump models assume   only a negligibly short stay in the transition state and a few conformational substates. Calculations have been performed with two substates, four substates in a tetrahedral arrangement or eight substates forming the corners of a cube. With barrier heights in the order of 20 kJ/mol these models explain the shape of the Mössbauer spectra but fail in the explanation of the temperature dependence of the absorption area.

Frauenfelder has proposed a model of free continuous jump diffusion with a distribution of barrier heights[18]. A hierarchy of conformational substates is assumed[23]. Jump diffusion measured in the Mössbauer spectroscopy is characterized by a diffusion constant $D = 1/2\ k_D\ d^2$ where d is the average jump distance. In the exponent the jump rate k_D contains the barrier height $H^{\#}$ between different substates. It is essential that for each individual molecule $H^{\#}$ is constant during more than 100 ns. However, within the sample a distribution of barrier heights exists. Fluctuations of a molecule into states with different $H^{\#}$-values occur slowly compared to 100 ns. In this model each molecule gives a Mössbauer doublet with a well defined line width depending on D. The experimentally obtained Mössbauer spectrum is a superposition of spectra with different line widths. This model allows to understand the shape of the Mössbauer spectra as a function of temperature. It fails to explain the temperature dependence of the $\langle x^2 \rangle$-values quantitatively. Nevertheless, the described model allows a comparison with results of flash photolysis of CO in CO liganded myoglobin[12,23]. The slow fluctuations between states with different $H^{\#}$ values can be correlated to the structural relaxations of type FIM1 while the fluctuations determining the Mössbauer spectrum of each molecule correspond to the relaxations of type FIM2.

An extended Brownian oscillator is a quite general model to describe diffusion-like motions of protein segments in restricted space. A segment of the molecule is either in a conformational substate or with the probability $p_t(T)$ in the transition state.

$$p_t(T) = (1 + \exp(G^{\#}/(RT)))^{-1} \tag{5}$$

$G^{\#}$ equals $H^{\#} - TS^{\#}$ and is the difference of the Gibbs free energy between substate and transition state. R is the gas constant. In the transition state diffusive motions occur around the average structure till the segment is trapped again in another conformational substate. The fluctuations are driven by the thermal bath represented by a stochastic force $F_t(T)$. The equation of motion with respect to a coordinate x_t looks as follows:

$$\ddot{x}_t + 2\beta_t \dot{x}_t + \omega_t^2 x_t = F_t(T) \tag{6}$$

A harmonic restoring force $\omega^2 x_t$ limits the diffusion in space and conserves the average structure of the molecule as found from X-ray analysis. The motion is damped due to the friction β_t which depends on temperature. In an extended Brownian oscillator the effective friction is determined by the depth of the conformational substates and the friction in the transition state. To understand the Mössbauer spectra one has to assume strongly overdamped oscillations. One obtains a superposition of Lorentzians with increasing width and decreasing area. If these Lorentzians are numbered by N the line N = 0 gives the sharp Lorentzians of Fig. 9. The broad line arises from contributions N = 0 where the line N has the width $N\alpha$ The parameter α is related to the parameters in Eq. (6) by $\alpha = \omega_t^2/(2\beta_t)$. Its reciprocal value characterizes the time in which a diffusive motion over the distance $\langle x^2 \rangle^{1/2}$ occurs.

Good fits of the Mössbauer data have been performed with the assumption that one segment jumps several times during 100 ns between conformational substates and the transition state[19]. The opposite assumption gives rather poor fits to the experimental data.

Recently the width $\Gamma^{(1)}$ of the additional broad lines was found to increase with decreasing temperature between 260 K and 200 K[20]. This complicates the data analysis because the Brownian oscillator predicts a controversial behaviour. The discrepancy can be removed by assuming a distribution of activation energies $H^{\#}$ for the different conformational substates. A distribution of activation energies is also necessary to understand the rebinding kinetics of CO in myoglobin after flash photolysis[12].

A special type of Mössbauer experiments supplies information on the size of collectively moving segments. Here, the radiation of a ^{57}Fe Mössbauer source is scattered by the sample in the same way as X-rays (Rayleigh Scattering of Mössbauer Radiation, abbreviated RSMR). The scattered radiation is transmitted through a Mössbauer absorber. This allows the determination of the fraction which was elastically scattered by the sample. An analysis gives the mean displacement, $\langle x^2 \rangle^R$, averaged over all atoms of the molecule. Since ^{57}Fe radiation is used one obtains the time resolution of Mössbauer absorption spectroscopy (compare[7]).

RSMR investigations on myoglobin have already given some important results. First of all the temperature dependence of the $\langle x^2 \rangle$-values is similar to that of the iron as shown in Fig. 10[21]. In the average all atoms of Mb behave almost like the heme iron. So, the iron appears to be a good marker for the investigation of protein dynamics!

Angular dependent RSMR experiments offer the possibility to determine the size of segments moving collectively on the time scale of 100 ns or faster. In the simplest picture one may imagine that the pro-

216

tein molecule is cut into independently moving parts with a characteristic length l_c. First RSMR experiments have shown that this length is
of the order of 5 A [22]. It should be emphasized, however, that a detailed
analysis of these data requires the knowledge of the water structure in
myoglobin crystals.

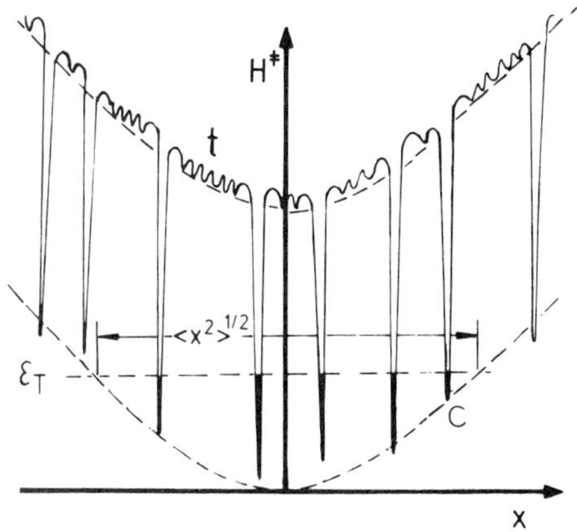

Fig. 11. Potential explaining the segmental fluctuations in myoglobin. In
a Mössbauer experiment x is the coordinate of the iron. For the
sake of simplicity the drawing is only one-dimensional. The conformational substates are occupied according to the Boltzmann
law. X-ray analysis measures the distribution $<x^2>^{1/2}$ of the
occupied substates. For details see the text.

COMPARISON OF RESULTS FROM X-RAY STRUCTURE ANALYSIS AND MÖSSBAUER
SPECTROSCOPY

Fig. 10 compares the $<x^2>$-values of the iron in Mb-crystals as determined by X-ray structure analysis and by Mössbauer spectroscopy. The discrepancy reflects the different time sensitivity of both
methods. Since slow processes and static disorder are included $<x^2>$-
values determined by X-ray analysis are always larger as those from Mössbauer spectroscopy. A potential as shown in Fig. 11 may help to visualize the situation. It consists of deep traps (conformational substates) and the transition state. The depth of the traps is distributed
around an envelope with a parabolic shape. The transition state has also

a parabolic envelope. Here, the ripples symbolize friction. The mean square displacements of X-ray structure analysis measure the distribution of populated conformational substates. Fluctuations between these states have to be fast enough to guarantee thermal equilibrium during the structure determination which takes several days. Mössbauer spectroscopy is insensitive to fluctuations between states that occur seldom compared to 100 ns. This is just the case below 200 K. Here, the $<x^2>$-values reflect only vibrations within one conformational substate. Only if the temperature is high enough to reach the transition state Brownian motions become possible and the Mössbauer $<x^2>$-values increase.

THE ROLE OF WATER

Mössbauer absorption spectroscopy can be used to compare myoglobin with a large hydration shell and freeze dried material. Fig. 10 shows that drying reduces drastically the intramolecular mobility. Nevertheless, the protein shows still some specific dynamics. The increase of the $<x^2>$-values changes at the temperature $T = 180$ K and $T = 250$ K. This may be a hint for a hierarchy of motions[23,20][57]. One can label the water mobility in Mb crystals by diffusing a $Fe(CN)_6^{4-}$ -complex into the crystal water. It becomes obvious from Fig. 10 that the mobility of the heme iron and the mobility of the crystal water is strongly correlated.

RSMR investigations on dry and wet myoglobin confirm the results discussed above. Moreover, there is evidence that the crystal reduces dynamics due to the close packing of the molecules[21].

DYNAMICS OF MEMBRANES

^{57}Fe Mössbauer spectroscopy allows to study the dynamics of model membranes and biological membranes. On the one hand, iron can be bind to the phosphate groups of lipids. On the other hand, membrane-bound proteins often contain iron. Growth of cells on a medium containing the isotope ^{57}Fe yields biological material suitable for Mössbauer spectroscopy. Fig. 12 compares some results[20]. The lecithin sample consists of DPPC double layers with iron bound to the phosphate groups. In the bacteriorhodopsin sample the iron is bound to the phosphatydyl-glycerophosphat of the lipid and the carboxyl groups of the protein part. In

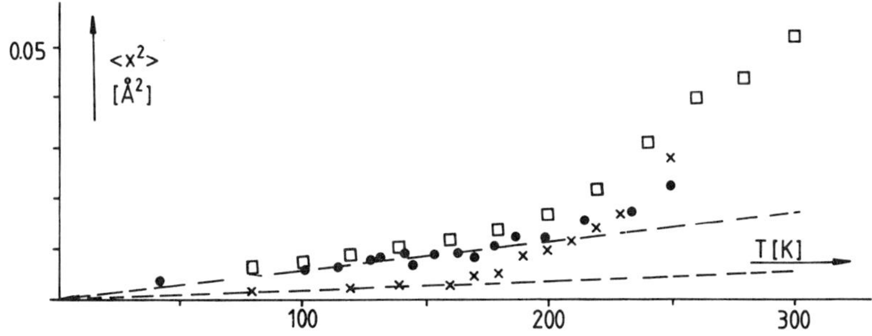

Fig. 12. Mean square displacements, $<x^2>$, of iron in membranes determined by Mössbauer spectroscopy. Squares: double lipid layer of leci-thin[25]; crosses: chromatophores of Rhodospirillum Rubrum[24]; circles: Fe^{3+} bound to purple membranes[26].

the chromatophores of Rhodospirillum Rubrum the iron of the reaction center and of cytochromes was used to label membrane dynamics. All samples show similar features as Mb. Above a certain temperature specific modes of motion arise which can be characterized as Brownian motions in restricted space. In the chromatophores these motions are already excited at rather low temperatures. It should be emphasized that in this sample a clear correlation between the mobility and the bio-chemical function has been found[24]. After photoabsorption a radiationless dissipation of a part of the energy occurs only if diffusive motions are excited. This dissipation makes fluorescence impossible and is essential for the mechanism of energy gain from light.

ACKNOWLEDGEMENT

This work was supported by the Deutsche Forschungsgemeinschaft, the Bundesministerium für Technologie and the Fond der Chemie.

REFERENCES

1. C.C. Phillips (1966) in: Advances in Structure Research by Diffraction Methods, Vol. 2, p. 75, ed. R. Brill, R. Mason, Vieweg Braunschweig

2. R. Huber (1988), Angew. Chem. 100, 79-89

3. Frauenfelder H., Petsko G.A., Tsernoglou D. (1979), Nature 280, 558-563

4. Frauenfelder H., Parak F. and Young R.D. (1988), Ann. Rev. Biophys. Chem. 17, 451-479

5. Topics in Applied Physics, Vol. 5, Mössbauer Spectrocopy, ed. U.Gonser, Springer Verlag (1975)

6. Topics in Current Physics, Mössbauer Spectroscopy II, ed. U. Gonser, Springer Verlag (1981)

7. Parak F. (1986), Methods in Enzymology, Vol. 127, ed.: L. Packer, Academic Press, pages 196-206

8. Parak F. and Reinisch L. (1986), Methods in Enzymology, Vol. 131, ed. C.H.W. Hirs and S.N. Timasheff, Academic Press, pages 568-607

9. Parak F., Hartmann H., Aumann K.D., Reuscher H., Rennekamp G., Bartunik H. and Steigemann W. (1987), Eur. Biophys. J. 15, 237-249

10. Huber R., Epp O., Steigemann W., Formanek H. (1971), Eur.J.Biochem. 19, 42-50

11. Hartmann H., Steigemann W., Reuscher H. and Parak F. (1987), Eur. Biophys. J. 14, 337-348

12. Austin R.H., Beeson K.W., Eisenstein L., Frauenfelder H., Gunsalus I.C. (1975), Biochemistry 14, 5355-5372

13. Case D.A. and and Karplus M. (1979), J. Mol. Biol. 132, 343-368

14. Ringe D. and Petsko G.A., Kerr D.E., Ortiz de Montellano P.R. (1984) Biochemistry 23, 2-4

15. Parak F., Knapp E.W. and Kucheida D. (1982), J. Mol. Biol. 161, 177-194

16. Singh G.P., Parak F., Hunklinger S. and Dransfeld K. (1981), Phys. Rev. Lett. 47, 685-688

17. Parak F., Heidemeier J., Knapp E.W. (1988), in: Biological and Artificial Intelligence Systems; ed. E. Clementi and S. Chin, ESCOM Science Publishers B.V., in press

18. Frauenfelder H. (1985), in: Structure and Motion: Membranes, Nucleic Acids and Proteins; ed. E. Clementi, G. Corongiu, M.H. Sarma, R.H. Sarma, Adenine Press, pages 205-217

19. Parak F. and Knapp E.W. (1984), Proc. Natl. Acad. Sci. USA 81, 7088-7092

20. Parak F., Heidemeier J. and Nienhaus G.U. (1988), Hyperfine Interactions 40, 147-158

21. Krupyanskii Yu.F., Parak F., Goldanskii V.I., Mössbauer R.L., Gaubmann E., Engelmann H. and Suzdalev I.P. (1982), Z. Naturforschg. 37 c, 57-62

22. Nienhaus G.U. and Parak F. (1986), Hyperfine Interactions 29, 1451-1454

23. Ansari A., Berendzen J., Bowne S.F., Frauenfelder H., Iben I.E.T., Sauke T.B., Shyamsunder E., Young R.D. (1985), Proc. Natl. Acad. Sci. USA $\underline{82}$, 5000-5004

24. Parak F., Frolov E.N., Kononenko A.A., Mössbauer R.L., Goldanskii V.I. and Rubin A.B. (1980), FEBS Lett. $\underline{117}$, 368-372

25. Parak F., Fischer M., Graffweg E. and Formanek H. (1987), in: Structure and Dynamics of Nucleic Acids, Proteins and Membranes, ed.: E. Clementi and S. Chin, Plenum Publish. Comp. New York, pages 139-148

26. Heidemeier J., Fischer M., Parak F., Engelhard M., Kohl K.D., Hess B. and Formanek H., submitted to Eur. Biophys. J.

ENZYME HYDRATION AND FUNCTION

John A. Rupley and Giorgio Careri

Department of Biochemistry
University of Arizona
Tucson, Arizona 85716

Dipartimento di Fisica
Universita di Roma I
Roma 00185, Italy

INTRODUCTION

The character of the surrounding solvent modulates
biological processes. Carboxylate ion attack on substituted
phenyl esters (Bruice and Turner, 1970), a reaction studied as
an enzyme model, changes in rate by four to six powers of ten
for change in solvent from water to a dioxane-water mixture.
The enthalpy of formation of a peptide hydrogen bond is small,
less than 1 kcal/mol, in water, but approaches the gas phase
value of 5 kcal/mol in solvents such as carbon tetrachloride
(Kresheck and Klotz, 1969). Hydrophobic interactions by
definition reflect the solvent environment. The effect of
solvent on rates and thermodynamics of reactions can be
understood to propagate through macromolecular processes to an
influence of solvent on higher levels of biological
organization.

The intent of the discussion that follows is to present the
basis for understanding the interaction between solvent and
protein, in the form of a picture of the water-protein
interface, and to use this picture in considering some of the
ways in which solvent influences protein and enzyme processes.
It will be seen that the description of the water-protein
interface is now reasonably detailed. The understanding of how
the properties of this interface modulate protein properties is
beginning to emerge.

METHODS OF MEASUREMENT OF HYDRATION

There is variety among the tools available to examine the
hydration of macromolecules (Table 1).

The measurements differ in the way they reflect the
following: time average versus dynamic properties; properties of

Table 1. Measurements of protein hydration

DIFFRACTION:	x-ray, neutron
SPECTROSCOPY:	IR, UV, CD, fluorescence, NMR, ESR, Mossbauer
SOLUTION THERMODYNAMICS:	Cp, V, preferential binding, compressibility, hydration forces, denaturation
SORPTION THERMODYNAMICS:	G, H, Cp, V, denaturation
RELAXATION:	NMR, ESR, dielectric
HYDRODYNAMICS:	sedimentation, viscosity, etc
SPECIAL RATE PROPERTIES:	enzyme activity, hydrogen exchange
SPECIAL TIME-AVG PROPERTIES:	non-freezing water
COMPUTER SIMULATION:	molecular dynamics and Monte Carlo, surfaces and packing, electrostatics

the solvent versus those of the protein; structure; powder or
solid samples versus solutions.

Particular attention is paid here to measurements on
partially hydrated protein powders. We believe that in order to
understand the interaction between water and protein, it is
necessary to vary the activity of the water over a wide range.
By working with protein powders, one can analyze changes in
thermodynamic properties and structure in much the same way that
one can analyze the complexes of proteins with substrates and
other molecules in solution, by varying the concentration of at
least one of the species participating in the complex.

A sorption isotherm, one of the earliest and simplest
measurements made on protein hydration, is obtained by
determining the amount of water adsorbed on the protein surface
as a function of the partial pressure of the surrounding water
vapor (the thermodynamic activity of the water). This reflects
a time-average, thermodynamic property, the free energy of the
system. The measurement is restricted, by definition, to the
solid phase. Another thermodynamic property, the heat capacity,
can be determined over the full range of system composition,
from the dry protein to the dilute solution (Yang and Rupley,
1979). Measurements of this kind serve to tie behavior of the
molecule in the solid phase to the behavior under conditions
more generally of interest, namely, the solution range.

For review discussions of measurements of protein
hydration, see Rupley et al. (1983), Kuntz and Kauzmann (1974),
and Edsall and McKenzie (1983).

SELECTED RESULTS DESCRIBING THE HYDRATION PROCESS

<u>Time-Average Properties</u>

Heat capacity measurements serve as framework upon which we
can develop a picture of protein hydration (Yang and Rupley,
1979). Fig. 1 gives such data for the lysozyme-water system as
a function of water content.

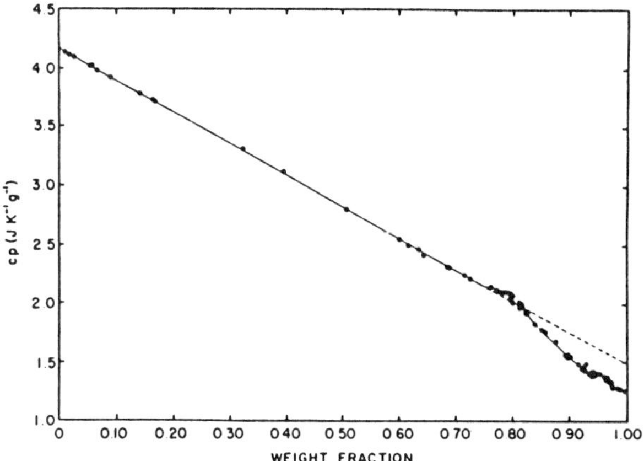

Fig. 1. Specific heat of the lysozyme-water system from 0 to 1.0
weight fraction protein. From Yang and Rupley (1979),
with permission.

The linear response of the heat capacity to change in
system composition, from the dilute solution to 0.73 weight
fraction protein, requires that in this range the only change in
the system be the addition or removal of bulk water. There is
no discontinuity at the transition from homogeneous solution to
the hydrated powder, at about 0.2 weight fraction protein.
Apparently, the protein in the powder, at hydration levels above
0.73 weight fraction protein, has the same thermal properties
and by inference the same characteristic hydration shell as
present in solution. The discontinuity in the thermal response,
in the water poor region at 0.73 weight fraction protein, which
is equivalent to 0.38 g of water/g of protein, or 300 mol of
water/mol of protein, corresponds to the end point of the
hydration process. We can understand the heat capacity
measurements as defining tightly the hydration end point.
Because these measurements are sensitive to both hydrogen-
bonding and hydrophobic interactions of water, they should
reflect all the principal time-average chemistry associated with
hydration.

The details of the heat capacity response in the water poor
region afford insight into the hydration process. Fig. 2 gives
the dependence of various time average properties of lysozyme on
the hydration level (h, g of water/ g of protein), over the
hydration range 0 to 0.4 h, from the dry protein to a level
slightly above the end point of the hydration process.

Curve (d) of Fig. 2, part I shows the dependence of the
apparent specific heat on hydration level. This function is
calculated from the data of Fig. 1. It is directly related to

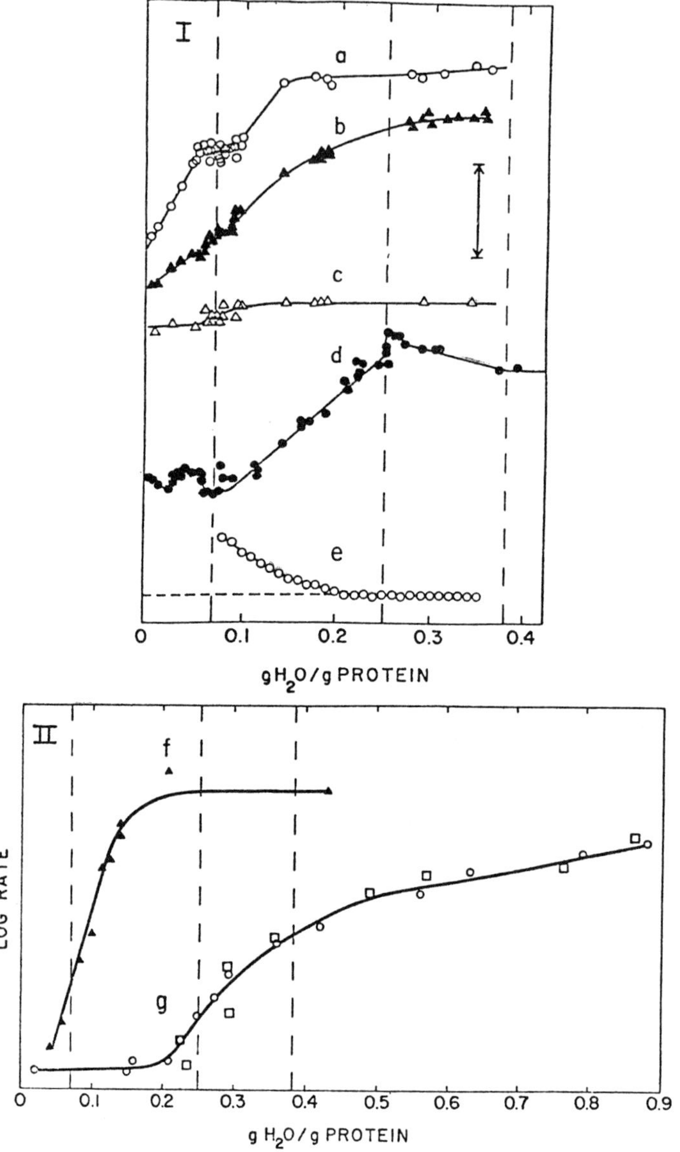

Fig. 2. Effect of hydration on lysozyme. Part I, time-average properties. Part II, dynamic properties. From top to bottom the curves are: (a) carboxylate absorbance (1580 reciprocal cm); (b) amide I shift (ca 1660 reciprocal cm); (c) OD stretching frequency (ca 2570 reciprocal cm); (d) apparent specific heat capacity; (e) diamagnetic susceptibility; (f) log rate peptide hydrogen exchange; (g) square symbols, log enzyme rate, and circles, log reciprocal rotational correlation time of an ESR probe. From Rupley et al. (1983), with permission.

the extent to which the thermal response of the lysozyme-water
system deviates from ideal behavior.

The nonideality of the system shows three discontinuities:
at 0.38, 0.25 and 0.07 h. The 0.38 h discontinuity corresponds
to the hydration end point, above which the nonideality per mol
of protein is constant. Condensation of water over the weakest
interacting regions of the protein surface begins at near 0.25
h, and proceeds until completion of the hydration shell at 0.38
h. This event is a typical condensation process. It occurs
within a narrow range of water activity, near unity, and is
associated with addition to the surface of about 1/3 of the
total hydration shell. The third discontinuity is at 0.07 h,
where the nonideality function shifts from generally downward-
trending to strongly upward-trending. It corresponds to another
2-dimensional condensation process, the formation of water
clusters from the dispersed set of water molecules present at
hydration levels below 0.07 h, where predominantly the water is
bound to ionizable protein surface groups. This transition of
the surface water at 0.07 h is seen also in the infrared
spectroscopic properties (curve (c), Fig. 2; see Careri et al.
(1979)).

The rise and fall in the apparent heat capacity between 0
and 0.07 h is a transition heat contribution, associated with
the transfer of protons from carboxylic acid to basic groups of
the protein. As the protein is dried to very low hydration
levels, below about 0.05 h, the system reacts to a high net
charge in the unfavorable low-dielectric environment of a
protein surface stripped of water. Charge is neutralized by
changing the effective pK order, so carboxylate groups become
among the most basic of the ionizable protein groups. Infrared
measurements show this transition in the carboxylate pK (curve
(a), Fig. 2; see Careri et al. (1979)).

Dynamic Properties

Protein rate processes are affected by hydration. Motion
at the protein surface was determined by electron spin resonance
(ESR) measurements of the rotational correlation time for a
nitroxide spin probe, TEMPONE, noncovalently bound at the
protein surface (Rupley et al., 1980). Below 0.2 to 0.25 h the
motion of the spin probe is frozen (curve (g), Fig. 2). Above
this hydration level the rotational motion grows explosively,
with an apparent 15th-order dependence on water activity.

These measurements suggest that the development of surface
motion is tied to the condensation event that completes the
hydration shell, seen in the heat capacity isotherm as the 0.25
h discontinuity. This correspondence appears physically
plausible. For rotational motion of a relatively large molecule
like TEMPONE at the protein surface, many protein-TEMPONE
interactions must be broken and reformed. The presence of a
substantially continuous water region should facilitate the
breaking and reforming of protein-TEMPONE interactions, through
the intermediate formation of protein-water and TEMPONE-water
interactions.

A series of nuclear magnetic resonance studies on the
hydration of lysozyme have been carried out by Bryant and his
collaborators. Water in partially hydrated powder samples of

lysozyme responds as a fluid, with motion increasing with
increase in hydration level (Hilton et al., 1977). The
characteristic time of the water motion is slower than the bulk
water value by less than a factor of 100. Study of a protein,
bovine serum albumin, labeled covalently with a nitroxide spin
probe, gave data consistent with a diffusion constant of the
surface water about a factor of 5 slower than the bulk water
value (Polnaszek and Bryant, 1984). The NMR results are in close
agreement with high-frequency dielectric relaxation measurements
of the water motion in protein powders, carried out by Harvey
and Hoekstra (1972). The discontinuity in the thermodynamic
properties at 0.25 h is reflected in the NMR and dielectric
properties of the interfacial water, although less strongly than
in the ESR measurements of the TEMPONE probe.

The rates of exchange of the labile amide hydrogens of a
protein are strongly affected by folding of the protein.
Folding can reduce the exchange rate by more than 8 orders of
magnitude. The hydration dependence of the amide exchange is
complete at below half coverage of the surface with water (curve
(f), Fig. 2). Analysis of the hydration dependence shows that
the exchange is only 3rd order in water activity (Schinkel et
al., 1985). It is possible that the hydration dependence
reflects entirely the participation of water in the underlying
chemistry of the exchange process, rather than a special effect
characteristic of protein hydration. At any rate, the exchange
process is relatively independent of the hydration process,
which suggests that the internal motions of the protein that are
sensed by hydrogen exchange are uncoupled from the surface and
its hydration.

PICTURE OF THE FULLY HYDRATED PROTEIN

Analysis of sorption isotherms, which are closely similar
for globular proteins (Kuntz and Kauzmann, 1974), shows that the
free energy of the surface water differs from the bulk solvent
by only 0.5 kcal/mol of water, i.e., by about the thermal
energy, RT. Apparently, the values for the thermodynamic
properties of the hydration shell are slightly but not strongly
different from the bulk water values (Table 2). From the
nonideality in the heat capacity at the end point, the change in
heat capacity associated with hydration of the protein is
estimated as 2.5 cal/K-mol of water. Other properties of the
surface water, for example, the partial molar volume and
enthalpy (Table 2; see Rupley et al. (1980)), show also
significant but small differences (10-15 percent) from the bulk
water values.

About 300 water molecules are sufficient to cover the
lysozyme surface. This is a remarkably small number. A
calculation based on the crystal structure of lysozyme shows the
surface is about 6000 square angstroms in area (Shrake and
Rupley, 1973). Thus each water covers, on average, about 20
square angstroms, which is twice the "projection" of a water
molecule packed as in the liquid. The arrangements of the water
on the protein surface must be special, and there can be no
multilayer water. Moreover, whatever arrangements there are must
integrate simply into the bulk water. It appears likely that
the protein surface selects local arrangements of water from

Table 2. Time Average Properties of Interfacial Water
for Lysozyme at Full Hydration

Amount of water:

305 mol of water/mol of lysozyme

Surface area covered per water molecule:

21.6 sq. angstroms/molecule

Thermodynamic differences between interface and bulk water:

free energy	=	-0.5 kcal/mol
enthalpy	=	-1.4 kcal/mol
entropy	=	-3.0 cal/k-mol
heat capacity	=	2.5 cal/k-mol
volume	=	-1.7 ml/mol

Picture consistent with observations:

interface water similar to bulk water

monolayer interface

interaction with charged and polar surface
groups selects locally ordered water
arrangements

arrangements selected mesh with bulk solvent
and cover large area per water

fluctuations between many instantaneous
arrangements, as in bulk water

water motion about 10 times reduced from
bulk water

among the many possible in bulk water, so satisfying the
requirement of large area covered per hydration water and also
the requirement of meshing with the surrounding bulk water.
Hydrogen-bonded water layers present in ice I, orthogonal to the
c-axis, show a projected area of 20 square angstroms, with 4.5
angstrom separation between water molecules. Infrared
spectroscopic and other measurements suggest that the polar and
charged groups are the primary sites of hydration and are filled
at hydration levels below 0.25 h. The average spacing of polar
and charged groups on a protein surface is 4-5 angstroms.

Blake et al. (1983) analyzed high resolution x-ray
diffraction data for human lysozyme (HL) and tortoise egg white
lysozyme (TEWL), locating 143 molecules of ordered water per HL
molecule, and 122 molecules per TEWL molecule. The ordered

water covers about 75 percent of the available surface of the
the proteins. Thus the estimate of surface water is in good
agreement with the 300 molecules of water seen by the heat
capacity measurements. The area covered per water molecule is
estimated as 19-21 square angstroms. The ordered water is in a
monolayer about the protein surface. The average number of
hydrogen bonded neighbors is 2-3. If the interaction between
the ordered water in the monolayer and the bulk solvent is taken
into account, then on average the number of hydrogen-bonded
neighbors for water in the surface monolayer is as for bulk
water. Ice-like or other regular structures were not seen,
perhaps a peculiarity of lysozyme. The waters are nearly all
bound to polar or charged protein atoms. Charged side chains
bind generally two ordered waters, polar groups one. There is
nothing surprising about being able to locate water at defined
(time-average) positions at the surface of a protein in the
crystal. Water seen about a protein surface group, by x-ray
diffraction, is similar to what would be seen by an observer
moving with the rotating coordinate system of a molecule of bulk
water and looking along one of the tetrahedral hydrogen-bonding
directions. For a review of water in protein crystals, see
Edsall and McKenzie (1983).

To summarize, the protein surface is covered by a monolayer
of water, differing only slightly in thermodynamic properties
from the bulk water, and organized in local arrangements that
are a subset of the local arrangements found in bulk water.
These local arrangements, like those in bulk water, fluctuate,
but have a defined time-average structure. The pictures of the
hydration water derived from the thermodynamic and diffraction
measurements are in notably close agreement.

The picture given above is a first-order description.
Clearly, water outside the monolayer region at the protein
surface participates in some protein processes. Rate processes
involving cooperative motions of water and larger ligands may
sense "multilayer" water: the change in TEMPONE motion with
hydration of lysozyme is largely but not completely saturated at
0.38 h. The "hydration forces" described for assemblies of
proteins, nucleic acids and membranes (Parsegian et al., 1985)
are cooperative phenomena, for which the energies of interaction
per single water molecule are much below the thermal energy, RT.

HYDRATION AND FUNCTION

Folding

The folding of the protein appears to produce a surface
that minimizes perturbation of the surrounding water. This
property can be seen as complementary to another thermodynamic
aspect of protein folding, the maximum withdrawal of hydrophobic
side chains from solvent. As suggested above, the minimal
perturbation of the solvent reflects the arrangement of charged
and polar groups on the protein surface. Not all biological
macromolecules are like globular proteins in minimally
perturbing their environment. Nucleic acids require, by weight
or surface area, about twice the amount of water required by
globular proteins for their hydration (Rupley and Siemankowski,
1986).

Chemistry of Transition States and Intermediates

There is a considerable literature on the participation of solvent in organic reactions, and we cite here only several of many such studies related to enzyme processes. The work of Bruice and collaborators on carboxylate attack on phenyl esters has already been noted (Bruice and Turner, 1970). Young and Jencks (1977) show that lysozyme must stabilize the putative oxocarbonium ion intermediate by 5-7 kcal/mol, compared with the species formed in the non-enzymic hydrolysis of glycosides and acetals, that reacts rapidly with water. Bell et al. (1974) have analyzed the reaction of chymotrypsin in several solvents, concluding that the substrate undergoes significant desolvation in the enzymatic activation process.

Enzyme Reactions in Partially Hydrated Powders

The reaction of lysozyme with its hexasaccharide substrate (Rupley et al., 1980) develops sharply at 0.2-0.25 h, and exactly parallels the hydration dependence of the surface motion of the ESR probe TEMPONE (curve (g), Fig. 2). The lysozyme reaction is understood to involve movement of the substrate more deeply into the active site cleft, during the passage from the last equilibrium complex through the transition state of the rate-determining step to form the oxocarbonium ion intermediate (Banerjee et al., 1975). Thus the development of enzymatic activity in the partially hydrated powder, at the hydration level at which surface motion becomes unfrozen, is in accord with the mechanism of the catalysis, which requires mobility of the substrate on the enzyme surface.

Protonic Conduction

There is another correlation between the development of the enzymatic activity of lysozyme and the chemistry of the protein surface. Dielectric measurements (Careri et al., 1985; Careri et al., 1986) on partially hydrated powders of lysozyme have shown that there is protonic conduction associated with a percolation transition involving water clusters. The binding of hexasaccharide substrate at the active site reduces the proton flux by half. It also shifts the hydration level of the percolation transition. We understand this to mean that a significant proportion of the protonic conduction paths pass through the active site. Lysozyme, like many other enzymes, exhibits general catalysis as an element of its mechanism of reaction with substrate. It is an attractive possibility that the active site may be special in facilitating proton movement.

Substrate Binding

Various enzyme studies have addressed the possible importance of changes in hydration associated with substrate binding and catalysis. For example, Sloan and Velick (1973) measured the buoyant density and preferential hydration of yeast glyceraldehyde phosphate dehydrogenase and its complex with NAD. They concluded that there is a 15 percent decrease in preferential hydration, corresponding to a 6 percent volume contraction, when the coenzyme is bound. X-ray diffraction studies have shown that conformational changes are associated with binding of cofactor and substrate to this class of

dehydrogenase. A cooperative release of water could give a
large cumulative effect, even though the free energy change per
water molecule transferred to bulk solvent may be small. If the
water in the active site has a special chemistry, such as a
large free energy of transfer from the bulk solvent, compared to
the average water in the hydration shell, then contributions of
this sort would be expected to be even more substantial.

In this regard, crystallographic studies, e.g., those of
Blake et al. (1983) on lysozyme, provide examples of water in
the active site of an enzyme being located at the position
occupied by a substate atom in the enzyme-substrate complex.
This situation should be entropically favorable for binding,
because the substrate, with a single conformation in the
complex, is replacing a water or a water region with also only
one "conformation". If this situation is common, it would help
explain the tight binding of substrate obtained through
relatively few interactions with the protein, i.e., apparently
large free energies of individual hydrogen-bonding and other
interactions.

Fluctuations and Catalysis

Proteins exhibit motions with widely varied characteristic
times (Careri et al., 1975). Suggestions have been made
concerning the coupling of protein fluctuation events and
catalysis (e.g., Careri and Gratton (1986)). The hydration
shell must play a role, if only in coupling the enzyme
fluctuations with the surrounding thermal bath. A succession of
rearrangements (isomerizations) of the geometry of the enzyme-
substrate complex is perhaps a universal aspect of enzyme
reactions. The hydration shell should facilitate optimization
of the active site environment during traversal of the reaction
path, in effect acting to damp oscillations (Rupley et al.,
1983).

Tapia and Eklund (1986) carried out a Monte Carlo
simulation of the substrate channel of liver alcohol
dehydrogenase, based on the x-ray diffraction structure for this
enzyme. The addition of substrate and the associated
conformation change induce an order-disorder transition for the
solvent in the channel. A solvent network, connecting the
active-site zinc ion and the protein surface, may provide the
basis for a proton relay system.

PROSPECT

Although by no means complete, the description of the
interface between protein and solvent is at a reasonably
satisfactory level. A principal motivation for devoting
attention to the interface has been the expectation that an
understanding of it will contribute importantly to the
understanding of protein function, in particular enzyme
function. From the preceding section, it is clear that the
correspondences that can be drawn at this time are, more often
that one would like, speculative. There are enough trails and
signs, however, to indicate the direction to a more complete
analysis of the relationship between hydration and function.
Where a sharp question can be formulated and so suggest a test

of a mechanism of solvent participation, site-directed
mutagenesis provides an experimental tool, which properly may be
guided by computer simulation of the protein-solvent system.
The participation of fixed or random (percolative) network
structures in proton movement is a particularly attractive
possibility for function of the surface water, and such
suggestions would appear to be now open to test.

REFERENCES

Banerjee, S. K., Holler, E., Hess, G. P., and Rupley, J. A.,
    1975, Reaction of N-acetylglucasamine oligosaccharides with
    lysozyme. Temperature, pH, and solvent deuterium isotope
    effects. Equilibrium, steady state, and presteady state
    measurements, J. Biol. Chem., 250:4355-67.
Bell, R. P., Critchlow, J. E., and Page, M. I., 1974, Ground
    state and transition state effects in the acylation of, J.
    Chem. Soc., Perkin Trans. 2:66-70.
Blake, C. C. F., Pulford, W. C. A., and Artymiuk, P. J., 1983,
    X-ray studies of water in crystals of lysozyme, J. Mol.
    Biol., 167:693-723.
Bruice, T. C., and Turner, A., 1970, Solvation and
    approximation. Solvent effects on the bimolecular and
    intramolecular nucleophilic attack of carboxyl anion on
    phenyl esters, J. Amer. Chem. Soc., 92:3422-8.
Careri, G., and Gratton, E., 1986, The statistical time
    correlation approach to enzyme action: the role of
    hydration, in: ''The Fluctuating Enzyme,'' G. R. Welch,
    ed., pp. 227-262, John Wiley, New York.
Careri, G., Fasella, P., and Gratton, E., 1975, Statistical time
    events in enzymes: a physical assessment, CRC Crit. Rev.
    Biochem., 3:141-64.
Careri, G., Giansanti, A., and Gratton, E., 1979, Lysozyme film
    hydration events: an IR and gravimetric study, Biopolymers,
    18:1187-203.
Careri, G., Geraci, M., Giansanti, A., and Rupley, J. A., 1985,
    Protonic conductivity of hydrated lysozyme powders at
    megahertz frequencies, Proc. Natl. Acad. Sci. U. S. A.,
    82:5342-6.
Careri, G., Giansanti, A., and Rupley, J. A., 1986, Proton
    percolation on hydrated lysozyme powders, Proc. Natl. Acad.
    Sci. U. S. A., 83:6810-14.
Edsall, J. T., and McKenzie, H. A., 1983, Water and proteins.
    II. The location and dynamics of water in protein systems
    and its relation to their stability and properties, Adv.
    Biophys., 16:53-183.
Harvey, S. C., and Hoekstra, P., 1972, Dielectric relaxation
    spectra of water adsorbed on lysozyme, J. Phys. Chem.,
    76:2987-94.
Hilton, B. D., Hsi, E., and Bryant, R. G., 1977, Proton nuclear
    magnetic resonance relaxation of water on lysozyme powders,
    J. Am. Chem. Soc., 99:8483-90.
Kresheck, G. C., and Klotz, I. M., 1969, Thermodynamics of
    transfer of amides from an apolar to an aqueous solution,
    Biochemistry, 8:8-12.
Kuntz, I. D., Jr., and Kauzmann, W., 1974, Hydration of proteins
    and polypeptides, Adv. Protein Chem., 28:239-345.
Parsegian, V. A., Rand, R. P., and Rau, D. C., 1985, Hydration
    forces: what next?, Chem. Scr., 25:28-31.
Polnaszek, C. F., and Bryant, R. G., 1984, Self-diffusion of

water at the protein surface: a measurement, J. Am. Chem. Soc., 106:428-9.

Rupley, J. A., and Siemankowski, L., 1986, in: ''Membranes Metabolism and Dry Organisms,'' C. Leopold, ed., pp. 259-272, Comstock Publishing Associates, Cornell Univ. Press, Ithaca.

Rupley, J. A., Yang, P. H., and Tollin, G., 1980, Thermodynamic and related studies of water interacting with proteins, ACS Symp. Ser., 127:111-32.

Rupley, J. A., Gratton, E., and Careri, G., 1983, Water and globular proteins, Trends Biochem. Sci. (Pers. Ed.), 8:18-22.

Schinkel, J. E., Downer, N. W., and Rupley, J. A., 1985, Hydrogen exchange of lysozyme powders. Hydration dependence of internal motions, Biochemistry, 24:352-66.

Shrake, A., and Rupley, J. A., 1973, Environment and exposure to solvent of protein atoms. Lysozyme and insulin, J. Mol. Biol., 79:351-71.

Sloan, D. L., and Velick, S. F., 1973, Protein hydration changes in the formation of the nicotinamide adenine dinucleotide complexes of glyceraldehyde 3-phosphate dehydrogenase of yeast. 1. Buoyant densities, preferential hydrations, and fluorescence-quenching titrations, J. Biol. Chem., 248:5419-23.

Tapia, O., and Eklund, H., 1986, On the molecular mechanism of liver alcohol dehydrogenase: Monte Carlo simulations and x-ray studies of water in the substrate channel, Enzyme, 36:101-14.

Yang, P., and Rupley, J. A., 1979, Protein-water interactions. Heat capacity of the lysozyme-water system, Biochemistry, 18:2654-61.

Young, P. R., and Jencks, W. P., 1977, Trapping of the oxocarbonium ion intermediate in the hydrolysis of acetophenone dimethyl ketals, J. Am. Chem. Soc., 99:8238-48.

PERCOLATION PROCESSES

John A. Rupley and Giorgio Careri

Department of Biochemistry
University of Arizona
Tucson, Arizona 85716

Dipartimento di Fisica
Universita di Roma I
Roma 00185, Italy

INTRODUCTION

Measurements of the MHz-frequency dielectric properties of
partially hydrated lysozyme powders have demonstrated protonic
conduction at the protein surface (Careri et al., 1985). The
dependence of the dielectric response on hydration level shows,
at a threshold value of the hydration, the explosive growth
characteristic of a phase transition (Careri et al., 1986). This
behavior will be seen to follow closely the percolation model.

The following discussion will have several foci. The
overall intent is to present the percolation model. This model
is not widely known among biochemists. It is applicable to
disordered systems for which processes dependent on long-range
connectivity are important. Movement of ions and other species
within membranes might be expected to show percolative behavior.
Before turning attention to the percolation phase transition, we
will consider, first, as background, some general aspects of
phase transitions, and second, the dielectric measurements that
led us to apply the percolation model.

We assume familiarity with the companion contribution in
this volume, by Rupley and Careri, on "Enzyme Hydration and
Function", which we will refer to by the acronym, JR-GC.

PHASE TRANSITIONS

Phase transitions, order-disorder transitions, the critical
region, and related events occupy an honorable place in physics
and chemistry, and they consume a significant amount of space in
texts on the statistical properties of matter (e.g., Hill
(1960), Careri (1983), and Landau and Lifshitz (1977)). It would
seem to be useful to introduce several concepts at this point.

Intuitively one associates a change in phase, such as the condensation of a gas to a liquid, with a change in the extent of order of the system. Under some conditions of measurement, such as room temperature and standard pressure for the system of water vapor and liquid water, two phases are present. The thermodynamic and structural properties of the two phases differ. Because the phases are in equilibrium, the partial molar free energies must be equal. For "first order" phase transitions, such as for the water vapor-liquid equilibrium, the first derivatives of the free energy, the volume and the entropy (or heat), will differ for the two phases.

There is not always a discontinuity in the volume or entropy across a phase transition. For the water vapor-liquid system there is a pair of values of the temperature and pressure that define the critical point, at which there is an order-disorder transition, but no discontinuity in the volume or entropy. For "second order" phase transitions of this kind, the second derivatives of the free energy, most typically the heat capacity, show a discontinuity.

The region about the critical point is particularly interesting by way of theory and experiment. Near the critical point properties of the system, such as the heat capacity, change similarly for very different materials, following a power law in the difference between the value of a thermodynamic variable and its critical value. The "critical exponents" are the exponents of these very general power law relationships:

$$\text{property} = (T - T_c)^{\text{crit. exp.}}$$

Some properties, again for example, the heat capacity, diverge near the critical point. The divergence would be described by a negative value of the power law exponent. We will return to critical exponents when considering percolation and the percolation threshold, which will be seen to be analogous to the critical point in the water vapor-liquid system.

In order to analyze phase transitions, the system must be described by use of a mathematically tractable model. The Van der Waals model, hard sphere molecules with interactions, provides relationships that are very useful but do not give insight into the underlying physics and chemistry. Lattice models, typically an "Ising model", have been applied for more than fifty years to the problem of phase transitions, with striking successes. The model consists of a regular lattice of sites in one or more dimensions. Each site can be occupied in one of two ways, A or B (for some systems, one way of occupancy would be a filled site, the other an empty site). There are potential energies of interaction between sites (AA, BB, AB). Statistical methods are used to compute the thermodynamic properties of the system by properly averaging all configurations, taking into account the energies of interaction.

Perhaps the application of the Ising model most familiar to biochemists is the treatment of the helix-coil transition (for a review discussion, see Scheraga (1963)). The familiar Zimm-Bragg constants characterizing the transition for a particular polypeptide are extracted by use of such an analysis. The greater usefulness of treatments of this kind, compared with classical thermodynamic treatments, comes from the insights the

Ising model approach gives into the way in which helix and coil
regions are distributed within a single chain, and the way in
which this distribution changes with temperature and the model
parameters.

A variety of lattice models have been applied to the
problem of adsorption of gas on a surface. One, by Hill (1949),
applies directly to the problem of protein hydration. Hill
(1949) treated the case of adsorption on a heterogeneous
surface, with the further restrictions that the adsorption be
localized, i.e., that the interaction between the surface and
the adsorbate be relatively strong, and that the adsorption be
unimolecular. Lateral interactions between gas molecules
adsorbed on the surface were included in the description. This
model applies in detail to protein surfaces and the interaction
of water with them: the strong localizing interactions with the
surface and the lateral interactions are provided by water
hydrogen bonds; the hydration shell is a monolayer; and the
surface is heterogeneous, consisting of charged, polar, and
nonpolar sites.

In view of this correspondence, the predictions of the
lattice model should be observed in the protein data. The Hill
treatment predicts two phase transitions: one at low coverage of
the surface, which would be the 2-dimensional condensation of
water dispersed about the surface into clusters; the second at
high coverage, where adsorbate condenses to complete the
monolayer. Both these transitions are found for globular
proteins (see JR-GC).

We will return to the subject of lattice models when
entering into the discussion of percolation.

DIELECTRIC MEASUREMENTS

Dielectric loss and capacitance were measured for powdered
samples of lysozyme, as a function of hydration level (h, g of
water/g of protein) and frequency (10 kHz to 10 MHz) (Careri et
al., 1985; Careri et al., 1986). Fig. 1 shows the
characteristic frequency of the dielectric dispersion, as a
function of hydration level, for several types of sample. These
data show clearly that protonic conduction is the dominant
contribution to the dielectric relaxation: the addition of 10
mol of NaCl/mol of protein has no effect; deuteration reduces
the dielectric relaxation time by exactly the square root of 2,
the expected change for a rate process involving protons. The
dielectric response is strongly dependent on pH. At $h = 0.3$,
the dielectric relaxation time is 7th order in the proton
concentration.

Fig. 2 shows the effect of blocking the active site with a
substrate, which increases the dielectric relaxation time by a
factor of 2. The conductivity of the sample is proportional to
the reciprocal of the dielectric relaxation time. Thus the
substantial effect of substrate binding suggests that a large
part, perhaps half, of the proton flow over the protein surface
passes through the active site. Considering the importance of
acid-base catalysis for enzyme processes, this channeling of the
proton flow very possibly plays a role in function.

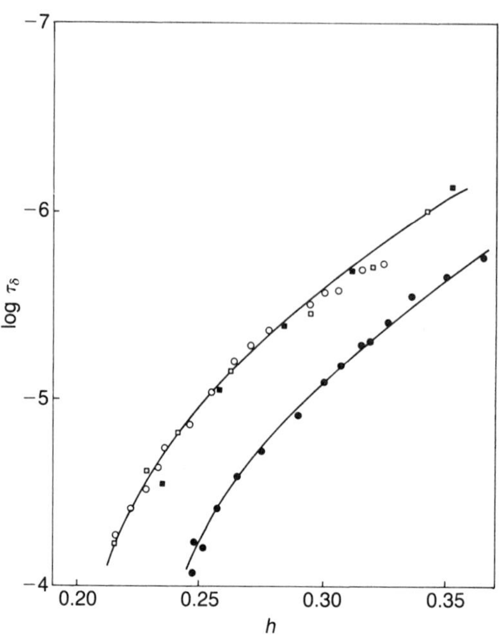

Fig. 1. The dielectric relaxation time versus hydration h for different lysozyme powder samples of pH near 7. Filled and open squares are for H_2O-hydrated samples from protein preparation 1, of pH 7.0 with 0 and 10 mol of NaCl per mol of protein, respectively. Open circles, data for two different runs on H_2O-hydrated samples from protein preparation 3, of pH 7.35. Closed circles, D_2O-hydrated samples, run under the conditions of the open-circle data. From Careri et al. (1985), with permission.

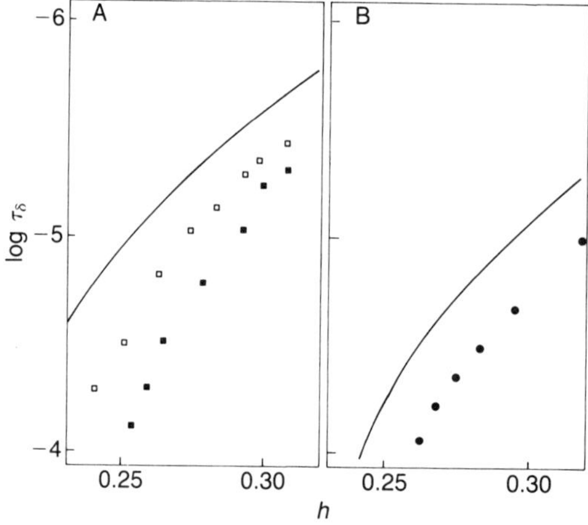

Fig. 2. Data obtained as for Fig. 1. (A) H_2O-hydrated samples: filled symbols, lysozyme with equimolar GlcNAc tetrasaccharide at pH 7.0; open symbols, with 3x molar GlcNAc tetrasaccharide at pH 6.5. (B) D_2O-hydrated sample with equimolar GlcNAc tetrasaccharide at pH 7.0. The curves of A and B are from the data of Fig. 1 for lysozyme without substrate. From Careri et al. (1985), with permission.

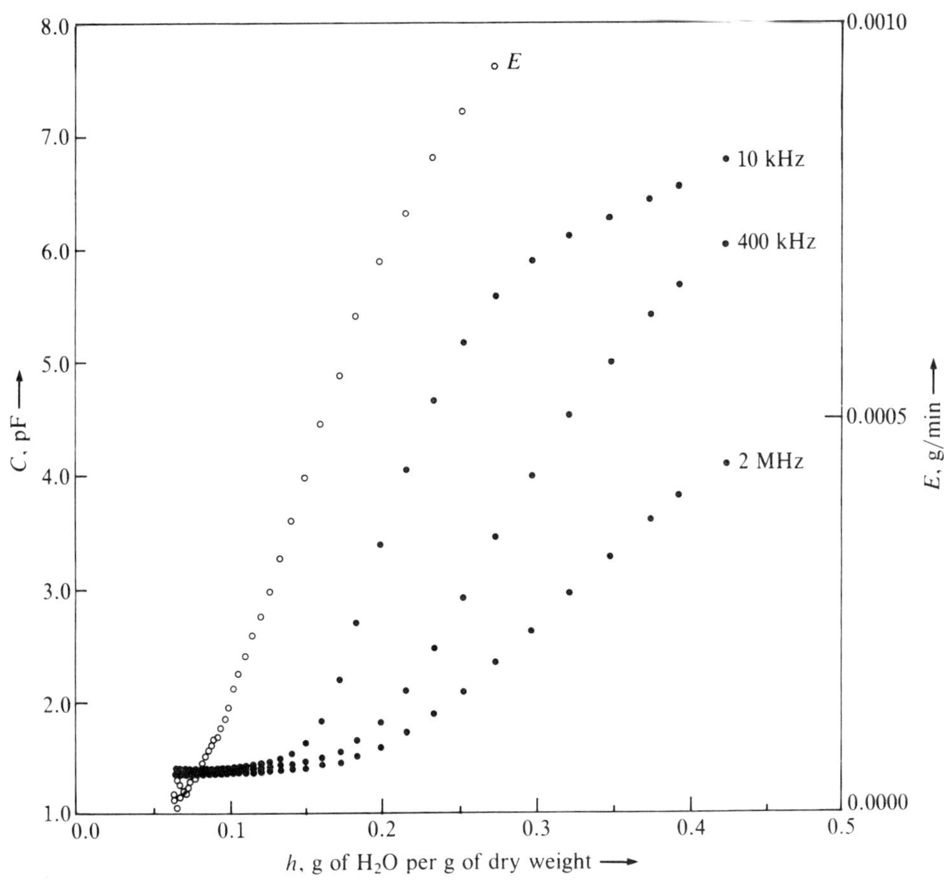

Fig. 3. Capacitance, closed symbols, for the composite capacitor
containing a sample of lysozyme powder of pH 3.11, as a
function of hydration level of the protein. The
capacitance data are given for three frequencies. The
hydration level was decreased from the high hydration
limit of >0.35 to the low hydration limit of near 0.07 h
by passage of dry air through the apparatus. The
evaporation rate E decreases to zero at the low hydration
limit. From Careri et al. (1986), with permission.

The above measurements were of the dielectric response at
near full hydration. We turn now to the initiation of the
protonic conduction, at low levels of hydration.

Fig. 3 shows that the capacitance grows explosively above a
hydration level of 0.15. This behavior is independent of
conditions of preparation of the sample (varied pH; deuterium
substitution) that affect the high hydration dielectric
response. The explosive growth is characteristic of a phase
transition. However the hydration dependence of the heat
capacity shows that there is no thermal phase transition between
0.1 and 0.25 h. We believe that the onset of protonic
conductivity can be explained as percolative behavior. Indeed,
the data of Fig. 3 and other results for lysozyme agree closely
with predictions of the percolation model. This general
physical model applies to a broad range of phenomena rooted in
spatially random events and topological disorder.

PERCOLATION THEORY

The central aspects of the percolation model can be
illustrated by a simple example, the "vandalized grid", taken
from Zallen (1983). Consider a square 2-dimensional lattice of
resistors, with two opposite edges in series with a battery and
ammeter. A random vandal stochastically severs resistors. We
ask, how does the current vary with the fraction of uncut bonds
(p)? and at what fraction of uncut bonds does no current pass?
In this case, intuition is not necessarily a good guide. But
percolation theory is. Fig. 4 shows what would be observed: an
approximately linear increase in the conductivity for increase
of p above the threshold, at p_c = 0.5; a sharp transition at the
threshold; no current passed below the threshold. At the
threshold, the net falls into pieces -- long-range connectivity
is lost.

To look at the above example differently, let us start with
an empty lattice, and throw resistors at it randomly and one at
a time (we are lucky, so each resistor sticks properly to the
lattice). Even though the resistors are distributed randomly,
some resistors will fall next to another, to form clusters.
These clusters of connected resistors will grow in size, until
at the percolation threshold (p_c = 0.5), one finds a cluster
infinite in extent, and current flows. Intuition again may not
be a good guide (but percolation theory is), as to how the
clusters vary with p. Fig. 5 shows various cluster properties
as a function of p. Just below the percolation threshold, the
average cluster size grows explosively, becoming infinite at p_c.
Just above p_c, the percolation probability, P(p), the
probability that any filled bond is part of the infinite
cluster, similarly grows explosively. The growth of the
conductivity, which reflects the richness of the
interconnections within the resistor network, lags behind P(p).

It should be apparent that the percolation model is a
special case of the Ising model. For the percolation model,
there is no energy of interaction between occupied sites. The
transition reflects the stochastic character of the system.
Fig. 5 shows clearly that at p_c there is an order-disorder
transition. Because there is no interaction energy, the
temperature second derivative, the heat capacity, is invariant
through the transition. The percolation transition is second
order, with a discontinuity in a derivative of the entropy. A
central aspect of the percolation transition, is that it is
detected as a discontinuity in a <u>process</u>. This process reflects
only the long-range connectivity within the system. In
contrast, the more familiar phase transitions, even though they
may show characteristic changes in long-range order, are
grounded in interactions between near neighbors and generally
exhibit changes in chemistry of the system.

The use of a lattice is a construct to enable analytical
treatment of a percolative system. When describing a
percolation process we wish to be independent of choice of
lattice. Zallen (1983) has shown that percolation can be
described in a lattice-independent manner by identifying p as
the fraction of the volume (or surface) filled (covered) with
the elements that support the percolation. In effect, lattice
points are expanded into close-packed spheres, and the lattice-
dependent packing is taken into account by calculating p as the

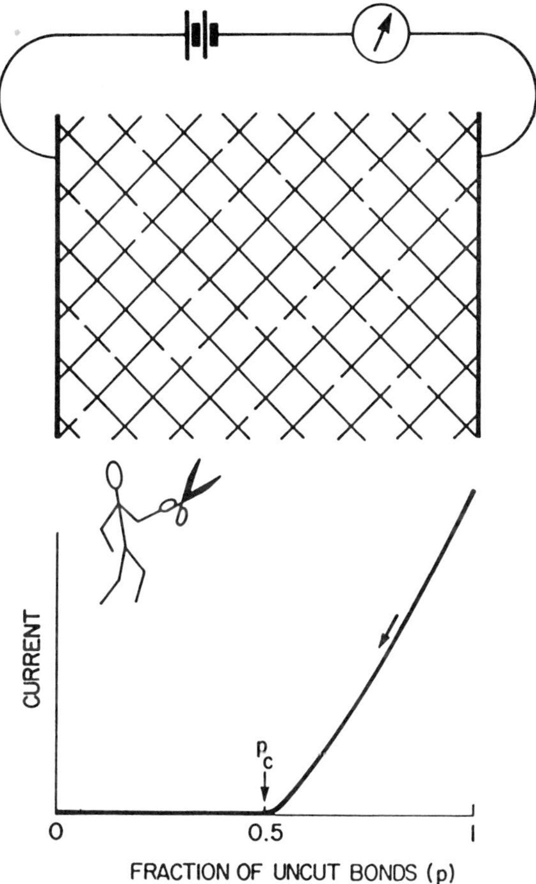

Fig. 4. The randomly cut network as an example of percolation. From Zallen (1983), with permission.

sum of the volumes of the "conducting" spheres over the total
volume of the system (spheres and interstices). With this
definition of p, the threshold for 2-dimensional percolation is
$p_c = 0.45$. The threshold depends on the dimensionality of the
system; in 3 dimensions, $p_c = 0.16$, a lower fractional occupancy
than in 2 dimensions, as should be anticipated.

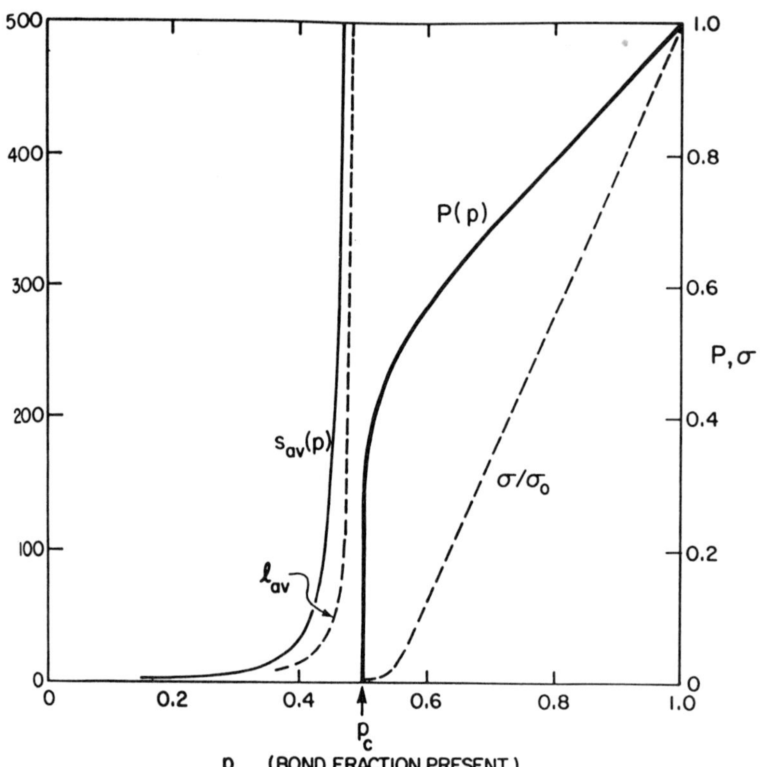

Fig. 5. The behavior, as a function of the fraction (p) of filled
bonds, of key properties that characterize bond
percolation on the square lattice in two dimensions.
P(p) is the percolation probability, $s_{av}(p)$ is the
average cluster size, l_{av} is the mean spanning length of
the clusters, and σ is the random-network conductivity.
From Zallen (1983), with permission.

The curves of Fig. 5, showing strong changes in cluster and
system properties near the percolation threshold, suggest that
there should be power laws with critical exponents that
characterize such changes. These would be analogous to the
critical exponents that characterize the more familiar phase

transitions near the critical point. Indeed such relationships
are exhibited by percolation processes. The critical exponent,
t, for the growth of the conductivity near p_c, has a value
approximately 1.3 for percolation in two dimensions and 2.0 for
percolation in 3 dimensions.

PROTONIC PERCOLATION AND PROTEIN HYDRATION

   One can now understand why the capacitance data of Fig. 3
are in agreement with the percolation model. The 2-dimensional
percolation threshold is predicted to be at a hydration level of
0.45 x 0.38 = 0.17 for a transition dependent on the protein
surface water. The observed value, estimated from capacitance
data for pH 3 to 8, is 0.15. The invariance of the threshold to
change in pH or deuterium substitution is expected, because the
onset of conduction depends only on the fraction of the surface
filled by conducting elements. In contrast, the magnitude of
the conductivity at high hydration should depend strongly on the
chemistry of the surface, which reflects the pH, etc., as is
observed.

   Fig. 6 shows that the growth of protonic percolation on
lysozyme shows the expected dependence on hydration level. The
critical exponent, t, describing the change in d.c. conductivity
close to p_c has the expected value of 1.3 for a 2-dimensional
percolative process. At larger values of $(h - h_c)$, the critical
exponent is characteristic of a 3-dimensional process. At these
higher hydrations there may be a contribution from 3-dimensional
networks, established through contacts between neighboring
protein molecules.

   Complexation of the active site with a substrate, the
GlcNAc tetrasaccharide, as noted above, reduces the conductivity
by half. It has no effect on the critical exponent, t,
indicating that the percolation remains 2-dimensional. However,
the system is more complex in its behavior than the free
molecule: the percolation threshold is shifted to $h_c = 0.22$.
Thus the onset of enzymatic activity coincides both with the
onset of protonic percolation and with the unfreezing of surface
motion (JR-GC). It is not possible to say which if either of
these correspondences is central to catalysis.

   To summarize, we picture the protonic process as being
transfer along threads of hydrogen-bonded water molecules
adsorbed on the protein surface. At the percolation threshold,
there is a sharp change in the characteristic length that
describes the connectivity of the surface water. Above h_c the
mean free path of the protons is long, and the threads act as
shorts bypassing the local details of the protein surface.
Between 0.1 and 0.25 h the surface water is in clusters (see
above and JR-GC), and these are mobile. Thus the hydrogen-
bonded conduction threads must also fluctuate. The stochastic
basis of the percolation model is consistent with fluctuating
conduction paths. It is possible that some regions of the
surface contribute more importantly to proton flow, as suggested
by the effect of complexation with substrate saccharide.

   This picture should hold for other globular proteins.
First, the hydration behavior of proteins in general, judged by

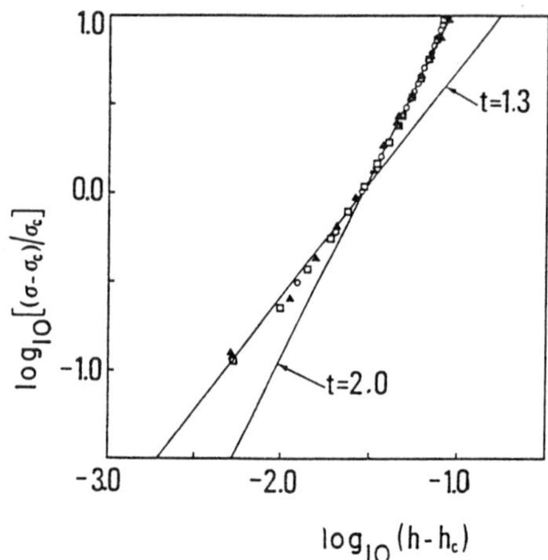

Fig. 6. Hydration dependence of the dc conductivity of lysozyme
powders of near pH 7.0, near the percolation threshold,
plotted according to the equation:

$$\sigma(p) - \sigma(p_c) \sim (h - h_c)^t$$

Open circles, native lysozyme hydrated with H_2O. Open
squares, native lysozyme hydrated with D_2O. Closed
triangles, 1:1 complex of GlcNAc tetrasaccharide with
lysozyme hydrated with H_2O. The lines are drawn for
values given by theory for the critical exponent for
two-dimensional percolation (t = 1.3) and three-
dimensional percolation (t = 2.0). From Careri et al.
(1988), with permission.

the sorption isotherms, is closely similar. Second, the
percolation transition is essentially independent of system size
and the details of the chemistry of the surface.

OTHER APPLICATIONS OF THE PERCOLATION MODEL

One of the more attractive features of the percolation
model is its wide applicability. It has provided a well-defined
and tractable model for analyzing an exceptional variety of
processes (Table 1). The theory of polymer gelation (Flory,
1941), although not recognized until recently as being based on
a percolation model, was the first such problem to be solved.

There should be a correspondence between proteins at low
hydration and and proteins in membranes, which have a portion of
their surface withdrawn from solvent. A significant amount of
water may be bound to the protein surface internal to the
membrane (Rupley and Siemankowski, 1986). Dielectric
measurements on the purple membrane of <u>Halobacterium Halobium</u>
(Rupley et al., 1988), similar to those carried out for
lysozyme, have demonstrated 2-dimensional protonic percolation,
with somewhat more complex properties than found for lysozyme.
The percolation threshold is at the hydration level at which

244

                  Table 1. Applications of percolation theory
                            (see Zallen (1983))

Phenomenon or System   Transition

Flow of Liquid in a porous medium   Local/extended wetting

Spread of disease in a population   Containment/epidemic

Communication or resistor networks   Disconnected/connected

Conductor-insulator composite   Insulator/metal
materials

Composite superconductor-metal   Normal/Superconducting
materials

Discontinuous metal films   Insulator/metal

Stochastic star formation in   Nonpropagation/propagation
spiral galaxies

Quarks in nuclear matter   Confinement/nonconfinement

Thin helium films on surfaces   Normal/superfluid

Dilute magnets   Para/ferromagnetic

Polymer gelation, vulcanization   Sol/gel

The glass transition   Liquid/glass

Variable-range hopping in   Resistor-network analog
amorphous semiconductors
----

changes in the d.c. photoresponse have been observed (Varo and
Keszthelyi, 1983).

     Percolation behavior is based on randomness of the
arrangements, here conducting elements distributed about the
protein surface or the membrane. The percolation model suggests
that it may not be necessary to have a rigid geometry and
definite pathway for conduction, as implied by the "proton-wire"
model of membrane transport. Statistical assemblies of
conducting elements need only partially fill a surface or pore.
For a 3-dimensional pore, in principal only 1/6 of the
conduction sites need be filled. Conduction could be switched
on or off by addition or subtraction of a few elements,
reflecting shift in the fractional occupancy across p_c. For
proton pumps, the fluctuating random percolation networks would
serve for diffusion of the ion across the water poor protein
surface, to where the active site would apply a vectorial
"kick". In this view, the special non-random structure would be
limited in size, to a dimension commensurate with that found for
proteins such as enzymes.

Connectivity has a central role in biochemistry and
biology, and one imagines the percolation model, with its focus
on connectivity, should have wide application. At the molecular
level, percolative behavior might be found, for example, for
systems of proteins that function coordinately, as in metabolic
pathways; for functional interactions between proteins embedded
in a membrane; for the interaction between domains in the
folding of a polypeptide.

For those interested in learning more about the percolation
model and percolative transitions, we recommend the following
books: Zallen (1983); Stauffer (1985); and a collection of
papers edited by Deutscher et al. (1983).

REFERENCES

Careri, G., 1983, ''Order and Disorder in Matter,''
    Benjamin/Cummings, Menlo Park.
Careri, G., Geraci, M., Giansanti, A., and Rupley, J. A., 1985,
    Protonic conductivity of hydrated lysozyme powders at
    megahertz frequencies, Proc. Natl. Acad. Sci. U. S. A.,
    82:5342-6.
Careri, G., Giansanti, A., and Rupley, J. A., 1986, Proton
    percolation on hydrated lysozyme powders, Proc. Natl. Acad.
    Sci. U. S. A., 83:6810-14.
Careri, G., Giansanti, A., and Rupley, J. A., 1988, Critical
    exponents of protonic percolation in hydrated lysozyme
    powders, Phys. Rev. A, 37:2703-2705.
Deutscher, G., Zallen, R., and Adler, J., eds., 1983, ''Annals
    of the Israel Physical Society, Vol. 5: Percolation
    Structures and Processes,'' Adam Hilger, Bristol.
Flory, P. J., 1941, J. Am. Chem. Soc., 63:3091.
Hill, T. L., 1949, Statistical mechanics of adsorption. VI.
    Localized unimolecular adsorption on a heterogeneous
    surface, J. Chem. Phys., 17:762-771.
Hill, T. L., 1960, ''An Introduction to Statistical
    Thermodynamics,'' Addison-Wesley, Reading.
Landau, L. D., and Lifshitz, E. M., 1977, ''Statistical Physics,
    3rd Edition Part 1,'' Course of Theoretical Physics, Vol.
    5, Pergamon Press, Oxford.
Rupley, J. A., and Siemankowski, L., 1986, in: ''Membranes
    Metabolism and Dry Organisms,'' C. Leopold, ed., pp. 259-
    272, Comstock Publishing Associates, Cornell Univ. Press,
    Ithaca.
Rupley, J. A., Siemankowski, L., Careri, G., and Bruni, F.,
    1988, Two-dimensional protonic percolation on lightly
    hydrated purple membrane, Proc. Natl. Acad. Sci. U. S. A.,
    in press.
Scheraga, H. A., 1963, Intramolecular bonds in proteins. II.
    Noncovalent bonds, in: ''The Proteins, Vol. 1. 2nd Ed,'' H.
    Neurath and R. L. Hill, eds., pp. 477, Academic, New York.
Stauffer, D., 1985, ''Introduction to Percolation Theory,''
    Taylor and Francis, Philadelphia.
Varo, G., and Keszthelyi, L., 1983, Photoelectric signals from
    dried oriented purple membranes of Halobacterium halobium,
    Biophys. J., 43:47-51.
Zallen, R., 1983, ''The Physics of Amorphous Solids,'' John
    Wiley, New York.

STATICAL   AND   DYNAMICAL   PROPERTIES   OF   MACROMOLECULAR
SOLUTIONS

F. Wanderlingh[*]
and
R. Giordano

[*]Inst. of Spectroscopic Techniques - CNR - University- Messina
Inst. of Phys. - University - Messina

1) Introduction

In the present lesson we would like to present some experimental results
that, in our opinion, could be relevant for an over-all comprehension of the
whole mechanism of enzymatic catalysis.

The puzzle of describing the enzymatic action is indeed very complicated,
because it encompasses a wide range of space and time scales.

At each time scale some peculiar behaviour takes place (or must take
place) in order to explain the entire process. However a common view seems
to arise from different investigations: some coherence, some correlation   is
needed between usually uncorrelated events,   for the machinery to work out.
In the works we present here, we focus our attention onto large scale
phenomena, involving a population of proteins in a macromolecular solution.
In a sense we are looking for the behaviour characterizing energy exchanges
between the thermal bath and the proteins.

Our goal is to find out if such an exchange (the "input" energy for a protein)
is in fact a random process. If this is the case, the scale of interest starts
from characteristic times and lengths involved in a single macromolecule (a
few tenths of Å).

If however correlated behaviour also takes place in the solution as a whole,
the above mentioned limit would be noticeably enlarged.

Now, after a large number of experimental results carried out with different
techniques, we believe that a collective, coherent, correlated behaviour
occurs in macromolecular solutions.

In addition, recent neutron scattering experiments showed that a "microscopic" modification in the density of states is strictly connected to the "macroscopic" correlated behaviour.

In the following sections, we sum up the fundamental experimental results of our investigation.

The second part will be devoted to a tentative explanation of observed phenomena, i.e. we try to construct a semi-phenomenological model that should be considered in describing the energy exchanges between proteins and thermal bath.

2) Thyxotropic structures

A thyxotropic structure is an ordering that turns out to be stable under small enough perturbation. However such a structure would be suddendly destroyed if the perturbation exceedes the prescribed limit.

If a thyxotropic structure arise in a fluid, the system will show unusual properties, as far as its rheological behaviour is concerned.
Namely:

1)- The viscosity becomes very large and strongly depends on the applied strain.

2)- There is possibility of a nonvanishing shear stress at zero shear rate (yield point).

3)- The above mentioned properties suddenly disappear if a mechanical disturbance destroys the structure.

The experiments described in the present section, show that a thyxotropic structure is built up, in a macromolecular solution in the course of time. The general behaviour of the observed phenomena is always the same, irrespectively of the kind of macromolecules involved. In addition the phenomenon is hardly affected by small changes in the external parameter, like temperature or pH values.

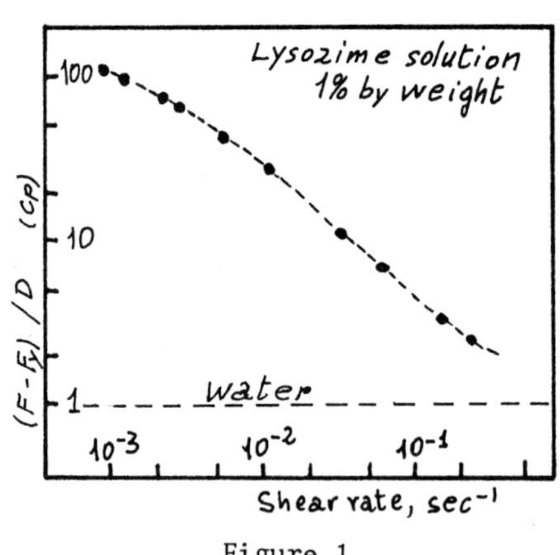

Figure 1

248

Our group performed this kind of measurement in several macromolecular solutions[1,2,3] (e.g. Lysozime, BSA, Haemoglobin, DNA) that all behave in the same way, apart from some scaling factor (that seems to be related to the size of the macromolecule involved).

A relevant common feature of all the experiments consist in the very large difference between the obtained values for the viscosity η_0 measured in an usual (capillary-flow) viscometer and the viscosity η measured at an extremely slow shear rate (Rotating cylinders viscometer). Even at very small concentrations, when η_0 does not differ appreciably from the pure water value (10^{-2} poise), η can be as large as some poises.

If one defines an "apparent viscosity" simply as the ratio between shear stress and shear rate, experimental results appear as shown in fig.1.

The viscosity decreases as the shear rate increases. The η_0 values is recovered, only at a large enough rate. In fig.2 the very long time needed for the shear stress to relax after the stopping of the shear rate, is clearly shown, together with the existence of a residual non zero shear stress (yeld point).

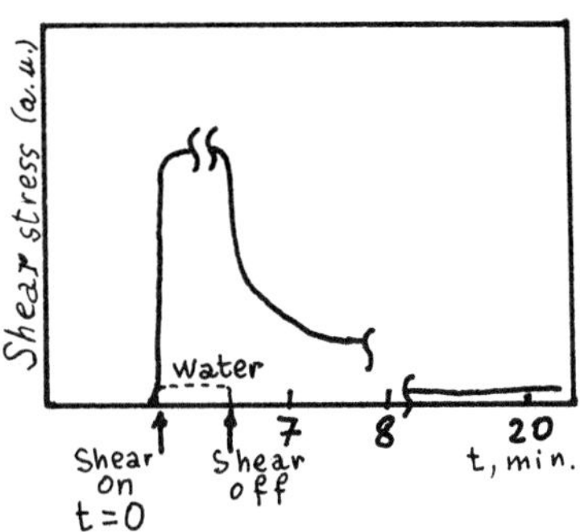

Figure 2

The rheological properties of a thyxotropically structured solutions, can be described by the phenomenological equations:

$$F = (\eta_0 + c\,\lambda)\,D + F_y$$
$$\frac{d\lambda}{dt} = a - (a + bD)\,\lambda \tag{1}$$

where: F = shear stress; D = shear rate; η_0 = viscosity in absence of the structure. The parameter λ ($0 \leq \lambda \leq 1$) measures the "degree" of development of the structure, that builds up in the course of time according to the second equation above. Here a is the rate at which the structure develops in absence of any disturbance (i.e. at zero shear rate), while b measures the "damage" introduced by the shear rate itself.

The above equations can be manipulated in order to predict the time behaviour of the "apparent viscosity" of the system. In fig.3 such a theoretical prediction is fitted with experimental results.
Such a fit, in turn, allows the evaluation of the implied parameter. Typical

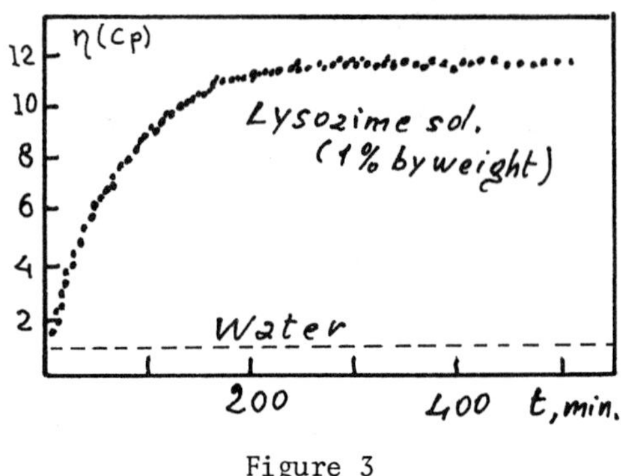

Figure 3

obtained values are, e.g. (in a lysozime solution) a yeld point $F_y = 10^{-3}$ dyn/cm², and c = 1.4 poise (the "viscosity" of a fully developed undisturbed structure). The time constant "a" as a function of concentration is shown in fig.4.

Notice the existence of a rather large plateau, that indicates some unsensitivity of the structure at different concentrations.

The peculiar behaviour of the structured solutions can also be shown by acoustic measurements[4]: the absorption coefficient (α / f^2) turns out to be frequency dependent, indicating the existence of some relaxational process. In addition the acoustic behaviour of the solution changes in time: the acoustic absorption coefficent decreases until a steady value, is reached which is only slightly larger than that of pure water. In such a condition the coefficient itself no longer depends on the

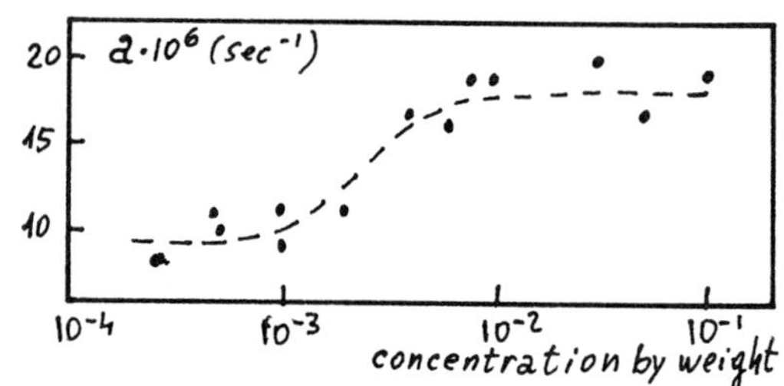

Figure 4

frequency and the relaxation process seems to disappear (see fig.5). The time constant that characterizes such a phenomenon turns out to have the same value implied in viscosity measurements.

At first sight the _decrease_ of acoustic absorption seems quite surprising if compared with the _increase_ of viscosity.

After all, viscosity is one of the main sources of the acoustic dissipation. However the phenomenon can be easily understood in terms of a growing thixotropic structure that gives rise to a diffusional relaxation process, until the fully developed structure prevents any diffusional motion, behaving like

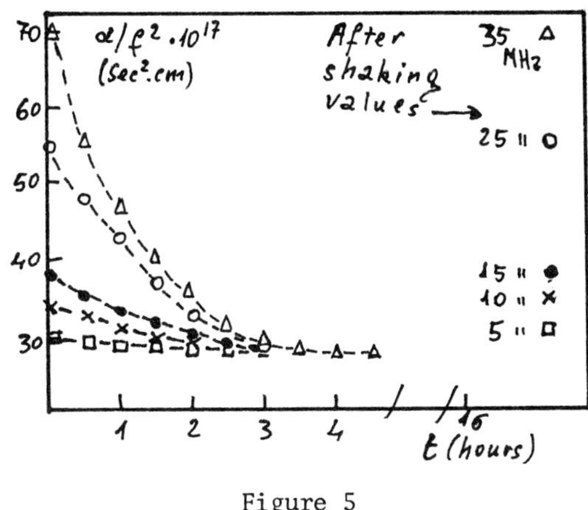

Figure 5

a rigid system.

In a relaxation process, the acoustic absorption can be written as

$$\frac{\alpha(\omega)}{\omega^2} = A + \frac{B}{1 + (\omega\tau)^2} \quad (2)$$

where τ is the relaxation time, that in the case of a diffusion process is given by[5]

$$\tau = \frac{m}{6\,\pi\,\eta\,R} = \frac{2}{9}\,\frac{\rho}{\eta}\,R^2 \quad (3)$$

Here m is the mass of the diffusing objects of radius R and density ρ. Increasing R, τ also increases, i.e. the relaxation frequency τ^{-1} shifts towards lower values, and the acoustic absorption, measured at higher frequencies ($\omega > 5$ MHz) decreases.

In eq. 3, R can be interpreted as a correlation length characterizing a cluster of spatially ordered molecules. From experimental results it turns out that R grows in time, becoming larger and larger. Ultimately such a correlation length becomes macroscopic (percolation) and the system behaves more or less like a solid lattice embedded in the solvent. Viscosity reaches its maximum value, while the acoustic relaxation frequency goes to zero.

3) <u>Light-scattering experiments</u>

Light scattering experiments furnish rather direct evidence of the existence of long-lived, long ranged structures in macromolecular solutions.

In this kind of experiment, structural information can be obtained both from static and dynamic light scattering .

In the first case the experiment gives the so-called static structure factor S(k)

$$S(k) = \int d\omega S\,(k,\omega) \quad (4)$$

In the second case the power spectrum of the scattered light gives information about the characteristic times involved in the "diffusional" processes of the system. In turn the variation of the "diffusion coefficient" as a function of the exchanged wave vector k can again be related to structural properties[6]:

$$D(k) = \frac{D_0}{S(k)} \qquad (5)$$

Eq. 5 suggests a "dynamical" interpretation of the structural properties: those preferred fluctuations, characterized by a wave vector k corresponding to a maximum in the structure factor, are also characterized by a longer decay time. In a sense they are more probable because they live for a longer time.

It is to be stressed that the S(k) we experimentally found is to be considered as a product:

$$S_{exp}(k) = F(k) \, S(k) \qquad (6)$$

where F(k) is the so called "form factor" characterizing the single scatterer.

Actually in order for a form factor to be distingushable in a light scattering experiment, the size of the single scatterer must be comparable with the wavelength of light, i.e. some thousand of Å. Clearly such a role cannot be played by a single macromolecule. In addition a cluster of macromolecules can be considered as a "single scatterer" if and only if a strong dynamical correlation exists between the molecules belonging to the cluster: in a sense the clusters would behave more or less like a rigid object in order to give rise to a well defined form factor.

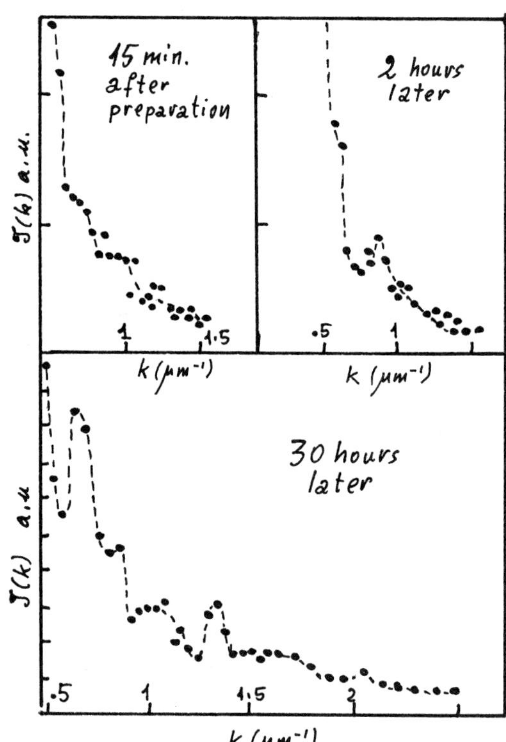

Scattered intensity as a function of exchanged wave vector. (DNA solution)

Figure 6

We performed a number of experiments of this kind, by using the correlation spectroscopy technique.[7,2,8,9]

The obtained results can be summarized as follows:

i) Elastic light scattering shows from the beginning the existence of large scatterers,indicating that clusterization takes place in a very short time.

ii) In the course of the time, the angular dis-tribution of the scat-tered light evolves in a structured diffraction pattern, indicating the building up of a spatial ordering among clusters. Such an order disappears if a mechanical distur-bance destroys the spatial structure.

As an example we show such a behaviour in a DNA solution[9] in fig.6.

As far as the dynamical properties are concerned, the correlation spectro-scopy furnishes the time autocorrelation function $C(k)$ of scattered light. In the case of a freely diffusing object (Brownian scatterers) such a function behaves like a single exponential decay:

$$C(k,t) = A(k) \, e^{-\Gamma t} \qquad (7)$$

being

$$\Gamma = D \, k^2 \qquad (8)$$

In eq. 8 the diffusion coefficient D is a constant. In our experiments such a simple behaviour is found only at the very beginning or just after shaking the sample.

Again, in the course of time, the $C(k,t)$ changes, and cannot longer be fitted with a single exponential (see fig.7). We are faced with a structured system, and eq.5 must be used.

A k-dependent "diffusion coefficient" can be defined as

$$D_{eff}(k) = \, <S(k,t) \, \dot{S}(k,t)> \, = \left. \left\{ \frac{d}{d\tau} <S(k,t) \, S(k,t + \tau)> \right\} \right|_{\tau = 0} \qquad (9)$$

and experimentally founded as the derivative at the origin of the normalized autocorrelation function. In turn, according to eq.5, the inverse of the diffusion coefficient will be proportional to the structure factor. In fig.8 the results obtained in the DNA solution are reported. A comparison with fig.6 shows that dynamical measurement actually reproduces the same structured behaviour directly found from static measurements.

Light scattering experiments therefore furnish direct evidence of a space-time correlated behaviour for the macromolecular solutions. Among the spatial Fourier components of the fluctuations of concentration, preferred wavenumbers exist which correspond to longlived components that, as a consequence, dominate in the spectrum of the fluctuations. The system is no longer spatially homogeneous, and long ranged, long lived structures take place.

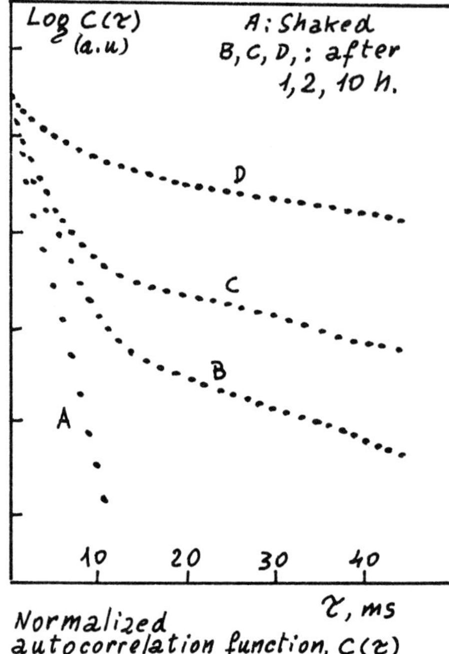

Normalized autocorrelation function, C(τ) (Lysozime sol.)

Figure 7

4)   Neutron scattering experiments

Inelastic neutron scattering is an almost unique experimental technique in order to study dynamical properties of physical systems. The spectrum of inelastically scattered neutrons is, in fact, related rather directly to the vibrational density of states, in contrast, for instance, to Raman spectroscopy in which the scattering efficiency of the system can change drastically at different excitations.

However neutron fluxes with energy up to 1 eV (~8000 cm^{-1}), have only recently become available, so that up to now only few systems have been investigated with such a tecnique.

In this section we report some preliminary results[10] concerning a 12% by weight solution of lysozyme in the 300-600 meV range, that comprises the two intramolecular stretching modes of OH and CH bonds. Neutron spectra of water[11] and supercooled water[12] in the above mentioned range has been reported in literature.

We performed measurements using the TFXA spectrometer of the Spatial Neutron Source ISIS at the Rutherford Appleton Lab. (Oxfordshire- U.K.).

A very important peculiarity that must be taken into account in such a kind of experiment, is the large exchanged momentum, in comparison, e.g. with light scattering in which such a quantity is nearly zero.

Intramolecular vibrational modes cannot, in themselves, carry actually out any momentum. If the scatterers are freely moving objects, the exchanged momentum is piked-up by the recoil of the scatterer, and the required energy appears as a shift toward higher frequency of the corresponding vibrational band. If however the system is more or less structured, the situation becomes more complicated. For instance, in ref.11 a shifted OH contribution is interpreted as the excitation of the intramolecular vibration together with the breaking of an H-bond: In order for a hydrogen bonded water molecule to recoil, the bond must in fact be broken.

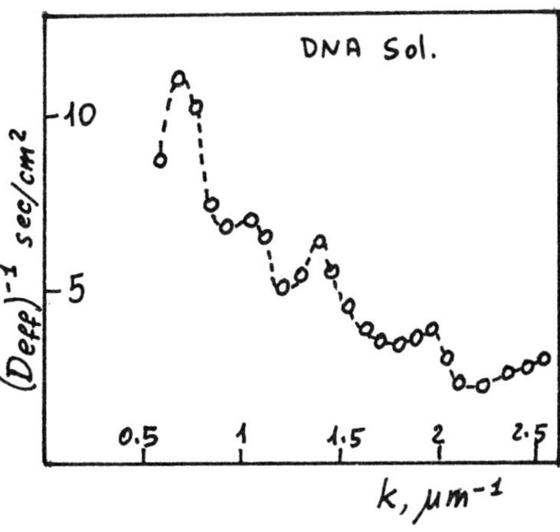

Figure 8

Generally speaking we are faced with multi-mode excitations, one of which at least must be able to carry out the exchanged momentum.

The very existence of this kind of modes therefore act as a "selection rule" for the scattering process to occur. (In our language the free recoil of the scatterer or the breaking of a bond are to be considered as extreme cases of such a kind of "modes").

Due to the above mentioned circumstance it turns out that a collective behaviour of the system, i.e. the existence of collective excitation ("phonons"), can also be revealed in an incoherent neutron scattering experiment, through the modifico tion of the vibrational intramolecular bands. As an example in fig.9 we show the spectrum of water. The dashed area reproduces the shape of the usual Raman spectrum, while the arrows indicate the literature assignment concerning the OH

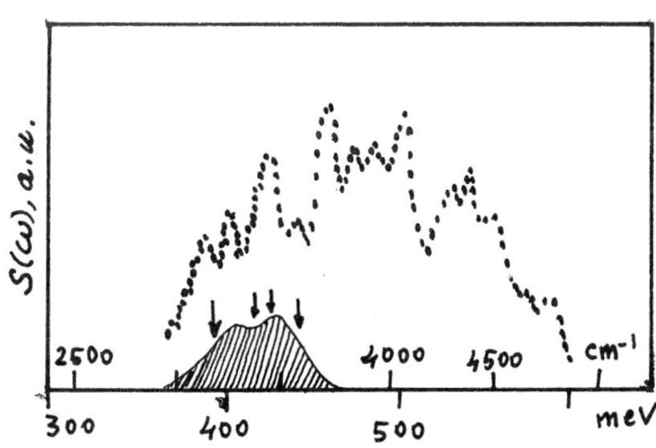

Anelastic neutron scattering in water.

Figure 9

stretching mode. Roughly speaking the neutron spectrum could be viewed as the superposition of two additional replicas of the OH band, shifted towards higher frequency by $\sim$ 70 meV and $\sim$ 140 meV respectively.

Now 70 meV correspond to a lower frequency mode of water, namely to the libration. It therefore seems that we are in fact concerned with combination bands.

Processes of this kind that seem to give negligible contributions in the case of optical spectroscopy, become relevant in the neutron scattering because of the large exchanged momentum.

As far as macromolecular solutions are concerned, we report in fig.10 the following spectra of the 12% by weight lysozyme solution:

a)- Just after the preparation of the sample

b)- The same solution, ten hours later

c)- A sample of the same solution, periodically shaken in order to avoid the
      building up of the clusters structure.

In spectrum a) the OH contribution is clearly recognizable, although partially merged in a new intense band centered at the frequency corresponding to the CH stretching mode (370 meV).

The spectrum b) shows an even more dramatic modification. Although the fundamental OH band could more or less still be recognized, the intense CH band seems to have disappeared and the entire spectrum behaves like a broad and almost continuous distribution of modes. As mentioned before, it therefore seems that a collective behaviour takes place in the structured solution. Such an indication is supported by the shape of the spectrum c). Apart from any detailed consideration, the very noticeable modification of the vibrational spectrum, solely due to the shaking of the sample, makes unescapable the conclusion that the long-range ordering among large-sized clusters deeply modifies the vibrational behaviour of a single molecule, as revealed by neutron scattering.

It is worth noticing that the same kind of experiment performed in pure

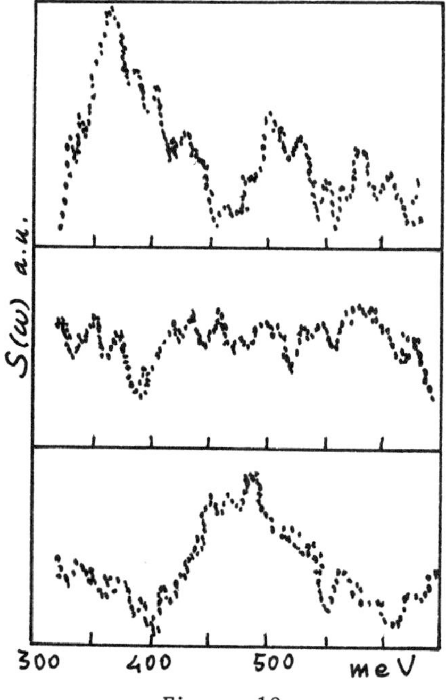

Figure 10

water does not show any modification after the shaking of the sample.

In addition we recall that optical Raman spectra of the OH stretching in the same solution do not show any modification with respect to pure water apart from some minor changes in the shape.[13]

Obviously we do not believe that the vibrational properties of the water molecule, can itself be so deeply modified by the presence of the macroscopic ordering. On the other hand Raman measurements exclude such a possibility.

In our opinion the collective properties of the structured solution are reflected on the observed spectra, by means of the above mentioned selection rule required by the momentum conservation law.

5) A possible stochastic model. The structure factor

In this section we try to construct a stochastic model for a structured macromolecular solution, from which both statical and dynamical behaviour can be recovered.

The properties of the solutions seem mainly concerned with a long-range order, in which aggregates of molecules (clusters) are implied.

The advantage of our model is twofold: it is also able to quantitatively explain the experimental results mentioned in the previous sections. In addition it focusses the attenption onto few fundamental ansatz, whose physical implication can be further investigated.

Let us briefly sketch the formalism that is implied in discussing light scattering experiments.

In principle we consider our system as a density distribution $\rho(\vec{r})$ of point - like scatterers (single molecules or even segments of a molecule). Then the light scattered by a volume element $d\vec{r}$ will be proportional to $\rho(\vec{r})\,d\vec{r}$.

For the field scattered at a given exchanged wave vector $\vec{K}$ we have

$$E(\vec{k}) = \int_V e^{i\vec{k}\cdot\vec{r}}\,\rho(\vec{r})d\vec{r} \tag{10}$$

V being the scattering volume. Let us now consider a frame of reference with the x axis parallel to $\vec{k}$. Putting

$$\overline{\rho}\ (x) = \int \rho(x,y,z)\,dydz$$

we have

$$E(\vec{k}) = \int e^{ikx}\, \overline{\rho}(x)\, dx$$

and for the intensity:

$$I(\vec{k}) = \int e^{ik(x-x')}\, \overline{\rho}(x)\, \overline{\rho}(x')\, dxdx'$$

Calling $\xi = x-x'$ we obtain

$$I(\vec{k}) = \int e^{ik\xi}\, C(\xi)\, d\xi \tag{11}$$

being

$$C(\xi) = \int \overline{\rho}(x)\, \overline{\rho}(x+\xi)dx \tag{12}$$

the spatial autocorrelation function of the "planar density" $\overline{\rho}(x)$.

Eq.11 shows that the scattered intensity reproduces the spatial Fourier transform of the "planar density" autocorrelation function i.e. by the Wiener-Kintchine theorem, the power spectrum of the density distribution. It can also be shown that $I(\vec{k})$ will be sizeably different from zero only if the autocorrelation function $C(\xi)$ is of finite range, the latter being not too different from k^{-1}.

Therefore the appearance of peaks in the experimental values of $I(\vec{k})$ indicate that spatial periodicity in the distribution of density occurs.

Because of the above mentioned circumstance we are therefore left with a one-dimensional model, and we recall that experimental results indicate the existence of more or less regularly distributed clusters.

We indicate with $R_i(t)$ the center of mass position of a cluster. Then the density distribution will be the convolution of the "shape" of a cluster (the form factor) with an array of delta-functions $\delta\ (x-R_i)$.

Our calculation is mainly concerned with the structure factor, deriving from the delta-functions, its product with a form factor being quite trivial. Therefore we write

$$\rho(x,t) = \sum_i \delta\big[r-R_i(t)\big] \tag{13}$$

The Fourier transform of the spatial autocorrelation function, i.e. the time dependent structure factor will be given by:

$$S(k,t) = \sum_{ij} \int\int \delta\left[x-R_i(t)\right]\delta\left[x+\xi-R_j(t)\right] e^{ik\xi} \, dx \, d\xi = \sum_{i,j} e^{ik\left[R_i(t)-R_j(t)\right]}$$

Such a quantity is proportional to the instantaneous value of the scattered intensity, so that the experimental time autocorrelation function will be proportional to

$$\left\langle S(k,0)\, S(k,t)\right\rangle = \left\langle \sum_{n,m} e^{ik(R_n(0)-R_m(0))} \sum_{p,q} e^{-ik(R_p(t)-R_q(t))} \right\rangle \tag{14}$$

where stationary conditions are supposed to exist.

Now we state that the center-of-mass position of a cluster will be given by

$$R_n(t) = nL + \sum_{i=1}^{n} \epsilon_i(t) = nL + \Delta_n(t) \tag{15}$$

In eq (15) L is a fixed distance, while $\epsilon_i(t)$ are stochastic variables describing the displacement of a cluster from the fixed distance L between nearest neighbours.

The time evolution of the stochastic variable $\epsilon_i(t)$ is described by a Langevin-like equation. Let us consider a small time interval τ. From a physical point of view we can think of τ as a mean collision time. Then:

$$\epsilon(t+\tau) = \epsilon(t)\, e^{-\alpha\tau} + \delta \tag{16}$$

In eq.16 the first term on the RHS describes a systematic evolution that tends to reestablish a perfect periodic array, while δ is a random variable described by a Gaussian distribution:

$$p(\delta) = \frac{1}{\sqrt{\pi\mu^2}}\, e^{-\frac{\delta^2}{\mu^2}} \tag{17}$$

Consequently, after a time $t=\nu\tau$ we have

$$\epsilon_i(t) = \epsilon_i(0)\, e^{-\nu\alpha\tau} + \sum_{j=1}^{\nu} \delta_{ij}\, e^{-(\nu-j)\alpha\tau} \qquad (18)$$

being δ_{ij} the random displacement of the i/th cluster, ocurred at the time $j\nu$. Then the displacement $\Delta_n(t)$ of the n/th cluster from its "regular" position nL, will be given by

$$\Delta_n(t) = \sum_{i=1}^{n} \epsilon_i(t) = \Delta_n(0)\, e^{-\nu\alpha\tau} + \sum_{j=1}^{\nu} \Delta_j^{*}\, e^{-(\nu-j)\alpha\tau} \qquad (19)$$

where

$$\Delta_j^{*} = \sum_{i=1}^{n} \delta_{i,j} \qquad (20)$$

In the same way:

$$R_p(t) - R_q(t) = (p-q)\, L + \Delta_{qp}(t) \qquad (21)$$

with

$$\Delta_{qp}(t) = \Delta_{qp}(0)\, e^{-\nu\alpha\tau} + \sum_{j=1}^{\nu} \Delta_{qp,j}^{*}\, e^{-(\nu-j)\alpha\tau} \qquad (22)$$

being

$$\Delta_{qp,j}^{*} = \sum_{i=q+1}^{p} \delta_{i,j} \qquad (23)$$

Now Δ_{qp}^{*} is the sum of the random displacements δ_{ij} that take place at each time interval τ for each R_i between p and q.

As a consequence Δ_{qp}^{*} is also a Gaussian variable, with r·m·s $|p-q|\mu^2$:

$$p(\Delta_{qp}^{*}=x) = \frac{1}{\sqrt{\pi|p-q|\mu^2}}\, e^{\dfrac{-x^2}{|p-q|\mu^2}} \qquad (24)$$

Let us consider as a first step the asymptotic behaviour of the structure factor

$$\left\langle S(k,t\to\infty) \right\rangle = \left\langle \sum_{p,q} e^{ik\left[R_p(t)-R_q(t)\right]} \right\rangle_{t\to\infty} \tag{25}$$

where the bracket indicates an average over the probability distribution.

In the frame of the model sketched above we have

$$\left\langle S(k,t) \right\rangle = \sum_{p,q} e^{ik(p-q)L} \left\langle e^{ik\Delta_{pq}(t)} \right\rangle$$

being:

$$\left\langle e^{ik\Delta_{pq}(t)} \right\rangle = \left\langle \exp\left[ik\Delta_{pq}(0)e^{-\nu\alpha\tau} + \sum_{j=i}^{\nu} \Delta^*_{pq,j}\, e^{-(\nu-j)\alpha\tau} \right] \right\rangle =$$

$$= e^{ik\Delta_{pq}(0)e^{-\nu\alpha\tau}} \left\langle \prod_{j=1}^{\nu} e^{ik\Delta^*_{pq,j}\, e^{-(\nu-j)\alpha\tau}} \right\rangle \tag{26}$$

Now the time evolution of the random variables is supposed to be a Markovian process, so that the mean can be factorized.
Then we have

$$\left\langle e^{ik\Delta^*_{pq,j}\, e^{-(\nu-)\alpha\tau}} \right\rangle = \frac{1}{\sqrt{\pi|p-q|\mu^2}} \int e^{ikxe^{-(\nu-j)\alpha\tau}} e^{-\dfrac{x^2}{|p-q|\mu^2}}\, dx =$$

$$= e^{-\dfrac{k^2\mu^2}{4}|p-q|\, e^{-2(\nu-j)\alpha\tau}}$$

and therefore

$$\left\langle \prod_{J=1}^{\nu} \exp \Delta^*_{pq,j}\, e^{-(\nu-j)\alpha\tau} \right\rangle = \left\langle \exp -\frac{k^2\mu^2}{4}|p-q| \sum_{l=1}^{\nu-1} e^{-2l\alpha\tau} \right\rangle$$

The sum of the geometrical series furnishes

$$\sum_{l=0}^{\nu-1} e^{-2l\alpha\tau} = \frac{1-e^{-2\nu\alpha\tau}}{1-e^{-2\alpha\tau}} = \left(\frac{1}{1-e^{-2\alpha\tau}}\right)(t\to\infty)$$

In the same limit $\Delta_{qp}(0)\, e^{-\nu\alpha\tau}$ goes to zero, so that

$$\lim_{t\to\infty} \left\langle e^{ik\Delta_{qp}(t)} \right\rangle = e^{-\frac{k^2\mu^2}{4}\,|p-q|\,\cdot\,\frac{1}{1-2\,e^{-2\alpha\tau}}}$$

Such a result indicates that in equilibrium conditions the displacements Δ_{qp} behave like random variables, with a Gaussian distribution:

$$p(\Delta_{qp}=y) = \frac{1}{\sqrt{\pi|q-p|\sigma^2}}\, e^{-\frac{y^2}{|q-p|\sigma^2}} \qquad (27)$$

being

$$\sigma^2 = \frac{\mu^2}{1-e^{-2\alpha\tau}} \qquad (28)$$

Notice that if $\alpha\to0$ (i.e. the system does not show any tendency to ordering), then $\sigma\to\infty$: the system becomes fully disordered as a consequence of the superposition of an infinite number of random displacements; its dynamic properties being those characteristic of Brownian motions.

On the contrary, if $\alpha\to\infty$ (i.e. any displacement is cancelled before the next one occurs), then $\sigma^2=\mu^2$. In the system a stable configuration exists, and the actual position of each object is

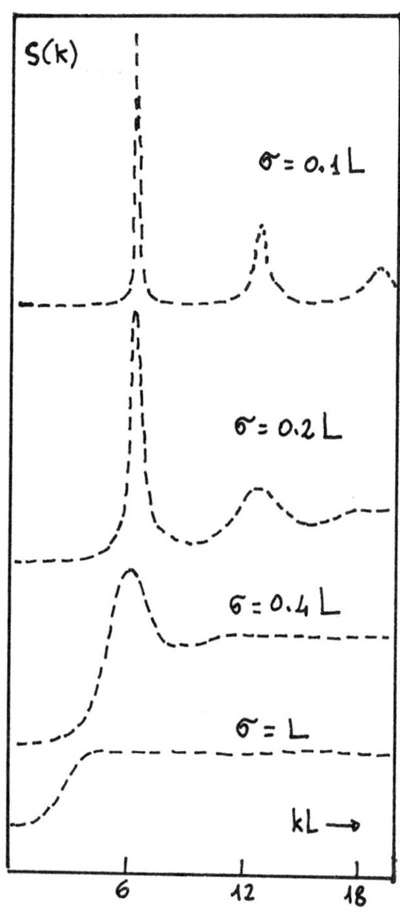

Figure 11

in a "thermal cloud" around the preferred position, like in a crystal.

A thixotropically structured solution is in an intermediate condition: the degree of order can be measured by α, while μ plays the role of some effective temperature.

With the distribution given by eq.27 the average in eq.25 can be performed and gives the following result for the (normalized) equilibrium structure factor:

$$\langle S(k) \rangle = \frac{1 - e^{-\frac{k^2\sigma^2}{4}} \cos kL}{1 - 2\,e^{-\frac{k^2\sigma^2}{4}} \cos kL + e^{-\frac{k^2\sigma^2}{2}}} \tag{29}$$

In fig.11 we show a plot of such a function for some different values of σ. The increasing disorder of the system is evident as σ increases, i.e. the "temperature" μ increases or the "restoring force" α decreases.

6) <u>Fluctuations in the system and autocorrelation function</u>

The autocorrelation function will be given by

$$\langle S(k,0)\, S(k,t) \rangle = \sum_{m,n,p,q} e^{ik(m-n)L} e^{-ik(q-p)L} \left\langle e^{ik\Delta_{nm}(0)} e^{-ik\Delta_{qp}(t)} \right\rangle \tag{30}$$

According to the preceedingly obtained results the mean is to be performed with the probability distribution given by eq. 27 (with eq.28):

$$\left\langle \exp(ik\Delta_{nm}) \cdot \exp\left[-ik\left(\Delta_{qp}\,e^{-\nu\alpha\tau} + \sum_{j=1}^{\nu} \Delta^*_{qp,j}\,e^{-(\nu-j)\alpha\tau}\right)\right] \right\rangle =$$

$$= \left\langle e^{ik\Delta_{nm}}\, e^{-ik\Delta_{qp}e^{-\nu\alpha\tau}} \right\rangle e^{-\frac{k^2\sigma^2}{4}|p-q|(1-e^{-2\nu\alpha\tau})} \tag{31}$$

Notice that for $t \to \infty$ (i.e. $\nu \to \infty$) the second factor into the mean disappears so that eq.31 becomes

$$\left\langle e^{ik\Delta_{nm}} \right\rangle e^{-\frac{k^2\sigma^2}{4}|p-q|} = e^{-\frac{k^2\sigma^2}{4}|m-n|} e^{-\frac{k^2\sigma^2}{4}|p-q|}$$

In such a case the sum in eq.30 factorizes and we obtain the usual result:

$$\left\langle S(k,0)\, S(k,t) \right\rangle_{t \to \infty} = \left\langle S(k) \right\rangle^2 \tag{32}$$

On the contrary, for finite values of t, the mean appearing in eq.31 cannot be factorized unless the two intervals nm and pq are disjointed. If, in fact, there is an overlap as e.g. in the following sequence of indices:

$$n < q < m < p$$

the common part m-q correlates the probability distribution of the two displacements Δ_{nm} and Δ_{pq}.

In such a case we can write

$$\Delta_{nm} = \Delta_{nq} + \Delta_{qm}$$
$$\Delta_{qp} = \Delta_{mp} + \Delta_{qm}$$

so that

$$\Delta_{nm} - \Delta_{qp}\, e^{-\nu\alpha\tau} = \Delta_{nq} - \Delta_{mp}\, e^{-\nu\alpha\tau} + \Delta_{qm}\,(1-e^{-\nu\alpha\tau}) \tag{33}$$

Now the segments nq, qm and mp are disjointed, therefore their probability distribution can be factorized, and the mean values calculated.
We get

$$\left\langle e^{ik\Delta_{nq}} \right\rangle = e^{-\frac{k^2\sigma^2}{4}|q-n|}$$

$$\left\langle e^{ik\Delta_{mp}e^{-\nu\alpha\tau}} \right\rangle = e^{-\frac{k^2\sigma^2}{4}|p-m|\, e^{-2\nu\alpha\tau}}$$

$$\left\langle e^{ik\Delta_{qm}(1-e^{-\nu\alpha\tau})} \right\rangle = e^{-\frac{k^2\sigma^2}{4}|m-q|\,(1-e^{-\nu\alpha\tau})^2}$$

Accordingly, the variable part of the autcorrelation function becomes (taking into account eq.30 and 31)

$$\left\langle S(k,0)\ S(k,t) \right\rangle_{var} =$$

$$= \sum_{n<q<m<p} e^{ik(m-n)L}\ e^{-\frac{k^2\sigma^2}{4}|q-n|}\ e^{-ik(p-q)L}\ e^{-\frac{k^2\sigma^2}{4}|p-m|}\ e^{-\frac{k^2\sigma^2}{2}|m-q|\vartheta} \qquad (34)$$

being $\vartheta = (1 - e^{-\nu\alpha\tau})$.

The effect of the intersection (m-q), that couples the two oscillating terms, also reduces the effect of the disorder by subtracting from the rms $|m-n|\sigma^2$ and $|p-q|\sigma^2$ the contribution of the common part $|m-q|$.

Notice that such a subtraction is complete at t=0 and disappears for t→∞: in such a limit we are left with uncorrelated terms, the sum in eq.34 can be factorized and we get the result corresponding to $\left\langle S(k) \right\rangle^2$.

Now, in order to perform the sum, we select a fixed value $\xi=|m-q|$. Then the sums over n and p reduce to the sums over s=(q-n) and s'=(p-m) respectively

$$\sum_{q} \sum_{\xi} e^{-\frac{k^2\sigma^2}{2}|\xi|\vartheta} \sum_{s} e^{ik(s+\xi)L}\ e^{-\frac{k^2\sigma^2}{4}|s|} \sum_{s'} e^{-ik(s'+\xi)L}\ e^{-\frac{k^2\sigma^2}{4}|s'|}$$

The sums over s and s' can be extended up to infinity because of the presence of the decaying real terms.

We recall, however, that the sum in eq. 34 refers to a selected sequence of indices. All the possible combinations sum up to take into account the following combination of signs for s and s':

1) s > 0 , s' > 0   3) s > 0 , s' < 0

2) s < 0 , s' < 0   4) s < 0 , s' > 0

As a consequence, calling

$$z = \sum_{s=0}^{\infty} e^{iksL}\, e^{-\frac{k^2\sigma^2}{4}s} = \frac{1}{1 - e^{ikL}\, e^{-\frac{k^2\sigma^2}{4}}} \qquad (35)$$

we are concerned with the four contributions:

$$z \cdot z^+ + z^+ \cdot z + z^2 + z^{+2} = (z + z^+)^2$$

so that

$$\langle S(k,0)\, S(k,t) \rangle_{var} = \sum_{q} \sum_{\xi} e^{\frac{k^2\sigma^2}{2}\xi\vartheta} (2zz^+ + 2\operatorname{Re} z^2\, e^{2ik\xi L})$$

The sum over q is simply a sum over the total number of N scatterers and only gives rise to a normalization factor. However as far as the sum over ξ is concerned we cannot consider the limit $N \to \infty$ because for $t = 0$ ($\vartheta = 0$) the sum no longer converges. The sum itself gives rise to a very complicated expression. However the leading term can be extracted and one gets:

$$\langle S(k,0)\, S(k,t) \rangle_{var} \simeq \frac{1}{N}\, \frac{1 - e^{-\frac{k^2\sigma^2}{2}N\vartheta}}{\left[1 - e^{-\frac{k^2\sigma^2}{4}}\cos kL + e^{-\frac{k^2\sigma^2}{2}}\right]\left[1 - e^{-\frac{k^2\sigma^2}{2}\vartheta}\right]} \qquad (36)$$

In the limit $t \to 0$ ((i.e. $\vartheta = 0$) eq. 36 becomes

$$\langle S(k,0)^2 \rangle_{var} = \frac{1}{\left[1 - e^{-\frac{k^2\sigma^2}{4}}\cos kL + e^{-\frac{k^2\sigma^2}{2}}\right]}$$

independent from the value of N. On the contrary for $t \to \infty$ (i.e. $\vartheta = 1$) we get:

$$\langle S(k) \rangle_{var}^2 = \frac{\left[1 - e^{-\frac{k^2\sigma^2}{4}N}\right]}{N\left[1 - e^{-\frac{k^2\sigma^2}{4}}\cos kL + e^{-\frac{kl^2\sigma^2}{2}}\right]\left[1 - e^{-\frac{k^2\sigma^2}{2}}\right]}$$

Notice that the latter quantity goes to zero as $N \to \infty$, and the same happens for any arbitrary value $\vartheta > 0$ ($t > 0$) in eq.(36) : in a truly infinite system the effect of fluctuations becomes undetectable, in comparison with the mean value.

Actually in a spatially correlated system, if the sample (i.e. the scattering volume) becomes much larger than the correlation length λ, the system behaves like an ensemble of $M = N/\lambda$ uncorrelated parts, fluctuating independently. Because, in such a case, the r.m.s. of the fluctuation increases like $\sqrt{M}$ while the total scattered intensity increases as M, the fluctuating contribution, normalized to the total intensity, tends to vanish for all values of t except t = 0.

A quite correct evaluation of the $t \simeq 0$ (i.e. $\vartheta \sim 0$) behaviour can be made by treating ϑ as a continuous variable. Then we can calculate the time derivative and we found a zero value for $\vartheta \to 0$.

More precisely, the autocorrelation function stays nearly constant until $\vartheta \ll 2 / N \ k^2 \sigma^2$.

Such a result, besides the above mentioned role of N, clearly shows a relevant feature of the correlation function: the function at the origin "round off" as a consequence of the permanence of the memory of the system before the next collision.

7) <u>Experimental check</u>

In order to make a comparison with the stochastic model described in the previous sections, we performed a number of experiments,[14] in lysozime solutions. A short account of these will be presented here.

In performing a correlation-spectroscopy experiment at different angles of scattering in a macromolecular solution, the following circumstances are of paramount importance:

i)- The dynamical properties cover a wide range of times. Previous results and the numerical evaluation of the model sketched above show this peculiarity.

ii)-Great care must be taken in comparing measurements performed at different exchanged wave vector (different angles) because of some unavoidable noise that could be different in the various measurements.

We overcome the first difficulty by using a specially designed single-clipped autocorrelator, that performs the temporal autocorrelation of the incoming scattered intensity in 88 channels divided into 11 groups of 8. The time delay between two successive channels is doubled at each group, so that while in the first group two channels are spaced by a time interval of τ_{min}, the spacing in the last group is $2^{10}\,\tau_{min}$.

Such a configuration allows the detection of the autocorrelation function on an enormous scale of time, with a delay of 16376 τ_{min} between the first and the last channel.

In such a way the uncorrelated part of the signal can also easily be detected and properly subtracted.

In our correlator we can select a value as small as 200 ns for τ_{min}.

The second above mentioned problem was overcome by finding an "internal standard" in the sample, given by free lysozyme molecules, i.e. molecules not involved in the structure. At very short times we find a small contribution that can be identified as due to freely diffusing lysozyme molecules. Such a contribution in fact corresponds to an exponential decay whose time constant is Dk^2, where D turns out to be independent from the scattering wave vector and reproduces with great accuracy the diffusion coefficient attributed to the lysozyme ($D = 10^{-6}$ cm^2 sec.$^{-1}$).

Now, because of the small size of the lysozyme molecule ($\sim$15 Å as compared to the wavelength of light (6328 Å) the intensity of such a contribution must also be independent from k, and can therefore be used as an internal standard for the normalization of results taken at different angles.

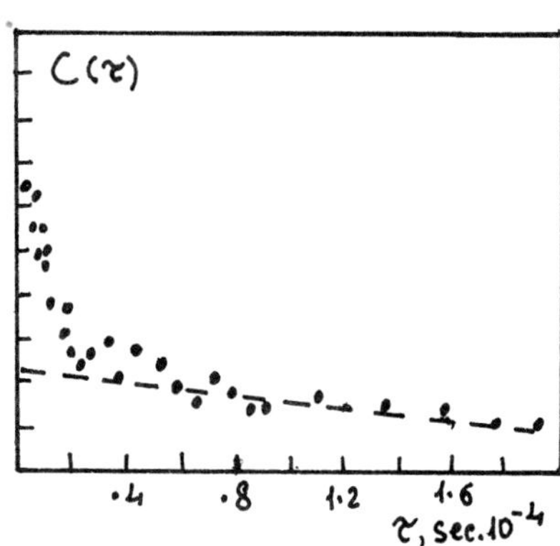

Figure 12

As far as the sample is concerned, we use a 10 % by weight solution of highly purified lysozyme dissolved in pure water.

The temperature is held constant at 20 °C. All measurements are performed several hours after the preparation in order to concern a fully developed structure.

In fig.12 we reproduce a typical result for the autocorrelation function at a very short time. The sharp contribution of the free molecules are clearly shown, while on such a scale the main correlated contribution nearly behave like a constant.

As a check for our stochastic model we can consider both the time-dependence of the autocorrelation function and the structure factor as measured by the $t\to\infty$ behaviour. Two typical results are reported in the next figures.

In fig.13 the experimental decay is shown, together with a fit with eq.36. The values of the implied parameter turn out to be $\alpha = 270$ sec^{-1} $\frac{\sigma}{L} = 0.036$ We want to stress the accuracy with which the theoretical curve fits the experimental data, notwithstanding the very large scale of time implied (from less than 10μs to more that 1msec), and the approximations involved in the use of eq. 36 (leading term).

In fig.14 the structure factor, as calculated from the theory is shown together the experimental point.

To do this we need to multiply the S(k) as given by eq.29 by a suitable form factor F(k) due to the finite size (and shape) of a cluster.

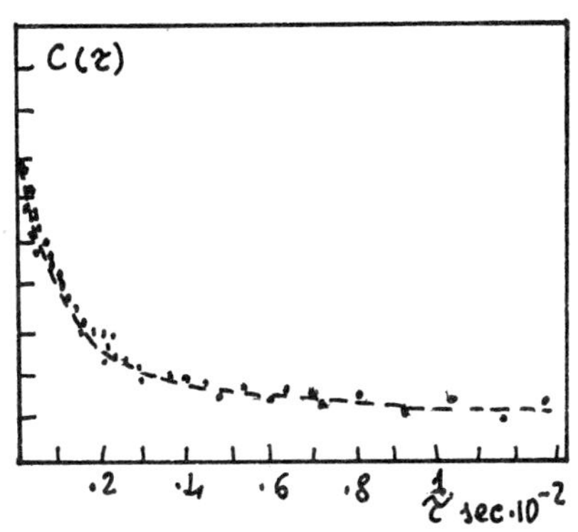

Figure 13

An evaluation of the form factor can be experimentally found by a static light scattering experiment, performed just after the preparation of the sample, before the thixotropic structure builds up. In fig.14 such an evaluation is shown as a dashed line (normalized to the height of the first peak in the S(k)), while, the continuous line represents the product F(k)·S(k). The agreement again seems to be quite good. The parameter implied in the fit furnishes:

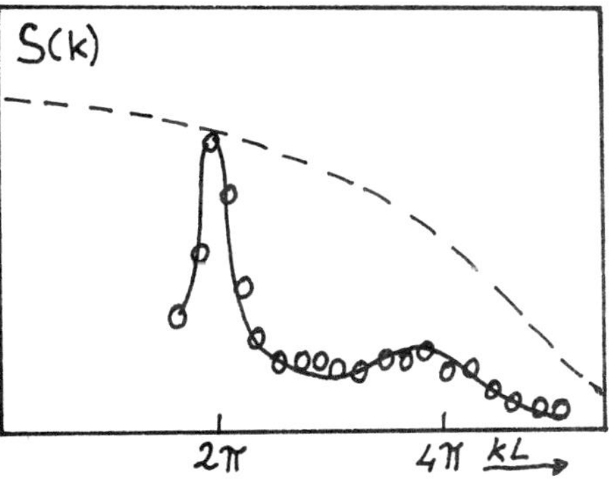

Figure 14

L = intercluster spacing $\simeq 6000$ Å

Δ = cluster size $\simeq$ 850 Å$<\Delta<$1200 Å

$\sigma/L = 0.2$

The exact evaluation of Δ depends to some extent on the "detailed" shape of the cluster. Here Δ is evaluated simply as $(\Delta k)^{-1}$, Δk being the half-width of the experimentally found form factor.

Some comments come from the comparison between the values of σ obtained respectively from the dynamical behaviour of the system ($\sigma/L = 0.036$) and from the static structural properties ($\sigma/L = 0.2$). It seems that, from a dynamical point of view the system is more ordered than in a static description. Actually, in our model there is no room for a static disorder, the only disordering process allowed being dynamical in character, and coming from collisions, like a Debye-Waller factor.

Experimental results however, seem to indicate that, in addition, there is some static disorder, as if quite reasonable to suppose.

8) <u>Concluding remarks</u>

From the body of experiments, some of which have been presented here, a collective correlated behaviour is recognizeable in the properties showed by macromolecules in a solution.

The number of widely different techniques used in the various experiments, that comprehend rheological properties, acoustic behaviour, light scattering, neutron scattering and Raman spectroscopy, makes unescapable the conclusion that a macromolecular solution does not behave like an amorphous disordered system.

As mentioned in sec.1, such a circumstance means that some care must be taken for the physical description of the "thermal bath", that, in our opinion, cannot be treated as a simple heat reservoir.

In a more or less ordered structure the distribution of modes can be deeply modified in comparison with a unstructured system. As an example, thermal waves ("Brillouin" waves) for which the ordered clusters positions correspond to nodal loci are presumably less damped than others.

In addition the single macromolecules are stuk together in a cluster, that in itself seems to be an ordered structure behaving more or less like a rigid body.

Why such a hierarchy of ordering (molecules in a cluster, cluster in ordered array...) takes place, and the kind of interaction (collective in character in our opinion) needed for such an ordering, are questions that we plan to investigate in the near future.

REFERENCES

1) R.Giordano, M.P.Fontana, F.Wanderlingh
   J.Chem.Phys. **74**, 2011 (1981).

2) R.Giordano, G.Maisano, F.Mallamace, N.Micali, F.Wanderlingh
   J.Chem.Phys. **75**, 4770 (1981).

3) R.Giordano
   Il Nuovo Cim. **1D**, 540 (1982).

4) R.Giordano, F.Mallamace, F.Wanderlingh
   Il Nuovo Cim., **2D**, 1272 (1983).

5) S.Temkin, R.A.Dobbins
   J. Acoust. Soc.of Am. **40**, 5 (1966)

6) J.C.Brown, P.N.Pusey, J.W. Goodwin, R.M. Otterwill
   J.Phys. A **8**, 664 (1975).

7) R.Giordano, A.Salleo, S.Salleo, F.Mallamace, F.Wanderlingh
   Optica Acta **27**, 1465 (1980).

8) R.Giordano, N.Micali
   "Correlation spectroscopy and structure properties of macromolecular solution"; in "The application of laser light scattering to the study of biological motion"; J.C.Earnshaw and M.W.Steer ed. Plenum Pub. Co. (1983).

9) R.Giordano, F.Mallamace, N.Micali, F.Wanderlingh, G.Baldini, S.Doglia
   Phys. Rev. A **28**, 3581 (1983).

10) G.Salvato, F.Wanderlingh, U.Wanderlingh
    "Inelastic Neutron Scattering and Macromolecular solutions"- To be published.

11) M.A.Ricci et al.
    Physica **136B**, 190 (1986).

12) S.M.Chen et al.
    Phys. Rev.Lett. **53**, 1360 (1984).

13) F.Aliotta, M.P.Fontana, R.Giordano, P.Migliardo, F.Wanderlingh
    J.Chem.Phys. **75**, 4307 (1981).

14) F.Farsaci, M.E.Fontanella, R.Giordano, G.Salvato, F.Wanderlingh, U.Wanderlingh
    "Dynamical behaviour of structured macromolecular solutions" - To be published.

(C) <u>Reactivity and Catalysis</u>

"There are agents in nature able to make the particles of joints stick together by very strong attractions and it is the business of experimental philosophy to find them out."

Isaac Newton

"Everything is made of atoms. That is the key hypothesis. The most important hypothesis in all of biology, for example, is that everything that animals do, atoms do. In other words, there is nothing that living things do that cannot be understood from the point of view that they are made of atoms acting according to the laws of physics."

Richard P. Feynman

INTRODUCTION TO THE BASIC CONCEPTS IN REACTION DYNAMICS

J.L. Houben

Istituto di Chimica Quantistica ed Energetica Molecolare
Consiglio Nazionale delle Ricerche
Via Risorgimento, I-56100 Pisa (Italy)

As has already been discussed (1), the efficiency and specificity of enzyme catalysis are currently interpreted as being due to a combination of factors, including:
- the interplay between the flexibility and the rigidity of the enzyme structure;
- the adaptation of the fluctuation time scale to the chemical reaction;
- the preparation of very specific transient species.
Enzyme catalysis is first of all a chemical reaction catalyzed by a macromolecule with a well-defined structure and dynamics. This aspect is discussed at length in the chapters dedicated to enzyme spectroscopy (2). Enzyme energy fluctuations are discussed in Cooper's chapter (3). In the present one, an attempt will be made to present a framework for the chapters on chemical reactions and molecular dynamics (4-10).

Two different approaches to the description of the chemical reaction in the condensed state are presented in this book. Warshel (5) in a self-contained chapter proposes a well-defined model applied to enzyme catalysis and does not need any further comment here. Hynes (5,6) together with Fonseca (7) instead offer an interesting opportunity to discuss enzyme catalysis within a more general framework. Before entering into the details of the theory of chemical reactions as it may be applied to enzyme catalysis, it may be helpful to recall that the serine protease mechanism involves various steps: (a) enzyme-substrate recognition; (b) preparation of a well-defined transition state; (c) proton tunneling and, finally (d) the peptide bond hydrolysis itself. Such a description involves molecular dynamics (9) in its full complexity - specifically, the movement of particles in potential wells (5,8,10), assisted by the energy flow, from various degrees of freedom to the "reaction coordinate" (10,11).

The enzyme turnover rate being typically about 1,000 events per second, a characteristic time on the order of 1 ms would seem to be suggested for the process; but since in a multi-step process, the overall rate is controlled by the slowest one, this 1 ms time scale requires some elaboration. From the nature of some of the individual steps, it appears clear that their characteristic times must be orders of magnitude shorter than this overall time scale. In particular, in the case of time-correlated statistical events the single event levels are expected to have characteristic times in the 1 ns time range or below (12). Consequently, molecular dynamics simulation in the ps range is probably quite fundamental to a full understanding of these chemical events (9,12).

The purpose of the present chapter is to introduce, through a very general and elementary discussion, the various topics relevant to the chemical reaction: energy distribution, migration and transfer; the potential energy surface; models for chemical reactions; and finally, the trajectory approach to chemical reactions. Many of these topics receive a more technical treatment in other chapters (3-12).

ENERGY DISTRIBUTION, MIGRATION, TRANSFER AND FLUCTUATIONS

At a macroscopic level - that is to say, for a sample containing a practically infinite number of molecules - the energy is distributed among the various degrees of freedom: translation, rotation and vibration, following the laws of statistical physics (13). The mass and the moment of inertia of a protein are fairly large and thus both the translational and rotational energy distributions can be treated as continua and to each of their degrees of freedom a mean energy of kT/2 can be associated. Vibrational energy levels are distributed from 0 to about 1eV and thus have quite different populations and contribute differently to the total energy of the protein.

This statistical description applies to a macroscopic sample where, due to the large number of molecules, we are interested in the average properties of the sample. At the microscopic level - i.e., at the level where the single event occurs - the situation is quite different as the kinetic energy is continuously redistributed among the nuclei. In other words if, given a collection of N molecules, n_k are in the kth vibrational state (13), at a molecular level this is equivalent to saying that among N consecutive instantaneous "photographs" of one single molecule, n_k pictures will show the molecule to be in this kth state. Such an experiment would also demonstrate that these n_k events are randomly distributed among the N snapshots if, with respect to the population relaxation time of this kth level, a single snapshot is short and two successive snapshots are separated by a sufficiently long time interval. This also means that each individual molecule will in the course of time reach all the allowed energy levels for each degree of freedom.

Indeed, since a molecule interacts with others, through collisions for example, it cannot be treated as an isolated thermodynamic system; i.e., the individual molecule is continuously pumped to, and relaxed from, a particular vibrational, rotational and translational state and the rate of exchange depends on the interactions of this particular degree of freedom with the remaining ones of the thermodynamics system. For a polyatomic molecule, at the intramolecular level two mechanisms of energy exchange between modes exist: (a) energy migration between equivalent oscillators, with the energy quanta hopping from one site to another, and (b) energy transfers where, if the total energy of the system is conserved, the energy distribution among the various degrees of freedom changes. The details of these mechanisms are discussed elsewhere (10,11). At the intermolecular level, these same mechanisms exist, but then the energy exchange requires an effective collision between the molecules.

In small molecules, the molecular energy fluctuations are large compared to their total internal energy content. For example, the vibrational temperature (14) of the hydrogen molecule is 6215 K, so that the difference in energy between a molecule in the vibrational ground state and one in a vibrational excited state is quite large. On the other hand, the vibrational states last only a short time - typical population relaxation times in the condensed state are less than 1 ps. As the molecular size increases, these fluctuations become relatively less and less important: the total energy is very roughly proportional to the number of modes, while the energy fluctuations are proportional to their root mean square. But their duration increases: the relaxation time increases linearly with the radius, since it is a volume to surface ratio (see refs. 1

276

and 3). For proteins, which have mesoscopic dimensions the fluctuations are long-lasting and still very large.

POTENTIAL ENERGY AND CHEMICAL REACTIONS

Consider, for the sake of simplicity, a colinear chemical reaction between the species XZ-Y and X-ZY:

$$XZ + Y \rightleftharpoons X + ZY \tag{1}$$

In this case, the space can be divided into two subspaces: one which contains the reactants (XZ and Y) and the other the products (X and ZY). The fundamental problem in quantum chemistry can be reduced to the determination of the potential surface in the whole space and to a discussion of what occurs at the frontier between these two subspaces. If for the non-interacting particles:
- the exact molecular wave functions are Ψ_{XZ}, Ψ_Y, Ψ_X and Ψ_{ZY}
- the Hamiltonians are H_{XZ}, H_Y, H_Y and H_{ZY}
the time-independent Schrodinger equations of the non-interacting systems are:

$$(H_{XZ} + H_Y)\,\Psi_{XZ}\,\Psi_Y = (E_{XZ} + E_Y)\,\Psi_{XZ}\,\Psi_Y \tag{2a}$$
$$(H_X + H_{ZY})\,\Psi_X\,\Psi_{ZY} = (E_X + E_{ZY})\,\Psi_X\,\Psi_{ZY} \tag{2b}$$

As the reactants approach, the interaction term between XZ and Y becomes less and less negligible; i.e., the full Hamiltonian contains new terms and the total wave functions are no longer solutions of the Schrodinger equations above. The potential curves in the crossing region are distorted; i.e., "avoided crossing" occurs. In the perturbation theory approach (15), the interaction energy is given by:

$$E_{XZ-Y} = \langle \Psi_{XZ}\,\Psi_{XZ-Y}\,|H_{XZ-Y}|\,\Psi_X\,\Psi_{ZY}\rangle \tag{3}$$

In reactive systems the interaction is strong, the energy splitting is large and the transmission coefficient - which measures the probability of crossing from the reactant to the product potential surfaces - approaches unity. If in this region of space, the potential surface is rather flat, the flight of the particle over the saddle will be slow enough to allow a full nuclear rearrangement of the reactants during the reaction event. If these interactions are small - if, for example, the reactants XZ and Y, and the products X and ZY, have different spin multiplicities then. (a) equations 2 are almost valid, (b) the splitting due to the curve-avoided crossing is small, and (c) when the X-Z-Y system is in the transition state region, there is only a small probability that it will move "from one potential curve to the other"; that is, the transmission coefficient is less than unity.

The first problem to be solved for a detailed description of a chemical reaction is thus to know the exact form of the potential surface. This implies - as indicated above - a knowledge of the dynamics of the solvation shell of the reactants, as the nuclear configuration of the reactants and consequently their electronic cloud will suffer drastic rearrangements in the course of the saddle crossing.

There is another, much more far-reaching consequence of these interactions: the potential curves become anharmonic for small displacements from equilibrium. Since normal modes (11) are defined by a symmetry reduction of the cartesian coordinates - an operation which requires a totally quadratic form for the Hamiltonian, and so a harmonic potential - a normal mode analysis cannot strictly be used in such cases. In practical terms, this means that energy can flow efficiently from mode to mode before energy relaxation takes place: i.e., the modes "communicate".

As discussed in the preceding section, a chemical reaction is simply the passage of a particle from one subspace to another. It has thus to be promoted to an energy level sufficiently high to overtake the relative potential barrier. As discussed in the first section, the average values of the translational, rotational and vibrational energies are described by statistical physics (13); but the instantaneous level of energy along one coordinate fluctuates continuously due to collisions and intermolecular interactions.

Eyring's Transition State Theory (TST)

In Eyring's transition state theory, the basic hypotheses are that the matrix-reactants energy exchange rate is: (a) slow compared with the collision time of the colliding reactants so that collisions occur without changes in the reactant total energy content, and (b) fast compared to the chemical reaction rate, so that the reactants can be considered to be in thermal equilibrium with the matrix throughout the course of the chemical reaction. During the reactive collision, the total energy (translational, rotational and vibrational) of the reactants is considered to be constant, but the energy distribution among the various degrees of freedom of the reactants may alter. In this respect, the TST is essentially a gas phase theory. The probability that a collision will lead to a reactive event is directly proportional to:
- the probability of the reactants having a total energy content which is at least as high as the potential barrier;
- the probability that a sufficient fraction of this energy will be channeled to the reaction coordinate during the collision time;
- the value of the transmission coefficient over the saddle.

Since the potential surface controlling the final fate as well as the overall rate of the reaction is defined, at least in part, by the solvent-reactant interactions, the solvent must play a fundamental role in the reaction process. According to the TST, the solvent is supposed to be in a relaxed state; that is, in thermal equilibrium with the thermal bath, so that a potential of mean-forces can be calculated along the reaction coordinate for each configuration of the colliding reactants.

Kramers' Theory and Subsequent Modifications

In its earliest version, the Kramers' theory described the movement of a particle Z of mass M, in a potential well V(Q) where Q is a generalized nuclear coordinate arising out of the action of various forces:
- one force, $- \partial V(Q)/ \partial Q$, derives from the potential V(Q) itself; as in the TST, this is assumed to be a potential of mean-forces;
- another describing the stochastic forces resulting from the random matrix-reactant interactions: F(t);
- a second deterministic force described by a friction coefficient also resulting from the matrix-reactant interactions : M γ v.

Therefore, the Langevin equations of motions are:

$$\partial Q/ \partial t \; = \; v \qquad\qquad (4)$$

$$\partial v/ \partial t \; = \; -1/M \; \partial V/ \partial Q \; + \; F(t)/M \; - \; \gamma \, v \qquad\qquad (5)$$

The stochastic forces F(t) and the friction coefficient, γ, resulting from the same matrix-reactants interactions must be directly related. This is given by the fluctuation-dissipation theorem:

$$< F(t) \; F(t+t) > \; = \; 2 \, \gamma \; M \, kT \, \delta(t) \qquad\qquad (6)$$

to which may be added without any loss of its general applicability:

$$< F(t) > = 0 \qquad (7)$$

The first condition states that stochastic forces have no memory on the time scale of the barrier crossing while the second simply ensures that no continuous force is included with the stochastic ones. These random forces represent in the Langevin model the energy exchanges discussed above.

Since it is assumed that the intensity of the solvent-reactant interactions are comparable, even on a short time scale, to the forces deriving from the potential, it can be said that the hypotheses which lie at the basis of the Kramers theory are exactly the opposite of those used to formulate the TST. This model of strong reactant-matrix interactions allows one to picture the flight over the barrier as the free diffusion of a Brownian particle (note that in this region, the forces deriving from the potential barrier are negligible). Therefore, in the Kramers theory there is a possibility of back-crossing: the particle's quantity of movement can change signs while it is on the saddle; in the TST, instead, the particle is considered to be experiencing a free flight over the potential barrier so that no back-crossing is possible.

Between the Eyring and the Kramers models, there are obviously many intermediate situations which depend upon the strength of the reactant-matrix interaction - weak, medium or strong - these concepts being defined with respect to the flight time over the barrier saddle. In more precise terms, it is possible to visualize three regions corresponding to low, medium and high values of the ratio between the frequencies linked to the friction coefficient and to the saddle of the potential barrier, respectively. At very low friction values, the stochastic forces are too weak to maintain the reactants at thermal equilibrium and, as a consequence, the macroscopic rate constant will go to zero as soon as the energy-rich fraction has reacted. At very high friction values, the reactants are efficiently pumped to a relatively high level of internal energy but, as soon as their energy increases, they are very quickly damped to a lower energy level so that once again, the reaction rate tends to go to zero. At intermediate friction values, the energy pumping will be sufficiently efficient and the energy damping sufficiently slow to allow an efficient crossing of the barrier. It should be noted that the Kramers' rate constant is always lower than or equal to the Eyring's constant.

Recently, some of the hypotheses at the basis of the Kramers' theory have been "relaxed":

- the friction coefficient is no longer expected to be represented by a delta function; it is now characterized by a memory function which accounts for the finite correlation time of the stochastic forces compared to the flight time over the barrier. The rate constants of the reaction obtained in this case have values intermediate between those generated by the TST and the Kramers models;
- the finite time response of the matrix to the changes in the reactants during the course of the reaction - for example, the dipole reorientation time - has been included in the Langevin equations of motion.

Hydrogen Tunneling

One of the most common reactions in biochemistry is the proton exchange between groups; a good example is presented by the serine proteases (1). In classical mechanics, a particle oscillates strictly along the potential curve, but in a quantum description it is allowed to probe forbidden space regions due to the Heisenberg uncertainty principle:

$$\Delta Q \cdot \Delta v_Q \geq h \qquad (8)$$

The proton being a light particle, a significant part of the wave function describing its movement in a potential well will extend a relatively long way into the forbidden region of the space. The probability of a particle being observed there decreases with the increasing mass, height and width of the potential barrier (15). In the case of a double well, a proton can therefore move from one well to the other without necessarily passing over the saddle - this being referred to as the "tunnel effect". The width and the height of the potential barrier depend upon the temperature, which in turn depends on: (a) a modulation of the separation between the potential minimae due to oscillations of the radicals around their equilibrium positions, and (b) a change in the population of the vibrational excited states. Therefore, the probability of tunneling through a barrier is expected to increase with the temperature.

<u>General comments</u>

Whatever the model one uses among the three described above, there are certain aspects which require further elaboration:
- the problem of the energy distribution, migration and transfer between the various degrees of freedom of the system;
- the description of the potential curve: this is crucial to a complete understanding of exactly which terms determine the saddle since: (a) the rate of escape is fixed by the ratio between the characteristic times associated with the friction coefficient and the saddle crossing, and (b) the potential curves obviously have a direct role to play in the chemical reaction;
- the description of the particle's movement in a complex potential well.

THE CLASSICAL TRAJECTORY DESCRIPTION

Consider, for the sake of simplicity, a linear triatomic molecule: XYZ. The symmetric and antisymmetric stretching modes can be visualized in terms of a particle moving in a potential well which, at the bottom, can be approximated by a harmonic potential. Such a particle will experience forces directed along the maximum slope of the potential surface and since on a purely harmonic potential surface, those forces are always parallel to the trajectories, if no perturbation is applied, the particle will oscillate indefinitely along the same trajectory.

If the particle is prepared in a higher energy state, the potential surface is no longer harmonic and the forces and the instantaneous trajectories are no longer co-linear. As a consequence, the particle prepared in such a state will describe a Lissajoux curve on the potential surface; that is it will not move for a long period of time along trajectories described by normal modes. If it is prepared in an even more energetic state, in the absence of energy dissipation it will after some time reach one of the dissociation regions and the molecule will dissociate. It should be noticed that in this region, the movement of the particle corresponds to very different states of motion for the external nuclei; the one escaping the well is in free flight while the other is more or less oscillating around an equilibrium position.

Using this classical treatment in computer simulation yields numerical results which are in surprising agreement with those obtained using a more sophisticated (and realistic) treatment by quantum description. A plausible explanation for this is to be found in the fact that, in classical terms, the particle escaping from the well passes a very large portion of its time in a region where the quantum effects are negligible due to the high density of states.

APPLICATION TO ENZYME CATALYSIS

The concepts discussed in the present chapter should theoretically be applicable to the enzyme catalysis process but when one tries to apply them a few problems arise:

- Energy distribution, migration, transfer and fluctuations: Enzymes are mesoscopic systems and their fluctuations can thus be estimated without any special difficulties (1,3,13) but an enzyme is a highly structured and thus anisotropic system; therefore one must ask two questions: (a) is there a tendency due to the three-dimensional structure to convey energy to the active site, and if so (b) how often and for how long can an enzyme concentrate a sufficient fraction of its internal energy "on the reaction coordinate" and so promote the reaction?

- Potential energy and the chemical reaction: As discussed above: it is necessary to know the exact form of the potential surface (4,5,7). This implies a knowledge of the dynamics of the active site which acts as the solvation shell of the reactants. The difficulties reside in the fact that on the one hand we have little more information than that provided by spectroscopy (2) and molecular dynamics (9) and on the other, it is unrealistic to assume that the whole system (substrates-active site and products-active site) is in thermal equilibrium during the course of the reaction.

- The choice between the TST and Kramers models: Since the enzyme-substrate complex lasts for about 1 ms, it is obvious that the system exchanges energy with its thermal bath during its lifetime, but this time interval covers many individual events which must be isolated in this theoretical treatment. There is little - or no - indication that the Eyring transition state theory cannot be applied.

- The classical trajectory approach: at this stage, this approach appears to be the only feasible one since as with a sufficient knowledge of the reaction mechanism and of the potential surface, it is possible to estimate with a high degree of precision the enzyme catalysis rate (4,8).

REFERENCES

(No attempt has been made in this chapter to present a complete bibliography of the subject since it is intended only to be a general introduction to the other chapters of the present volume. A few basic references not expected to be cited elsewhere have been included.)

(1) A. Cooper, J.L. Houben: Enzyme catalysis: A physical chemist's description, in the present volume.
(2) J.L. Houben: The spectroscopy of enzymes: Introductory remarks, in the present volume.
(3) A. Cooper, D.T.F. Dryden: Thermodynamic fluctuations in proteins, in the present volume.
(4) A. Warshel: Microscopic simulations of chemical reactions in solutions and protein active sites: Principles and examples, in the present volume.
(5) J.T. Hynes: The role of the environment in chemical reaction, in the present volume.
(6) D. Borgis, J.T. Hynes: Proton transfer reactions, in the present volume.
(7) T. Fonseca: The concept of the potential of mean-force in enzyme catalysis, in the present volume.
(8) G. Alagona, C. Ghio: Theoretical calculations on an enzyme catalyzed reaction mechanism, in the present volume.
(9) R.L. Somorjai: Proteins: interactions and dynamics, in the present volume.
(10) A. Lami: Highly excited vibrational states and chemical reactivity, in the present volume.
(11) F. Fillaux: Nonlinear coupling and vibrational dynamics, in the present volume.
(12) G. Careri: Enzyme catalysis: an overview from physics, in the present volume.
(13) D.A. McQuarrie:"Statistical Physics." Harper & Row, New York (1976).

(14) The vibrational temperature is the energy of the vibrational state expressed in kT units.

(15) P.W. Atkins: "Molecular quantum mechanics" 2nd ed. Oxford University Press, Oxford (1983).

(16) T. Fonseca, J.A.N.F. Gomes, P. Grigolini, F. Marchesoni: The theory of chemical reaction, p. 389 in: "Memory approaches to stochastic processes in condensed matter," M. Evans, P. Grigolini and G. Pastori Parravicini (eds.), in Adv. Chem. Phys., Vol. 62, I. Prigogine and S. Rice (Gen. eds.), J. Wiley, New York (1985).

THE ROLE OF THE ENVIRONMENT IN CHEMICAL REACTIONS

James T. Hynes

Department of Chemistry and Biochemistry
University of Colorado
Boulder, CO 80309-0215   USA

1. INTRODUCTION

The last decade has seen a renaissance in the study, both experimental and theoretical, of chemical reactions in solution[1,2]. Naturally, a focus of these studies is the determination of the role of the solvent in affecting the rate constant for the chemical process. Here we attempt to give a very brief glimpse at some of the ideas and results that have emerged. We restrict the discussion exclusively to reactions in solution (and even then, to a very few topics). But it is likely that, at least in a general way, similar ideas should ultimately prove useful in the context of understanding enzyme catalysis rates.

It is perhaps useful at the outset to draw draw a few rough analogies to connect the solution and enzyme rate problems. In solution a reaction can be schematically represented by

$$A + B \; \underset{k_{diss}}{\overset{}{\rightleftharpoons}} \; [AB] \; \overset{k}{\longrightarrow} \; P \quad ,$$

which shares some features in common with the Michaelis-Menten scheme

$$E + S \; \underset{k_s}{\rightleftharpoons} \; [ES] \; \overset{k_{cat}}{\longrightarrow} \; P + E \quad .$$

In the solution case, [AB] represents species "caged" by the solvent, with k_{diss} representing a physical diffusive transport step. The chemical reaction proper, which is our focus, is the final step, and is characterized by the rate constant k. One wants to know: what determines k and how does the solvent influence it? The corresponding questions for the enzyme problem would focus on k_{CAT} and would involve an investigation of the environment in the active site, as an analogue of the solvent.

Some of the reactions we will discuss are similar in type to reactions in enzymes. For example, the nucleophilic displacement

$$Cl^- + CH_3Cl \rightarrow ClCH_3 + Cl^-$$

in water is of the same type as phosphoryl transfers

$$\sim O^- + O_3PO\sim \; \rightarrow \; \sim OPO_3 + {}^-O\sim$$

described elsewhere in this volume[3]. Further, the model proton transfers

$$AH^+ + B \rightarrow A + H^+B$$

we will discuss in a separate chapter have strong analogies to proton transfers in many enzymatic systems, e.g. serine proteases.[4]

It is also useful at the outset to distinguish between equilibrium or static solvent effects and dynamic or nonequilibrium solvent effects. The former fall within the province of Transition State Theory, while the latter do not. In what follows we will focus mainly on the latter.

2. TRANSITION STATE THEORY

Without a doubt the most important theory for reactions in solution, both from conceptual and practical viewpoints, is Transition State Theory (TST)[5,6]. Here we briefly describe the ideas and ingredients of TST.

To focus the discussion, we consider a simple model _cis_-_trans_ isomerization in an inert solvent

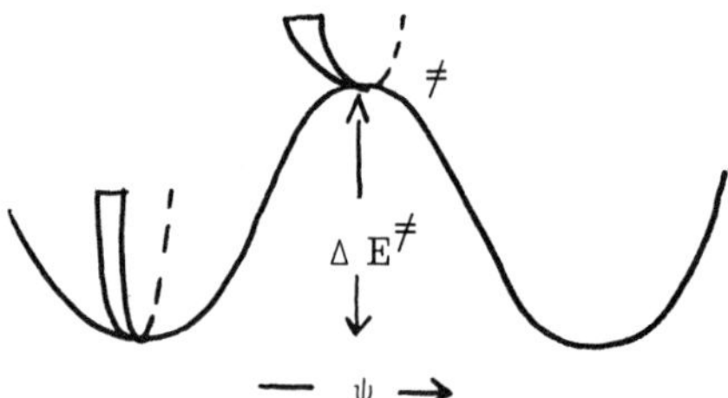

in terms of two coordinates: the XCCX angle ϕ and the CC bond length l. The schematic potential energy for this system is represented as Fig. 1

Fig. 1   Schematic illustration of the potential wells and barrier for the isomerization. The transverse vibrational wells for the reactant and transition state are displayed.

with the _cis_ and _trans_ wells in ϕ indicated. The curvature in e.g. the reactant _cis_ well is characterized by the frequency ω_R, i.e. the square root of the spring constant divided by the reduced mass.

The transverse wells in l represent bound CC stretching motions in the reactant, with frequency $\omega_{l,R}$, and at the transition state, with frequency $\omega_{l,\ddagger}$. Since a double bond will weaken near $\ddagger$ as the overlap of the p electronic orbitals on the C atoms diminishes when ϕ nears 90°, the latter frequency will be less than the former. For simplicity, we treat both the torsional angle and all vibrations classically in the following.

The TST expression for the rate constant k can be rationalized as follows. If we consider an equilibrium solution of _cis_ and _trans_ isomers dilute in solution, both forward and reverse reactions will be occurring. For example, if we imagined watching the trajectories for forward reactions, we would see _cis_ molecules being activated by collisions with the solvent molecules, passing over the barrier, i.e. through the transition state. These molecules would ultimately be deactivated, again by collisions with the solvent molecules, into the _trans_ product well. The microscopic details of these trajectories will all differ from each other due, e.g., to the differing details of the interaction with the solvent, that is, due to the differing forces exerted on the reaction system. If we now focus on those trajectories originating from the reactant _cis_ well, TST states that k is given by

$$k^{TST} = P^{\ddagger} \nu^{\ddagger}, \tag{1}$$

the product of the equilibrium probability $P^{\ddagger}$ of reaching the transition state and the frequency $\nu^{\ddagger}$ of passing through it from the _cis_ side to the _trans_ side. [A proper way to evaluate $\nu^{\ddagger}$ is to consider motion along the reaction coordinate ϕ as a translation[5], rather than as a vibration as is done in many texts]. This works out to be

$$k^{TST} = \frac{k_B T}{h} \frac{Q^{\ddagger}}{Q_R} e^{-\Delta E^{\ddagger}/k_b T}, \tag{2}$$

where k_B is the Boltzmann constant and h is Planck's constant. Q_R is the partition function for the _cis_ reactant. $Q^{\ddagger}$ is the partition function for the transition state species, but it is missing any contribution from the reaction coordinate, i.e. $Q^{\ddagger}$ is the partition function for the vibration in l transverse to the reaction coordinate. The reaction coordinate motion at $\ddagger$ has already been accounted for by the frequency factor $\nu^{\ddagger}$.

The partition function for a classical vibration is just $k_B T/\hbar\omega$, so eq. (2) works out to be

$$k^{TST} = \left(\frac{\omega_R}{2\pi} \right) \left(\frac{\omega_{l,R}}{\omega_{l,\ddagger}} \right) e^{-\Delta E^{\ddagger}/k_B T}. \tag{3}$$

But the ratio $(\omega_{l,R}/\omega_{l,\ddagger}) = \exp(\Delta S^{\ddagger}/k_B)$ is just an activation entropy factor reflecting the feature that for $\omega_{l,R}/\omega_{l,\ddagger} > 1$, there is a broader "pass" across the barrier favoring the reaction. In this simple model, there is no difference between enthalpy and energy, so that activation free energy is $\Delta G^{\ddagger} = \Delta E^{\ddagger} - T\Delta S^{\ddagger}$ and we have finally

$$k^{TST} = \left(\frac{\omega_R}{2\pi} \right) e^{-\Delta G^{\ddagger}/k_B T}. \tag{4}$$

The TST rate constant is then an "attempt" frequency factor along ϕ in the reactant times an exponential free energy factor. The latter includes the energy change between the bottom of the reactant well and the transition state, and the entropy change associated with changes in the confining potentials perpendicular to the reaction coordinate ϕ on passing from the reactant to the transition state.

The basic picture and predictions of TST are the same for atom or group transfer reactions $A + BC \rightarrow AB + C$ (Fig. 2). In this case, TST gives the rate constant as

$$k^{TST} = \left(\frac{\omega_{coll}}{2\pi} \right) e^{-\Delta G^{\ddagger}/k_B T}, \tag{5}$$

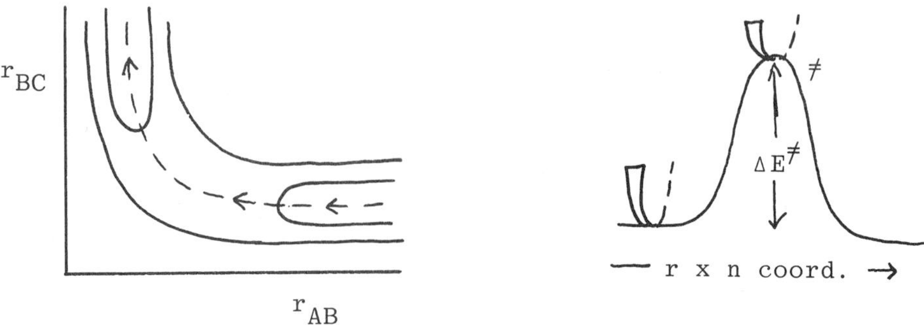

Fig. 2   Schematic illustration of the potential energy surface for a collinear atom transfer A + BC → AB + C (here taken to be symmetric, A=C). The potential barrier and transverse reactant and transition state wells are also shown.

where ω_{coll} is a collision frequency for the relative approach of A and BC. Here $\Delta G^{\ddagger}$ includes the energy cost $\Delta E^{\ddagger}$ for reaching the transition state and the entropy change $\Delta S^{\ddagger}$ associated with the differing vibrational frequencies in the reactant (AB vibration) and in the transition state (e.g. a symmetric ABC vibration in the symmetric case where A = C).

Solvent effects on reactant rate constants within the TST framework arise from a change in $\Delta G^{\ddagger}$ on passing from gas phase to solution, or from one solvent to another. In this framework, the influence of a solvent arises from a contribution to the solvation free energy which differs for the reactants and the transition state. With a few exceptions[7], these effects are only important for reactions involving the displacement of changes, and for those reactions, the effects can be enormous, representing many orders of magnitude for the rate. We return to this issue after we have discussed a different influence of the solvent on the rate.

3. DYNAMIC SOLVENT EFFECTS - KRAMERS THEORY

TST assumes that once the reacting system is at the transition state and heading from the reactant side of the barrier to the products side, it will always go on to form products rather than recrossing back towards reactants. But it is certainly possible that one or more of the surrounding molecules collide with the reaction system and induce a recrossing of the barrier back towards reactants. In this case, TST will be an overestimate of the rate, and it is conventional to characterize the deviation of the actual rate constant k from the prediction of TST by the transmission coefficient κ:

$$k = k^{TST}\kappa \; ; \qquad \kappa \leq 1. \qquad (6)$$

The more extensive is the solvent-induced recrossing the smaller will be κ.

The first and still very much vital theory of k for reactions in solution is due to Kramers.[8] In this description, a very simple and schematic description of both the reaction system and the solvent is assumed. In particular, the reaction is idealized as passage along a reaction coordinate x over a parabolic barrier (Fig. 3)

$$\Delta G(x) = \Delta G^{\neq} - \frac{1}{2} \omega_b^2 x^2 \; .$$

The dynamic interaction with the solvent is represented by a Langevin equation (LE) for the average motion:

$$\overline{\ddot{x}}(t) = \omega_b^2 \, \overline{x}(t) - \zeta \, \overline{\dot{x}}(t) \ . \tag{7}$$

Here the frictional term $-\zeta \, \overline{\dot{x}}(t)$ represents an average damping force proportional to the velocity. This force is supposed to represent, for example, the collisional influence of the solvent in inducing recrossings of the barrier. The friction constant ζ is related to the forces exerted by the solvent on the reaction system in the neighborhood of the transition state. In particular, ζ is

$$\zeta \propto \int_0^\infty dt \, \langle F(0) \, F(t) \rangle \ , \tag{8}$$

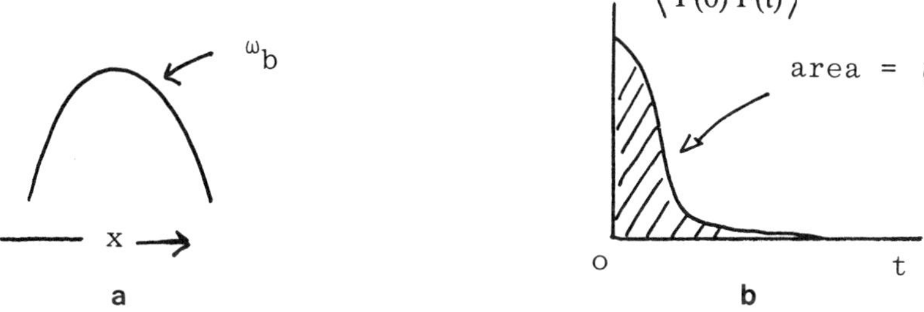

Fig. 3   (a) The free energy versus reaction coordinate in the transition state region.
(b) Schematic time behavior of the time dependent friction.

i.e., proportional to the time integral of the correlation function of the reaction system — solvent force F, as indicated in Fig. 3. In the simplest conception, ζ would be proportional e.g. to the solvent viscosity. We will subsequently return to the identity of the friction, but for now it suffices to say that the stronger the coupling of the solvent to the reaction system, the larger is the friction constant ζ.

With this model, Kramers found that the transmission coefficient is

$$\kappa_{KR} = [1 + (\zeta/2\omega_b)^2]^{1/2} - (\zeta/2\omega_b), \tag{9}$$

whose behavior is illustrated in Fig. 4. For small friction ζ, solvent-induced recrossing is negligible, and κ approaches unity so that TST applies. (At extremely low friction, there is another behavior[8], but this will not concern us here). But as ζ increases, there is increased barrier recrossing and κ drops. Ultimately the diffusive regime is reached in which $\kappa \propto \zeta^{-1}$, and the reaction is well described as a diffusive motion across the barrier, characterized by extensive recrossing, and a very small κ value.

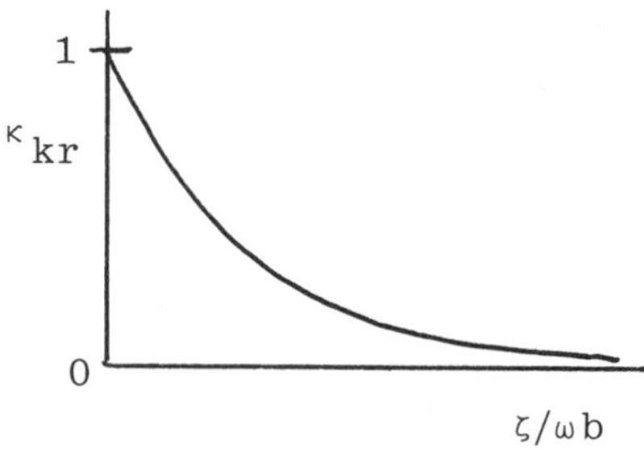

Fig. 4   The behavior of the Kramers Theory transmission coefficient as a function of the friction constant ζ (see text).

There have been a number of recent tests, both experimental and computational, of Kramers theory. At this stage, it seems fair to say that sometimes it works and sometimes it does not. We focus first on the former situation. The most extensive experimental tests have involved barrier crossing reactions in excited electronic state isomerizations. One example is the <u>trans-cis</u> isomerization in the first electronically excited state of stilbene. In this case, the Kramers predictions are fairly well borne out[9]. The rate declines with increasing friction more or less according to Eq. (9). Another example is provided by a molecular dynamics (MD) computer simulation of a model solvent-separated ion pair → contact ion pair reaction in a polar solvent,

$$A^+|B^- \rightarrow A^+B^-,$$

whose free energy curve as a function of separation r is sketched in Fig. 5.

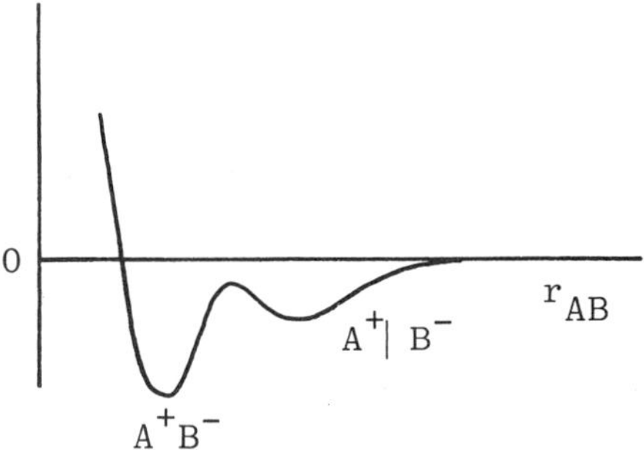

Fig. 5   Illustration of the potential of mean force for the ion pair association described in the text.

For this low barrier (~ 2.5 k_BT) reaction, the MD simulation yields $\kappa = 0.2$, in excellent agreement with the Kramers theory prediction.[10]

These reaction examples are special in the sense that the reaction barriers are low (a few kcal or less) as are the barrier frequencies ω_b, i.e. the barriers are fairly broad. It is more typical that reaction barriers are higher and sharper. In such cases, Kramers theory fares considerably less well, as we now describe.

4. GROTE-HYNES THEORY

In what follows, we will describe the theory which has been constructed to improve upon Kramers theory, and a test of that theory. But first it will be useful to introduce a qualitative characterization of the strength of reaction system-solvent forces.

For many nonionic isomerization and atom or group transfer reactions, the reaction system-solvent forces are of the non-electrostatic or van der Waals type, and are weak. By contrast, for transfers such as nucleophilic displacements $X^- + RY \rightarrow XR + Y^-$ in polar solvents, these interactions are electrostatic and/or hydrogen-bonding in character, and are comparatively strong. This disparity has important consequences for both the static and dynamic influence of the solvent on k, as we will now illustrate.

We consider first a weak interaction case reaction: a model atom transfer reaction A + BC in compressed rare gas solvents. The transmission coefficient κ has been determined[11] in an MD simulation for various barrier heights $\Delta E^{\ddagger}$ (cf Fig. 2). For $\Delta E^{\ddagger}$ values ranging from 20 kcal to 5 kcal in assorted liquid density rare gas solvents, κ is found to differ negligibly from unity; TST is essentially perfect. The fundamental reason for this is simply that the strong chemical forces -- i.e. the sharp barrier for which the energy drops rapidly as the transition state is left -- overwhelm any interaction forces due to the solvent, and the solvent is powerless to induce any recrossing of the barrier.

To set the stage for the key ideas of Grote-Hynes Theory, we consider a strong interaction case reaction: the nucleophilic displacement S_N2 reaction $Cl^- + CH_3Cl \rightarrow ClCH_3 + Cl^-$ in water. The strength of the reaction-system — solvent interaction forces is revealed in Fig. 6, which shows schematically the theoretical gas phase potential and the solution phase potential of mean force along the reaction coordinate.

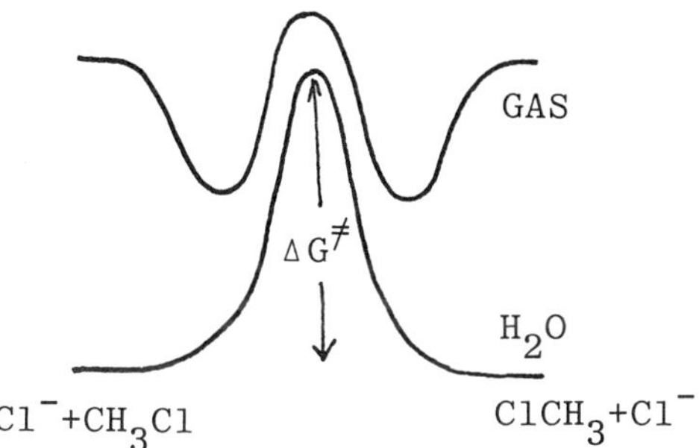

Fig. 6   Schematic gas phase and solution phase (free) energy diagrams for $Cl^- + CH_3Cl$ reaction in H_2O.

In the gas phase, the central Walden inversion barrier is flanked by two wells arising from ion-dipole $Cl^- \cdot CH_3Cl$ complexes. The solution mean potential, which is calculated[12] by allowing the water solvent to equilibrate to the reaction system at each value of the latter's reaction coordinate, is quite different. The equilibrium has erased the wells by lowering the energy, i.e., stabilizing, the charged localized reactants and products. The charge delocalized transition state $Cl^{\delta-}CH_3^{\delta+} Cl^{\delta-}$ is also stabilized, but to a lesser extent, by the solvation. The net effect is quite large for the rate: The gas phase central barrier of ~ 14 kcal/mol is raised to a barrier of ~ 26 kcal/mol by the differential solvation.

In the TST approach, this is the end of the story. Basically eq. (5) gives the rate constant and $\Delta G^{\ddagger}$ for the reaction is given by the appropriate solution phase free energy difference in Fig. 6. But there is a potential problem: it has been assumed that the solvent is always equilibrated to the reaction system at each stage of the latter's reaction coordinate. (More generally, it can be shown that TST corresponds to this assumption of equilibrium solvation[13]). The assumption will hold if the key time scale t_{rx} for the reactive motion is slow compared to the time scale t_{solv} for solvent relaxation or equilibration. Is this true?

What is t_{rx}? It is not, as one might first imagine, the inverse of the rate constant, k^{-1}. This is roughly the time to <u>reach</u> the transition state; it is very long — seconds to hours

— since k $\propto$ P‡ is very small due to the low probability P‡ of reaching the transition state. Instead t_{rx} is the characteristic time for <u>crossing</u> the barrier once the transition state neighborhood is reached. It can be shown[14] that t_{rx} is approximately ω_b^{-1}, where ω_b is the barrier frequency introduced in Sec. 3 which measures the curvature of the barrier. The higher and "sharper" the barrier, the higher is ω_b and the shorter is t_{rx}. This time can be quite short indeed for modest-to-high barrier reactions. For the the Cl$^-$ + CH$_3$Cl S_N2 reaction, it is 0.02 ps.[16]

Given this short t_{rx}, can the H$_2$O solvent keep up with the reaction system and provide equilibrium solvation? The characteristic solvation time for water involving, e.g., water molecule dipole reorientation is in the range $t_{solv} \sim 0.25$ - 1 ps.[15] Clearly the condition for equilibrium solvation does not apply and instead we must think about <u>nonequilibrium</u> solvation.

Van der Zwan and Hynes[13] have shown that such nonequilibrium solvation effects can be described by the Grote-Hynes (GH) Theory of Reaction Rates.[14] We will illustrate this feature in action presently, but first we describe the key elements of GH Theory. It is assumed that the average motion along the reaction coordinate x in the transition state neighborhood is governed by the Generalized Langevin Equation (GLE)

$$\overline{\ddot{x}}(t) = \omega_b^2 \, \overline{x}(t) - \int_0^t d\tau \, \zeta(\tau) \, \overline{\dot{x}}(t-\tau) \tag{10}$$

in which the time dependent friction coefficient $\zeta(t)$

$$\zeta(t) \propto \langle FF(t) \rangle \tag{11}$$

is proportional to the correlation function of the force exerted by the solvent on the reaction system along the reaction coordinate in the transition state neighborhood. The essential difference between the GLE and the LE eq. (7) is that the GLE allows a description of events occurring on short time scales. In particular, the GH equation[14] for the transmission coefficient,

$$\kappa = [\kappa + \omega_b^{-1} \int_0^\infty dt \, e^{-\omega_b \kappa t} \zeta(t)]^{-1} \, , \tag{12}$$

reflects this. It shows that when the barrier is high and sharp and ω_b is large — so that t_{rx} is small — only the short time aspects of the solvent forces, i.e., the time dependent friction $\zeta(t)$, are important in the reaction (when ω_b is large, the exponential in eq. (12) is non-zero only for small times t). In contrast, when the barrier is low and broad, ω_b is very small and the reaction time scale t_{rx} is long. Then the full friction, i.e., the friction constant ζ, eq. (8), is relevant (ω_b can be put to zero in the time integral in eq. (12). In this case the GH equation for κ reduces to the Kramers eq. (9)).

The importance of nonequilibrium solvation described by GH Theory can be illustrated via comparison with the results of Molecular Dynamics (MD) computer simulation of the Cl$^-$ + CH$_3$Cl reaction in a solvent composed of 64 H$_2$O molecules. The details of the potentials and techniques employed in the MD simulation are described elsewhere,[15] but one detail that is worth stressing here is the rapid charge shift from the attacking chloride ion to the leaving chloride ion in the neighborhood of the transition state. This is responsible for strong electrical coupling to the protic polar solvent in this heavy particle charge transfer reaction.

The value of κ determined in the MD simulation is $\kappa_{MD} = 0.54 \pm 0.05$, so that there is a serious departure from the equilibrium solvation TST theory. The GH Theory predicts $\kappa_{GH} = 0.57 \pm 0.08$, which agrees with the MD result within the error bars. Further, the most important part of the time dependent friction $\zeta(t)$ in the GH equation is its initial value $\zeta(t=0)$. This feature is a consequence of the time scale relation $t_{rx} \sim \omega_b^{-1} \ll t_{solv}$ — the solvent molecules hardly move during the passage over the barrier top. It tells us that the solvent-induced recrossing reflected in $\kappa_{MD} \sim 0.5$ is due to static solvent <u>configurations</u>, rather than to any solvent dynamics. Indeed the recrossings can be explained in detail in terms of static solvent water configurations which, for example, more effectively solvate the $Cl^{\delta-}$ in $Cl^{\delta-}-CH_3^{\delta+}Cl'^{\delta-}$ at the transition state (Fig. 7). As the Cl' species attempts to leave to form the products, this asymmetric solvation pulls the reaction system back to form the more favored reactants--a recrossing of the barrier.

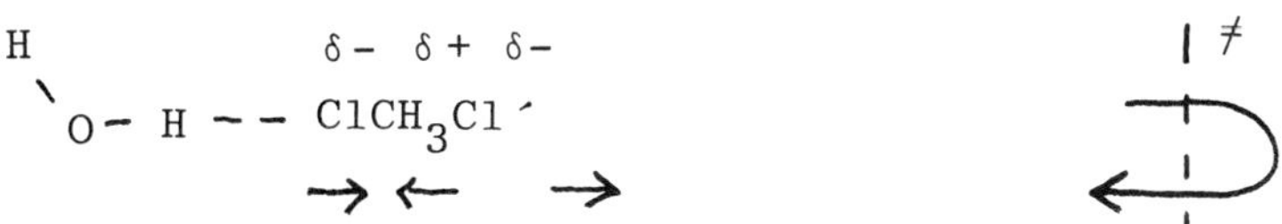

Fig. 7   Schematic of asymmetric solvation induced recrossing of the transition state in the Cl^- + CH_3Cl reaction in H_2O.

Thus the picture that emerges is a nonequilibrium solvation perspective that is far removed from the equilibrium solvation TST picture. It is also far removed from the Kramers picture. In fact, for this reaction eq. (9) predicts $\kappa_{KR} = 0.02$, far below the true value. The reason for this dramatic failure is that the long time scale frictional forces, which are measured by the friction constant ζ, are very large. The Kramers Theory states that it is these large forces which govern the barrier motion and cause extensive diffusive recrossing of the barrier and a resulting very low transmission coefficient. In reality, these long time forces are irrelevant on the short time scale t_{rx} of the reaction and the true relevant "frictional" solvent forces are considerably less. This crucial feature is accurately described by the GH Theory.

The basic picture above has been thoroughly tested in simulations of variants of the basic S_N2 reaction system.[15,16] The validity of the GH description is confirmed in every case. This validity extends to cases where the central barrier is artificially lowered thereby reducing the barrier frequency. In such cases, the solvent <u>dynamics</u> plays a crucial role in allowing the reaction to proceed. In particular, the short time rapid water librational motions provide the dominant dynamical component for the time dependent friction $\zeta(t)$ and determine the transmission coefficient.[16] This short time aspect is well described by GH Theory, but is completely missed in a Kramers description which incorrectly assumes that it is instead the longer time water molecule reorientational and translational dynamics that influence the rate. This dominance of the molecular scale reaction dynamics by faster solvent motions should also apply to different types of charge transfer reactions such as S_N1 dissociations[17] $RX \rightarrow R^+ + X^-$ and electron transfers.[18]

What happens when the reaction barrier is quite low and broad and the frequency ω_b is small? In this case, there is time enough for the full friction constant ζ to be relevant for the reaction and GH Theory reduces to Kramers Theory. One example of this is provided by MD simulation of an ion pair association mentioned in Sec. 3. For this low broad barrier reaction, MD, GH and Kramers all give the same result for κ.[10]

CONCLUDING REMARKS

In view of the above, one might ask: what is _right_ about Transition State Theory? The answer is that it correctly incorporates in the rate the factor

$$P^{\ddagger} \propto e^{-\Delta G^{\ddagger}/k_b T}$$

giving the probability that the transition state is achieved. This exponential factor is of course extremely important in determining the rate constant, and can be understood and calculated using equilibrium solvation ideas. What TST does not necessarily describe well is, in effect, the effective frequency for crossing the barrier in the transition state region, including recrossings. But as we have seen, the GH Theory is able to correct the TST Theory, both numerically and conceptually, to account accurately for such nonequilibrium solvation effects on the rate. As we have also seen, the latter can be quite significant for the rate, leading, e.g., to an order of magnitude correction.

Both equilibrium solvation and nonequilibrium solvation rate effects should also be present in enzyme catalytic rates. Here charged and polar groups in the active site provide the environment that is the analogue of the solvent. We expect that the molecular level questions of the influence on enzyme catalytic rates due to equilibrium and nonequilibrium "solvation" by the environment in the active site can now begin to be addressed using ideas, techniques and theories similar to those we have briefly described within.

ACKNOWLEDGMENTS

This work was supported in part by grants from The U.S. National Science Foundation, CHE 84-19830 and 88-07852, and the Petroleum Research Fund, administered by the American Chemical Society.

REFERENCES

1. J. T. Hynes, in "The Theory of Chemical Reactions," M. Baer, ed. Vol. IV (CRC Press, Boca Raton, FL 1985), p. 171.
2. J. T. Hynes, _Ann. Rev. Phys. Chem._ 36, 573 (1984).
3. H. Watson, this volume.
4. P. Carey, this volume.
5. S. Gladstone, K. J. Laidler and H. Eyring, "The Theory of Rate Processes," (McGraw-Hill, New York, NY, 1941); D. G. Truhlar, W. L. Hase and J. T. Hynes, _J. Phys. Chem._ 87, 2664 (1983).
6. T. Fonseca, this volume.
7. B. M. Ladanyi and J. T. Hynes, _J. Am. Chem. Soc._ 108, 585 (1986).
8. H. A. Kramers, _Physica_, 7, 284 (1940).
9. S. K. Kim and G. R. Fleming, _J. Phys. Chem._ 92, 2168 (1988).
10. G. Ciccotti, M. Ferrario, J. T. Hynes and R. Kapral, to be published.
11. J. P. Bergsma, J. R. Reimers, K. R. Wilson and J. T. Hynes, _J. Chem. Phys._ 85, 5625 (1986).
12. J. Chandrasekhar, S. F. Smith and W. L. Jorgensen, _J. Am. Chem. Soc._ 107, 154 (1985).
13. G. van der Zwan and J. T. Hynes, _J. Chem. Phys._ 76, 2993 (1982); _ibid._ 78, 4174 (1983); _Chem. Phys._ 90, 21 (1984).
14. R. F. Grote and J. T. Hynes, _J. Chem. Phys._ 73, 2715 (1980).
15. J. P. Bergsma, B. J. Gertner, K. R. Wilson and J. T. Hynes, _J. Chem. Phys._ 86, 1356 (1987); B. J. Gertner, J. P. Bergsma, K. R. Wilson, S. Lee and J. T. Hynes, _ibid._ 86, 1377 (1987).
16. B. J. Gertner, K. R. Wilson and J. T. Hynes, _J. Chem. Phys._, in press.
17. D. A. Zichi and J. T. Hynes, _J. Chem. Phys._ 88, 2513 (1988)
18. D. A. Zichi, G. Ciccotti and J. T. Hynes, to be submitted; J. T. Hynes, _J. Phys. Chem._ 90, 3701 (1986).

PROTON TRANSFER REACTIONS

Daniel Borgis and James T. Hynes

Department of Chemistry and Biochemistry
University of Colorado
Boulder, CO 80309-0215   USA

Proton (H^+) transfers play a central role in many enzymatic reactions.[1] A good illustration is provided by serine protease reactions.[2-4] In a recent theoretical study[4] of the serine protease tripsin-catalyzed hydrolysis of a specific tripeptide substrate, there are no less than four proton transfer steps, beginning with a serine to histidine H^+ transfer in the Michaelis complex (Fig. 1)

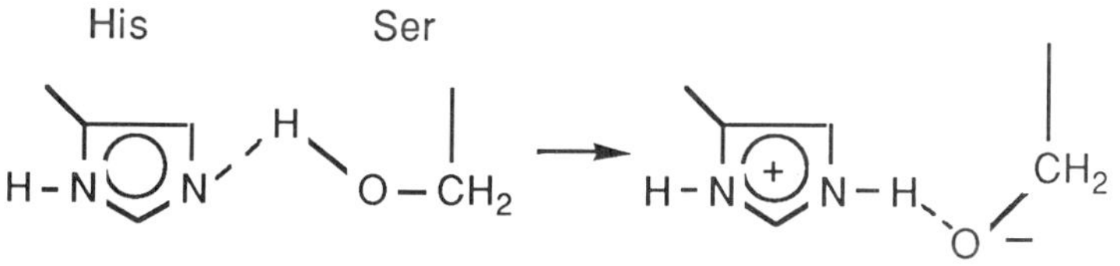

Fig. 1   The serine-histidine proton transfer.

Another example is furnished in a study[5,6] of the reversible isomerization of dihydroxyacetone (DHAP) to glyceraldehyde 3-phosphate (GAP) by the enzyme triose phosphate isomerase (TIM). A key step in the DHAP $\rightleftharpoons$ GAP reaction is the proton abstraction by a carboxylate group in Glu 165 of a CH proton in DHAP (Fig. 2).

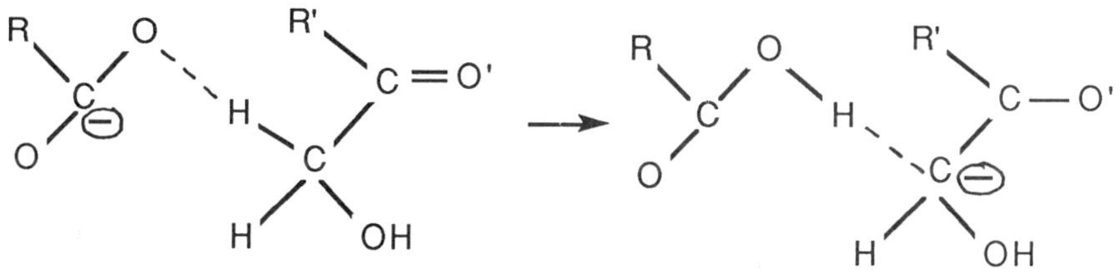

Fig. 2   Schematic of the proton transfer in TIM discussed in the text.

In order ultimately to understand such H^+ transfers theoretically, several key ingredients are required. The first ingredient is potential energy surfaces. These are now beginning to appear.[2-6] But knowledge of these surfaces is not sufficient, due to the special character of proton transfers, described below. A second ingredient, a theory for proton transfer rate constants, is necessary. Such a theory is only now being constructed for proton transfers in solution. We describe some aspects of that theory in this paper in a way which we hope indicates the relevance of these ideas to enzymatic proton transfers. A central theme will be the critical importance to the proton transfer of the fluctuations of the heavy atoms in its environment.

1. THE SPECIAL NATURE OF H^+ TRANSFERS

Three features make proton transfer reactions special and their theoretical treatment a challenge. First, the H^+ motion will be strongly coupled to internal or intramolecular vibrational modes in the reaction system. For example, the proton transfer barrier is strongly dependent on the O-C separation in a $RCOO + HCR'$ proton transfer in TIM[5,6](Fig. 2). Second, the proton may strongly electrically couple to the environment. In solution this will involve polar solvent molecules, and in enzymes it will involve, for example, charged groups in the active site, which strongly perturb H^+ transfer potential barriers.[5,6,8] Third, the light proton is a fundamentally quantum particle and may tunnel through a potential energy barrier rather than be activated over it.

The standard theory for interpreting H^+ transfers and H^+/D^+ isotope effects is Transition State Theory (TST),[9] sometimes corrected by tunneling factors.[10] The predictions of this theory are heavily used as indicators of mechanisms of proton transfers in enzymes.[11] It is, however, likely that TST and its underlying picture of H^+ transfers is wrong. (It is now known to be wrong for the related problem of H atom transfer in the gas phase[12]) The central difficulty with TST is that it grossly underestimates H^+ tunneling, which can completely dominate the actual H^+ transfer mechanism.

In the following, we sketch some key features of a theory due to Hynes and Borgis[7] incorporating all the aspects above for intramolecular H^+ transfers in polar solvents. While our focus is on reactions in solution, we anticipate that many qualitative features will also apply in enzymatic H^+ transfers. For example, the role of intramolecular vibrations described below for the solution case will be played by the motion of proton donor and acceptor groups in the active site of an enzyme. As a further analogy, the role of the polar solvent molecules in solution will be played by polar and charged groups in the active site of the enzyme. The importance of the latter for general enzymatic reactions has been emphasized by Perutz[16] and Warshel[17,18] among others.

The Borgis-Hynes theory has some features in common with earlier treatments of proton transfers;[13-15] similarities and differences are discussed elsewhere.[7] In the present paper, we first discuss the ideas and predictions of the theory in the solution context. Then we give a qualitative discussion of a particular enzymatic proton transfer and point out the strong general similarities to the solution case.

2. IMPORTANT COORDINATES FOR H^+ TRANSFER

The first important coordinate for an intramolecular proton transfer is q, the coordinate of the proton itself. There is a double well potential in q which is illustrated in Fig. 3 for the schematic symmetric transfer $AH^+ \, A \rightarrow A \, H^+A$, where the A's represent heavy atoms such as C, O, N, etc.

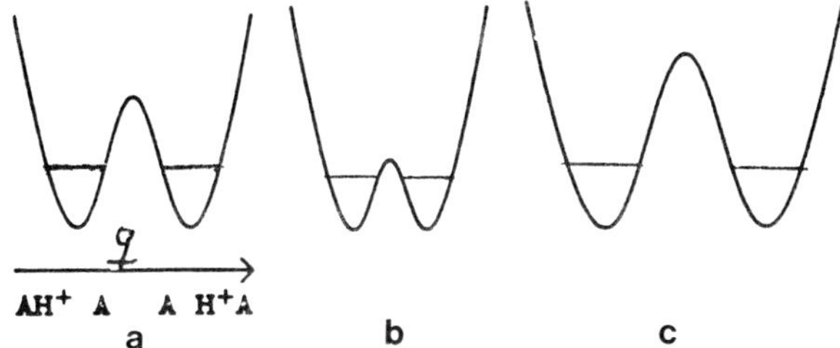

Fig. 3   Proton double wells for (a) zero, (b) negative, and (c) positive values of the intramolecular coordinate $Q = R-R_{eq}$. The proton ground vibrational levels for the reactant and product configurations are indicated.

The central barrier for the H^+ transfer arises when a sufficient portion of the energy cost for breaking the AH^+ bond must be paid before the product H^+A bond is formed. Tunneling through this barrier will occur if it is sufficiently low and narrow.

The second important coordinate $Q = R-R_{eq}$ is related to the intramolecular separation R between the two A's. The proton barrier in q can depend very strongly on Q, especially when there is a hydrogen bond (Fig. 4).

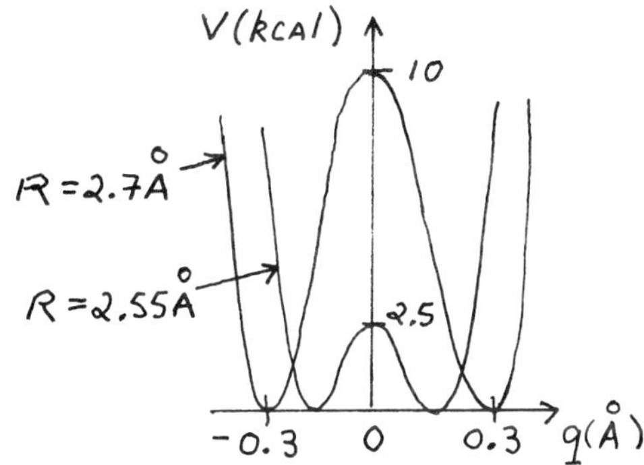

Fig. 4   The proton double well for various $Q = R-R_{eq}$ calculated in a hydrogen bonding system.[7]

Calculations[8,19,20] and experiments[21] indicate that the barrier height can drop very rapidly, at the rate of ~ 50 kcal/mol/Å, when the intramolecular separation Q decreases. This feature will lead to the possibility of a strong dynamical modulation of the tunneling rate as vibration in Q occurs since the tunneling probability increases exponentially as the barrier height and width drop(Figs. 3 and 4). We will call the fluctuations in Q "coupling fluctuations" since they modulate the *coupling* C (proportional to the square root of the tunneling probability[22]) allowing the tunneling to occur. Analogous coupling fluctuations are critical in understanding hydrogen *atom* transfer rates in matrices, where enormous tunneling H/D isotope effects are observed.[23] In brief, the vibrations in Q will assist the tunneling.

The third important coordinate is a collective solvent coordinate.[5] For present purposes, we can think of this as measuring the solvent orientational polarization, i.e., the alignment of the polar solvent molecules. This coordinate will couple electrically to the coordinate q of the charged proton. The most important effect here will be to modulate the *symmetry* of the proton transfer double well. For example, imagine that the solvent is equilibrated to the reactant AH⁺ A, i.e., there is an appropriate orientation of the solvent dipoles around AH⁺ A. Then the left-hand well and the proton vibrational level (Fig. 5) will be lowered while the right-hand well and the proton vibrational level will be raised in energy, introducing an asymmetry.

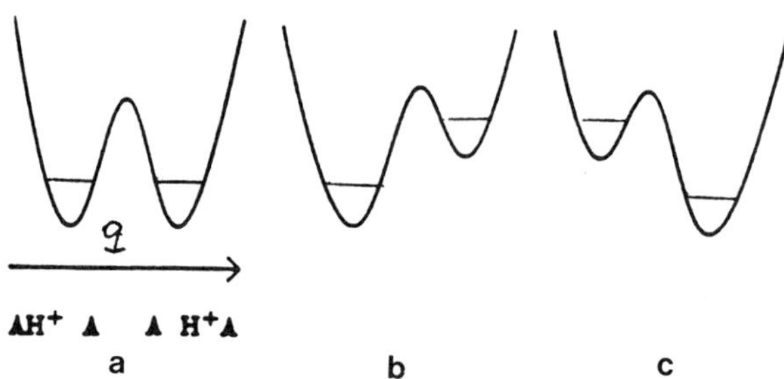

Fig. 5   Proton double wells in (a) the symmetric case, and (b) with solvent stabilization of AH⁺ A, and (c) with solvent stabilization of A H⁺A.

If instead the solvent is equilibrated to A H⁺A, then there is an asymmetry in the opposite sense.

Thus there is a differing stabilization of the "reactant" AH⁺ A and "product" A HA⁺ species for different values of the solvent coordinate.   Fig. 6 shows the resulting picture for the proton vibrational energy levels for the

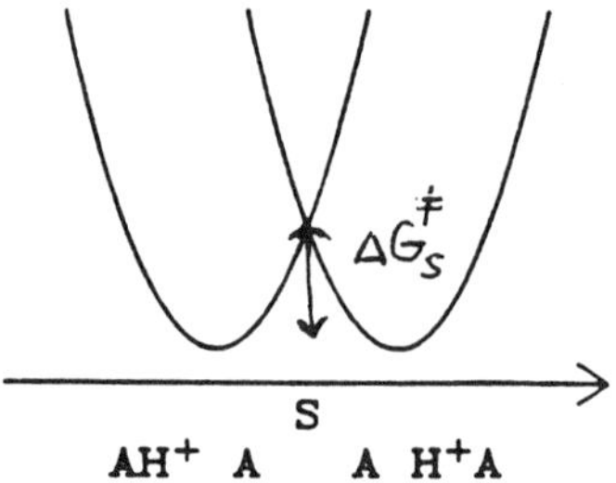

Fig. 6   The proton energies for the reactant and product configurations as a function of the solvent coordinate s.

reactant and product species, plotted as a function of the solvent coordinate. At any fixed value of s, the difference between the two curves is the *splitting* $\Delta\varepsilon$, equal to the difference between the proton vibrational energies in the product and reactant species. (This double well system should not be confused with the proton double well in q.)

Motion in the solvent coordinate is important for the following reason. Tunneling probabilities are very sensitive to the q double well symmetry. They decline sharply with increasing asymmetry, i.e. with increasing magnitude $|\Delta\varepsilon|$ of the splitting.[24] Fluctuations in the splitting are therefore required for tunneling to proceed; otherwise the proton is trapped in one of its two wells in q. Thus as solvent molecules rotate and reorient, there is motion in s in, e.g., the left-hand well of Fig. 6 if the proton is initially in the reactant AH^+ A configuration. Tunneling will not occur until the neighborhood of the intersection $s^{\ddagger}$ of the reactant and product curves in Fig. 6 is reached where the splitting $\Delta\varepsilon = 0$. At $s^{\ddagger}$, approximate symmetry for the *proton* double well in Fig. 5 holds, and tunneling can occur. Subsequent solvent motion then induces asymmetry and traps the proton in either the AH^+ A or A H^+A configuration. Reaching $s^{\ddagger}$ costs free energy, so the solvent will slow down the tunneling. The importance of such a factor has also been emphasized, for example, by Warshel[25] in some model computer simulations of H^+ transfer.

Splitting fluctuations are also important in short range electron transfer reactions.[15] One novel feature of proton transfer compared to those reactions is the importance of the coupling fluctuations discussed above. They are critical here because proton wave functions decay rapidly in space compared to electron wave functions; this leads to rapid variations in the coupling C -- which is related to the wave function overlap -- as vibrational motion in the intramolecular coordinate Q occurs.

We pause here to point out some differences that the perspective sketched above has with TST. First, in TST one pictures a smooth barrier occurring as one proceeds from reactants to products along a reaction path (Fig. 7);

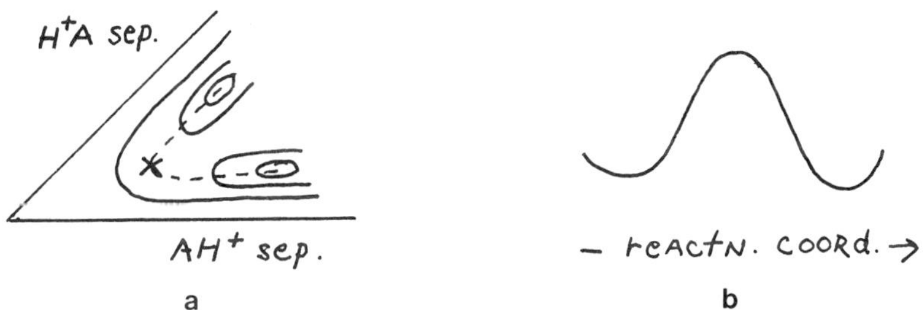

Fig. 7   (a) The schematic reaction path for H^+ transfer pictured in TST. (b) The potential energy along this path.

note that proton double well in q does not appear along this path. The activation energy for the reaction is supposed to be determined by the height of this barrier. We will see in Sec. 3 that the activation energy is in fact *not* determined by this barrier. Second, any tunneling that occurs in the TST view occurs through this barrier. In the current perspective, the tunneling occurs instead before the standard transition state region is even reached; this makes tunneling much more pronounced. Finally, in the TST view, kinetic isotope effects are primarily related to certain zero-point energy differences for the transition state and the reactants.[9,10] In contrast, in the present perspective they will be seen to arise predominantly from tunneling (Sec. 3.B).

3. H+ TRANSFER RATE CONSTANT

3-A. Formulation

To evaluate the rate constant k for the proton transfer, we use the time correlation function formula (tcf)[7,26]

$$k = \int_0^\infty dt \ \mathrm{Re} \ \langle j(0)j(t) \rangle .$$ (3.1)

Here j(0) is the initial value of the proton probability flux out of the reactant AH+ A and into the product A H+A well in q, and j(t) is its value at time t. The brackets denote an equilibrium average over both the proton transfer system in its reactant state and the solvent. Re denotes the real part. The larger is the flux correlation, the larger is its time integral, the rate constant.

Within the Born-Oppenheimer approximation that the solvent and the intramolecular vibration are slow on the very fast time scale of H+ motion (the limit in which Figs. 3-6 make sense), k can be written in terms of the fluctuations in the coupling C and the splitting $\Delta\varepsilon$:

$$k = \int_0^\infty dt \ \mathrm{Re} \ \langle C(0) \ e^{\frac{i}{\hbar}\int_0^t d\,\tau\Delta\varepsilon(\tau)} \ C(t) \rangle .$$ (3.2)

If there were no dynamics of Q or s, then the coupling and the splitting would be constant, say, at C_0, $\Delta\varepsilon_0$. In that case, the flux tcf would behave like $\cos(\Delta\varepsilon_0 t/\hbar)$ as in Fig. 8(a).

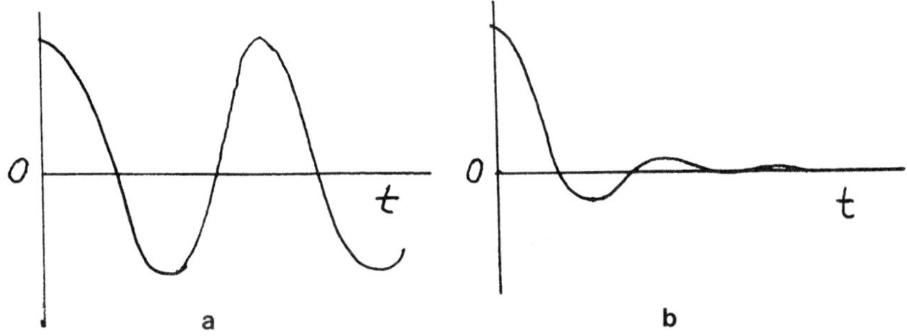

Fig. 8   The flux time correlation function (a) for the artificial case where the coupling and splitting are fixed and (b) for the reaction in solution where these quantities fluctuate.

This is coherent tunneling of H^+, as the probability oscillates back and forth between the wells. Such oscillatory coherent tunneling with a well defined frequency is responsible for gas phase spectroscopic features such as tunnel splittings in, e.g., malonaldehyde.[27] But there is in fact no irreversible rate process if the tunneling is coherent in this way; for example Eq (3.1) for k is nonsensically infinite in this limit and no rate constant exists.

Things are quite different in solution. The tunneling will now be an incoherent process. The fluctuations in the coupling C and especially the splitting $\Delta\varepsilon$ due to the Q and s motions destroy the coherence. This is illustrated in Fig. 8(b) for a model solution H^+ transfer to be described in more detail later. Here the key point is that there are no recurrent peaks in the tcf, thanks to the fluctuations. The tunneling is now an incoherent process, the time integral of the flux tcf is finite, and a rate constant k exists. It is also worth noting that the time scale of the events determining k is very short ($\sim 10^{-2}$ ps).

We have applied this formulation to a model symmetric intramolecular proton transfer $AH^+ A \rightarrow A H^+A$ occurring in a water solvent. The model reaction system is very roughly similar to the oxonium ion $[HOH \bullet \bullet \bullet \bullet OH]^-$. The double well potential for H^+ is constructed from Lippincott-Schroeder potentials[19] which well describe H bonding systems. The coupling C is determined by a semiclassical method.[22] The solvent coordinate and its electrical coupling to the H^+ transfer system is defined using a recent theory.[28] The solvent is described as a generalized dielectric continuum, some features of which are mentioned below. In what follows, we suppress most of the details and highlight the results and interpretation.

The splitting is predominantly determined by the solvent, i.e., $\Delta\varepsilon = \Delta\varepsilon(s)$, while the coupling is primarily determined by the vibration Q and is of the form

$$C(Q) = C_{eq}e^{-\alpha Q}. \tag{3.3}$$

Here C_{eq} is the value at the equilibrium position for Q ($Q = 0$; $R = R_{eq}$) and α measures the rate at which the coupling increases as the vibration contracts, i..e., when Q becomes negative. With the assumption that the fluctuations in Q and s are of a damped harmonic character, the tcf in Eq (3.2) can be expressed in terms of the tcf's $\langle Q(0)Q(t)\rangle$ and $\langle s(0)s(t)\rangle$. The former is evaluated in a model in which the vibration Q "dephases" i.e., loses coherence, as the solvent modulates its frequency by collisions of the solvent molecules with the A species. The latter is evaluated in a model in which s can both oscillate due to any nearly free rotational motion of solvent dipoles and decay due to diffusive Debye-like reorientation of the solvent dipoles.[29] It turns out that the *dynamics*, i.e., the time dependence of $\langle Q(0)Q(t)\rangle$ and $\langle s(0)s(t)\rangle$. is very important in destroying recurring peaks in the proton tcf and inducing incoherence, but has little effect on the short time dynamics of the remaining tcf peak that determines k.(cf Fig. 8)

3-B Analytic Results

We find that some fairly simple analytic formulas describe well the numerical results obtained for k. We now discuss these and the trends they illustrate for proton transfer rates.

In a number of H bonded systems the Q vibration frequency ω is low ($\lesssim 500$ cm^{-1}). If $\hbar\omega/k_BT \lesssim 2$, then we find that

$$k = \langle C^2\rangle \left[\left(\frac{\pi}{\hbar^2 k_BT(\Delta G_s^\ddagger + \Delta G_\alpha^\ddagger)}\right)\right]^{1/2} e^{-\left(\Delta G_s^\ddagger + \Delta G_\alpha^\ddagger\right)/k_bT} \tag{3.4}$$

well describes our results. The factor $\langle C^2\rangle$ is the thermal average of the square of the coupling; the larger the fluctuations in C are, the larger is the rate. The exponential in k involves the activation free energy $\Delta G_S^{\ddagger}$ for the solvent to reach the intersection point in Fig. 6 where the q double well is symmetric and tunneling can occur. This free energy is larger the less polar is the solvent and the larger is the distance over which the proton is transferred. The additional activation free energy $\Delta G_{\alpha}^{\ddagger}$ can be thought of as the cost of a fluctuation in Q to reach regions of stronger coupling and easier tunneling. The larger is the rate of variation of the coupling with changes in Q [cf. Eq. (3.3)], the larger is $\Delta G_{\alpha}^{\ddagger}$.

Eq. (3.4) is appropriate when $\hbar\omega/k_BT \lesssim 2$ (i.e., low frequency Q vibrations). At room T, this condition can be satisfied, as noted above, by systems with the A species involved in an H bond. But when the A's are far apart and negligible H bonding occurs, the frequency of the Q vibration will be set by other stronger bonds in which the A atoms participate (e.g., if A = O, bonds such as CO or NO). Then ω will tend to be higher: $\hbar\omega/k_BT \gtrsim 2$. In that case, we find that k is approximately given by

$$k = \langle C^2 \rangle e^{-\alpha^2 \langle Q^2 \rangle} e^{-\beta \Delta G_s^{\ddagger}} \approx C_{eq}^2 \, F e^{-\beta G_s^{\triangle \ddagger}} \; ; \quad \beta = 1/k_BT \, . \tag{3.5}$$

Here F is a temperature independent Franck-Condon factor. In particular, it is the square of the quantum average of $e^{-\alpha Q}$ in the ground vibrational quantum state for the Q vibration, the only state occupied at "low" temperatures $T < \hbar\omega/2k_B$.

The temperature dependence of the rates Eqs. (3.4) and (3.5) illustrates an important point made earlier. Consider first the high frequency rate Eq. (3.5). The activation energy $E_a = -k_B \, \partial \ln k/\partial(1/T)$ is just the activation free energy $\Delta G_S^{\ddagger}$ for the solvent; E_a is *not* determined by the barrier height in the proton q coordinate. For the low frequency rate Eq. (3.4) the temperature dependence is more complex, $k \sim \exp[BT - E_a/k_BT]$, with the exp $[BT]$ term arising from the growth of the average coupling fluctuations $\langle C^2 \rangle$ with T as the Q vibrational amplitudes increase. But again the activation energy $E_a = \Delta G_S^{\ddagger} + \Delta G_{\alpha}^{\ddagger}$ is not given by the barrier height of the potential in the proton's coordinate q. The lesson is clear: one *cannot* identify proton barrier heights by measured activation energies.

But the barrier through which the proton *tunnels* is definitely in q. The kinetic isotope effect (KIE) for H^+ and D^+ transfer, k_{H+}/k_{D+}, essentially arises from the difference for these isotopes of the coupling C allowing the tunneling. In particular the KIE arises from the thermally averaged square coupling $\langle C^2 \rangle$ in Eq. (3.4) when the Q vibration frequency satisfies $\hbar\omega/k_BT \lesssim 2$. In terms of the value of the coupling at the equilibrium separation $Q = 0$ and the fluctuations,

$$\langle C^2 \rangle = C_{eq}^2 \, e^{2\alpha^2 \langle Q^2 \rangle} = C_{eq}^2 \exp\left[\frac{2\alpha^2 k_B T}{m\omega^2} \right] , \tag{3.6}$$

where m is the effective mass for the Q vibration. C_{eq}^{H+} for a proton exceeds C_{eq}^{D+} for a deuteron, since the lighter proton can more easily tunnel through the double well barrier in q. On the other hand, the thermal fluctuations favor D^+ tunneling over H^+. This arises from the fact that $\alpha_{D+} > \alpha_{H+}$ since the coupling C(Q) and thus the tunneling probability increases more rapidly as Q contracts for D^+ than for H^+. (The wavefunctions for the heavier deuterons fall off more rapidly in space.) Since these coupling fluctuations become more important as the temperature increases, the KIE will decline.

We find that the KIE k_{H+}/k_{D+} is quite sensitive to the hydrogen bond potential. Table 1 presents the results of a model calculation which is based on a Lippincott-Schroeder H bonding potential and which accounts for the rapid drop of both the center barrier in q for the proton and the proton frequency in the reactant well. For smaller values of the equilibrium separation R_{eq} between the A species and thus smaller proton transfer distances Δq, the KIE k_{H+}/k_{D+} plummets. The hydrogen bonding leads to greater coupling C as R_{eq} diminishes.

Table 1. Kinetic Isotope Effects[7]

$R_{eq}(\text{Å})$	$\Delta q(\text{Å})$	k_{H+}/k_{D+} [a]	k_{H+}/k_{D+} [b]
2.7	0.7	7.7	170
2.8	0.8	3 0	530
2.9	0.93	1 90	2200
3.0	1.1	15 30	10^4

a) Hydrogen bond properties included.
b) Hydrogen bond properties ignored.

For comparison, we also present in Table 1 the KIE's calculated when the special features of hydrogen bonding -- dropping barrier height and proton frequency -- are simply ignored, as they are in some treatments[13-15] in the literature. It is seen that this oversight quite considerably overestimates the KIE.

All these features described above can also be studied for non-symmetric proton transfers to explore the role of reaction endothermicity and exothermicity.[7] The only idea from this that we require in what follows is that for endothermic reactions, such that the reaction free energy $\Delta G > 0$, the rate constant exponentially decreases with increasing ΔG.

4. RELEVANCE FOR ENZYMATIC H[+] TRANSFERS

While no applications to enzymatic H[+] transfers of the theory described above have yet been carried out, we would like to argue that these ideas could be highly relevant to the catalytic case. We first offer some general remarks and then consider briefly a specific case.

First, we have seen that proton transfer barriers can drop rapidly as the proton donor and acceptor groups approach each other and that the transfer rate is very sensitive to this. One obvious potential role of the enzyme is to facilitate this approach via its structure. This could allow small separations leading to enhanced tunneling rates in a way not possible in the analogous uncatalyzed *intermolecular* proton transfer in solution. Second, we have seen the key role in the solution case played by the electrical coupling of H[+] to the environment. This aspect, in which the symmetry of H[+] transfer barriers is modulated, is probably even more important in enzymes, given the greatly increased electrostatic interaction of H[+] with nearby polar ionic groups.[16-18] This enhancement arises from the low dielectric constant medium character of, and the comparatively immobile character of dipoles and ionic groups within, the inhomogeneous enzyme. The dramatic influence of ionic charges on proton transfer barriers has been illustrated by Scheiner.[8] We will now see how the ideas we have described ought to be relevant in one of the few enzymatic reactions for which detailed potential energy surface information is available.[5,6]

Triose phosphate isomerase (TIM) is one of the most efficient enzymes known, with a catalysis rate of the reversible isomerization of dihydroacetone (DHAP) to glyceraldehyde 3-phosphate (GAP) which is at least 10^9 times faster than the corresponding uncatalyzed reaction.[30] The enzyme has been characterized as "perfectly evolved" in the sense that the rate determining step for DHAP $\rightleftharpoons$ GAP isomerization is product dissociation from the enzyme.

A key step in the DHAP $\rightleftharpoons$ GAP reaction is the proton abstraction by a carboxylate group in Glu 165 of a CH proton in DHAP (Fig. 2). This was an early puzzle, given that the pKa of a carboxylate group is about 4 versus a pKa ~ 14 for CH. Knowles[31] suggested

a polarizing electrophile in the active site of TIM which could stabilize
the incipient anion. This basic picture is confirmed by Alagona et al.[5] in
ab initio studies: the imidazole NH group of His 95 and the side chain NH_3^+ of Lys
13 provide a positive electrostatic potential in the enzyme active site.

In a simplified but realistic model *ab initio* calculation, Alagona et al.[6] found that a
positive charge positioned near the O' oxygen in Fig. 2 - mimicking the electrophilic effect
of the His 95 and Lys 13 - converted the H^+ transfer potential from $\sim$ 30 kcal endothermic
to thermoneutral with a low barrier of $\sim$ 12 kcal. This is a striking example of the
environment modulating an H^+ barrier symmetry and height. Further, it was found that the
barrier is a very strong function of an "intramolecular" Q coordinate -- the distance between
the O and C atoms between which the H^+ is transferred. The barrier height drops $\sim$ 8 kcal
as this distance decreases by $\sim$ 0.3 Å.

Even with this brief description, it is clearly apparent that proton transfer energetics
in an enzyme system can be extremely sensitive to the distance between the heavy atoms. It
seems safe to predict then that the fluctuations in these distances should play a critical role in
affecting the rates of H^+ transfer reactions.

ACKNOWLEDGMENTS

This work was supported in part by grants from The U.S. National Science
Foundation, CHE 84-19830 and 88-07852. D. Borgis is a CNRS-NSF Postdoctoral
Fellow.

REFERENCES

1.   See, e.g., A. Fersht, "Enzyme Structure and Mechanism," W. H. Freeman, New
      York, 1985.
2.   D. M. Hayes and P. A. Kollman, in "Catalysis in Chemistry and Biochemistry,."
      B. Pullman, ed., Reidel, Dordrecht, 1979. p. 77.
3.   M. J. S. Dewar, *Enzyme*, 36, 8 (1986).
4.   S. J. Weiner, G. L. Seibel and P. A. Kollman, *Prof. Natl. Acad. Sci. USA*, 83,
      649 (1986
5.   G. Alagona, P. DesMeules, C. Ghio and P. A. Kollman, *J. Am. Chem. Soc.*, 106,
      3623 (1984).
6.   G. Alagona, C. Ghio and P. A. Kollman, *J. Mol. Biol.*, 91, (1986).
7.   D. Borgis and J. T. Hynes, to be published.
8.   S. Scheiner, *Acc. Chem. Res.*, 18, 174 (1985).
9.   R. D. Gandour and R. L. Schowen, "Transition States of Biochemical Processes,"
      Plenum, New York, 1978.
10.   F. H. Westheimer, *Chem. Rev.* 61, 265 (1961).
11.   I. A. Rose, in "The Enzymes," P. D. Boyer, ed. 3rd Ed., Vol. 2.
12.   See, e.g., B. C. Garrett, D. G. Truhlar, A. F. Wagner and T. Dunning, *J. Chem.
      Phys.*, 78, 4400 (1983).
13.   R. R. Dogonadze, A. M. Kuznetzov and V. G. Levich, *Electrochim. Acta*, 134,
      1025 (1968).
14.   E. D. German, A. M. Kuznetzov and R. R. Dogonadze, *J. Chem. Soc. Faraday II*,
      76, 1128 (1980).
15.   J. Ulstrup, "Charge Transfer Professes in Condensed Media," Springer-Verlag,
      New York, 1979.
16.   M. Perutz, *Proc. Roy. Soc.* B167, 448 (1967).
17.   A. Warshel, *Proc. Nat. Acad. Sci. USA* 75, 5259 (1978).
18.   A. Warshel, *Biochemistry*, 20, 3167 (1981).
19.   E. R. Lippincott and R. Schroeder, *J. Chem. Phys.*, 56, 1099 (1955).
20.   B. O. Roos, *Theor. Chim. Acta*, 42, 77 (1976).
21.   A. Novack, *Struct. and Bonding*, 18, 177 (1974).
22.   See, e.g., J. N. L. Connor, *Chem. Phys. Lett.*, 4, 419 (1969).

23. W. Siebrand, T. A. Wildman and M. Z. Zgierski, *J. Am. Chem. Soc.*, 106, 4083 (1984).
24. See, e.g., J. R. de la Vega, *Acc. Chem. Res.*, 15, 185 (1982).
25. A. Warshel, *J. Phys. Chem.*, 86, 2218 (1982).
26. S. Lee, D. Ali and J. T. Hynes, to be published.,
27. T. Carrington and W. H. Miller, *J. Chem. Phys.* , 84, 4364 (1986).
28. S. Lee and J. T. Hynes, *J. Chem. Phys.*, 88, 6853, 6863 (1988).
29. J. T. Hynes, *J. Phys. Chem.* 90, 32701 (1986).
30. W. J. Albery and J. R. Knowles, *Biochemistry*, 15, 5631 (1976).
31. J. R. Knowles, *Acc. Chem. Res.*, 10, 105 (1977).

MICROSCOPIC SIMULATIONS OF CHEMICAL REACTIONS
IN SOLUTIONS AND PROTEIN ACTIVE SITES; PRINCIPLES AND EXAMPLES

Arieh Warshel

Department of Chemistry
University of Southern California
Los Angeles, CA 90089

INTRODUCTION

The reason for the enormous catalytic power of enzymes is one of the most fundamental questions in molecular biophysics. Disregarding magical effects it is clear that enzyme catalysis must be based on some clear and probably simple physical concepts. This paper will examine the microscopic origin of enzyme catalysis by computer simulation approaches. Methods for simulation of chemical reactions in solutions and proteins will be reviewed, emphasizing the insight obtained from the simple Empirical Valence Bond (EVB) formulation. The reader will be introduced to simple approaches that should let him judge for himself which catalytic effects are important and how to formulate a mechanistic hypothesis in terms of well defined energy values.

Section (a) will explain how to obtain EVB potential surfaces for chemical reactions in solutions by both the simple Langevin Dipole model and the more elaborated all atom model. This section will also consider the relationship between potential surfaces and the corresponding reaction rate constants. Section (b) will extend the simulation approach of Section (a) to chemical reactions in proteins and will give several demonstrations of EVB treatments of enzymatic reactions, encouraging the reader to construct an EVB surface for a specific enzyme. Finally, some general conclusions will be drawn emphasizing the key role of electrostatic effect in enzyme catalysis.

(A) Chemical Reactions in the Gas Phase and in Solutions

The rate constants for chemical reactions can be written as

$$k = A \, exp\{-\Delta g^{\neq} \beta\} \tag{1}$$

here $\beta = (k_B T)^{-1}$ (where k_B is the Boltzmann constant). This section will consider the evaluation of rate constant from a reliable yet simple microscopic point of view. As will be argued at the end of this section, the preexponential factor A includes dynamical effects which are not drastically different for reactions in different environments. On the other hand, the changes in $\Delta g^{\neq}$ in different environments lead to huge changes in the rate constant. Thus we will concentrate in the main part of this section on the evaluation of the activation free energy $\Delta g^{\neq}$.

(1) <u>How to evaluate potential surfaces</u>

In order to evaluate activation free energies in a microscopic way one should be able to evaluate the forces between the reacting atoms at any given nuclear configuration. The relevant information about the internuclear forces is provided by the so called <u>Born Oppenhimer</u> potential surface of the reacting system. This potential surface (E(r)) can be obtained, at least in principle, by solving the quantum mechanical Schrodinger equation.

$$H(\mathbf{r},\mathbf{x})\psi(\mathbf{r},\mathbf{x})=E(\mathbf{r})\psi(\mathbf{r},\mathbf{x}) \tag{2}$$

where $\mathbf{x}$ and $\mathbf{r}$ are, respectively the electronic and nuclear position vectors. Unfortunately the rigorous solution of this equation is entirely unpractical for large molecules and one must search for approximated solutions. As much as bond breaking and bond making reactions are concerned it seems that the Valence Bond (VB) approach provides the best approximation. This method gives a simple and clear physical insight which is particularly important when one is interested in comparing reactions in different environments. This feature is exploited by the Empirical Valence Bond (EVB) method [1] which will be our method of choice. To outline this method let's consider first the heterolytic bond cleavage reaction.

$$X-Y \rightarrow X^--Y^+ \tag{3}$$

The wave function of this system can be described by the two resonance structures

$$\psi_1=X-Y \tag{4}$$

$$\psi_2=X^-\,Y^+$$

where we assume for simplicity that the resonance structure $X^-\,Y^+$ is of much higher energy than X-Y. The potential surface, U(r), for this system is the ground state energy obtained by solving the secular equation

$$\begin{vmatrix} H_{11}^0\text{-}E & H_{12}^0 \\ H_{12}^0 & H_{22}^0\text{-}E \end{vmatrix}=O \tag{5}$$

The diagonal matrix elements can be approximated by

$$H_{11}^0=\varepsilon_1^0=\overline{M}(R) \tag{6}$$

$$H_{22}^0=\varepsilon_2^0=(I-EA)-e^2/r+A\,exp\{-bR\}$$

where $\overline{M}$ is a Morse type potential for the covalent resonance structure, I and EA are the ionization potential of Y and electron affinity of X, respectively. The off diagonal element H_{12}^0 is obtained (following ref. 2) by using the fact that the ground state energy of the bond in the gas phase can be determined experimentally and approximated by a Morse type function, M, and that the solution of eq. (5) can be written as

$$H_{12}^0=((\varepsilon_1^0-E)(\varepsilon_2^0-E))^{1/2}\approx((\varepsilon_1^0-M)(\varepsilon_2^0-M))^{1/2} \tag{7}$$

Note that the Morse potential for the ground state energy, M, is different than the function $\dot{M}$ which describes a pure covalent state [1].

The above derivation might seem to some readers like a circular exercise, however, our point is not in reexpressing the gas phase potential surface but in preparing a reference Hamiltonian for calculations in solution. To see this point let's first consider the simple problem of the interaction of our reaction system with an external charge. This effect is easily introduced in the EVB formulations by adding to ε_2 the interaction of X^- and Y^+ with the external charge, using the expression

$$\varepsilon_2=\varepsilon_2^0-Q_Z/r_{XZ}+Q_Z/r_{YZ}\approx\varepsilon_2^0+\mu_2\xi_Z \tag{8}$$

where Q_Z is the charge on the Z atom and ξ_Z is the field on the X^- Y^+ dipole from the external charge.

The same approach used above to consider the effect of an external charge can be used to simulate the effect of many solvent molecules, as the solvent is the main effect in stabilizing the X^- Y^+ ionic state (see Fig. 1). Thus we can write

$$H_{11}=H_{11}^0 \tag{9}$$

$$H_{22}=H_{22}^0+U_{Ss}^{(2)}$$

$$H_{12}=H_{12}^0$$

where S and s represent the solute and the solvent respectively and $U_{Ss}^{(2)}$ is the solute-solvent interaction potential for the ionic state. This treatment is not much different than the approach taken in eq. (8). Now, however, we deal with, many solvent configurations and free energy, $\Delta g(r)$, that corresponds to the ground state surface obtained by mixing H_{11} and H_{22} at different solvent configurations. The simplest approach is to consider the potential of mean force for H_{22}

$$H_{22}\rightarrow\Delta g^{(2)}=H_{22}^0+\Delta g_{sol}^{(2)} \tag{10}$$

(where $\Delta g_{sol}^{(2)}$ is the solvation free energy of the ionic state) and to use the ground state obtained by mixing $\Delta g^{(2)}$ with H_{11} as the approximated ground state free energy.

$$\Delta g(r)\approx\tfrac{1}{2}[(H_{11}+\Delta g^{(2)})-((H_{11}-\Delta g^{(2)})^2-4H_{12}^2)^{1/2}] \tag{11}$$

Thus our task is reduced to estimating the solvation free energy of the ionic state. This solvation energy should be estimated, however, using the solvent polarization induced by the charge

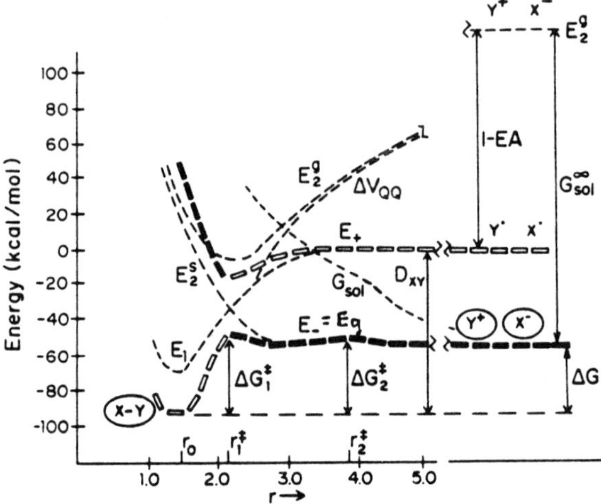

Figure 1 Ionic bond cleavage in solution. The figure shows how the gas phase ionic state becomes a solvated ground state due to the contribution of the solvation energy.

distribution of the ground state obtained by mixing ψ_1 and ψ_2 (rather than by the pure ionic state ψ_2). The simplest way to estimate this energy is to use a continuum model where the solvation energy of a dipolar state is given by the product of the field from the solvent and the solute dipole, $\mu^{(2)}$,

$$\Delta g_{sol}^{(2)} = -\xi^{(g)}\mu^{(2)} \tag{12}$$

where $\xi^{(g)}$ is the so called "reaction field" which is due in the present case to the ground state dipole of the solute and is given in kcal mol^{-1} Å^{-2} by

$$\xi = 166(\mu^{(g)}/a^3)((\varepsilon-1)/(2\varepsilon+1)) \tag{13}$$

Here a is the radius of the cavity around the solute (given in Å), μ is given in Å and electron charge units, and ε it the macroscopic dielectric constant of the solvent.

The crucial problem, however, is that the cavity radius is an arbitrary parameter which is not given by the macroscopic model, making the results of eq. (13) rather meaningless from a quantitative point of view. A much more quantitative model is provided by the semimicroscopic model described below.

(2) Potential surfaces can be obtained using the Langevin Dipole model

The evaluation of solvation energies of ionic states presents a major challenge; the traditional macroscopic models [e.g. refs. 3,4] depend on the unknown cavity radius of the solute (eq. 13). Fully microscopic models, on the other hand, require one to include explicitly a huge number of solvent molecules and involve major convergence problems. An alternative approach that overcomes some of the convergence problems of fully microscopic models, yet treats the solvent explicitly in a simplified dipolar model, is the Langevin Dipoles (LD) model [5]. The model represents

308

the time average polarization of the solvent molecules by a cubic grid of polarizable dipoles. The grid is placed around the solute atoms excluding each grid point which is a van der Waals distance from a solute atom. The remaining grid points are then replaced by point dipoles whose polarization should mimic the average polarization of the solvent molecules at the same region in space. This is accomplished by using a Langevin type relationship

$$\mu_i^n = (coth(X_{n-1}) - X_{n-1}^{-1})\mu_0 \xi_i^0 / \xi_i^0 \tag{14}$$

$$X = C\mu_0 \xi / k_b T$$

$$\xi_i = \xi_i^0 + \xi_{\mu,i}$$

where ξ_i^0 is the field from the solute charge and $\xi_{\mu,i}$ is the field on the ith dipole from all other dipoles. The index n indicates that we are dealing with an iterative procedure starting with $\xi_\mu^{(0)} = 0$. The parameters in this model (C and μ_0) were obtained by using an explicit all atom solvent model and molecular dynamics simulations to evaluate the field dependent on the polarization of the water molecules around ions, and then fitting eq. (14) to the corresponding results of the simulation. The free energy of the (LD) model is simply obtained by

$$\Delta G = -\frac{1}{2} \sum_i \mu_i \xi_i \tag{15}$$

Although this model is clearly a rough approximation, it seems to capture the main physics of polar solvents. The model eliminates the main problems associated with the macroscopic model of eq. 13; in particular since the model defines a microscopic cavity around the solute, it eliminates the dependence of the results on an arbitrary cavity radius and the need to use a "dielectric constant" which is not defined properly at a short distance from the solute. To appreciate the simplicity of the LD model it is recommended to evaluate the potential surface for an ion pair using both eq. (13) and the LD model. To perform such a calculations we have to evaluate the solvation energy of the ionic state, ψ_2, as a function of r. This is done by fixing r and building a grid around the given solute, excluding grid points which are within van der Waals contact from the solute atoms and evaluating the interaction between the ionic state to the self consistently polarized Langevin dipoles situated at the remining grid points. Now, however, we must introduce a complication; the Langevin dipoles should be polarized by the charge distribution of the ground state obtained by mixing ε_1 and ε_2, rather than by the purely ionic charge distribution of ψ_2. This is accomplished by solving the secular equation **HC** = **EC** of eq. (5) in an iterative way using

$$H_{22}^{(n+1)} = \varepsilon_2^0 + \sum_i q_i^{(2)} V_i^{(n)} - \frac{1}{2} \sum_i Q_i^{(n)} V_i^{(n)} \qquad H_{11}^{(n+1)} = \varepsilon_1^0 - \frac{1}{2} \sum_i Q_i^{(n)} V_i^{(n)} \tag{16}$$

$$Q_i^{(n)} = (C_i^{(n)})^2 q_i^{(2)} \qquad\qquad V_i^{(n)} = -\sum_j \mathbf{r}_{ij} \mu_j(\mathbf{Q}^{(n)}) / r_{ij}^3$$

where $V_i^{(n)}$ is the potential on the ith solute atom from the solvent dipoles, $q^{(2)}$ and Q are, respectively, the atomic charges of ψ_2 and of the ground state, $\mu(Q^{(n)})$ designates the solvent polarization due to the given $Q^{(n)}$, and $C^{(n)}$ is the eigenvector of the secular equation, obtained with $H_{22}^{(n)}$. The term $(-\frac{1}{2}\Sigma Q^{(n)} V^{(n)})$ reflects the energy invested in polarizing the solvent dipoles.

Another useful example of treating solvent effects is provided by considering the proton transfer reaction

$$RXH+Y \rightarrow RX^- +HY^+ \tag{17}$$

This reaction can be described by the three resonance structures

$$\Phi_1 = RX\text{–}H\ Y \tag{18}$$

$$\Phi_2 = RX^-\ H\text{–}Y^+$$

$$\Phi_3 = RX^-\ H^+\ Y$$

The electrons involved in the actual reaction, referred to here as the <u>active electrons</u>, can be treated according to the general prescription for the four electrons three orbitals problem with the VB wave functions [2].

$$\Phi_1 = N_1\{|XH Y\bar{Y}|-|\bar{X}HY\bar{Y}|\}\chi_1 = \phi_1\chi_1 \tag{19}$$

$$\Phi_2 = N_2|X\bar{X}H\bar{Y}|-|X\bar{X}\bar{H}Y|\}\chi_2 = \phi_2\chi_2$$

$$\Phi_3 = N_3|X\bar{X}Y\bar{Y}|\chi_3 = \phi_3\chi_3$$

where the N's are normalization constants and the χ's are the wave functions of the inactive electrons moving in the field of the active electrons. We have thus partitioned the molecular electronic space into active and inactive parts and we assume no interaction between these parts. This requires that the Hamiltonian be of the form $\mathbf{H} = \mathbf{H}_{act} + \mathbf{H}_{inact}$. The matrix elements between the basis wavefunctions are then as follows:

$$\mathbf{H}_{ii} = <\Phi_i|\mathbf{H}|\Phi_i> = <\phi_i|\mathbf{H}_{act}|\phi_i> + <\chi_i|\mathbf{H}_{inact}|\chi_i> \tag{20}$$

$$\mathbf{H}_{ij} = <\Phi_i|\mathbf{H}|\Phi_j> = <\phi_i|\mathbf{H}_{act}|\phi_j> + <\chi_i|\mathbf{H}_{inact}|\chi_j>$$

The quantum mechanical matrix elements of the active electrons are given in [2].

The matrix elements of the inactive electrons can be evaluated by expressing the corresponding potential surfaces as a quadratic expansion around the equilibrium value of the various internal coordinates.

$$<\chi_i|\mathbf{H}_{inact}|\chi_i> = 1/2\sum_{bonds} K_b^{(i)}(b-b_0^{(i)})^2 + 1/2\sum_{angles} K_\theta^{(i)}(\theta-\theta_0^{(i)})^2 + V_{nonbonded}^{(i)} \tag{21}$$

$$<\chi_i|\mathbf{H}_{inact}|\chi_j> = 0$$

where the b's and θ's are, respectively, the bond lengths and bond angles in the R fragment.

The usual EVB procedure involves diagonalizing this 3 x 3 Hamiltonian. However, here we wish to use a very simple model for the proton transfer reaction and will represent the potential surface and wavefunction of our reacting system using only two electronic states (mixing the high energy state ψ_3 with ψ_1 and ψ_2 by a perturbation treatment). Using a two-state system will preserve most of the important features of the potential energy surface while at the same time providing a simple model that will be more amenable to discussion than the three-state system (see ref. 7). The effective matrix elements of the two state model can be evaluated using standard quantum chemical methods. This evaluation is tedious and the earlier assumptions that we made will lead to errors in the matrix elements. On the other hand, we can use experimental information to approximate the diagonal matrix elements. That is, the rather complicated function H_{11} (see ref. 7) will be given, at the range where the H...Y distance is large compared to the X-H bond length, by a Morse potential function that depends on the distance R_{X-H}. When H approaches Y we have a repulsive van der Waals interaction between these atoms. We can describe both of these forces using analytical potential energy terms. The same argument applies to H_{22}. As far as H_{12} is concerned we can approximate it by an exponential term and fit the parameters in this term to the experimental information on the gas-phase potential energy surface of the reaction and, if needed to accurate gas phase calculations.

Thus we describe the gas phase potential by

$$H_{11}^0 = M(b_1) + U_{nb}^{(1)} + (K/2)(\theta_1 - \theta_1^0)^2 + U_{inactive}^{(1)} \tag{22}$$

$$H_{22}^0 = M(b_2) + U_{nb}^{(2)} + (K/2)(\theta_2 - \theta_2^0)^2 + \alpha^{(2)} + U_{inactive}^{(2)}$$

$$H_{12}^0 = A\, exp\{-\mu(r_3 - r_3^0)\}$$

where b_1, b_2 and r_3 are, respectively, the X-H, H-Y, and X$\cdots$Y distances, where θ_1 and θ_2 are, respectively, the R-X-H and X-H-Y bond angles.

The effect of the solvent of our reaction Hamiltonian is obtained by using eq. (16). The ground state potential surface is then obtained by solving the corresponding secular equation. A typical surface, for proton transfer in water, is shown schematically in Fig. 2. The approach described in this section provides a convenient way for estimating the energetics of chemical reactions in various solvents. However, equation (10) involves an approximation that can be justified only if the free energy of the actual ground state is proportional to $\Delta g^{(2)}$. The validity of such a linear free energy relationship cannot be justified without detailed studies using more rigorous models. Such models will be considered next.

(3) Potential surfaces in more realistic solvent models

In the previous section we considered a rather simple solvent model treating each solvent molecule as a Langevin-type dipole. Furthermore, in calculating activation free energies we assumed linear free energy relationships, and introduced free energies in the diagonal elements of the Hamiltonian (instead of evaluating the ground state free energy). Although this model appears to capture the key solvent effect it is crucial to examine more realistic solvent models. While doing this, however, one has to realize that many models which are in principle rigorous do not converge (and therefore do not give meaningful results).

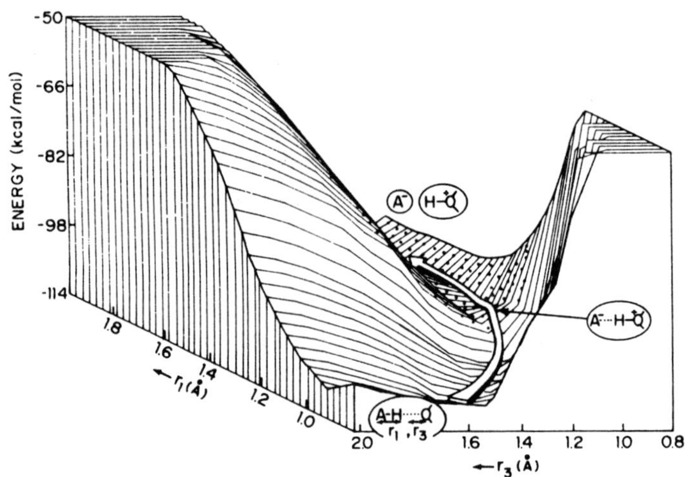

Figure 2 Potential surface for proton transfer between an acid $R''COO_AH$ and an RO_BR' molecule in solution. The independent coordinates r_1 and r_3 are respectively the distance from the proton to O_A and O_B. Regions of the potential surface that have more than 50% ionic character are dotted. The figure is based on the calculation of ref. 1.

The implementation of the EVB method with an all atom solvent model is in principle very simple (forgetting for a while about convergence problems). The Hamiltonian matrix elements are written as

$$H_{ii}=H_{ii}^0+U_{Ss}^i+U_{ss}+U_{ind}^{(i)} \tag{23}$$

$$H_{ij}=H_{ij}^0$$

where, as before, S and s represent respectively the solute and solvent molecules. The potential U_{ss} is the solvent potential surface, represented by a standard all atom potential of the form

$$U_{ss}=\sum_{ij}(U_{nb}^{ij}+U_{qq}^{ij}) \tag{24}$$

$$U_{nb}^{ij}=\sum_{k(i)k'(j)}(A_kA_{k'}/r_{kk'}^{12}-B_kB_{k'}/r_{kk'}^6)$$

$$U_{qq}^{ij}=\sum_{k(i)k'(j)}q_k q_{k'}/r_{kk'}$$

here k runs over the atoms of each water molecule and the intramolecular potential of each water molecule is kept constant for simplicity. The solute solvent potential U_{Ss} is written as

$$U_{Ss}=\sum_{i(S)j(s)}(U_{nb}^{ij}+U_{Qq}^{ij}) \tag{25}$$

$$U_{Qq}^{ij}=Q_iq_j/r_{ij}$$

where Q and q are the solute and solvent residual charges respectively. The three body induced dipole potential $U_{ind}^{(i)}$ is treated by the self consistent iterative approach which is described in detail elsewhere [5].

312

To see how we actually implement this all atoms EVB method let's consider as a test case the S_N2 type reaction which can be written as

$$X^+ + CH_3Y \rightarrow XCH_3 + Y^+ \tag{26a}$$

This reaction can be described by the three resonance structures

$$\psi_1 = [X^- \; C - Y]\chi_1 \tag{26b}$$

$$\psi_2 = [X - C \; Y^-]\chi_2$$

$$\psi_1 = [X^- \; C^+ \; Y^-]\chi_3$$

The relevant Hamiltonian for the gas phase solute molecules can be treated by the same three orbitals four electron model used in the previous section. Representing the system by the two lowest energy configurations we can write

$$\varepsilon_1^0 = H_{11}^0 = M(b_1) + U_{nb}^{(1)} + U_{ind}^{(1)} + 1/2\sum_m K_b^{(1)}(b_m^{(1)} - b_0)^2 + 1/2\sum_m K_\theta^{(1)}(\theta_m^{(1)} - \theta_0)^2 \tag{27}$$

$$\varepsilon_2^0 = H_{22}^0 = M(b_2) + U_{nb}^{(2)} + U_{ind}^{(2)} + \alpha^{(2)} + 1/2\sum_m K_b^{(2)}(b_m^{(2)} - b_0)^2 + 1/2\sum K_\theta^{(2)}(\theta_\theta^{(2)} - \theta_0)^2$$

$$H_{12} = A \; exp\{-\mu(r_3 - r_3^0)\}$$

in which the gas phase energies ε_i^0 are represented by a quadratic expansion of the actual contribution, b_1, b_2 and r_3 are the X-C, C-Y and X-Y distances, $M(b_j)$ is the Morse potential for the jth bond, the b_m are the C-H bond lengths, and the θ_m are the X-C-H, H-C-H and H-C-Y angles defined by the covalent bonding arrangement for a given resonance structure. For example, in ψ_1 the θ_m are defined by the H-C-Y and H-C-H angles, and the K_θ for the X-C-H angles are set to zero. The $U_{nb}^{(i)}$ are the nonbonded interactions between the nonbonded atoms in the ith resonance structure. For ψ_1 this includes the $X^- \cdots C$ repulsion, the $X^- \cdots H$ repulsion, and the $X^- \cdots Y$ interaction. These nonbonded interactions are described by either $Ae^{-\mu r}$ or 6-12 van der Waals potential functions. The induced dipole terms describe the attraction between the charges of the system (Q) and its induced dipoles. This term is evaluated for simplicity by neglecting the induced dipole-induced dipole interaction. Finally the $\alpha^{(2)}$ parameter is the energy difference between ψ_2 and ψ_1 with the fragments at infinite separation (this is simply the difference in the heat of formation of $X^- + CH_3Y$ and $Y^- + CH_3X$). As before we can determine the parameters in eq. (27) by fitting the corresponding ground state potential surface to ab-initio calculations and relevant experimental information. With the gas phase potential surface we can obtain the solvent Hamiltonians by adding the solvent-solute interaction to the "classical" part of the diagonal EVB matrix elements (eq. (25)).

Once the analytical ground state potential surface is constructed, we can start to explore the configurational space and evaluate the reaction free energy surface and the activation free energy $\Delta g^{\neq}$. The strategy for this involves the free energy perturbation approach [8,9] using a mapping potential which is composed of the EVB diagonal energies using [7]

$$\varepsilon_m = \lambda_1^m \varepsilon_1 + \lambda_2^m \varepsilon_2 - 2(\lambda_1^m \lambda_2^m)^{1/2}|H_{12}| \tag{28}$$

and changing the vector $\lambda^m = (\lambda_1^m, \lambda_2^m)$ in a systematic way from (1,0) to (0,1) in n small increments, where the index m is changed from 0 to n - 1. In this way we can drive the system from the reactant state through the transition state and into the product state. Note that ε_1 has a minimum at the reactant geometry and ε_2 has a minimum at the product geometry so that as we change m we force the system to move from the reactant state to the product state. Specifically, the solvent is forced to adjust its polarization to the changing charge distribution of the solute and we will be able to see how much free energy the solvent needs to adjust to the solute configuration by using the free energy perturbation formalism. The H_{12} term which is added to the weighted contributions of ε_1 and ε_2 minimizes the difference between the mapping potential and the actual ground state potential at the transition state region.

The free energy associated with changing ε_1 to ε_2 can be obtained by

$$\delta G(\lambda_m \to \lambda_{m'}) = (1/\beta) \ln[<exp(-\varepsilon_{m'} - \varepsilon_m)\beta>_m] \tag{29}$$

$$\Delta G(\lambda_n) = \Delta G(\lambda_0 \to \lambda_n) = \Sigma_{m=0}^{n-1} \delta G(\lambda_m \to \lambda_{m+1})$$

the free energy functional $\Delta G(\lambda)$ reflects both the electrostatic solvation effects associated with the changes of the solute charges, as well as the intramolecular effects associated with changing ε_1 to ε_2.

Obtaining $\Delta G(\lambda)$ with the perturbation procedure using the mapping potentials ε_m is not sufficient for evaluating the activation free energy, $\Delta g^{\neq}$, which reflects the probability of being at the transition state on the actual ground state potential surface. The quantity $\Delta G(\lambda)$ is basically used to find the free energy associated with changing ε_1 to the potential $\varepsilon_{m\neq}$ that forces the system to spend the longest time near the actual transition state. However, one still needs to find the probability of reaching the transition state for trajectories that move on E_g. The corresponding formulation defines the reaction coordinate X^n in terms of the energy gap $\varepsilon_2 - \varepsilon_1$. This is done by dividing the configuration space of the system into subspaces, S^n, that satisfy the relationship $\Delta\varepsilon(S^n) = X^n$, in which the energy differences X^n are constants and $\Delta\varepsilon = \varepsilon_2 - \varepsilon_1$. The X^n are then used as the reaction coordinate, giving

$$exp\{-\Delta g(X^n)\beta\} = exp\{-\Delta G(\lambda_m)\beta\} <exp\{-(E_g(X^n) - \varepsilon_m(X^n))\beta\}>_m \tag{30}$$

This expression relates the probability of finding the system at X^n on the ground state E_g to the probability of being on the mapping potential ε_m (which keeps X around X^n). The evaluation of $\Delta g(X^n)$ for an exchange reaction is described in Fig. 3. In addition to $\Delta g(X^n)$, we can obtain the probability of being at X^n on ε_1. This probability defines the free energy functionals, Δg_i ($\Delta g_i(X^n) = \int exp\{-\varepsilon_i(X^n)\beta\}dS^n$), which can be obtained through the same derivation that led to eq. (30), and are given by

$$exp\{-\Delta g_i(X^n)\beta\} = exp\{-\Delta G(\lambda_m)\beta\} <exp\{-(\varepsilon_i(X^n) - \varepsilon_m(X^n)\beta\}>_m \tag{31}$$

The evaluation of the Δg_i functionals for an exchange reaction is demonstrated in Fig. 4.

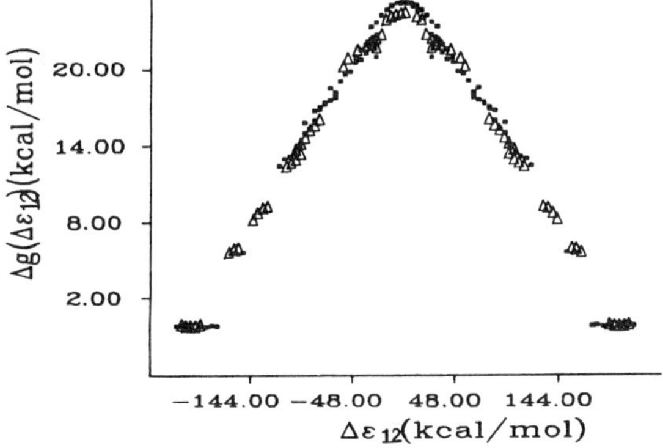

Figure 3 Actual free energy profile for the ground-state surface as a function of the energy gap $\Delta\varepsilon$. The calculations are done for the $Cl^-+CH_3Cl \rightarrow ClCH_3+Cl^-$ exchange reaction.

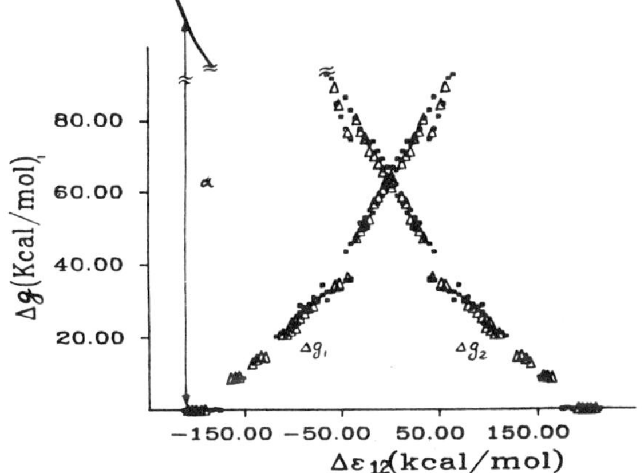

Figure 4 Free energy functionals, Δg_1 and Δg_2 as a function of the energy gap $\Delta\varepsilon$.

The Δg_1 and Δg_2 curves can be used to estimate the dependence of $\Delta g^{\neq}$ on the reaction free energy, ΔG_0, as will be shown in Section 5.

(4) Examining Dynamical Effects

In cases of reactions with significant activation barriers the reactant system spends most of its time encountering ineffective collisions. However, once in a long time the kinetic energy stored in the many degrees of freedom of the system is converted to kinetic energy along the reaction coordinate, producing a reactive trajectory that reaches $X^{\neq}$. Thus, the rate constant can be expressed in terms of the probability that the system will be at a range $\Delta X^{\neq}$ around $X^{\neq}$ and the average time, $\bar\tau$ it takes a reactive trajectory to cross that region, as:

$$k=\bar\tau^{-1}(P(X^{\neq})dX^{\neq})/\int P(X)dX = (<\dot{X}>^{\neq}/\Delta X^{\neq})exp\{-\Delta g^{\neq}\beta\} \qquad (32)$$

where $<\dot{X}>$ is the average velocity along X for trajectories that arrive at $X^{\neq}$ from the reactant and continue to the product region without being deflected backward. Note that the time $\bar\tau$ is not different than the time it takes a productive trajectory to move from the reactant region to the transition state.

The inverted time, $\bar{\tau}^1$, is given by

$$\bar{\tau}^{-1} = F(k_b T/h) \approx F(T/300)6.2 \bullet 10^{12} sec^{-1} \tag{33}$$

where F is the <u>transmition factor</u> which expresses the fraction of trajectories which continue to the product state after arriving at $X^{\neq}$.

Eq. (32) is frequently expressed in the form

$$k = \frac{1}{2}F<|\dot{X}|>P(X^{\neq})/\int P(X)dX \tag{34}$$

When F is equal to unity this equation reduces to the rate expression of the well known Transition State Theory. In order to examine possible dynamical effects it is important to evaluate the time τ of eq. (33) by microscopic simulation. This can be done, at least in principle, by generating equilibrated trajectories of the system at the reactant state and evaluating the velocity of a reaction trajectory. Unfortunately, such an approach is entirely unpractical for processes with large activation barriers, where it takes an extremely long time until the kinetic energy of the system is converted to the reaction coordinate and generates a productive trajectory that reaches the transition state. However, one can use a practical trick by preparing an equilibrated system at the transition state region and then running "<u>downhill</u>" trajectories. Since classical dynamics is invariant to time reversal, we can use the time reversal of the downhill trajectories to construct the corresponding very rare "uphill" productive trajectory. This is demonstrated in Fig. 5 which describes the time reversal of downhill trajectories that are used to construct a productive trajectory for an S_N2 reaction. The downhill trajectory approach is useful not only for evaluating the preexponential factor τ but for examining the relationship between the dynamics of the solute and solvent contributions to the reaction coordinates. Some reactions can be considered as <u>solvent driven reactions</u> when the solute coordinates approaches their transition state configuration only after the solvent polarization reaches a configuration that stabilizes the product state. In other cases the reation is a "<u>solute driven</u>" process. Downhill trajectories allow one to explore the nature of the reactive trajectories and the corresponding dynamics of the solute and the solvent coordinates.

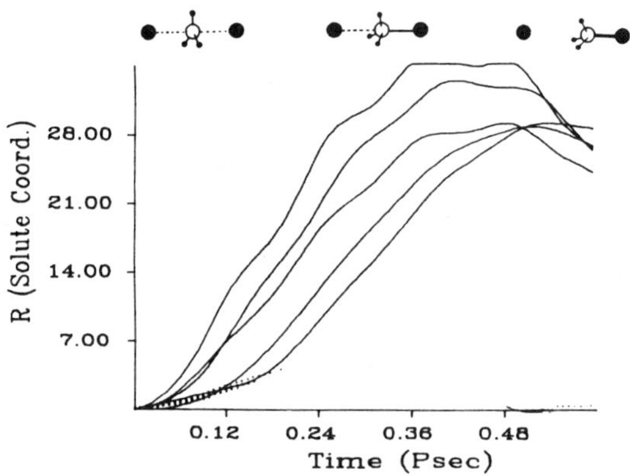

Figure 5 Time-dependent solute coordinate, R in typical downhill trajectories.

316

(5) <u>Linear free energy relationships</u>

Examination of the functionals Δg_1 and Δg_2 in Fig. 4 gives an interesting insight about the relationship between the reaction free energy, ΔG_0, and the activation free energy $\Delta g^{\neq}$. That is the transition state free energy can be approximated by

$$\Delta g^{\neq} \approx \Delta g_1(X^{\neq}) - \bar{H}_{12}(X^{\neq})　\tag{35}$$

where $\bar{H}_{12}(X^{\neq})$ is the average of H_{12} and we use the fact that $\varepsilon_1 = \varepsilon_2$ at $X^{\neq}$ so that $E_g(X^{\neq}) = \varepsilon_1(X^{\neq})-H_{12}(X^{\neq})$. We can further manipulate eq. (35) by assuming that Δg_1 and Δg_2 can be approximated by paraboli with the same curvatures (see Fig. 6). With this assumption we can evaluate Δg at the intersection of the two paraboli by

$$\Delta g_1(X^{\neq}) = (\alpha - \Delta G_0)^2/4\alpha \tag{36}$$

where α, which is called the <u>reorganization energy</u>, is defined by Fig. 6 and the reader can verify this equation in the simple case where $\Delta G_0 = 0$ and $\Delta g_1(X^{\neq}) = \alpha/4$. Now we can write the actual activation free energy as

$$\Delta g^{\neq} \approx (\Delta G_0 - \alpha)^2/4\alpha - \bar{H}_{12}^2(X_0)/\alpha - \bar{H}_2(X^{\neq}) \tag{37a}$$

$$|\Delta G_0| < 4\alpha$$

Outside the range of $(|\Delta G_0| < 4\alpha)$ we have

$$\Delta g^{\neq} \approx \Delta G_0 \qquad \Delta G_0 > 4\alpha \tag{37b}$$

$$\Delta g^{\neq} \approx 0 \qquad \Delta G_0 > -4\alpha$$

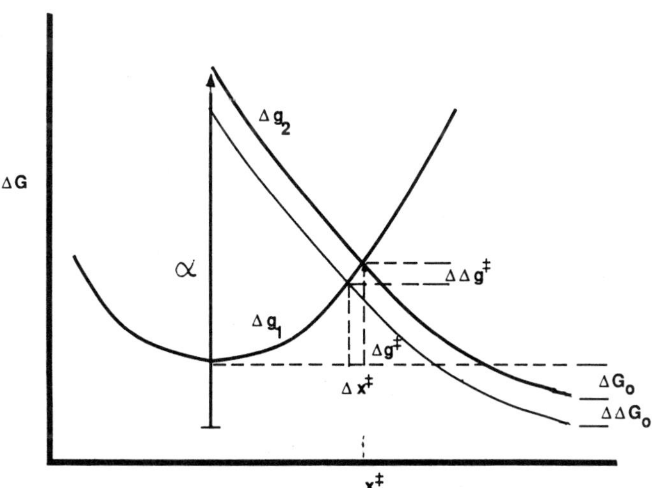

Reaction Coordinate

<u>Figure 6</u> The relationship between α, ΔG_0, and $\Delta g^{\neq}$ in harmonic free energy surfaces.

Eq. (37) without the perturbation term H_{12}^2/α is identical to the Marcus' equation for methyl transfer reactions [10]. This equation predicts, at the range ($|\Delta G_0|<4\alpha$), a linear relationship between $\Delta\Delta G_0$ and $\Delta g^{\neq}$ by

$$\Delta\Delta g^{\neq} \approx \theta\Delta\Delta G_0 = [(\Delta G_0 - \alpha)/2\alpha]\Delta\Delta G_0 \tag{38}$$

Such a linear relationship dates back much further to Brönsted [11] and Hammond [12] and played a major role in physical organic chemistry [13]. The problem is, however, that eq. (38) is not exact nor based on a fundamental microscopic principle. In fact, the Δg curves will not be harmonic for systems that do not obey the linear response approximation. Furthermore, the parameter α is not reproduced correctly by macroscopic estimates. Nevertheless, even the approximate validity of eq. (38) can provide a powerful way of avoiding the actual evaluation of $\Delta g^{\neq}$.

A microscopic examination of the linear free energy relationship for S_N2 reactions is described in Fig. 7.

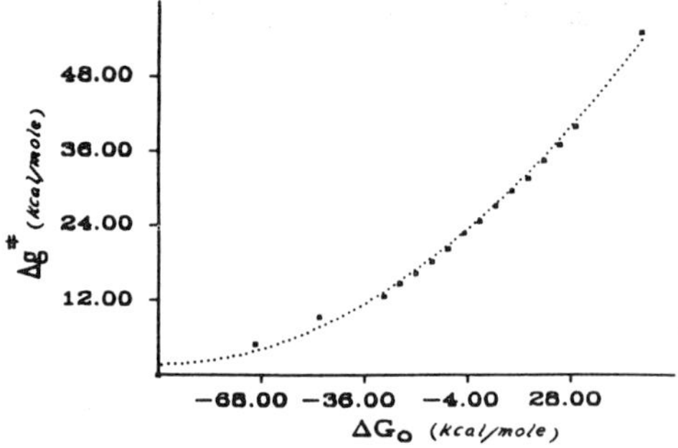

Figure 7 The relationship between the activation free energy $\Delta g^{\neq}$ and the reaction free energy ΔG_0 for S_N2 type reactions.

The figure presents the calculated dependence of $\Delta g^{\neq}$ on ΔG_0 for an hypothetical system where all the potential parameters are set at their values for the $Cl^- + CH_3Cl \rightarrow ClCH_3 + Cl^-$ except $\alpha^{(2)}$ of eq. (27), which is varied in a parametric way. This corresponds to a variation of the electronegativity of the atom while keeping its radius constant. The calculated results do give support to eq. (38) where θ changes from one for very endothermic reactions with a product like transition state to 0.5 for $\Delta G_0 \approx$ 0 and then to zero for very exothermic reactions with a reactant like transition state. The approximated validity of the linear free energy relationship is also supported by many experimental studies, however, the experimental verification of such relationships is not conclusive since it is hard to find a common denominator for experiments with different X's and Y's (the changes in many parameters (eg. H_{12}) makes a direct comparison to a macroscopic models far from simple).

(B) Potential Surfaces for Chemical Reactions in Proteins

The same approach used in the previous section for simulation of chemical reactions in solutions can be exploited in studies of chemical reactions in enzymes. In fact, the solution studies provide a unique way for calibrating the enzyme calculations. That is, one can simply view the enzyme as another "solvent" concentrating on the change in the EVB energies moving from the reference solvent (i.e. water) to the active site. For example, if we use the average protein structure as obtained from X-ray crystallography and represent the solvent by the LD model (such a model is

318

referred to as the Protein Dipoles Langevin Dipoles (PDLD) model), we can write:

$$\varepsilon_i^p = \varepsilon_i^s + (\Delta\Delta g_{sol}^{(i)})_{s \to p} \tag{39}$$

where $\Delta\Delta g_{sol}^{(i)}$ is the change in the solvation free energy of the indicated resonance structure and s and p designates solvent and protein respectively. With these diagonal energies we can now express the effect of the enzyme active site in terms of the changes in solvation free energies. To appreciate this simple and powerful concept we will consider the class of proton transfer reactions in a protein. As we saw before one can calibrate the reference reaction in solution using the pK_a's of the proton donor and acceptor. With the calibrated surface and the PDLD approach we can write

$$\varepsilon_i^p = H_{ii}^0 \tag{40}$$

$$\varepsilon_2^p = \varepsilon_2^s + (\Delta g_{sol,p}^{(i)} - \Delta g_{sol,s}^{(i)})$$

This leads to the surface shown in Fig. 8.

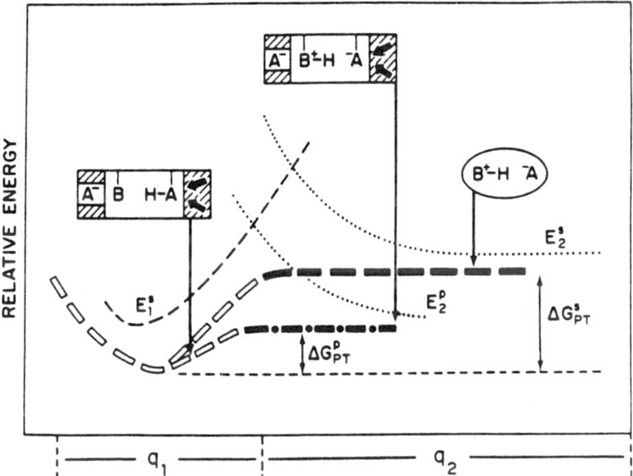

Figure 8 Comparing potential surfaces for proton transfer reactions in solution ▬ ▬ and in an enzyme ▬•▬. The figure considers the energy of the covalent (BH-A) and ionic (BH⁺A⁻) resonance forms. The energy of the covalent resonance form is the same in solution and in the enzyme, while the ionic resonance form is strongly stabilized by the enzyme. This makes the free energy of proton transfer in the enzyme, ΔG_{PT}^p, much smaller than the corresponding free energy in solution, ΔG_{PT}^s.

As seen from the figure there is an almost linear correlation between the stabilization of ionic state and the activation free energy for the reaction. An enzyme which is designed to accelerate the proton transfer step will simply stabilize the (B⁺H A⁻) state more than water does.

One can argue, however, that the use of eq. (40) is based on an ad-hoc assumption and requires a more rigorous treatment. This can be readily provided by the use of an all atom model for the protein-solvent system and the free energy perturbation approach. That is, we now represent the two diagonal energies by potential functions rather than free energies (Fig. 9) Now we have an analytical potential surface for the solvated reacting system (E_g in Fig. 9) and we can use the free energy perturbation (FEP) approach (section b) to evaluate the reaction free energy. Such a study is described in Fig. 10 for the proton transfer step in the catalytic reaction of serine proteases [14].

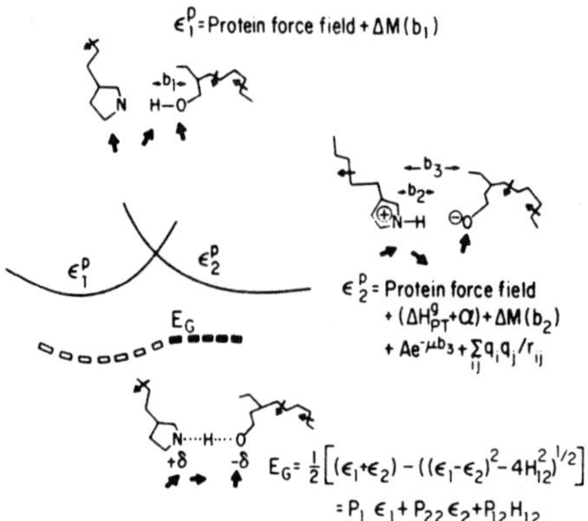

Figure 9 A schematic description of the way we obtain the EVB potential surface for simulating the dynamics of proton transfer reactions. The figure considers ϵ_1, ϵ_2 and E_G for a proton transfer from serine to histidine in a protein active site.

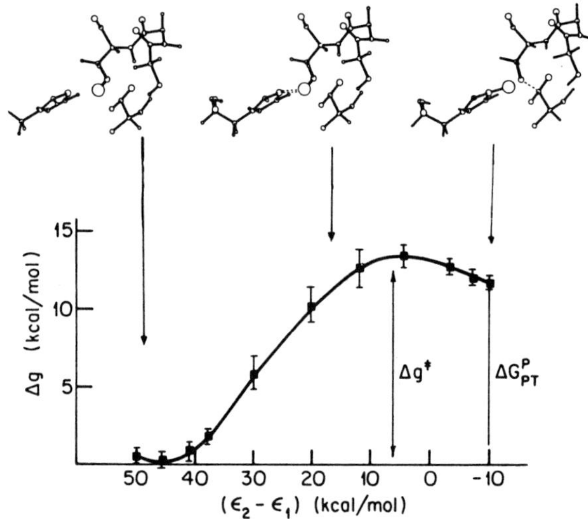

Figure 10 The free energy surface for the proton transfer step in the catalytic reaction of trypsin. The figure also presents some configurations obtained by a "downhill" trajectory.

The calculation gives converging results and can be used to examine the linear free energy relationship assumed in using the diagonal energies of eq. (40) in solving the secular equation and obtaining the ground state free energy. In fact, test calculations in solutions (section 1) and proteins (Fig. 11) indicate that one can use linear free energy relationships in proteins and more importantly that the substitution of the solvation free energy in the diagonal elements of eq. (40) gives similar results to those obtained with all atom Free Energy Perturbation (FEP) calculations.

The main take home lesson from the above analysis is that the activation free energy is strongly correlated with the stabilization of the ionic resonance structure by the protein active site. The generality of this concept will be examined below.

320

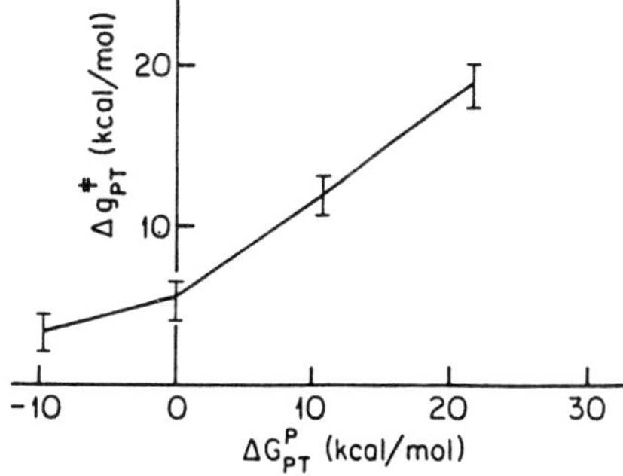

Figure 11 The relationship between $\Delta g^{\neq}$ and ΔG for proton transfer reactions in the active site of trypsin.

(C) Electrostatic Stabilization and Enzymatic Reactions

After the introduction given above we are ready to examine specific enzymatic reactions. Several key systems will be considered below; both as a demonstration of our method and as test cases of the origin of the catalytic power of enzymes.

(1) Lysozyme

Hen egg-white lysozyme catalyzes the hydrolysis of various oligosaccharides, especially those of bacterial cell walls. This enzyme with several of its saccharide substrates was the first enzyme-substrate complex to be solved by X-ray crystallography (15). This provided the first glimpse of the details of the active site of an enzyme and led to proposals on how catalysis might take place. To review these proposals and their more recent modifications let's start from the commonly accepted mechanism of this reaction. This mechanism involves the so called general acid catalysis

$$ROR'+AH \quad ROH^+R'+A^- \quad R^++A^- \, {}^+R'OH \tag{41}$$

The rate limiting step of this reaction is described in Fig. 12.

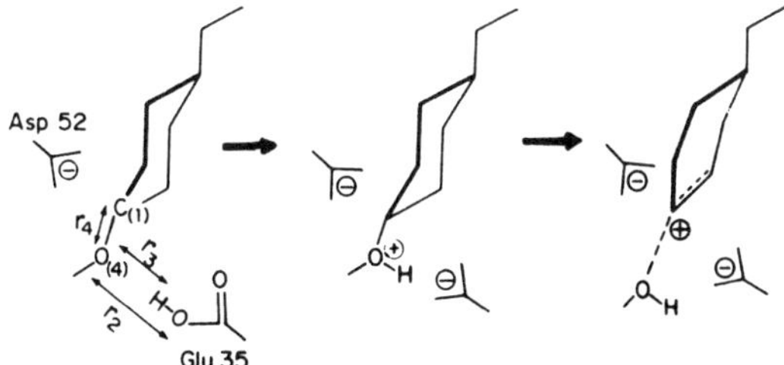

Figure 12 The catalytic reaction of lysozyme. The reaction involves proton transfer from Glu-35 to O_4 of the sugar residue in subsite D and a cleavage of the protonated C_1-O_4 bond.

This step consists of a proton transfer from Glu-35 to O_4, and a cleavage of the protonated C-O_4 bond forming a carbonium ion-like transition state. The sugar ring assumes the planar "sofa" form, while the ground state is characterized by a chair form. Model building studies that used a rigid framework indicated that the planar transition state geometry "fits" better to the active site of the D residue than the chair ground state [15]. This led to the "strain hypothesis", that argued that enzyme reduces the activation free energies by applying steric strain on the ground states of the reacting substrates. Subsequent studies that took into account the enzyme flexibility indicate that the strain effect cannot be a major one [16]. Another possiblility that one should consider is the electrostatic stabilization of the C^+ carbonium center by the neighboring ionized Asp 52. However, early studies with model compound in <u>solutions</u> [17] indicated that the effect is unlikely to be important. Nevertheless, the dielectric effect in solution might be very different than the corresponding effect in proteins. Thus it is crucial to examine which mechanism is really important by the actual microscopic model. This can be done by the EVB method, writing the resonance structures that correspond to the general acid catalysis mechanism of Fig. 12.

$$\psi_1=(A\!-\!H\ R\!-\!OR') \quad \psi_3=(A\!-\!H\ R^+\ {}^-OR') \tag{42}$$

$$\psi_2=(A^-\ R\!-\!OH\ {}^+R') \quad \psi_4=(A^-\ R^+\ H\!-\!OR')$$

The corresponding potential surface for the reaction in solution was calibrated by using pK_a's and other experimental information [16]. For example, the diagonal energies of ψ_1 and ψ_2 are written in the EVB+PDLD formulation as

$$\varepsilon_1^s=M_{OH}(r_1)+U_{nb}^{(1)}+M_{CO}(r_4) \tag{43}$$

$$\varepsilon_2^s=M_{OH}(r_3)+U_{QQ}^{(2)}+U_{nb}^{(2)}+\alpha^{(2)}+\Delta g_{sol}^{(2)}$$

where $\alpha^{(2)}$ can be uniquely determined by using the pK_a's difference between the acid and the sugar oxygen. The detailed evaluation of the rest of the matrix elements is given in ref. (17). With parameters calibrated on the reference reaction in solution we can try to evaluate the potential surface in the protein. The corresponding EVB+PDLD calculations are outlined in Fig. 13.

The calculations produce in a qualitative way the difference in activation free energy between the reaction in the enzyme active site and a reference reaction in a solvent cage. The main reason for this catalytic effect is the compliment of the change in charge distribution upon formation of the transition state and the electrostatic potential from the protein. An instructive way for examining this effect is to compare the contribution of solvation free energy to $\Delta g^{\neq}$ for the (Asp$^-$ C$^+$ Glu$^-$) system in water and in the enzyme active site [18]. The active site appears to solvate the (- + -) configuration much better than water does. The reason for the larger solvation by the enzyme appears to be associated with the fact that the enzyme provides dipoles (e.g. hydrogen bonds) which are already oriented to stabilize the transition state while the corresponding reference solvent cage have randomly oriented water molecules at the ground state [18]. The polarization of these water molecules towards the (- + -) configuration requires an investment of significant energy to overcome the dipole-dipole repulsion (about half of the energy gained by dipole-charge interaction is lost in dipole-dipole repulsion).

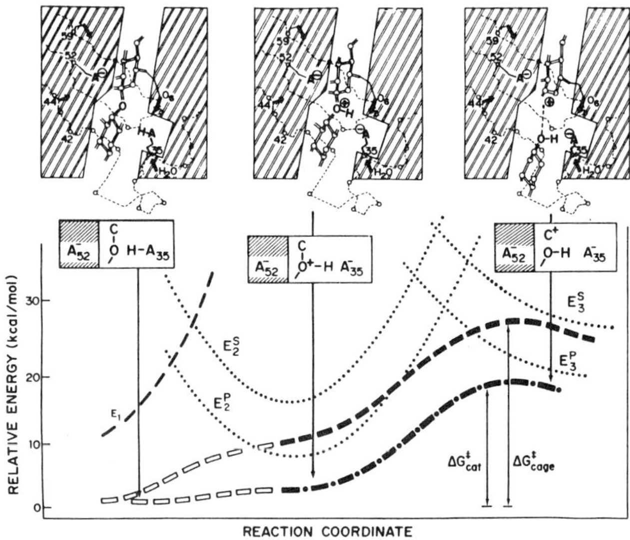

Figure 13 Comparison of the potential surfaces for general acid catalysis in lysozyme (━•━) and in solution (━). The figure describes the energetics of the reaction as crossing between the indicated resonance forms. The extra stabilization in the ionic resonance forms in the enzyme relative to solution is almost equal to the reduction of $\Delta G_{cat}^{\ne}$ relative to $\Delta G_{cage}^{\ne}$.

(2) <u>Serine Proteases</u>

The serine proteases are the most extensively studied class of enzymes. Kinetic studies of their catalytic activity made use of their broad substrates specificity to investigate the effect of varying types of substrates and/or inhibitors upon kcat and K_M. This class is characterized by the presence of a unique serine residue in an environment that makes it very effective in the hydrolysis of aminoacids. Here we will consider the three related members of this family, (<u>Cheymotrypsin</u>, <u>Trypsin</u> and <u>Subtilisin</u>). All three catalyze the hydrolysis of peptide bonds in proteins and peptides

$$H_2O+ \ -NH-\overset{R}{\overset{|}{C}}H-\overset{O}{\overset{\|}{C}}-NH-\overset{R'}{\overset{|}{C}}H- \ \rightarrow \ -NH-\overset{R}{\overset{|}{C}}H-CO_2H \ + \ H_2N-\overset{R'}{\overset{|}{C}}H-$$

$$(44)$$

They differ, however, in striking ways in their preference for amino acid side chains at the position R. For example, trypsin cleaves bonds only after Lys and Arg residues. The three enzymes are characterized by active sites with the so called <u>Catalytic triad</u> ($Asp_e His_c Ser_c$) (see Fig. 14) where c designates a catalytic residue. The elucidation of the X-ray structure of these enzymes [19-24] leads to detailed mechanistic proposals which are, despite their frequent appearance in text books, far from being established (see below). The problem is of course that the X-ray structure [19,20] tells us about the position of the catalytic residues but does not provide direct information about the energetics of the transition state. To correlate structure and energetics we will use here our EVB approach and try at the same time to demonstrate its effectiveness in examining different alternative mechanisms. We start by considering in Fig. 14 two possible mechanisms for the reaction of serine proteases. The mechanism in pathway (b) is a double proton transfer mechanism which is presented in many text books that discuss enzyme mechanisms. As will be seen below, we find clear evidence that this mechanism is uneffective in serine proteiases. Thus we start discussing mechanism (a) and only at the end discuss mechanism (b). To explore a given mechanism, we have to start by defining the key resonance structures and calibrating their energies on the relevant experimental information for the reference system in solution. The relevant resonance structures are

$$\psi_1 = (A^- \ Im \ H{-}O \ C{=}0)$$

(45)

$$\psi_2 = (A^- \ Im^+{-}H \ ^-O \ C{=}0)$$

$$\psi_3 = (A^- \ Im^+{-}H \ O{-}C{-}0^-)$$

$$\psi_4 = (A{-}H \ Im \ O^- \ C{=}0)$$

$$\psi_5 = (A{-}H \ Im \ O{-}C{-}0^-)$$

where A, Im, H-O and C=0 indicates, respectively Asp_c, His_c, Ser_c and the carbonyl of the substrate. The details of the parameterization are given elsewhere [22] and we only present in Fig. 15 the EVB+PDLD comparison of mechanism (1) in the active site and in a reference solvent cage.

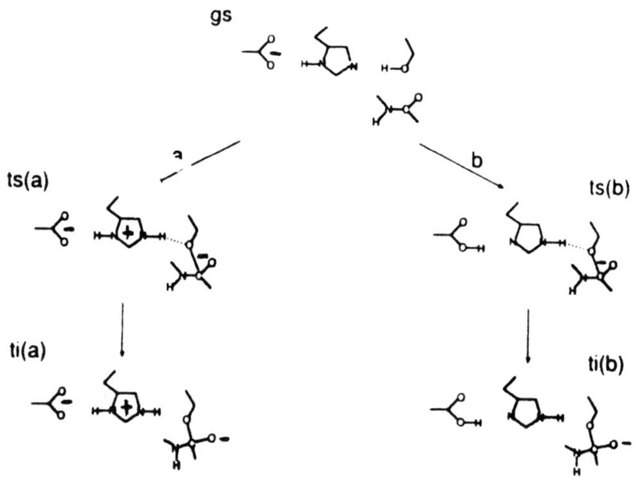

Figure 14 Two possible pathways for the catalytic reaction of serine proteases.

As in the lysozyme case it appears that the enzyme stabilizes the (- + -) transition state better than water does and that this accounts qualitatively for the observed catalytic effect [22].

While the EVB calculations of mechanism (a) seem to account for the catalytic rate enhancements, it it crucial to examine the double proton transfer mechanism (mechanism (b)). The trick, however, is to arrive at a unique conclusion which is not dependent on the uncertainty associated with quantum mechanical calculations of large molecules. To do this we exploit the fact that the energetics of the double proton transfer mechanism can be obtained from the difference between the activation barriers of route (b) and route (a) by $\Delta G_b^{\neq} = \Delta G_a^{\neq} + \Delta\Delta G_{ab}^{\neq}$. If $\Delta\Delta G_{ab}^{\neq}$ is positive then the double proton transfer mechanism is not important. The free energy $\Delta\Delta G_{ab}^{\neq}$ is basically the free energy associated with a proton transfer from His_c to Asp_c at the transition state.

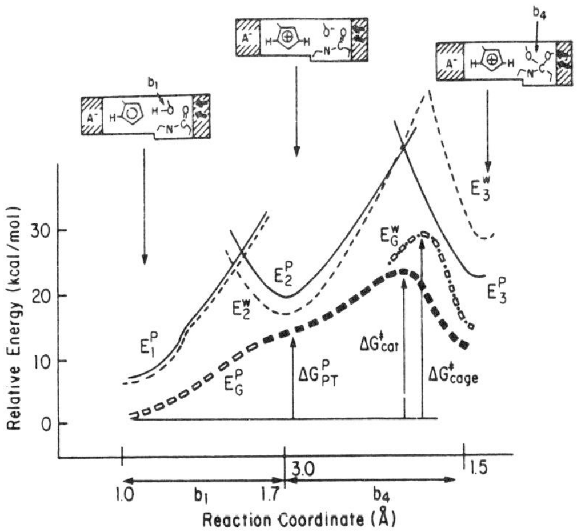

Figure 15 EVB potential surfaces for the rate limiting step in the hydrolysis of amides in the active site of trypsin (ε^P) and in the corresponding reference reaction in a solvent cage (ε^W).

This free energy, can be evaluated in two steps. First we estimate the free energy difference for the reference reaction in water. This is done using the experimental pK_a values for Asp and His in water (that gives about 4 kcal/mol for a proton transfer between His to Asp at infinite separation) and the change in Coulomb's interaction for the [Asp$^-$...His$^+$...t$^-$] $\rightarrow$ [Asp...His...t$^-$] process in water (which is estimated to be around 2 $\pm$ 1 kcal/mol [22]), giving an overall estimate of 6 kcal/mol for $\Delta\Delta G_{ab}^{\neq}$ in water. This indicates that the double proton transfer mechanism is quite unfavorable in an aqueous medium. Replacing the solvent cage by the protein environment <u>increases</u> the energy loss by an additional 6 kcal/mol (as estimated by the PDLD calculations), yielding a value of 12 kcal/mol for $\Delta\Delta G_{ab}^{\neq}$. This estimate indicates that, the double proton transfer mechanism is strongly unfavorable as compared to the electrostatic one. To realize this from a simple intuitive point of view it is important to recognize that the ionized form of Asp_c is even more stable in the protein active site than in water, as is apparent from its observed pK_a being equal to 3 in chymotrypsin [22]. As the stability of the negative charge of Asp_c increases, the probability of a proton transfer from His_c to Asp_c decreases.

The emergence of genetic engineering has revolutionized the studies of the catalytic action of serine proteases. Recent studies have examined the effect of substitutions of all the groups in the catalytic triad of subtilisin [23] and the effects of major substitutions of the active site groups in trypsin [24]. These studies provide direct experimental information about the storage of catalytic free energy. For example, substitutions of Asn-155 in subtilisin [25] resulted in a major decrease in $\Delta G^{\neq}$ (3-5 kcal/mol). This Asn residue is a part of the "oxyanion hole" which provides hydrogen bond stabilization to the negative charge of the oxyanion resonance structure ψ_3 (see the dipoles in Fig. 15). However, even the information from genetic engineering cannot tell us how to correlate in an <u>a priori</u> way structure and catalysis (which is different than the experimental determination of the contribution of a given group in a given structure). A true structure function correlation must be based on some theoretical model. Here we may find the real value of the genetic engineering experiments; they provide clear tests for various theoretical models. That is, up to now one could only calculate the changes in $\Delta G^{\neq}$ between the reaction in the enzyme and the reference reaction in solution. Unfortunately, the definition of a proper reference state (e.g. our solvent cage [18]) is too complicated to be generally accepted. With the genetic engineering experiments the reference state is uniquely defined and the name of the game is very clear; a theoretical approach that is supposed to explain enzymatic reactions should account for the experimental effects of site specific mutagenesis. In this respect the EVB approach is quite encouraging. Several major simulation studies of the EVB+FEP method [26] and in some cases the EVB+PDLD method [27] reproduced the effects of genetic engineering experiments in a semiquantitative way. The agreement between some calculated and observed results is demonstrated in Fig. 16.

(3) Staphylococcal Nuclease

Staphyloccal nuclease (SNase) is a single peptide chain enzyme that catalyzes the hydrolysis of both DNA and RNA at the 5' position of the phosphate diester bond, yielding a free 5'-hydroxyl group and a 3'-phosphate monoester. The enzyme requires one Ca^{2+}-ion for its action and shows little or no activity when Ca^{2+} is replaced by other divalent cations. The crystal structure of SNase has been obtained in complexes with the inhibitor pdTp and Ca^{2+} (28). pdTp has an extra negative charge on the 5'-phosphate group as compared to the substrates, but serves as a model for binding and also resembles the Michaelis E-S complex in solution.

Cotton et al.(28) have postulated the following three step mechanism (see Fig. 17) for the catalytic reaction: (1) General base catalysis by Glu43 which accepts a proton from a (crystallographically observed) water molecule in the active site, yielding an hydroxide ion. (2) Nucleophilic attack by the OH^- on the phosphorous atom in line with the 5'-O-P ester bond leading to the formation of a trigonal bipyramidal transition state or metastable intermediate. (3) Breaking of the 5'-O-P bond and formation of products. The overall rate constant k_{cat} for the wildtype enzyme is $95s^{-1}$ (at 25°C and pH 7.4) [29] corresponding to a total activation barrier of 15 kcal/mol, where the second step is the rate limiting one. As the pseudo-first-order rate constant for nonenzymatic hydrolysis of phosphodiester, where the attacking species is a water molecule is about $2x10^{-14}$,[30] the enzyme rate acceleration is 10^{15}-10^{16}. A more convenient reference reaction is provided by an OH^- nucleophilic attack of the phosphate group, as in the proposed mechanism for SNase. The rate constant for this reaction is $\sim7x10^{-12}$.[30]

In order to understand the origin of the enormous catalytic effect we use the EVB+FEP method to simulate the action of SNase. T This is done by describing the rate limiting step of the enzymatic reaction by the three resonance structures of Fig. 17. The free energy surface for the reaction in the protein, Δg_p, is obtained by changing the mapping parameter to drive the system between the three resonance structures; the general base catalysis (step1) is mapped by $\lambda=(1,0,0) \rightarrow \lambda=(0,1,0)$ and the rate limiting step, the formation of the doubly negatively charged pentacoordinated phosphate group (step 2), is mapped with $\lambda=(0,1,0) \rightarrow \lambda=(0,0,1)$. The selection of the proper reference state in solution is not unique. One possibility is to consider a reference system consisting of a glutamic acid, a water molecule and the substrate, all in the same solvent cage. The calculated free energy profile for this reaction is designated by $(\Delta g_s)_I$ in Fig.17. Another reference system is an OH^- ion and the substrate in a solvent cage. The free energy surface for the reaction of this system is designated by $(\Delta g_s)_{II}$ and the corresponding activation barrier is $(\Delta g_s^{\neq})_{II}$.

As seen from the figure we reproduce the observed trend in the change of the catalytic free energy: $[(\Delta g_s^{\neq})_{II} - \Delta g_p^{\neq}]_{obs} \sim 18$kcal/mol, $[(\Delta g_s^{\neq})_{II} - \Delta g_p^{\neq}]_{calc} \sim 19$kcal/mol. Perhaps more instructive is the

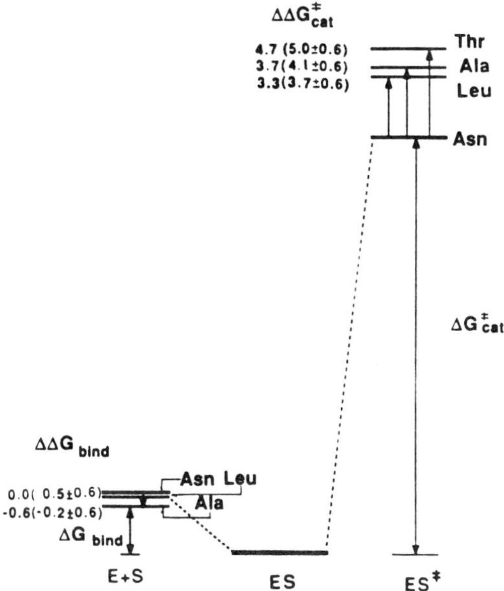

Figure 16 Calculated and observed free energy changes for the Asn-155 → Thr, the Asn-155 → Leu, and the Asn-155 → Ala mutations. The calculated free energies are given in parentheses near the corresponding observed values.

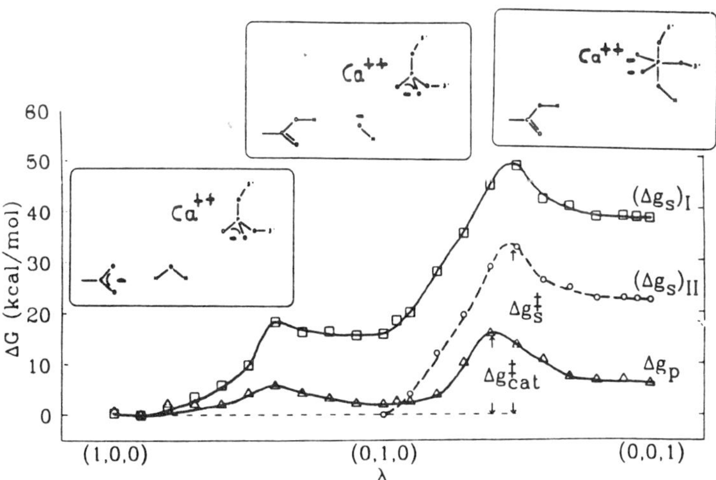

Figure 17 The reaction profile for the SNase reaction and the corresponding reference reactions in solution (see text for definitions). The calculations are taken from ref. 31.

comparison of $(\Delta g_s)_l$ and Δg_p, as we are basically dealing with the same system in the protein and in the reference solvent cage. As can be seen from the figure, the protein eliminates the large contribution associated with the proton transfer step. This effect is mainly due to the stabilization of the hydroxide ion by the calcium ion. The large barrier reduction, compared to the solution reaction, for the second step is accomplished by partial neutralization of the negative charge on the phosphate group by the Ca^{2+}-ion and the two arginine residues 35 and 87 which are closely interacting with the 5'-phosphate. In the SNase case as in the cases of lysozyme and serine proteases we find that the main reason for catalysis is the electrostatic complimentarity between the enzyme potential and the change in charge distribution during the course of the reaction.

(4) Concluding Remarks

The demonstrations given in this chapter are not just selective cases. In fact, the reader is encouraged to examine his specific enzymatic system and try to formulate the assumed mechanism by the EVB formulation. Although this can be done automatically by our program package MOLARIS, it is very instructive to take a "do it yourself" approach and to convert mechanistic hypothesis to valence bond structures and to clear energetic statements. As was shown in the case of the double proton transfer mechanism there is no need to accept our or other's suggestions of key catalytic effects. The conceptual EVB approach lets one judge for himself.

Although different catalytic effects might be important in different enzymes, we believe that most systems with a significant rate acceleration will be found to have an electrostatic factor as the key contribution to the reduction in $\Delta g^{\neq}$. We also believe that for any system where structural information is available one will find that the active site potential stabilizes the transition state charge distribution and the corresponding transition state resonance structures.

Acknowledgment

This work was supported by NIH Grant GM-24492 and by ONR Contract N000014-87-K-0507.

References

(1) Warshel, A. and Weiss, R.M. (1980). J. Am. Chem. Soc., 102, 6218-6226.

(2) Coulson, C.A., & Danielsson, U. (1954). Part II. Ark. Fys., 8, 245-255.

(3) Born, M. (1920) Phys. I, 45-48.

(4) Kirkwood, J.G. (1934). J. Chem. Phys., 2, 351-361.

(5) Warshel, A. and Russell, S. (1984) Quart. Rev. of Biophys. 17 283-427.

(6) Russell, S. and Warshel, A. (1985) J. Mol. Biol.

(7) Hwang, J-K., King, G., Creighton, S., and Warshel, A. (1988) J. Am. Chem. Soc. (in press).

(8) Valleau, J.P. and Torrie G.M. (1972) in <u>Modern Theoretical Chemistry</u> ed. Berne, B. (Plennum, New York) Vol 5 pp. 137

(9) (a) Warshel, A., J. Phys. Chem. (1982) <u>86</u>, 2218.

   (b) Hwang, J.-K. and Warshel A. (1987) J. Am. Chem. Soc. **109**, 715.

(10) Marcus, R.A. (1968) J. Phys. Chem. **72**, 891.

(11) Bronsted, J.N. and Pederson, K. (1924) Z. Phys. Chem. **108**, 185.

(12) Hammond, G.S. (1955) J. Am. Chem. Soc. **77**, 334.

(13) Albery, J.W. and Kreevoy, M.M. (1978) Adv. Phys. Org. Chem. **16**, 87.

(14) Warshel, A., Russell, S.T. and Sussman F.(1987) Israel J. Chem. **27**, 217.

(15) (a) Blake, C.C.F., Mair, G.A., North, A.C.T., Philips, D.C., Sarma, V.R. (1967) Proc. R. Soc. London, Ser. B, 167, 365.

   (b) Philips, D.C. (1966) Sci. Am. <u>215</u> 78-90.

(16) Warshel, A. and Levitt, M. (1976). J. Molec. Biol., <u>103</u>, 227-249.

(17) Dunn, B.M., Bruice, T.C., Adv. Enzymol. Relat. Areas Mol. Biol., (1973), 37, 1-59.

(18) Warshel, A. (1981) Biochemistry **20**, 3167.

(19) Blow, D.M., Birktoft, J.J., and Hartley, S.S. (1969) Nature **221**, 337-340

(20) Kraut, J., Annu. Rev. Biochem, (1977) **46**, 331-358.

(21) Stroud, R.M., Kossiakoff, A.A., Chambers, J.L., (1977) Ann. Rev. Biophys. Bioeng. **6**, 177-193 (1977).

(22) Warshel, A. and Russell, S., (1986a) J. Am. Chem. Soc. **108**, 6569-6579.

(23) Carter, P., and Wells, J.A., Nature, (submitted).

(24) Craik, C.S., Roczniak, S., Largeman, C., and Rutter, W.J., (1987) Science **237**, 909-913.

(25) Wells, J.A., Cunningham, B.C., Craycar, T.P., and Estell, D.A. (1986) Philos. Trans. R. Soc. (London) Ser. A No. 317, 415-423.

(26) Warshel, A., and Sussman, F. (1986) Proc. Natl. Acad. Sci. (U.S.A.) **83**, 3806-3810.

(27) Naray-Szabo, G., Sussman, F., Hwang, J-K, and Warshel, A. (submitted).

(28) Cotton, F.A. Hazen, E.E., and Legg, M.J. (1979) Proc. Natl. Acad. Sci. $\underline{76}$ 2551-2555.

(29) Serpersu, E.H., Shortle, D. and Mildvan, A.S. (1987) Biochemistry, $\underline{26}$, 1289-1300.

(30) Guthrie, J.D. (1977), J. Am. Chem. Soc. $\underline{99}$, 3991-4001.

(31) Aqvist, J and Warshel, A. (1988) in preparation.

THE CONCEPT OF THE POTENTIAL OF MEAN-FORCE

IN ENZYME CATALYSIS

Teresa Fonseca

Chemistry Department
Colorado State University
Fort Collins, Co 80523, USA

When trying to understand the microscopic details of any chemical
reactive process in a condensed phase it is always very useful to introduce the concept of a potential of mean-force. As we shall see below from
a number of examples, this concept will tell us how the reaction coordinate
(we assume here for simplicity a one-dimension reaction coordinate)is
affected on the average by its environment. This concept is particularly
useful when applied to enzyme catalysis as it can make more transparent
and explain, at least in part, how enzymes are so successful in speeding up
specific chemical reaction processes.

However, the evaluation of the potential of mean-force is a very
difficult one and it is only now becoming possible, for simple systems, with
the use of supercomputers. Obviously, it is still in the future when it will
be possible to completely determine the potential of mean-force for an
enzyme reaction, but using several sensible simplifications reasonably good
pictures are already available and some of the traditional ideas of how
enzymes work are being tested.

The outline of this paper will be as follows:in section 1 several
examples of potentials of mean-force will be given and in section 2 the
connection to thermodynamics and traditional reaction rate theories will be
presented. Section 3 will be devoted to a short discussion of a few examples
of potentials of mean-force for enzymatic reactions and section 4 presents
the ending remarks.

1- THE POTENTIAL OF MEAN-FORCE: DEFINITION

Let us first introduce the concept of reaction coordinate. For any
chemical activated process, i.e., any process where for the conversion of
reactants into products the passage across a state of higher energy is
unavoidable, the reaction coordinate is defined as the path of minimum
energy. In most chemical reactions the reaction coordinate can be defined
using a one-dimensional variable, but there are a number of cases where a
higher dimensionality is required.

Now, in the gas-phase, we can consider that the reactants proceed in
their conversion into products in the absence of interactions with the
remaining particles that constitute the chemical system. This is not true,

however, for reactions in solution, where due to the more condensed packed
system the interactions among the particles that constitute the reactive
system will influence not only the approach of the reactants (in the case
of non-unimolecular reactions) but also the energy profile of the reaction
coordinate. The potential of mean-force is defined as the energy profile of
the reaction coordinate in solution assuming that, at any value of the
reaction coordinate, the surroundings have time enough to equilibrate
themselves with respect to the assigned evolution of the reactive event.
As we shall see in the examples given below, the potential of mean-force
can be substantially different in solution with respect to gas-phase, but
even in solution it can depend strongly on the kind of solvent involved.
So, we can interpret the potential of mean-force as a "metric" for solvent
effects. We must emphasize that, the solvent effects discussed in this way
should be considered equilibrium, or static efffects, and that they have
to be distinguished from dynamical solvent effects. However, this last
class of effects is beyond the scope of this paper and will not be treated
here.

Our first example refers to the case of solvolysis reactions,

$$R^+X^- \;\rightleftarrows\; R^+//X^- \;\rightleftarrows\; R^+ + X^- \qquad ; \qquad (1.1)$$

in the equation above, R^+X^- refers to the contact ion pair, $R^+//X^-$ to the
solvent separated ion pair and R^++X^- refers to the free solvated ions. It is
well known that for the same reaction in the gas-phase the solvent separated ion pair does not exist and the reaction proceeds directly from the
contact ion pair to the free ions, requiring an energy comparable to the
electrostatic interaction between the two ions (see figure 1)[1]. However, in
solution the reaction mechanism is believed not to be so simple and the
experimental results are interpreted in terms of an intermediate species,
the solvent separated ion pair. Although there have been for long a controversy on the real existence of this species, recent Monte Carlo (MC)[2]
simulations of the potential of mean-force for the solvolysis of t-Bu$^+$Cl$^-$
in water, and Molecular Dynamics (MD)[3] simulations for the solvolysis of
a model system in a polar solvent, support the existence of at least two
distinct species in the ion pair region, whose interconversion implies
the surmounting of an energy barrier. This result is substantially different
from the one predicted using a continuum picture for the solvent, where the
conversion of the contact ion pair in the free ions does not involve any
subsequential barrier, and reveals the importance of the solvent structure.

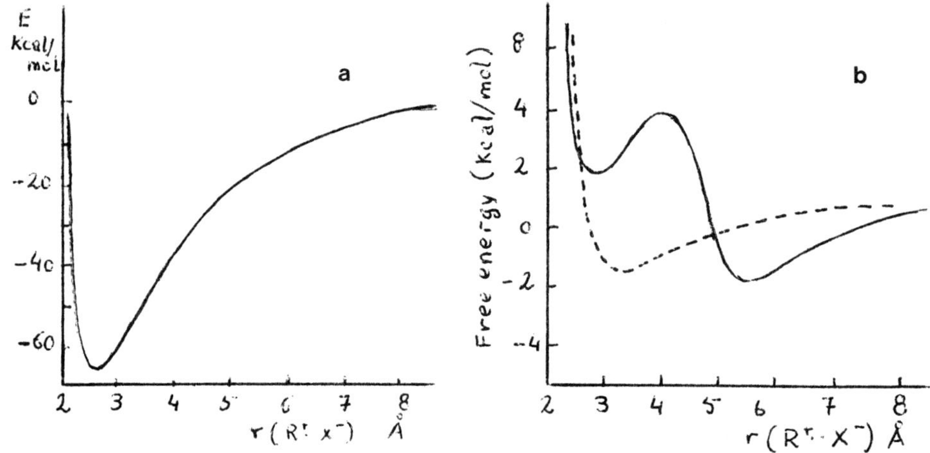

Fig. 1- Different energy profiles for the solvolysis reaction.
a) gas-phase b) —— solution; - - - continuum picture.

Our second example is provided by bimolecular nucleophilic substitution reactions (S_N2)[4]:

$$X^- + RY \rightarrow Y^- + RX \qquad . \qquad (1.2)$$

The energy profile for this kind of reaction in the gas-phase is well settled[5] and looks like as it is pictured in figure 2. There the two wells represent two stable intermediates, $[X^-RY]$ and $[XRY^-]$, that occur before and after the transition state, $[X^{\delta-}..R^{\delta+}..Y^{\delta-}]$.

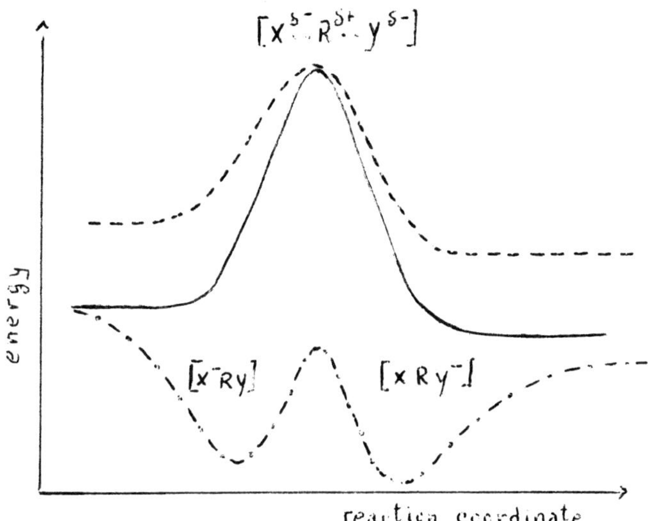

Fig. 2- Energy profiles for the S_N2 reaction. Solution : —— protic solvent, - - - aprotic solvent ; -.-.- gas-phase.

It is worthwhile to discuss here in some detail the mechanism of this type of reaction in the gas-phase as it was recently used to justify an hypothesis about how enzymes work. We will comment about that hypothesis later on. In the gas-phase a naïve interpretation of the energy profile depicted in figure 2 would lead to the wrong conclusion that the reaction rate should be determined by the collision frequency between the ion and the molecule. This is not the case as is proved by most of the reactions studied up until now. The reaction rates reported are usually smaller, and sometimes much smaller, than the collision rate constants. So, given that the energy profile is qualitatively accurate, how can a central barrier lower than the barrier to give the reactants slow down the overall reaction? The mechanism of the reaction is believed to be a unimolecular one between the ion-molecule complex, $[X^-RY]$, and the transition state, $[XRY]^{-\dagger}$,

$$[X^-RY] \; \rightleftarrows \; [XRY]^{-\dagger} \qquad ; \qquad (1.3)$$

Once the ion-molecule complex is formed, the energy released in that process is very quickly redistributed randomly among the degrees of freedom of the complex and keeps flowing among them. The reaction will proceed further, leading to products or reconversion in reactants, when the necessary energy of activation will accumulate in the normal mode (or normal modes) that lead to the corresponding transition state. So the concept of a unidimensional reaction coordinate is misleading here as all or almost all of the normal modes participate actively in the reaction. The reaction rate depends strongly on the density of energy states of the transition states. Intuitively

speaking, the more dense are the energy levels of the transition state the
more easy is to achieve the necessary energy accumulation. A detailed
study[5] of the transition states involved in the reaction under study leads
to the conclusion that the density of states on the transition state leading
to reconversion of reactants is larger than the density of states on the
transition state leading to products. This implies that, once the ion-
molecule complex is formed it has a higher probability of transforming back
in reactants than to transform in products. This explains how the smaller
barrier can slow down the reaction.

Now, before discussing the energy profile in solution, let us give a
look to the values of various absolute rate constants displayed in table I.

Table I – Absolute rates in various media of S_N2 reactions at 298k.
(log k in $M^{-1}s^{-1}$)a

Reactants	Solvent			
	H_2O	CH_3OH	DMF	gas-phase
$OH^- + CH_3F$	−6.2			10.2
$OH^- + CH_3Cl$	−5.2			12.0
$OH^- + CH_3Br$	−3.9			12.0
$Cl^- + CH_3Cl$				9.6
$Cl^- + CH_3Br$	−5.3	−5.2[b]	−0.4[b]	9.9
$Cl^- + CH_3I$	−5.5	−5.5[b]	0.5[b]	
$CN^- + CH_3Br$				10.3
$CN^- + CH_3I$	−3.2[b]	−3.2[b]	2.5[b]	
$F^- + CH_3Cl$	−7.8			11.7
$F^- + CH_3Br$	−6.5			11.6
$Br^- + CH_3I$	−4.4[b]	−4.1[b]	0.1[b]	

a) values from reference 5, b)reference 4.

The most striking difference in the rates in solution is their absolute
magnitude: there is a rate acceleration of several orders of magnitude on
switching from protic solvents, like water and methanol, to the polar aprotic solvent dimethylformamide (DMF). The comparison between any of the
solvents displayed and the gas-phase value is even more dramatic. It is
clear that the form of the potential of mean-force depends on the structure
of the solvent and that going from gas-phase, through dipolar protic solvents
to protic solvents, there is a considerable increase in the height of the
activation barrier. Why is that so? The main reason is again the solvation
of the charged species and here the differential solvation of the reactant
anion relative to the transition state. Comparing to the gas-phase, in
solution, the energy of the reactants drops significantly with respect to
that of the transition state*, because of the more localized charge on the
reactant anion. A negative charge on small atoms, especially first-row atoms,
which do not have electron-withdrawing substitutents attached to them,
produces anions which are very strong hydrogen-bond acceptors. Such anions
have strong general hydrogen-bonding interactions with protic solvents. On
the other hand, this interaction is absent in dipolar aprotic solvents.
Therefore, the differential solvation of reactants and transition state,
where the charge is delocalized, is greater for protic solvents. This is
the main origin of the great rate enhancement in polar aprotic solvents.
We are thus led to the form of the potential of mean-force schematized
in figure 2. This form was recently confirmed after several MC simulations
of the potential of mean-force of various S_N2 reactions in protic and aprotic
solvents[6,7].

---

*In solution the ion-molecule complex is almost energetically indisting-
uishable from the separated reactants.

What we start to learn from here is that the microscopic details of a
reactive event are substantially different in the gas-phase and in solution.
We have also seen that, even in solution, those details are very dependent
on the particular structure of the environment of the chemical reaction.
We will postpone a discussion about potentials of mean-force for enzymatic
reactions until section 3 but, we can already advance that, when looking
to chemical reactions in enzymes, we will have to compare the typical
solution environment for the non-catalyzed reaction with the environment
provided by the protein matrix to the same reaction, and analyse (if we
are able to!) the reasons for the catalytic power of the enzyme.

2- CONNECTION TO THERMODYNAMICS AND RATE THEORIES

Up til now we have not been very precise in the definition of the
potential of mean-force. We had always used the word energy without specifying what kind of energy we were talking about. In the gas-phase it is
customary to use the potential energy surfaces to picture the energetics
of the reaction. The connection to thermodynamic quantities and experimental
parameters is made on the basis of the traditional rate theories. As we
are concerned only with equilibrium effects we shall introduce the
Transition State Theory (TST)[8].

The TST involves two basic assumptions:
-that the activated complex (or transition state complex) is in equilibrium with the reactants and,
-that once an activated complex is formed the conversion into products
 is ensured.
The reaction rate, v, is then determined as

v= probability of having the activated complex *
   * frequency with which the complexes pass over the barrier;

Now, the probability of having the activated complex is given (under the
equilibrium assumption) by the equilibrium concentration of the activated
complex. In terms of statistical mechanics that reads

$$[X]^{\dagger} = [\text{Reactants}] * \frac{Q^{\dagger}}{Q^{R}} \exp(-U^{\dagger}/RT) \tag{2.1}$$

where $U^{\dagger}$ is the molar potential energy at the transition state assigning
the zero of energy to the reactants. $Q^{\dagger}$ and Q^{R} are the partition functions
per unit volume of the transition state and reactants, respectively, including zero-level vibration energies. To proceed on the derivation it is
convenient to express the motion over the barrier as a translational motion
on an arbitrary length δ. Then, the partition function for the activated
complexes may be written as

$$Q^{\dagger} = \frac{(2\pi m^{\dagger} K_B T)^{\frac{1}{2}}}{h} \delta \; q_{\dagger} \tag{2.2}$$

where $(2\pi m^{\dagger} K_B T)^{\frac{1}{2}} \delta/h$ is the translational partition function for the motion
of a particle of mass $m^{\dagger}$ in a one-dimensional box of length δ. We have now
to calculate the frequency with which the complexes pass over the barrier:
according to kinetic theory this frequency is given by the ratio between the
average speed of particles moving into the direction of products over the
barrier,

$$\langle v_+ \rangle = \left(\frac{K_B T}{2\pi m^{\dagger}} \right)^{\frac{1}{2}} , \quad (\text{+ stands for the direction of motion}) \tag{2.3}$$

and the length δ.

Puting all these equations together we get for v:

$$v = [\text{Reactants}] \; \frac{(2\pi m^\dagger K_B T)^{1/2} \delta}{h} \; \left(\frac{K_B T}{2\pi m^\dagger}\right) \; \frac{1}{\delta} \; \frac{q_\ddagger}{Q^R} \; \exp(-U^\dagger/RT)$$

$$= [\text{Reactants}] \; \frac{K_B T}{h} \; \frac{q_\ddagger}{Q^R} \; \exp(-U^\dagger/RT) \quad , \tag{2.4}$$

and the rate constant can be immediately derived as being given by,

$$k = \frac{K_B T}{h} \; \frac{q_\ddagger}{Q^R} \; \exp(-U^\dagger/RT) \tag{2.5}$$

This last equation deserves some comments. First of all we see that the arbitrary length δ cancels out in the final expression for the rate. But one must be careful when interpreting the partition function $q_\ddagger$: this is not the <u>total</u> partition function for the activated complex because the motion over the col has been excluded. This point will be important while making the connection to thermodynamics but it seems to have been often overlooked over the years. The importance of this point to thermodynamic studies in enzyme catalysis has been stressed only very recently by Kraut[9].

The thermodynamic formulation of the TST rate constant is always obtained by identifying

$$\frac{q_\ddagger}{Q^R} \; \exp(-U^\dagger/RT) \tag{2.6}$$

with the "equilibrium" constant, $K_{eq}'^\dagger$, for the formation of an activated complex from the reactants. Then k is given by

$$k = \frac{K_B T}{h} \; K_{eq}'^\dagger \tag{2.7}$$

We must stress again that $K_{eq}'^\dagger$ is not a true equilibrium constant because the motion over the col is missing on the partition function, $q_\ddagger$, for the activated complex. If we want to express k in terms of the true equilibrium constant then

$$k = \frac{\langle v_+ \rangle}{\delta} \; K_{eq}^\dagger \tag{2.8}$$

The frequency appearing on this expression is the actual frequency (under the non-recrossing assumption) with which the complexes pass over the barrier.

Expressing $K_{eq}^\dagger$ in terms of $\Delta G^\dagger$, the change in the standard Gibbs energy when the activated complexes are formed from the reactants, the result for k is

$$k = \lambda \, \exp(-\Delta G^\dagger/RT) \tag{2.9}$$

where we have substituted λ for $\langle v_+ \rangle/\delta$. The quantity $\Delta G^\dagger$ is known as the standard Gibbs energy of activation. It can be split into $\Delta H^\dagger - T \Delta S^\dagger$ to give

$$k = \lambda \, \exp(+\Delta S^\dagger/R) \, \exp(-\Delta H^\dagger/RT) \tag{2.10}$$

where $\Delta S^\dagger$ is the standard entropy of activation and $\Delta H^\dagger$ is the standard enthalpy of activation. Note that the expression above for k is usually used to determine the thermodynamic activation parameters. Some problems can occur on the evaluation of $\Delta S^{\dagger}$ as the frequency λ can be substantially different from the one commonly used, that is $K_B T/h$.

The thermodynamic version of the TST expression for the rate constant (eq.(2.9)) was derived without making any distinction between gas-phase and solution. Under the assumptions of the TST how can the solvent affect k with respect to the gas-phase? There are two ways in which the solvent can affect k. The first concerns the equilibrium constant $K_{eq}^{\dagger}$ as the solvent can influence the relative stabilization between the reactants and the activated complex. The second concerns the frequency λ. As the length δ is arbitrary, when comparing two distinct rate constants we can assume that it is the same for both, but $m^{\dagger}$ is the effective mass for motion over the barrier, and this can be affected by the solvent. In the case of S_N2 reactions we already argue that the interactions between the activated complex and different solvents are expected to be similar, the main effect of the solvent being attributed to its different interactions with the reactants. For these reactions it is probably a reasonable assumption to assume that λ is preserved when changing solvents and the effect of the solvent can be interpreted in terms of differences in the free energies.

With respect to enzymes a word of caution should be put on approaches that make use of thermodynamics and rate constants to conclude that enzyme catalysis can be simply understood in terms of a greater stabilization of the activated complex in the presence of the enzyme[10]. Although there are various indications that that is an important contribution to catalysis, the derivation generally used to support that conclusion is formally incorrect. The derivation starts from the thermodynamic cycle:

$$
\begin{array}{ccccc}
 & & K_N^{\dagger} & & \\
E + S & \underset{\leftarrow}{\overset{\rightarrow}{}} & E + S^{\dagger\prime} & \rightarrow & E + P \\
K_S \;\updownarrow & & \updownarrow K_T & & \updownarrow \\
ES & \underset{\rightleftharpoons}{} & ES^{\dagger} & \rightarrow & EP \\
 & K_E^{\dagger} & & &
\end{array}
\qquad (2.11)
$$

In the scheme above K_S is the equilibrium constant for association of the substrate, S, with the enzyme, E; $K_N^{\dagger}$ and $K_E^{\dagger}$ are equilibrium constants for the formation of the transition states of the nonenzymatic and enzymatic reactions, $S^{\dagger\prime}$ and $ES^{\dagger}$, respectively. K_T is the equilibrium constant for the binding of $S^{\dagger\prime}$ to E to form $ES^{\dagger}$. It is easily seen that the following relation holds between the different equilibrium constants:

$$
\frac{K_E^{\dagger}}{K_N^{\dagger}} = \frac{K_T}{K_S} \qquad \bullet \qquad (2.12)
$$

Now, using the transition state theory, one can relate the equilibrium constants for the formation of the transition states, with the rate constants to the same processes, remembering though that $K_E^{\dagger}$ and $K_N^{\dagger}$ are true equilibrium constants. A simple manipulation gives the following relation between rate constants and association equilibrium constants:

$$
\frac{k_e}{k_n} = \left(\frac{\lambda_n}{\lambda_e}\right)\frac{K_T}{K_S} = \left(\frac{m_n^{\dagger}}{m_e^{\dagger}}\right)^{\frac{1}{2}}\frac{K_T}{K_S} \qquad (2.13)
$$

Here k_e and k_n are the rate constants for the catalyzed and noncatalyzed reaction, respectively, and $m_e^{\dagger}$ and $m_n^{\dagger}$ are the effective masses for motion over the barrier for the catalyzed and noncatalyzed reaction, respectively. The relation usually used in the literature misses the factor $(m_n^{\dagger}/m_e^{\dagger})^{\frac{1}{2}}$ as true equilibrium constants are mixed with non regular equilibrium constants.

Equation (2.13) above, without the factor $(m_n^{\dagger}/m_e^{\dagger})^{\frac{1}{2}}$, has been frequently used to argue that the transition state must bind enormously more

strongly to the enzyme than does the substrate in its ground state, by a
factor equal to the enzymic rate acceleration. There is no indication about
the relative values of the effective masses for motion over the barrier and
so we can not predict for the moment what is the influence of that factor
on catalysis.

Let us finish this section with a word more about potentials of mean-
force. These potentials are actually free energies and can be calculated by
a variety of methods. Calculations have been performed by now for a large
number of systems and our interest will be foccus only on the ones refering
to proteins. The results have been used to predict rate constants for some
activated processes that take place inside proteins, like the well studied
rotational isomerization of tyrosine sidechains in the interior of the protein BPTI[11], and also to predict relative affinities of different ligands
for a protein[12]. Some of these methods make use of special tricks to make
calculations more easy. When the obtained free energies are used in predicting rate constants one must be sure that the calculated free energy is the
one wanted for the application of TST.

3- FREE ENERGY CALCULATIONS IN ENZYMES

In recent years the great improvement in computational resources and in
the determination of X-ray structures of proteins, have led to a large number of simulations of activated processes and real chemical reactions occuring in proteins and in particular in enzymes. These studies, although
always invoking a number of simplifying assumptions, provide at least semiquantitative ideas about the energetics and the dynamics of those processes.
Thanks to them it is becoming possible to explain how enzymes work so well.

In this section we will present some of the recent studies of enzymatic
reactions that have appeared in the literature. Our intent while doing so
will be to try to interpret those results in terms of similar reactions
occuring in solution.

3.A- <u>Dihydroxyacetone Phosphate (DHAP)-Glyceraldehyde Phosphate (GAP)</u>
       <u>Isomerization Catalyzed by Triosephosphate Isomerase (TIM)</u>

Triosephosphate isomerase (TIM) is a very efficient enzyme and it has
been characterized as a "perfectly evolved" enzyme[13] as the rate determining
step of the catalyzed reaction, reversible conversion of dihydroxyacetone
phosphate (DHAP) in glyceraldehyde phosphate (GAP), is the product dissociation from the enzyme. The mechanism of the TIM catalyzed reaction is reasonably well understood. One of the elementary steps during catalysis is the
enolization of DHAP. For this process it is known that the rate at the active site of TIM is more than 10^9 times faster than the rate in aqueous buffer.

Alagona et al.[14,15] have done extensive studies on this elementary
process, schematized in figure 3 below, using a combination of molecular
and quantum mechanics. Calculations for this process in vacuo were preformed
showing that the proton-transfer reaction is mainly uphill and energetically
very unfavourable, with the products and the transition state nearly equal
in energy. The results of molecular mechanical calculations for the same
process in yeast TIM showed that in the enzyme the products are instead stabilized with respect to the reactants. Alagona et al.[14,15] performed then a
number of quantum-mechanical calculations where they mimic the influence of
the amino-acids that are believed to participate in catalysis (Hys 95, Lys13
and Glu97), by including partial charges. Analysing the results it is seen
that the main effect of these charges is to bring the products at an energy
level compared to that of the reactants. But, even then, it is seen that the

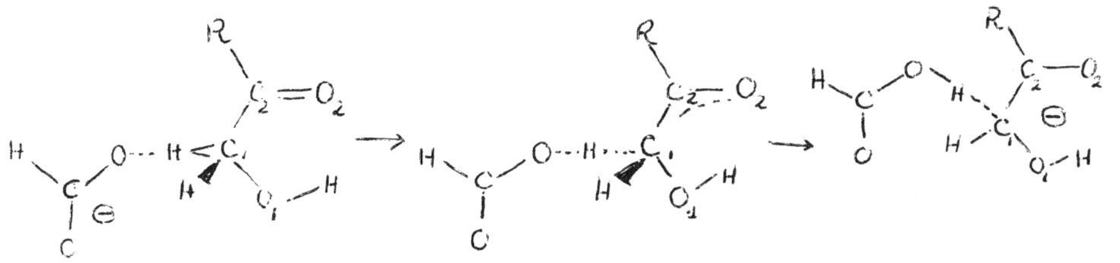

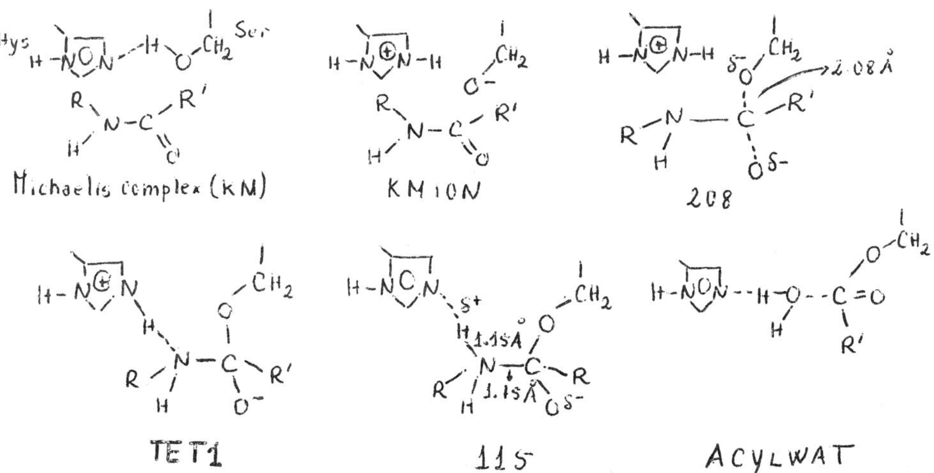

Fig. 3- Enolization of DHAP in TIM.

activation barrier for the proton-transfer is about 26-28 Kcal/mol, a value too high to explain the acceleration provided by the enzyme.

A combination of molecular and quantum mechanics was further applied to determine the variations in the energy barrier for the proton-transfer with the C..O and C..H distances. They then saw[14] that, although there is a cost in strain energy, the effective barriers for the proton-transfer actually decrease to about 9-15 Kcal for C..O distance of 3Å and to 6-12 Kcal for C..O equal to 2.9 Å. The role of the enzyme in speeding up the DHAP enolization seems to be double: it stabilizes the products to an energy comparable to that of the reactants and, probably in a concerted way, it shortens the distance between the heavy atoms involved in the proton-transfer. Recent advances in the theory of proton-transfer reactions in solution[16] indicate that these are the key steps to make a proton-transfer reaction fast.

3.B- Amide Bond Cleavage in the Active-site of a Serine Protease – Trypsin

The nature of enzyme catalysis in trypsin was recently studied by Weiner et al.[17]. The elementary step, believed to be the rate limiting step, consists on the break of the amide bond in the peptidic substrate. Weiner et al. preformed quantum mechanical simulations associated with molecular mechanics simulations, in a way similar to the one used by Alagona et al. in the example above, for a number of structures of the trypsin-substrate system that are believed to occur as intermediates during the reaction (see figure 4).

Fig. 4- Intermediates occuring in the trypsin-catalyzed hydrolysis of a specific tripeptide substrate.

The most interesting feature of this work for our purposes is that
Weiner et al. have also performed calculations on an analagous reaction,

$$OH^- + H_2NCHO \longrightarrow HCOO^- + NH_3 \quad ,$$

taking place in the gas-phase and in water[18]. That allows comparisons between
different media. We report on figure 5 their energy profiles along the
reaction.

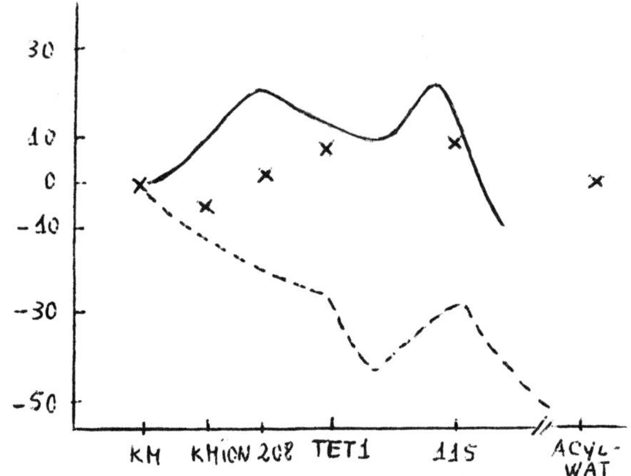

Fig. 5- Reaction profile for formamide hydrolysis
in solution (——), in the gas-phase (- - -)
and in trypsin active-site (x).

Despite the uncertainties in the energy values for the enzyme reaction
it is clear from figure 5 that the trypsin catalyzed reaction is quite different from the corresponding solution reaction and is somewhat intermediate
between solution and gas-phase profiles. Let us analyse in detail these profiles and try to understand how the enzyme can be working. In the gas-phase
there is no barrier for the formation of the tetrahedral intermediate; the
breakdown in formate and ammonia can occur via direct donation of the internal proton to the nitrogen with concurrent C-N bond cleavage or, if a water
molecule is available, the breakdown of the tetrahedral complex is water
catalyzed by water mediated proton donation to the nitrogen. The barrier for
the internal proton transfer was estimated in 35 Kcal/mol[19] and for the
water-catalyzed transfer the barrier, as one can see from figure 5, is about
12 Kcal/mol. In solution there is a solvent induced barrier to OH^- attack,
~22 Kcal/mol, due to the more favorable solvation of OH^- than the more diffuse anions that conduce to the tetrahedral complex. The breakdown of this
complex involves a water mediated proton-transfer from the carbonyl end of
the molecule to the nitrogen end, in a way similar to the water-catalyzed
gas-phase step, and it is not surprising that the energy barrier is about the
same, ~13 Kcal/mol.

What happens in the enzyme? From the energy profile depicted in figure5
it is not obvious if the formation of the tetrahedral complex is an uphill
process or if intermediate barriers exist between each of the steps reported
in figure 4. One can though argue that, for the proton-transfer reactions,
the barriers involved should be low as they occur between structures of
similar energies and, the groups between which the proton transfers are close
enough. The most important role of the enzyme is to isolate the O_γ of serine
from the solvent and thus facilitate the attack on the carbonyl carbon. This
is the process that in solution involves a high solvent induced barrier. The
remaining groups in the protein stabilize the ionic structures along the

catalytic pathway, with little energy being required in that process because
these groups are mainly preoriented and fixed.

3.C- Acid Catalysis in Lysozyme

Warshell[20] used a complete different approach from the ones cited above
to simulate the acid catalysis in lysozyme. This approach, called the empirical valence bond method, represents the reacting system as a superposition
of ionic and covalent resonance forms. Then, the effect of the solvent,
water or protein, is incorporated in an empirical way in the diagonal element
of the ionic resonance form. Although it is hard to evaluate the accuracy
of this method, the results obtained clearly point out some of the factors
present in the two examples above.

The catalytic reaction in lysozyme involves the cleavage of a peptidic
bond on the substrate. Again, it is believed that the first step of the
reaction is a proton-transfer between one of the amino-acids of the protein
and the substrate, creating in this way an ionic intermediate structure that
it is better stabilyzed by the enzyme than by water, mainly because the
remaining groups in the active-site of the protein provide a better electrostatic environment. That structure facilitates the break of the peptidic bond
and this last step also involves an ionic transition state. From the energy
profiles provided by Warshell[20] (see figure 6) we can see that one of the
important roles played by the enzyme is to stabilize, mainly via electrostatic interactions, the ionic structures occuring during the catalytic
reaction.

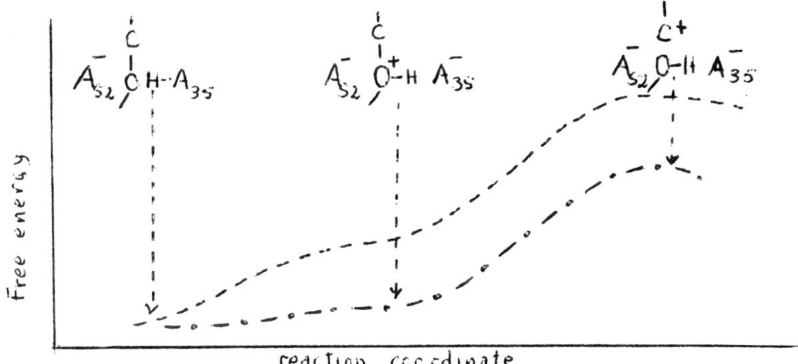

Fig. 6- Free energy profiles for acid catalysis in lysozyme,
(-.-.-) and in solution (- - -).

4- CONCLUDING REMARKS

We have tried here, without advocating any particular theory, to analyse
what makes enzymes so powerful in speeding up chemical reactions. From the
examples given above, and puting reliability on the proposed mechanisms and
on the methods employed in the calculations, we can point out some common
features among them. In all of them the formation of the Michaelis complex,
enzyme-substrate complex, involves the exclusion of the free water solvent
molecules from the active-site and partial desolvation of groups in the
protein. In our opinion, it is however erroneous to use this fact to infer
that enzymatic reactions are analogous of gas-phase reactions.[21] Even though
water is excluded, the protein itself acts as the solvent to the chemical
reaction. It is of course a very particular kind of solvent and, as we saw,
it seems to be particularly well designed to provide, at least for those
reactions involving ionic intermediates, the necessary electrostatic complement to stabilize those structures. This is a support to the traditional
idea that enzymes work so well because they bind more tightly the transition

state complex compared to the enzyme-substrate complex, but probably this
is not the only factor that counts.

Although we must look at the protein as a very special solvent to the
chemical reactions taking place in the active-site, we think that understand
how reactions occur in solution, with water or some other solvent, is a valuable step to understand enzyme catalysis. We focused here only on equilibrium effects of the solvent, but there is a rich variety of dynamical
effects by the solvent, that emerge from the study of the microscopic dynamics of chemical reactions in solution, and they can play an important role
too on enzyme catalysis. To assess the importance of the protein fluctuations on the catalytic reaction it would be desirable to have dynamical
simulations other than static simulations. However, these last ones are very
important as they can provide the starting point for the first ones.

Genetic engineering is now proving to be a very powerful tool to understand enzyme catalysis. It makes it possible to selectively modify the
protein groups in the active-site and thus determine the influence of
distinct groups on the chemical reaction. We believe that a combination of
these studies with theoretical ones, like the ones described in section 3,
can give real insights on the way enzymes work and pave the way to the
design of new catalysts.

Acknowledgments

This work was supported in part by a NSF grant to B. Ladanyi and a
NHI grant to M. Fixman. I acknowledge J.T. Hynes for many helpful
discussions and for calling my attention to the work of J. Kraut.

References

1. D.J. Raber , J.M. Harris and P.v.R. Schleyer in " Ions and Ion Pairs in
   Organic Reactions ", ed. M. Szwarc, Wiley, New York (1974),vol.2 p.247.
2. W.L. Jorgensen, J.K. Buckner, S.E. Huston and P.J. Rossky, J. Amer. Chem.
   Soc. 109: 1891 (1987).
3. G. Ciccotti, M. Ferrario, J.T. Hynes and R. Kapral, Chem. Phys., in
   press.
4. A.J. Parker, Chem. Rev. 69: 1 (1969).
5. W.N. Olmstead and J.I. Brauman, J. Amer. Chem. Soc. 99: 4219 (1977).
6. J. Chandrasekhar, S.F. Smith and W.L. Jorgensen, J. Amer. Chem. Soc.
   107: 154 (1985).
7. J. Chandrasekhar and W.L. Jorgensen, J. Amer. Chem. Soc. 107: 2974 (1985).
8. S. Glasstone, K.J. Laidler and H. Eyring, "The Theory of Rate Processes",
   McGraw-Hill, New York (1941).
9. J. Kraut, Science, to be published.
10. G.E. Lienhard, Science 180: 149 (1973).
11. J.A. McCammon and S.C. Harvey, " Dynamics of Proteins and Nucleic Acids",
    Cambridge, London (1987) , and references therein.
12. C.F. Wong and J.A. McCammon, J. Amer. Chem. Soc. 108: 3830 (1986).
13. J.R. Knowles and W.J. Albery, Acc. Chem. Res. 10: 105 (1977).
14. G. Alagona, P. Desmeules, C. Ghio and P.A. Kollman, J. Amer. Chem. Soc.
    106: 3623 (1984).
15. G. Alagona, C.Ghio and P.A. Kollman, J. Mol. Biol. 191: 23 (1986).
16. D. Borgis and J.T. Hynes, to be published in this volume.
17. S.J. Weiner, G.L. Seibel and P.A. Kollman, Proc. Nat. Acad. Sci. USA
    83: 649 (1986).
18. S.J. Weiner, U.C. Singh and P.A. Kollman, J. Amer. Chem. Soc. 107:
    2219 (1985).

19. G. Alagona, E. Scrocco and J. Tomasi, J. Amer. Chem. Soc. 97: 9876
    (1975).
20. A. Warshell, Acc. Chem. Res. 14: 284 (1981); ibid. Proc. Natl. Acad.
    Sci. USA 81: 444 (1984).
21. M.J.S. Dewar, Enzyme 36: 8 (1986).

THEORETICAL CALCULATIONS ON AN ENZYME CATALYZED REACTION

MECHANISM

Giuliano Alagona and Caterina Ghio

Istituto di Chimica Quantistica ed Energetica Molecolare, del C.N.R.
Via Risorgimento 35, I-56126 Pisa, Italy

INTRODUCTION

Because of the continuous expansion of genetic engineering techniques,
the availability of a theoretical approach for predicting the effect of site-
specific mutagenesis could be extremely useful. Therefore, we are interested in
the application of theoretical methods, mainly *ab initio*, in order to elucidate
the enzyme catalysis process. While most of the classical mechanical tools are
presented elsewhere in this book, a discussion of the quantum mechanical ones
falls outside its scope. Thus, we will focus more on the methodological aspects of
our research, rather than on its results, these nonetheless are thoroughly
discussed in the source papers [1-3].

The reversible isomerization of dihydroxy acetone phosphate (DHAP) to
glyceraldehyde-3-phosphate, as catalyzed by TIM, a mechanism of obvious
interest in itself, will be used as an example to illustrate the method employed.
Of course, a number of adjustments should be made to this method if one is
studying other processes involving different kinds of mechanisms.

First, we would like to show how the combination of *ab initio* quantum
mechanics and molecular mechanics enables one to study a chemical reaction
in the presence and absence of an enzyme, providing insight as well into how
the enzyme achieves its catalytic power.

Secondly, we will try to explain why certain mutant enzymes are less
effective or, even, not effective at all with respect to the wild type enzyme in
catalyzing the normal reaction, sometimes in contradiction with biochemical
intuition.

Finally, based on the understanding of this mechanism we can model the
enzyme environment in a very simple way, obtaining a full quantum
mechanical description of the "enzyme catalyzed" reaction.

In addition, we will present an overview of another method, recently
proposed by Chandra Singh and Kollman [4], also referring the interested
reader to the book by McCammon and Harvey [5] for a comprehensive review of
the theoretical approaches presently being used to study the dynamics of
proteins and nucleic acids.

The most commonly employed method to study reaction mechanisms is the potential energy surface scan, in which the calculations are performed along a putative reaction coordinate. There are, however, several levels of sophistication within this framework.

We will not enter into a discussion of the different quality *ab initio* calculations in use from the point of view of the basis set extension and the level (SCF, SCF+CP, MP2, CI, etc.), since this topic is rather technical and its interest is limited to theoreticians. Suffice it to state that in every case one should use the *ab initio* calculations best suited to the chemical problem under scrutiny, taking into consideration the extension and the characteristics of the system, the reliability of the results, and the available computer resources.

The first and least expensive (in terms of computer time) method usually exploits chemical intuition and simplified, but still adequate, models of the system. For example, chemical intuition suggests that a given reaction proceeds through a number of steps (see Scheme 1): in I-III, a base approaches the ketone and abstracts a proton; in III-V, a rearrangement occurs, together with the proton shuttling; and in V-VII, the base delivers the proton back to the aldehyde (this part of the mechanism is essentially the reverse of I-III). If this hypothesized sequence of steps is sound we can try to evaluate the energy changes involved in each step of the mechanism by resorting to simplified models of the interacting species.

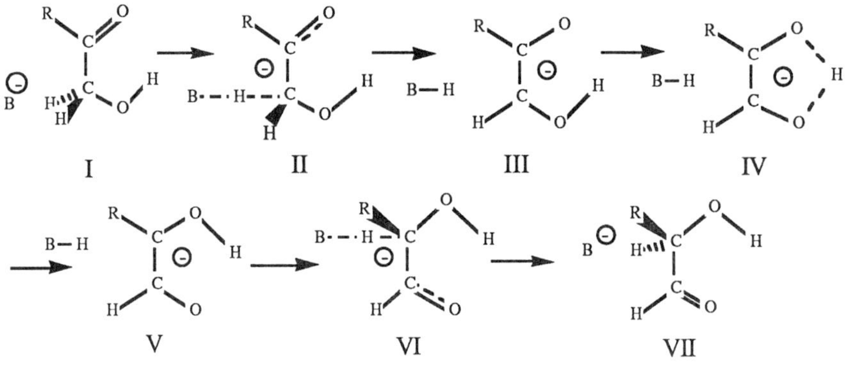

I   II   III   IV

V   VI   VII

Scheme 1

A good number of assumptions are, however, necessary. In the real biological system R = CH_2-O-PO_3^{--}, but of course, we cannot use this group in our calculations. Since -CH_2- has proven itself to be a good insulator, we can use a methyl group as a reasonable approximation to R. In some cases the system may still be too large, so we should resort to an H in place of R. The base B^- is a glutamate anion (Glu 165) in TIM; however, in our model we replace it with a formate anion ($HCOO^-$) or, sometimes, with an OH^-.

As far as the intramolecular proton transfer (III-V) is concerned, its energy barrier (Fig. 1) is fairly low even *in vacuo* and thus it is not likely that the enzyme exerts its catalytic effect in this step.

Let us examine the first three steps (I-III), i.e., the intermolecular proton transfer, that seems to require the inclusion of the environment. In fact, it is difficult to explain how a carboxylate ($pK_a \sim 4$) could abstract a proton

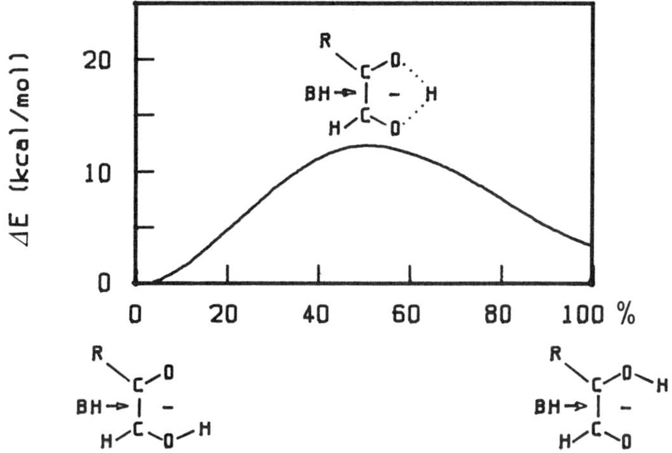

Fig. 1. Energy involved in the intramolecular proton transfer (STO-
3G/SCF level) at various percentages of proton transfer. The
geometries have been linearly interpolated.

from DHAP ($pK_a \sim 14$) without a polarizing electrophile. The *in vacuo* results
(Fig. 2) along the reaction path (three pairs of $R_{C...O}$ and $R_{C...H}$ values,
determined using model systems) confirm this, showing high barriers
inconsistent with a fast reaction.

Before passing to the inclusion of the environment in our calculations,
let us look at the results obtainable with a better description of the reaction
path. This description depends on a more complete geometry relaxation of the
relevant parameters $R_{C...O}$ and $R_{C...H}$, which allows us to draw a potential
energy hypersurface (Fig. 3). The energy profile along the minimum energy
path (Fig. 4) gives a high energy difference between I and III: upon internal
geometry optimization II probably is just a point along the straight uphill path
to III. We used in the various regions only three different geometries of the
partners (fully optimized on model systems), namely: in the reactant region,
HCOO$^-$····ketone; in the TS region [HCOO···H···interm.]$^-$; and in the enediolate
region HCOOH····enediolate.

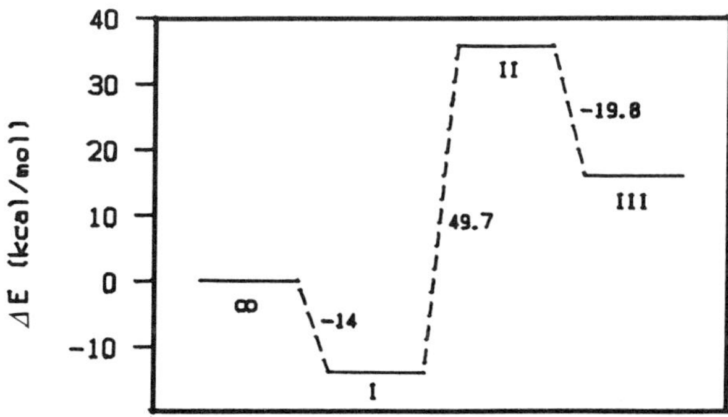

Fig. 2. Energy involved in the steps leading from I to III (4-31G/SCF
level). oo corresponds to the energy of the isolated reactants.

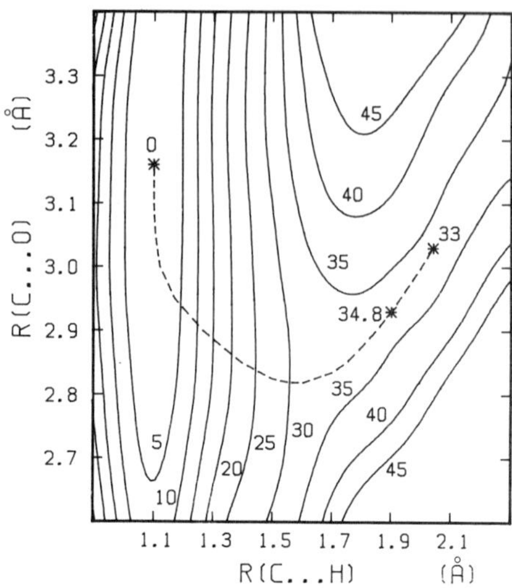

Fig. 3. Potential energy hypersurface for the *in vacuo* reaction mechanism (I-III) at the 4-31G/SCF level for the model with R=CH$_3$ (from Ref. 3, with permission of The HUMANA Press).

In so-called "adiabatic mapping", on the contrary, one keeps the ruling parameters fixed, while all the others are allowed to change so as to minimize the potential energy of the system at each position of the reaction coordinate (determined from a previous calculation), or at each position of a grid on the hypersurface. The shape of the resulting surface, however, should not be much different from that in Fig. 3.

In conclusion, even employing the most appropriate calculations, *in vacuo*, the reaction appears to be very slow, since we obtain either a large barrier or a large energy difference between I and III.

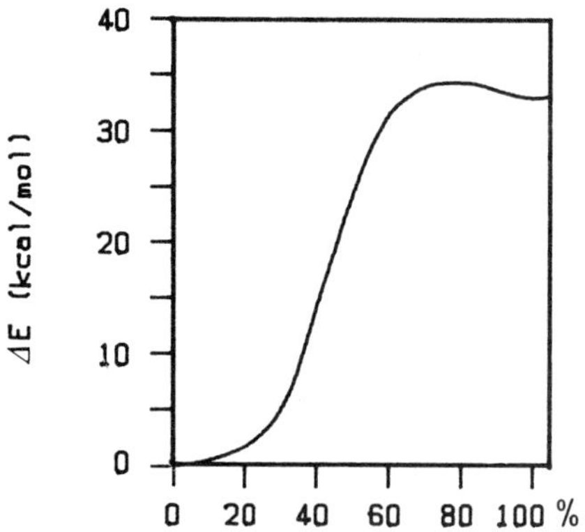

Fig. 4. Energy profile along the reaction coordinate of Fig. 3.

ENVIRONMENTAL EFFECTS

In the active site of TIM there are charged groups (Lys 13, His 95 and Glu 97), which are supposed to participate in the catalytic effect.

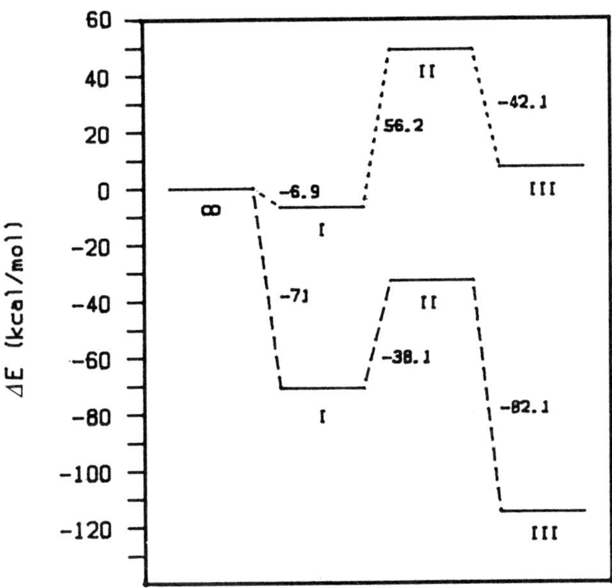

Fig. 5. Comparison between the energy changes involved in the steps leading from I to III for the model with R=CH$_3$ in the presence (----) and in the absence (·····) of an NH$_4$$^+$ group polarizing the carbonyl group (STO-3G/SCF calculations).

Either Lys 13 or His 95 (or both of them) could be the polarizing electrophile necessary to make proton abstraction easier. Before introducing the charges of all these residues in the *ab initio* calculations we wanted to assess the viability of this approach. So we added an NH$_4$$^+$ group alongside one of the carbonylic oxygen lone pairs, in order to simulate the effect of Lys 13. The relative energies of the various species are shown in Fig. 5, along with the consistent ones obtained in the absence of the NH$_4$$^+$ group. The effect that charged groups can have in stabilizing species III with respect to species I turns out to be quite dramatic from this comparison. Thus we had to include in the calculations a more realistic description of the key active site groups, derived from the known structures of TIM.

MODEL BUILDING AND MOLECULAR MECHANICAL SIMULATIONS

X-ray structures of TIM were available both for the chicken liver TIM and for the yeast TIM (YTIM). From the TIM-DHAP complex model built using the PS2 (Evans & Sutherland) we sorted out all of the residues (32) within 8 Å from any atom of DHAP to be included in the molecular mechanical refinement. This refinement makes use of an analytical potential function of the form:

$$E_{tot} = \sum_{bonds} k_r \left(r - r_{eq}\right)^2 + \sum_{angles} k_\vartheta \left(\vartheta - \vartheta_{eq}\right)^2 + \sum_{dihedrals} \frac{V_n}{2} \left[1 + \cos\left(n\Phi - \gamma\right)\right]$$

$$+ \sum_{nonbonded} \left[\frac{A_{ij}}{R_{ij}^{12}} - \frac{B_{ij}}{R_{ij}^{6}} + \frac{q_i q_j}{\varepsilon R_{ij}}\right] + \sum_{H\,bonds} \left[\frac{C_{ij}}{R_{ij}^{12}} - \frac{D_{ij}}{R_{ij}^{10}}\right] \qquad (1)$$

to compute the total energy of the system and minimizes this functional by varying all the parameters in order to bring them as close to the reference parameters as possible, given the imposed constraints. We decided to restrain all C_α carbons at the termini of the chain segments with a 10 kcal/(mol·Å^2) penalty function, to simulate the presence of the discarded segments of the backbone.

First of all, we had to insert into the parameter file the specific parameters of the various species (DHAP and enediolate (ENE)), using Glu 165 (TIM) or the protonated Glu 165 (TIMH$^+$), depending on the complex considered. After a very short minimization to eliminate the worst steric and electrostatic contacts, we model built various structures.

In all of our subsequent work, however, we used the structure model built according to the geometry proposed by Alber and Petsko [6] based on a difference density analysis of the YTIM/DHAP complex and the native enzyme, in which His 95 and Lys 13 are placed above and below the plane containing the open ring O=C-C-O-H. In this arrangement they are able to move from one oxygen to the other, consistent with the 'reversible' character of the isomerization, under the influence also of the Glu 97 located just in front of them.

In the molecular mechanical simulation the inclusion of the residues in the active site of TIM makes the complex between the protonated YTIM and the enediolate (III) to be more stable than the complex between YTIM and DHAP (I) by 23 kcal/mol (Fig. 6), thus confirming the marked effect of polarizing electrophiles on the interaction.

For the sake of clarity we report in Table 1 some of the numerical results obtained at the various stages of the research. It is worth noting that the molecular mechanical refinement gives a stabilization of the same order of magnitude as that from the *ab initio* calculations with the inclusion of the environment that we will discuss in the next section. The stereo pictures of a few energy refined structures are reported in Fig. 5 of Ref. 1.

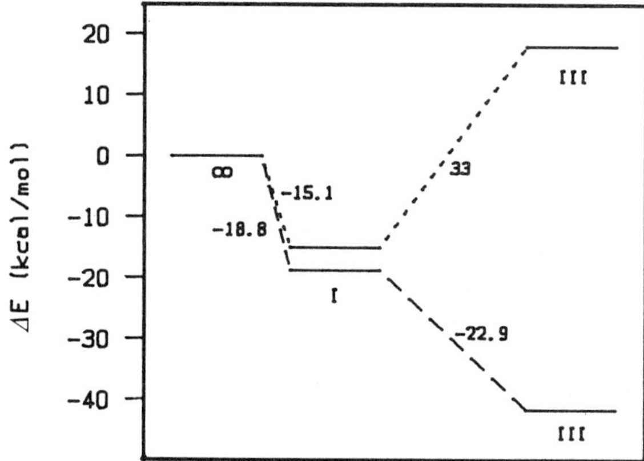

Fig. 6. Comparison between the energy changes obtained from the *ab initio* calculations *in vacuo* (.....) and the molecular mechanical refinement of the substrate-bound enzyme (----).

Table 1. Quantum mechanical energies obtained with the 4-31G basis set and molecular mechanical refinements of the various adducts (in kcal/mol).

| Complex | SCF | (MP2) | SCF | (MP2) | SCF | Mol. | mechanics |
	in vacuo		H,L,G[a]		G,L,G[b]	WT[c]	Mutant[d]
I	-14.0	-14.1	-45.9	0	-44.2	-465.4	-470.8
II	35.7	21.6	-3.6	28.3	-0.9	-	-
III	17.9	14.8	-40.5	2.4	-35.9	-488.3	-543.3
I → III	31.9	28.9	5.4	2.4	8.3	-22.9	-72.5

[a] Calculation perturbed by the charges of His, Lys, and Glu;
[b] calculation with the charges of Gln in place of those of His;
[c] wild type enzyme; [d] the mutant has Gln instead of His 95.

ENVIRONMENTAL EFFECTS ON THE *AB INITIO* REACTION MECHANISM

We included in the one-electron Hamiltonian the charges of those residues expected to most influence the isomerization reaction, i.e., His 95, Lys 13, and Glu 97. However, it was a rigid perturbation, because the charges were kept fixed in the position determined around DHAP. A better description would have been attained if at least three different positions of the system of charges had been considered, one for each different substrate. Even in our far from ideal situation it appears evident (Table 1) that the role of the enzyme is to stabilize III with respect to I. The estimated energy difference at the MP2 level *in vacuo* is ~29 kcal/mol, whereas after the inclusion of the charges of the three residues it is 2.4 kcal/mol, with a stabilization of about 26 kcal/mol.

MUTANT ENZYMES

Acting on the suggestion of G. Petsko, who was attempting to assess the need for a general acid in the catalyst, we included in both of our calculations (*ab initio* quantum mechanical as well as molecular mechanical) glutamine instead of histidine. Gln, in fact, like His, may act as a proton donor and acceptor but, due to its larger pK_a, it cannot act as a general acid. From both calculations (Table 1) it turned out that the mutated enzyme can be as effective as the native one in stabilizing III over I.

However, when we examined the minimized structures using the Picture System we realized that Gln, due to its higher flexibility in comparison to His, had moved during the simulation to a location different from His, and had formed an H bond with Glu 165. Therefore, Gln can reduce the catalytic rate either by inhibiting the proton transfer to Glu 165 or by moving to another position not suited to stabilizing III.

Our quantum mechanical calculations, on the contrary, in this specific case were not able to give a correct view of the process, because the Gln charges had been placed at a position matching as far as possible those of His, and were kept fixed in this arrangement. From the failure of this method to describe the outstanding features of the mechanism the need of a more efficient method, one which takes into account the mutual interplay between the quantum mechanical and the molecular mechanical parts of the system, becomes obvious. Of course, any method with these characteristics would be much more expensive, in terms of computer time, than the one employed here (roughly, it is like performing several geometry optimizations rather than as many single point calculations).

A COMBINED QUANTUM AND MOLECULAR MECHANICAL METHOD

As was hinted earlier, the approach described above is not completely satisfactory because it is still a rigid perturbation that we are introducing in each Hamiltonian. Much more promising is a method recently proposed by Chandra Singh and Kollman [4], which we will discuss in connection with the TIM mechanism even though it has not been applied to that particular mechanism yet. As will be seen, the net separation that we have between the quantum mechanical (QM) part of the system - DHAP (or its model) and COO^-, where the bond breaking or formation occurs - and the molecular mechanical (MM) part, i.e. the environment, is not going to be maintained.

The method proposed is actually an attempt to effectively couple *ab initio* quantum mechanical calculations with molecular mechanical and molecular dynamical calculations within one system of programs. The analytical gradients computed in both methods are combined in such a way as to geometrically optimize the structure of the system and to evaluate its global quantum mechanical and molecular mechanical energy.

The process starts with an energy refinement using molecular mechanics (in our case we could start with the model built refined structure, in which the DHAP parameters, not originally contained in the parameter set, are added from both the *ab initio* results or by inference based on similar known structures), specifying which atom in the following process will be in the QM set and which in the MM one. This step generates the initial gradient, which is the base gradient for the method. Then one begins the quantum mechanical optimization on the QM atoms with the MM atoms kept fixed, adding to the QM gradients the MM gradients on those atoms. These combined gradients are used to optimize the geometry of the QM part. In any case, at the very beginning, after a few cycles it is advisable to allow an adjustment of the MM atoms and to compute appropriate forces on the QM atoms, because in the meantime the QM atoms have changed their position. The geometry optimization of the QM part is carried out until the r.m.s. gradient on these atoms is less than the specified threshold. At this point one turns to optimize the MM atoms, keeping the QM ones fixed, and so on till the required convergence is reached.

Besides the molecular mechanical minimization, additional runs of molecular dynamical calculations can be useful, "shaking" the MM atoms in order to search for other local minima.

Then the total energy of the system must be evaluated:

$$E_T = E_{QM} + E_{MM} + E_{QM,MM} \tag{2}$$

where E_{QM} is the energy of the QM part and E_{MM} the energy of the MM part (from Eqn. (1)). $E_{QM,MM}$ is derived from an expression equal to (1), but which involves atoms from both sets, and where the sum over i runs on the QM atoms and that over j on the MM ones.

As with our earlier model, it is possible to include the MM charges in the one-electron Hamiltonian, thus polarizing the QM atoms. Of course, this is an alternative procedure, requiring a different formula to evaluate the total energy:

$$E'_T = E'_{QM} + E_{MM} + E'_{QM,MM} + E_{pol} \qquad (3)$$

where E'_{QM} is the QM energy, containing the polarization of the QM atoms by the MM charges, E_{MM} is the same as before, $E'_{QM,MM}$ is evaluated as above but with no electrostatic interactions, since these have already been included in the Hamiltonian, and E_{pol} contains the polarization of the MM atoms by the QM ones.

Before studying the enzyme catalysis process using this procedure, we would like to gain some experience employing it to simulate environmental effects in simpler systems. This method, in fact, is well suited to the study of solvation effects, phenomena usually investigated using either Monte Carlo simulations (if we consider the structure of the solution as an important result) or using the continuum model of the solvent (if the stress is more on the effect of the solvent on the electronic distribution of the solute). Since in this framework both kinds of information are obtainable with ease, a direct comparison with the results from the other two methods is possible. We can thus evaluate its performance (and those of the other methods as well) from the point of view of computational efficiency and cost, reliability, accuracy and convergence.

Going back to the purely quantum mechanical methods, a typical application of them to the study of environmental effects is discussed in the following section.

A SIMPLE MODEL OF THE ENVIRONMENTAL EFFECT

Exploiting the finding that the role of the enzyme is to stabilize species III with respect to species I, we designed a very simple model of the enzyme, putting a unit positive point charge along one of the carbonylic oxygen lone pairs in such a position as to equalize the energies of I and III.

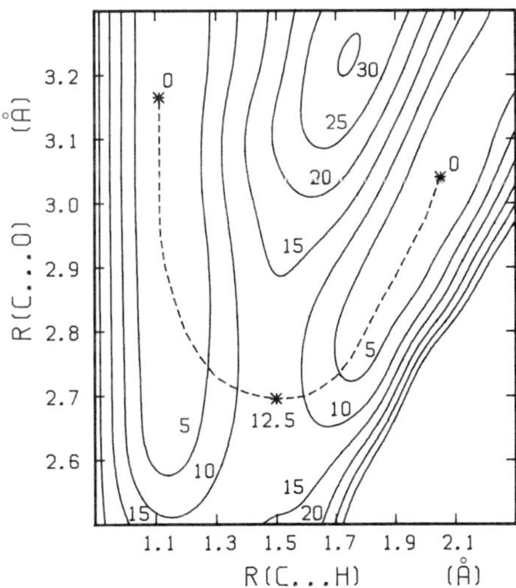

Fig. 7. Potential energy hypersurface at the 4-31G/SCF level in the presence of the polarizing charge (from Ref. 3, with permission of The HUMANA Press).

We then recomputed the whole potential energy hypersurface in the presence of this charge (Fig. 7), obtaining a barrier (Fig. 8) that, in terms of simple Arrhenius theory, is perfectly consistent with the observed catalytic rate. This finding may be fortuitous, but it does further support the idea that this is the rate limiting step.

In addition, because of this good agreement, we did not have to invoke proton tunneling to explain the mechanism, even though we cannot exclude a mechanism of that kind.

There already exists an experimental evidence concerning the probable rate limiting step. As a matter of fact, substituting Glu 165 with Asp, Straus *et al.* [7] obtained a reduction of $\sim 10^3$ in the catalytic rate, such that the intermolecular proton transfer becomes the rate limiting step in the catalysis. Therefore, we used our hypersurface to determine the likely effect of removing a single $-CH_2-$ group from the side chain. Examining the hypersurface it is obvious that the energies of the reactants and "products" are rather insensitive to changes in $R_{C...O}$, so the mutation would have little effect on them. The transition state, on the contrary, is very sensitive to $R_{C...O}$ and a shortening by 0.3 Å is sufficient to account for the 4 kcal/mol corresponding to the observed decrease in the catalytic rate. We were not able to determine whether the protein allows Asp to move to the same position as Glu by straining itself or if it keeps Asp at least 0.3 Å from that position. In any case, both of these explanations are consistent with the apparently innocuous change taking place, which has such a dramatic influence on the catalytic rate.

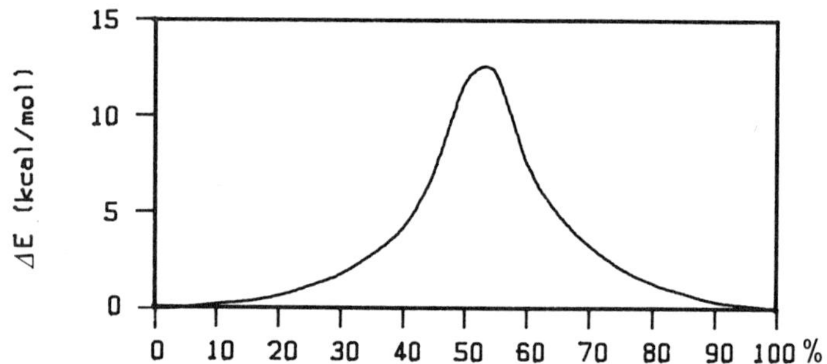

Fig. 8. Energy profile along the reaction coordinate of Fig. 7.

ACKNOWLEDGMENT

We are grateful to Prof. P. A. Kollman for having directed our attention to enzyme catalysis and for granting his permission to use joint and private material.

REFERENCES

1. G. Alagona, P. Desmeules, C. Ghio, and P. A. Kollman, Quantum Mechanical and Molecular Mechanical Studies on a Model for the Dihydroxyacetone Phosphate - Glyceraldehyde Phosphate Isomerization Catalized by Triosephosphate Isomerase (TIM), *J. Am. Chem. Soc.* 106: 3623 (1984).

2. G. Alagona, C. Ghio, and P. A. Kollman, Simple Model for the Effect of Glu 165 -> Asp165 Mutation on the Rate of Catalysis in Triosephosphate Isomerase, *J. Mol. Biol.* 191: 23 (1986).
3. G. Alagona, C. Ghio, and P. A. Kollman, Computational Approaches to the Study of Protein-Ligand Interactions, *in*: "Macromolecular Biorecognition", I. Chaiken, E. Chiancone, A. Fontana, and P. Neri, eds., p. 13, The Humana Press, Clifton (1987).
4. U. Chandra Singh and P. A. Kollman, A Combined *Ab Initio* Quantum Mechanical and Molecular Mechanical Method for Carrying out Simulations on Complex Molecular Systems: Applications to the $CH_3Cl + Cl^-$ Exchange Reaction and Gas Phase Protonation of Polyethers, *J. Comp. Chem.* 7: 718 (1986).
5. J. A. McCammon and S. C. Harvey, "Dynamics of Proteins and Nucleic Acids", Cambridge University Press, Cambridge (1987).
6. T. Alber and G. A. Petsko, private communication.
7. D. Straus, R. Raines, E. Kawashima, J. R. Knowles, and W. Gilbert, Active Site of Triosephosphate Isomerase: in Vitro Mutagenesis and Characterization of an Altered Enzyme, *Proc. Natl. Acad. Sci. U.S.A.* 82: 2272 (1985).

HIGHLY EXCITED VIBRATIONAL STATES AND CHEMICAL REACTIVITY

Alessandro Lami

Istituto di Chimica Quantistica ed Energetica Molecolare
Via Risorgimento 35, 56100 Pisa (Italy)

INTRODUCTION

The movement of nuclei on the potential energy surface created by electrons has a fundamental role to play in the "becoming" of matter in the animate and inanimate world. In fact, every chemical reaction happens just because, at a given instant and due to particular circumstances, a certain nucleus (or group of nuclei) may receive sufficient energy to escape from the potential well in which it exists, vibrating around an equilibrium position. It may then leave the rest of the molecule or begin to oscillate around a new equilibrium position.

The study of the way nuclei move is thus a central topic for understanding chemical reactivity. However, from a historical point of view the main impulse to study nuclear motion was given by the development of infrared spectroscopy, by which scientists were able to detect small amplitude vibrations around the molecular equilibrium geometry. Unfortunately the theoretical methods suitable for studying small amplitude oscillations are quite different from the ones needed to tackle the large amplitude motions characteristic of highly excited vibrations. The small amplitude problem is amenable to a very elegant and well established treatment in terms of harmonic normal modes (1) whereas the large amplitude problem is harder to solve, even if the general equations of motion can be easily expressed mathematically. To this intrinsic difficulty one must add the fact that, whereas for the normal mode problem there has been (and still is) a continuous interplay between experimental data from infrared spectroscopy and theoretical models for describing the force field in which nuclei move, nothing comparable has existed for many years for highly excited vibrational states due to the lack of spectroscopic tools for exciting and detecting such states. Of course, one could test a theory against data from thermal reactions rates, but the dynamical content of the latter is quite low, due to the averaging required to take into account the thermodynamic equilibrium.

In the last ten years the rapid development of powerful and tunable laser sources, as well as of techniques for studying molecules under collisionless conditions, where the dynamical features are not obscured by interactions with the surroundings (molecular beam, low-pressure cells, etc.), has made possible the direct observation of highly excited vibrational states.

At the same time theoretical and numerical studies using coupled oscillator systems have shed new light on the problem of motion in the high energy region (2) and *ab-initio* quantum chemical calculations have reached a stage in which they can be used to obtain reliable potential energy surfaces for a wide range of nuclear configurations, at least when one is dealing with small or medium-size molecules (3).

Various concepts which had been developed to explain the behaviour of nuclear motions at high energy rapidly spread to the contiguous field of the theoretical interpretation of chemical reactions and thus are now relevant not only for physical chemistry, but also for organic and biochemistry. In this paper I will try to present, in a non-specialist way, some of the basic ideas regarding highly excited vibrational states which underlie the most recent dynamical interpretations of chemical reactions.

This introduction will be followed by a brief description of a few selected experiments which have made important contributions to this field; this will be followed by a section devoted to their interpretation, based on classical as well as quantum dynamics. A short conclusion will summarize the main ideas presented and enumerate some of the problems to be solved in the future.

HOW TO EXCITE AND DETECT HIGH-ENERGY VIBRATIONAL STATES

Before illustrating some of the experiments which in recent years have helped to modify our way of looking at high-energy vibrational states, let me briefly recall what the previous ideas were. The normal mode picture, with anharmonic corrections for the first few overtones, furnished a satisfactory framework for interpreting the spectroscopic data available at the time (4). The energy regions where chemical reactions may take place could not be explored spectroscopically. Scientists agreed that at such energies, any independent-oscillator model would break down due to the very rapid energy exchange involved and the randomization taking place among the huge number of nearly degenerate coupled states (corresponding to the independent-oscillator model).

In this context various groups began to use highly sophisticated experimental setups to directly excite and study high vibrational states; they found that, despite the high density of states, some of them were excited preferentially. This indicated that the idea that above a certain energy threshold one should find only a very dense manifold of levels in which any independent-oscillator characteristic is lost, is not quite correct. At least some of the states should conserve their identity for a physically significant time interval (5).

The technique used successfully by Zare and co-workers (6) to populate highly excited vibrational states is based on the internal conversion by which a molecule undergoes a redistribution transition from an electronically excited state, say S_1, to the degenerate vibrational manifold of S_0 (Fig. 1).

The highly excited states populated in this way are then probed by absorption spectroscopy. In their study of the $CH_3-CO-CO-CH_3$ molecule, the above authors found sharp absorption spectra for a vibrational excitation energy of $\sim 22,000$ cm^{-1} in a region where the density of states can be estimated to be $\sim 10^8$ states/cm^{-1}.

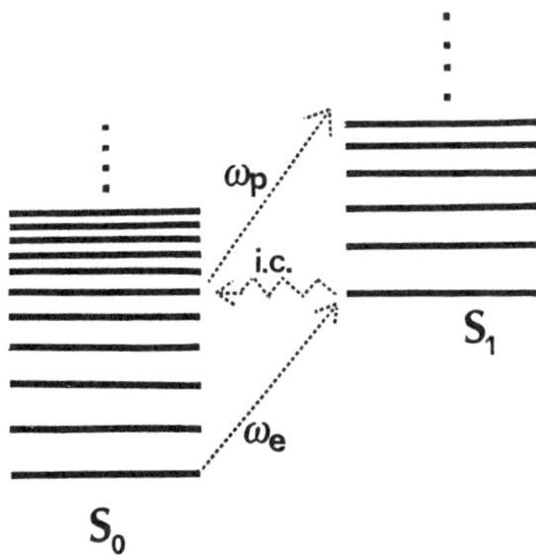

Fig. 1   The excited electronic state S_1 is populated by absorption of ω_e. The probe laser ω_p is used to detect high vibrational states of S_0 reached by internal conversion.

Another technique utilizes spontaneous fluorescence from an electronically excited state to populate selected vibrational states of the electronic ground state. Since the resulting population is very small, one needs to use a very sensitive detection technique, such as multiphoton ionization, by which the various excited vibrational and rotational states can be seen as peaks in the photoelectron spectrum (Fig. 2). In this way Smalley and co-workers (7) studied alkylated benzenes and found that the vibrational relaxation rate increases rapidly with the length of the chain.

Another variation on the same theme is the technique used by Field, Kinsey and co-workers (8). In their experiments the intially excited state of the S_1 manifold is forced to emit a photon by a tunable laser source. This stimulated emission determines a loss of population from the initial state, whose extent reaches its maximum when the laser is tuned to the transition frequency between the initial state and certain vibrational states of S_1. The effect is detected by looking at the spontaneous fluorescence intensity as a function of the stimulating frequency.

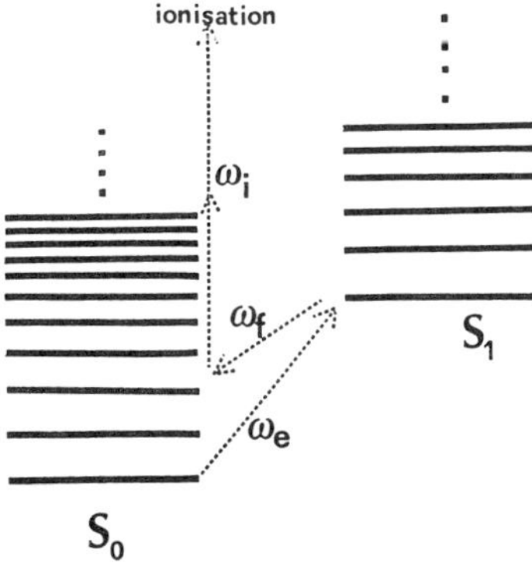

Fig. 2   After the excitation of S_1 by ω_e, spontaneous fluorescence populates the vibrational states of S_0, which are detected by multiphoton ionisation induced by the laser ω_i.

The direct absorption of many infrared photons is a low probability process, detectable only by using very powerful infrared laser sources such as the CO_2 laser. This has been used extensively in recent years to study the multiphoton excitation and dissociation of molecules (9). The information obtained in this way constitutes very stimulating material for the study of highly excited vibrational states.

The most direct way of preparing selected vibrational states is by high overtone excitation, which is however essentially limited to X-H stretchings (X = C, N, O, ...). In a classic work on benzene, Reddy, Heller and Berry (10) were able to obtain the excitation spectrum of C-H stretching overtones, up to v = 9. As explained in greater detail in (1), the position of the overtone peaks can be fitted by considering each C-H as an isolated Morse oscillator illustrating the high-energy breakdown of the normal mode picture.

The energy one can put into the X-H stretching through overtone excitation is in many cases sufficient to permit a chemical reaction; in fact, overtone excitation has been used (11) to verify the possibility of promoting chemical reactions at a rate different from that predicted by statistical theories, such as the transition state theory. The fundamental question is, again, whether the selected excited states we are able to prepare among the millions of others present, have a sufficient lifetime to play a role in chemical reactions, or are they just a kind of spectroscopic curiosity, being rapidly destroyed and absorbed in the sea of indistinct levels mentioned earlier. The actual experimental data tends to support the more pessimistic point of view, but some new ideas concerning the best excitation conditions for promoting selectivity certainly merit experimental investigation (12).

THE THEORETICAL APPROACH

Classical Mechanics

The first step in the treatment of the nuclear motion problem consists in abandoning the 3N orthogonal cartesian coordinates (N being the number of atoms in the molecule) in favour of a new system of coordinates in which the Hamiltonian (i.e. the energy as a function of the coordinates and conjugate momenta) becomes separable into translational, vibrational and rotational parts (I don't consider here the vibration-rotation interaction, which is usually negligible for large molecules). I will now focus on the vibrational Hamiltonian as a function of 3N-6 internal coordinates, which permit one to define the relative position of nuclei. The absolute position needs 6 other coordinates; these may include, for example, the 3 cartesian coordinates of the center of mass in a space-fixed frame and 3 Euler angles to specify the orientation of a second cartesian frame rotating with the molecule, in relation to the space-fixed one (13).

Before writing a general expression for the vibrational Hamiltonian in terms of the 3N-6 internal coordinates, let us consider some general features of the above transformation from cartesian to internal coordinates. The first interesting thing to notice is that when one combines the cartesian coordinates of different atoms one must also combine in the same way their masses. This means that a particle moving along the new coordinate has a mass which depends on the masses of all the atoms that are undergoing displacement. For example, when we combine the cartesian coordinates of two atoms to obtain the bond length stretching in a diatomic molecule, the mass to be associated with the new coordinate is the reduced mass: $\mu = m_1 m_2/(m_1 + m_2)$. When the internal coordinate can be obtained simply by a linear combination of the cartesian coordinates (as for stretching) there is nothing more to add. If, however, one wants to build up a coordinate describing, for example, the variation of a $C \diagdown C\text{-}H$ angle (bending) for arbitrary amplitudes, then a more general non-linear transformation is needed. As a consequence one finds that the new masses become variable with the position! To be more precise let us write the vibrational Hamiltonian for a diatomic molecule (where q is the stretching and p its conjugate momentum):

$$H = \frac{1}{2\mu}\ p^2 + V(q)$$

(1)

and for a general polyatomic molecule:

$$H = \frac{1}{2}\sum_{i=1}^{N-6} G_{ii}(q_1, \ldots q_{3N-6})\,P_i^2 + \sum_{i>j}^{N-6} G_{ij}(q_1, \ldots q_{3N-6})\,P_i P_j + V(q_1, \ldots q_{3N-6})$$

(2)

Comparing equations (1) and (2) one sees that the diagonal term G_{ii} represents the inverse of the masses and, as anticipated, depends on the internal coordinates. The extradiagonal term G_{ij} also depends on the q's and couple different momenta.

The standard normal mode treatment refers to small amplitude vibrations and is applied in the following way (1, 13):

i)   expand V in a Taylor series around the equilibrium position and take only the second order terms;

ii)   take G as fixed by assuming that all the q's are at their equilibrium values:

iii)   find a linear transformation for the coordinates (and momenta) in such a way that the Hamiltonian (2) becomes the sum of 3N-6 harmonic oscillator Hamiltonians;

iv)   solve the trivial 3N-6 separated Hamiltonian equation of motion. The new coordinates $Q_1, \ldots Q_{3N-6}$ are the normal coordinates or normal modes and involve the simultaneous harmonic motion of many nuclei.

The classical motion can be examined by looking at the behaviour of the various P's and Q's as a function of time, i.e. by looking at their trajectories in phase space. For a diatomic molecule (i.e., a single harmonic oscillator) one has periodic motion and the trajectory for a given energy is an ellipse, as shown in Figure 3.

For a collection of 3N-6 harmonic oscillators the trajectories do not describe general closed curves in a phase space of dimension 2 x 3N-6. This can be easily understood by considering two harmonic oscillators with frequencies v_1 and v_2, i.e. periods $T_1 = 1/v_1$ and $T_2 = 1/v_1$. In order to have a periodic trajectory in the four-dimensional phase space P_1, Q_1, P_2, Q_2, one should find a time t such that:

$$t = m_1 T_1 = m_1 T_1 \qquad\qquad (m_1 \text{ and } m_2 \text{ being integers})$$

(3)

(i.e., t should be simultaneously an integer multiple of both periods).

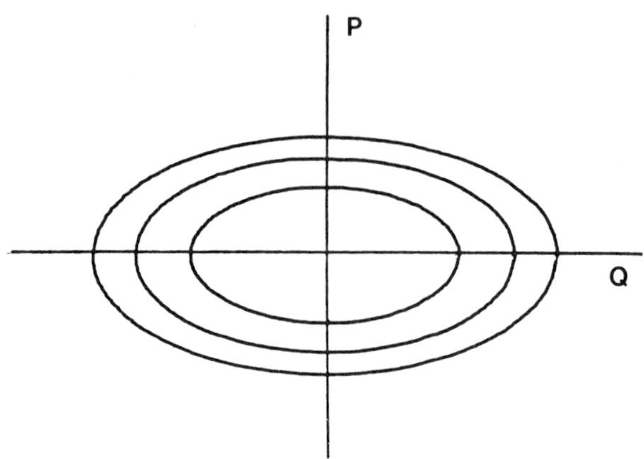

Fig. 3   Classical phase-space trajectories for a diatomic molecule in the
harmonic approximation. The three curves refer to energies in
the ratio 1:2:3.

Equation (3) leads to:

$$\frac{m_1}{m_2} = \frac{T_2}{T_1} = \frac{v_1}{v_2} \tag{4}$$

which is only possible in the case in which the ratio of the two frequencies
is a rational number (a zero-probability event, according to mathematics).
Thus the nuclear motion of a polyatomic molecule in the low energy region
is not exactly periodic, but rather quasi-periodic, since one can demonstrate
that its trajectories are nearly closed curves.

The above mentioned quasi-periodicity can be traced back to the fact
that the motion does not utilize all the phase space, but rather a subspace of
dimension 3N-6 (i.e., only one half). In fact, since we were able to show that
in the low-energy region (i.e. for small amplitudes) the molecule behaves
like a collection of independent harmonic oscillators, then the energy of
each one of them must be conserved over time. This immediately puts 3N-6
constraints on the motion and confines it to a 3N-6 subspace. For example,
in the case of a single harmonic oscillator the trajectories do not fill the bi-
dimensional P, Q space, but cover only the ellipse (Fig. 3) corresponding to:

$$E = \frac{1}{2\mu} P^2 + \frac{1}{2} k Q^2 = a\ constant \tag{5}$$

Hence, the main characteristic of nuclear motion in the low-energy
region is, according to classical mechanics, the existence of regular, quasi-
periodic orbits that cover just a small part of the phase space (2).

What happens as the energy increases?   First of all, small amplitude
approximations must be avoided, which means that:

i)   higher order terms in the expansion of V must be considered;
ii)   the dependence of the matrix G on the coordinates must be taken
      into account.

As far as normal modes are concerned, points (i) and (ii) mean that they become coupled by the potential energy terms, as well as by terms involving simultaneously the coordinates and the momenta and coming from the Taylor's series development of G. From a physical point of view this translates into the fact that normal modes tend to exchange energy at a rate which rapidly increases with increasing energy.

It can be shown that for a certain range of energies the independent-oscillator model can be maintained, abandoning the normal mode model and using more sophisticated mathematical techniques to find the appropriate coordinates which minimize the inter-mode coupling (2). As the energy level increases, one observes that certain trajectories remain regular and quasi-periodic, while others become erratic and begin to spread more and more in the phase-space. The fraction of regular orbits decreases progressively until, after an energy threshold is reached, essentially all the trajectories become chaotic; this means in effect the complete destruction of all the rules of conservation (except that of total energy conservation!) or, in other words, the loss of any independent-oscillator characteristic.

<u>Quantum Mechanics</u>

The above considerations were based on a classical mechanics analysis, while in principle quantum mechanics should be used. The correspondence principle states that in the high energy region (when the actions involved are large with respect to Planck's constant, h) the two approaches should eventually agree. In practice we are interested in an energy region in which the results of the two approaches may differ. In quantum mechanics one must start with the quantum translation of the Hamiltonian in equation (2), which can become quite complicated (14). However, an accurate analysis of the terms involved shows that by taking only the most relevant ones, one can write (15):

$$H = -\frac{\hbar^2}{2} \sum_{i=1}^{3N-6} G_{ii}(q_1, \ldots q_{3N-6}) \frac{\partial^2}{\partial q_i^2} - \hbar^2 \sum_{i>j}^{3N-6} G_{ij}(q_1, \ldots q_{3N-6}) \frac{\partial^2}{\partial q_i \partial q_j} + V(q_1, \ldots q_{3N-6}) \tag{6}$$

The quantum analysis proceeds in a manner which parallels the classical one; first of all one searches for coordinates that allow a zero-order description in terms of separate oscillators, plus a perturbation coupling different oscillators:

$$H = H_0 + H_1 \tag{7}$$

Then, with the help of the time-dependent Schroedinger equation, one may study how a given state, corresponding to the excitation deposed in a given mode (which can then be described as an eigenstate of H_0, eq. 6), evolves in time and what is the time interval, if any, after which the population becomes essentially equidistributed among all the quasi-degenerate eigenstates of H_0. This is the quantum analog of the classical spreading of trajectories in the whole phase space and is an important test for the applicability of statistical theories (such as the transition state theories) to chemical reactions.

By the above kind of analysis one may recognize the main pathways for vibrational energy redistribution after the localized excitation. For example, the decay of a C-H overtone is thought to begin with the transformation of one quantum of stretching into two quanta with the C-C-H bending (16), whereas the vibrations of the carbon skeleton seem to play only an indirect role, being quite inefficient in carrying out the initial excitation, at least for saturated hydrocarbon chains (17). The main mechanism for the above 1:2 Fermi resonance between stretching and bending seems to be the mixed coordinate-momenta coupling mentioned earlier.

CONCLUSIONS

We are beginning to understand some of the peculiar features of highly excited vibrational states. Present knowledge is the result of an interplay between the output of very careful experiments conceived in order to explore the high-energy region of the electronic potential energy surface on which nuclei move, and theoretical and computational studies of coupled oscillator systems mimicking molecules. The current view is that at those energies at which reactive channels begin to be open, there are a few regular states corresponding to vibrational excitations localized in particular modes (usually local modes) coexisting with a sea of states in which the independent-oscillator model is completely meaningless (i.e., energy is exchanged very rapidly among the modes).

This brings up some interesting theoretical and practical questions. Does the existence of these regular states demand, at least in some cases, a completely new approach to chemical reactions? Would it be possible to utilize local mode excitations to promote certain reactions which, according to statistical theories, should have only a negligible yield? Do particular local modes, or other long-lived excitations, play a role in enzymatic reactions? Only research in the coming years will give a definite answer to these very intriguing questions.

ACKNOWLEDGEMENTS

I would like to express my thanks to Ms. Lisa Chien for having proofread and typed the text for publication in this book.

REFERENCES

1. See the chapters by F. Fillaux in this book, and the references therein.
2. See e.g., S. A. Rice, Dynamics of intramolecular transfer of vibrational energy, in "Advances in Chemical Physics", J. Jorner, R. D. Levine and S. A. Rice, eds., Vol. XLVII, 117 (1981) .
3. See e.g., D. G. Truhlar, R. Steckler and M. S. Gordon, Potential energy surfaces for polyatomic reaction dynamics, in "Chemical Reviews," 87:217 (1987).
4. G. Herzberg, "Infrared and Raman Spectra," Van Nostrand Reihnold Company, NY (1945); P. Gans, "Vibrating Molecules. An Introduction to the Interpretation of Infrared and Raman Spectra," Chapman and Hall, London (1971).
5. For a review on vibrational relaxation, see: V. E. Bondybey, Relaxation and vibrational energy redistribution processes in polyatomic molecules, Ann. Rev. Phys. Chem. 35:591 (1984).
6. R. Naaman, D. M. Lubman and R. N. Zare, Vibrational energy redistribution in glyoxal following internal conversion, J. Chem. Phys. 71:4192 (1979).

7. J. B. Hopkins, P. R. R. Langridge-Smith and R. E. Smalley, Ground state vibrational randomization in alkylbenzenes, <u>J. Chem. Phys.</u> 78:3410 (1983).

8. P. H. Vaccaro, J. L. Kinsey, R. W. Field and H. L. Dai, Electric dipole moments of excited vibrational levels in the $\tilde{X}^1A_1$ state of formaldehyde by stimulated emission spectroscopy, <u>J. Chem. Phys.</u> 78:3659 (1983); C. E. Hamilton, J. L. Kinsey and R. W. Field, Stimulated emission pumping: new methods in spectroscopy and molecular dynamics, <u>Ann. Rev. Phys. Chem.</u> 37:493 (1986).

9. T. B. Simpson, J. G. Black, I. Burak, E. Yablonovitch and N. Bloembergen, Infrared multiphoton excitation of polyatomic molecules, <u>J. Chem. Phys.</u> 83:628 (1985).

10. K. V. Reddy, D. F. Heller and M. J. Berry, Highly vibrational excited benzene: overtone spectroscopy and intramolecular dynamics of C_6H_6, C_6D_6 and partially deuterated or substituted benzenes, <u>J. Chem. Phys.</u> 76:2814 (1982).

11. For a review, see F. F. Crimm, Selective excitation studies of unimolecular reaction dynamics, <u>Ann. Rev. Phys. Chem.</u> 35:657 (1984).

12. A. Lami, Enhancement of reactivity through near-resonant radiative excitation of an overtone: a model study of an isomerization by hydrogen tunneling, <u>Chem. Phys.</u> 115:399 (1987); A. Lami and G. Villani, Control of the yield of competing unimolecular reactions through double-resonance coherent trapping, <u>J. Phys. Chem.</u> 92:4348 (1988).

13. E. B. Wilson, Jr., J. C. Decius and P. R. Cross, "Molecular Vibrations. The Theory of Infrared and Raman Vibrational Spectra," McGraw-Hill Book Company, London (1955).

14. See e.g., E. C. Kemble, "The Fundamental Principles of Quantum Mechanics," p. 237, Dover, NY (1958).

15. E. L. Sibert III, J. T. Hynes and W. P. Reinhardt, Fermi resonance from a curvilinear perspective, <u>J. Phys. Chem.</u> 87:2032 (1983).

16. E. L. Sibert III, W. P. Reinhardt and J. T. Hynes, Intramolecular vibrational relaxation and spectra of CH and CD overtones in benzene and perdeuterobenzenes, <u>J. Chem. Phys.</u> 81:1115 (1982); Photoacoustic spectroscopy of vibrational overtones in gas-phase CH_3OH and CH_3OD, H. L. Fong, D. M. Meister and H. L. Swofford, <u>J. Phys. Chem.</u> 88: 405 (1988).

17. A. Lami and G. Villani, Excitation and decay of a C-H overtone coupled to a linear hydrocarbon chain: a simple quantum-mechanical model, <u>J. Chem. Phys.</u> 88:8 (1988) and unpublished results.

(D) <u>Practical and Industrial Applications</u>

"As gold which he cannot spend will make no man rich, so knowledge which he cannot apply will make no man wise."

Samuel Johnson

"If we possessed a thorough knowledge of all the parts of the seed of any animal (e.g. man), we could from that alone, by reasons entirely mathematical and certain, deduce the whole conformation and figure of each of its members..."

Rene Descartes

MICROCALORIMETRY OF PROTEIN-LIGAND INTERACTIONS

Alan Cooper

Department of Chemistry
Glasgow University
Glasgow G12 8QQ
Scotland, U.K.

INTRODUCTION

The energetics of enzymic processes and the interactions
responsible for protein behaviour have been central to our discussions
so far. Several indirect methods are available for probing the
thermodynamics of protein interactions, but the most unambiguous and
direct experimental approaches are based on calorimetry. Most reactions
have an associated heat effect and, apart from its intrinsic interest in
terms of energetics, this heat can serve as a useful general probe of
biomolecular processes. My purpose here is to describe the basis and
applications of isothermal microcalorimetry to the study of interactions
at protein binding sites, illustrating the range of information that may
be obtained from such studies. But before doing this, we should perhaps
consider the nature of the problem and why an empirical, rather than
theoretical approach is necessary.

A crucial first step in any catalytic event is the binding of
substrates to the catalyst in such a manner as to facilitate formation
of the appropriate transition state for the reaction. In the case of
enzymes, much of their power as catalysts - both in terms of specificity
and rate enhancement - arises from the precise stereochemistry of their
active sites which have been designed under evolutionary pressures to
allow formation of stable complexes with specific substrate molecules
with the correct disposition of functional groups. To understand and,
eventually, manipulate this process to our own ends, we must explore the
details of the non-covalent forces responsible for both the folding of
the polypeptide chain to create the active site conformation and the
subsequent binding of ligands to these sites. This is particularly
important now because of the intense efforts directed towards molecular
recognition, drug design, and related problems using computer-based
molecular mechanics and graphics techniques. With few exceptions, such
approaches currently use empirical or semi-empirical intermolecular
potential functions derived from gas phase studies or crystallographic
data bases of small molecules, thereby neglecting the crucial role of
solvent interactions that dominate these processes under physiological
conditions. But the seminal work of Kauzmann(1959) showed how the unique
properties of water can not only give rise to subtle, non-pairwise

bonding in the hydrophobic interaction, but also how these can modify the thermodynamic properties of even the apparently simple electrostatic and hydrogen bond interactions. For example, given the amount of emphasis placed on the hydrogen bond in descriptions of protein conformations and interactions, one might be forgiven for assuming that H-bonds might be responsible for much of the stability of such structures. But this is at odds with simple physical chemistry. Consider, for example, the protein folding process. As we know, given the correct amino acid sequence, a polypeptide chain will spontaneously fold into a more compact, unique conformation. This implies that the folded structure has a lower free energy than the more open, solvated random coil. To be sure, intermolecular hydrogen bonds are formed in the process, but this is at the expense of breaking hydrogen bonds with solvent water. There is unlikely to be any nett change in the number of H-bonds in the process, and any stabilising effect must arise because peptide-peptide bonds, for instance, are stronger than peptide-water bonds. But, as far as we know, this is not the case, at least with small molecules. N-methyl-acetamide, $CH_3CONHCH_3$, for example might be taken as a good small molecule analogue of the peptide group. Yet this molecule is extremely soluble in water, and shows no tendency to aggregate in solution. It is generally true that for small polar, hydrogen bonding molecules or functional groups the aqueous phase is normally the thermodynamically preferred state, contrary to what we observe in both the folded protein and in its complexes with other molecules.

This is the core of the problem. We really have no sound understanding of the thermodynamic forces active in biological macromolecules, and we must tackle the problem experimentally. One problem with studying protein folding is that the multitude of intermolecular interactions makes it difficult to estimate individual contributions. Things are a little easier with ligand binding on which we shall concentrate, both for its intrinsic importance and for the light it might eventually shed on protein interactions in general.

BASIC THERMODYNAMICS - a reminder

Any enzyme-substrate or ligand binding process is, at least in the first instance, a dynamic equilibrium process:

$$E + L \rightleftharpoons EL$$

which may be described by an equilibrium dissociation constant:

$$K_{Diss} = [E][L]/[EL]$$

with a standard Gibbs free energy (ΔG^0) given by:

$$\Delta G^0 = RT.\ln K_{Diss} = \Delta H^0 - T.\Delta S^0$$

where ΔH^0 and ΔS^0 are, respectively, the standard enthalpy and entropy changes in the entire system during the binding process. Both of these parameters are, in principle, dependent on temperature in the following manner, because of heat capacity effects:

$$\Delta H^0(T_2) - \Delta H^0(T_1) = \int_{T_1}^{T_2} \Delta C_P^0 . dT$$

$$\Delta S^0(T_2) - \Delta S^0(T_1) = \int_{T_1}^{T_2} \frac{\Delta C_P^0}{T} . dT$$

where: $\quad \Delta C_P^0 = \dfrac{d\Delta H^0}{dT}$

A useful rule that follows from the above definition is that K_{Diss} can be visualised as the concentration of free ligand, [L], at which the enzyme binding sites are just 50% occupied. Another way of writing the thermodynamic expressions emphasizes the statistical nature of the process. Thus:

$$K_{Diss} = \frac{w_{Free}}{w_{Bound}} . exp(\Delta H^0/RT)$$

with: $\quad \Delta S^0 = -R.ln(w_{Free}/w_{Bound})$

and where w_{Free}/w_{Bound} represents the ratio of the number of ways that the enzyme:ligand system (including solvent, etc.) may exist in either the free or bound states, respectively. In these terms, K_{Diss} can be seen to describe the statistical probability of a ligand molecule being dissociated from the enzyme at any one time, weighted by a classical Boltzmann factor in the enthalpy. Thus binding affinity is controlled not only by the balance of intermolecular forces, as measured by the enthalpy or heat of reaction, but also by the relative statistical factors (i.e. the entropy changes) which are much harder to visualise.

There are many ways in which K_{Diss} and, therefore, the free energy of ligand binding may be measured. These include direct binding methods such as equilibrium dialysis or filter binding assays, as well as indirect methods based on spectroscopic or other changes in the protein or ligand proportional to the extent of complex formation. And, to the extent that the Michaelis-Menten approximation is valid, the K_M determined from kinetic studies can be a good measure of the substrate dissociation constant. But information about the separate entropic and enthalpic contributions requires detailed temperature variation studies coupled with the Van't Hoff relation for the enthalpy per binding site:

$$\Delta H^0 = RT^2 . \frac{d \, ln \, K_{Diss}}{dT}$$

Unfortunately the range of temperatures covered and the precision of the available techniques are rarely adequate to yield accurate values, and direct determination by calorimetry is much more satisfactory.

Here we use the fact that when enzyme and ligand (say) are mixed, the observed heat effect (Q) is proportional to the amount of complex formed, i.e. $Q = -\Delta H^0 [EL]/C_E$ per mole of enzyme, where C_E is the total

enzyme concentration. In the general case of a protein with n identical
binding sites, use of the equilibrium equation leads to a hyperbolic
expression for the observed heat as a function of total enzyme (C_E) and
ligand (C_L) concentration (Bjurulf & Wadsö,1972; Cooper & Jenkins,1973;
Cooper,1974):

$$Q = -\Delta H^0(nC_E+C_L+K_{Diss})\{1 - [1-4nC_EC_L/(nC_E+C_L+K_{Diss})^2]^{1/2}\}/2C_E$$

An alternative and simpler double-reciprocal expression may be used in
situations where binding is so weak that the free ligand concentration
approximates the total:

$$-n\Delta H^0/Q = 1 + K_{Diss}/[L]$$

so that a plot of 1/Q versus 1/[L] is linear, with the slope and
intercept giving K_{Diss} and the enthalpy, respectively. Examples of data
plotted according to both these expressions are given in Fig.1. In
practise the hyperbolic binding equation, though more complex, is more
satisfactory for data analysis since the double-reciprocal plot puts
heavier weighting on the lower concentration data which are
intrinsically less accurate.

Thus calorimetry can yield not only the heat of reaction but also
the dissociation constant and, hence, the free energy and entropy of
binding as well. Note that in the general case described here, what is

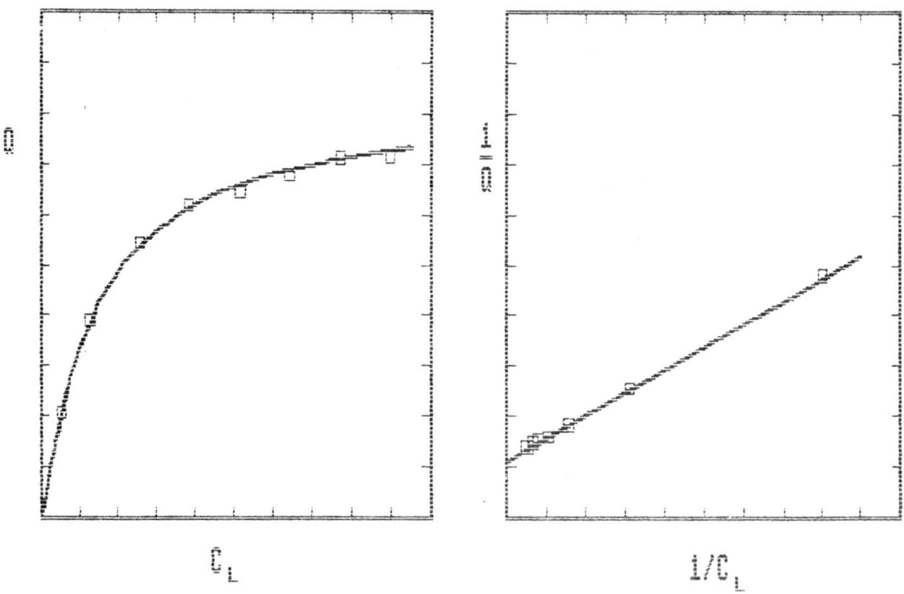

Fig. 1. Examples of calorimetric ligand binding data plotted according
to (A) the hyperbolic thermal titration equation, and (B) the
double-reciprocal expression.

actually obtained is nΔH^0, that is the enthalpy of ligand binding per mole of protein. To get the enthalpy per site we need independent data on the number of binding sites, n. Alternatively, we may use the observation that the Van't Hoff enthalpy determined from the temperature dependence of K_{Diss}, though less accurate, gives the enthalpy per site, ΔH^0. Comparison of this with the calorimetric enthalpy will therefore give additional information about the number of available binding sites on an enzyme (Cooper & Jenkins,1973). An example of this will be mentioned below.

PRACTICAL MICROCALORIMETRY

When a ligand binds to a protein, the enthalpic part of the interaction free energy gives rise to a heat of reaction proportional to the extent of binding. But, although calorimetry is an old technique, it is only relatively recently that it has been possible to apply this to biological systems envisaged here. There are two major practical problems: firstly, we are usually dealing with non-covalent interactions with intrinsically small heat effects and, secondly, biological materials are usually available in such small amounts or are so insoluble that only very small quantities - typically 1 μmole or less - can be used in any one experiment. Consequently, the temperature changes induced in the predominantly aqueous sample (if, as is normal, it is in solution) are very small - maybe just a few millionths of a degree - and cannot be measured accurately, as would be required in classical calorimetry.

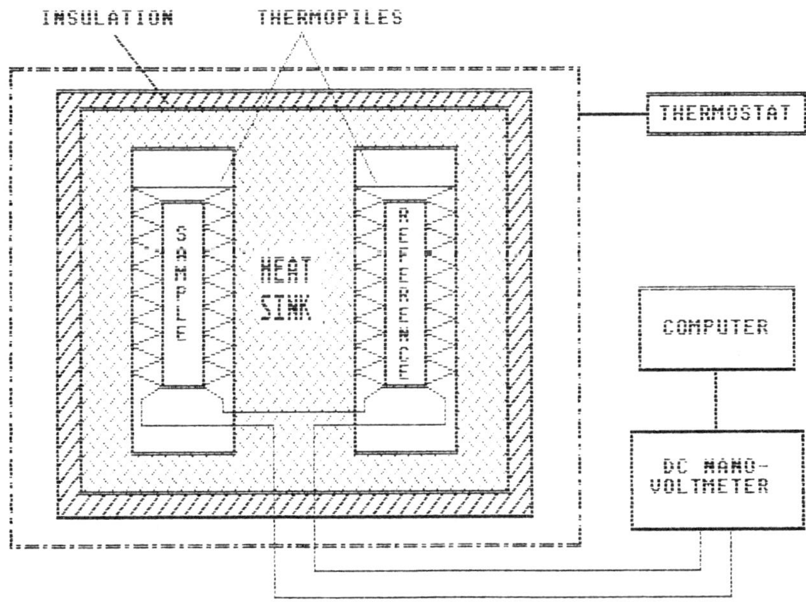

Fig.2. Basic layout of a typical isothermal reaction microcalorimeter. Note that the sample and reference thermopiles are connected back-to-back in series with each other so that differential measurements may be made.

Various techniques and applications of microcalorimetry have been reviewed by various authors (Wadsö,1974,1983; Sturtevant,1974; Langerman & Biltonen,1979; Biltonen & Langerman,1979; for example). The basic layout of a typical isothermal or "heat-leak" microcalorimeter used for most biomolecular work is shown in Figure 2. [These should not be confused with the equally important differential scanning calorimeters for the study of thermally-induced processes such as protein denaturation, lipid phase transitions, etc. Ligand binding effects can also be studied by these techniques in certain circumstances (Hu & Sturtevant,1987), but will not be considered further here.] At the heart of the system is the sample (or reference) contained within a thermally conducting cell and in good thermal contact with a thermopile, the other face of which is in contact with a large heat sink. Any reaction within the sample may give rise to a transient increase (or decrease) in temperature (ΔT), but this immediately begins to dissipate by conduction to the heat sink. For small temperature changes the rate of heat energy dissipation is proportional to the temperature differential across the thermopile:

$$\frac{dQ}{dt} = a.\Delta T \qquad (a = constant)$$

But any temperature gradient across the thermopile generates a voltage (V) which is itself proportional to ΔT. Consequently:

$$\frac{dQ}{dt} = k.V \qquad (k = constant)$$

This voltage will, of course, decay back to baseline as the heat in the sample dissipates, but the total heat generated (or absorbed) during the reaction may be obtained by integration of the resulting voltage/time response, i.e.

$$Q = k.\int V.dt$$

The value of the calibration constant, k, is usually determined electrically or by use of standard reactions.

The thermopiles are the critical components in calorimeters of this type. Early versions of such instruments used ingeniously wire-wound devices of up to 10,000 junctions, but these proved unsatisfactory for high-sensitivity work because their high electrical resistance resulted in too much electrical noise and poor impedance matching to available dc amplifiers. Significant advances in sensitivity were made possible with the introduction of solid-state thermopile devices (actually Peltier cooling devices, or heat pumps, used in reverse) which combine high sensitivity, good thermal conductivity and low internal resistance. Measurements down to less than 0.1 μvolt (corresponding to temperature differentials of less than 10^{-5}C) are now routinely obtainable, and measurements are limited solely by the temperature stability of the thermostat and the intrinsic heat effects associated with sample mixing or initiation of the reaction.

Various types of calorimetric reaction cell are available, and some are illustrated in Figure 3. Most of the work that is of relevance to us here involves mixing of two solutions - enzyme + ligand, say - for

which the batch, flow, or titration cells are mostly used. More
specialised cells have been devised for specific applications, including
the photocalorimeter cell for light-induced reactions to be mentioned
later.

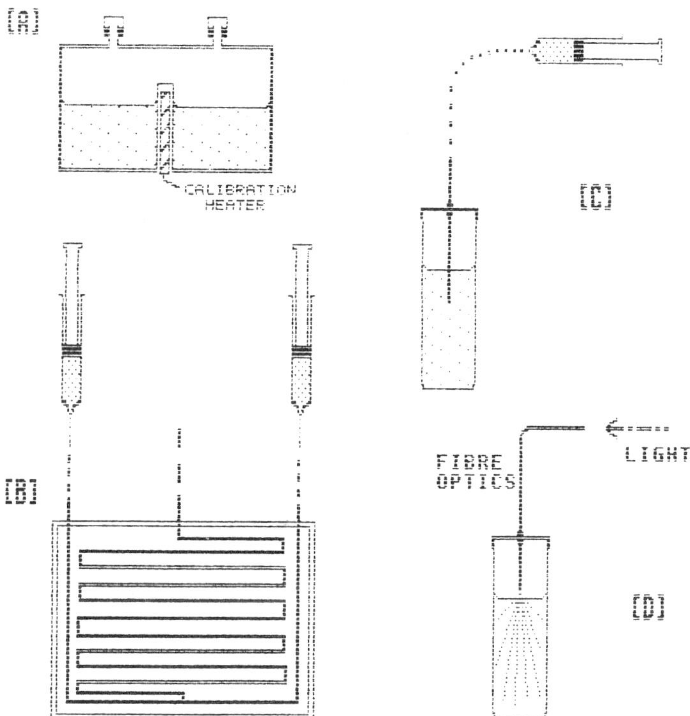

Fig.3. Calorimetric reaction vessels: (A) Batch cell for single mixing
experiments. Reactant solutions are loaded either side of the
dividing wall and, after thermal equilibration, are mixed by
rotation of the entire calorimetric unit. (B) Flow cell in
which continuous or stopped-flow mixing experiments may be
performed by continuous injection of sample. (C) Titration cell
similar to the batch cell but for in situ thermal titrations by
microsyringe injection of reagent. (D) Photocalorimeter cell in
which samples may be irradiated with light via flexible fibre
optics light guides. Note that all vessels will normally
include electrical calibration heaters, shown here only in (A).

APPLICATIONS

There are numerous reports of calorimetric studies of ligand
interactions in a diverse range of proteins from which some very general
trends may be ascertained. Binding of small molecules to protein sites

is usually exothermic (ΔH negative) and, in cases where the temperature dependence has been measured, ΔC_P is found to be usually negative, i.e. the heat of reaction gets more exothermic at higher temperatures. Similarly, the ΔS for ligand binding is usually negative and, because of the ΔC_P effect, decreases with temperature. A general explanation for the ΔC_P behaviour has been given by Sturtevant (1977) in terms of hydrophobic interactions and changes in low frequency vibrational modes in the protein, as discussed earlier in this volume. But unfortunately, despite the diversity of proteins studied so far, we have no clear predictive understanding of the magnitudes of the various interactions involved in the ligand binding process. The problem is that any one protein-ligand complex involves a potentially large number of intermolecular contacts and, apart from the occasional and fortuitous use of chemical modification (Cooper, 1974), no attempt has been made to dissect out the individual contributions to the binding free energy. Nor has it, in most cases, been feasible hitherto. Modern techniques of site-directed mutagenesis are changing this, and our current work involves a systematic study of the effects of individual amino acid mutations on enzyme-substrate binding parameters, but this work is at too early a stage to report here. Rather than attempt to review all the previous studies, I will concentrate here on some examples from our own past work which throw light on some of the problems that are of interest to us here: that is, the role of hydrogen bonds in stabilising protein interactions, the binding of substrates and transition state analogues to enzymes, and the energetics of structural fluctuations within protein molecules. These examples will also serve to illustrate some of the additional information that can be obtained from calorimetric experiments.

(i) Lysozyme-Inhibitor Complexes

Much of the early microcalorimetric work was done on this system (Bjurulf & Wadsö,1972), stimulated by the detailed crystallographic studies of this enzyme (Blake et al.,1967; Beddell et al.,1970). Our own work (Cooper,1974) involved the use of flow microcalorimetry to investigate the binding of the monomeric inhibitor, N-acetyl-glucosamine (NAG). The interest here lies in the fact that NAG in solution mutarotates quite rapidly to give the anomeric mixture of α and β saccharides, which differ only in the orientation of the 1-hydroxyl group and which both bind competitively to the enzyme, and we were able to study the binding of each separately. (Experiments were done at 5°C to slow down mutarotation during the course of the measurements.) Crystallographic and nmr experiments (Blake et al.,1967; Dahlquist & Raftery,1968) indicated that the two anomers bound in two slightly different orientations in the lysozyme active site cleft, with β-NAG residing in essentially the same way as in the oligosaccharide substrate and making important contacts with tryptophan-62, but with α-NAG oriented somewhat lower in the cleft, avoiding interaction with Trp-62 and making one less hydrogen bond to the protein. Paradoxically, in view of this, α-NAG appears to bind slightly more strongly than the β-anomer, with a two-fold difference in K_{Diss}, though with a somewhat lower enthalpy.

The situation revealed by calorimetry of binding of the two NAG anomers is slightly more complex, but we were fortunate here in that it was possible to chemically modify Trp-62 in such a way as to allow rational interpretation of the data. Careful treatment of lysozyme with N-bromosuccinimide under appropriate conditions results in specific oxidation of Trp-62 to the oxindole and dramatically reduces binding of β-NAG, as would be expected from the structural data. More surprisingly, this modification also resulted in a two-fold increase in K_{Diss} for binding of α-NAG. Our interpretation of this, supported by careful re-examination of the crystallographic data, is that α-NAG actually binds to the unmodified enzyme in two, mutually exclusive orientations: one identical to the β-anomer binding orientation (and therefore inhibited by Trp oxidation), and the second slightly lower in the cleft as described above. Consequently the α-NAG/lysozyme complex is rather more dynamic than originally visualised, with the small saccharide ligand rocking back and forth between two equilibrium orientations with approximately equal affinities. Thus the dissociation constants for both β-NAG and α-NAG in either of its two orientations are all approximately equal, and the apparent two-fold decrease in K_{Diss} for the α-anomer is solely a statistical consequence of the multiplicity of available binding orientations.

As a further consequence of these observations we can now extract data for the binding of NAG in two unique orientations making different numbers of hydrogen bonds with the protein:

	H-bonds	K_{Diss} (mM)	ΔG^0 (kJ mol^{-1})	ΔH^0	ΔS^0 (J K^{-1}mol^{-1})
β:	4	22.4	-8.9	-25.3	-59.0
α:	3	26.5	-8.5	-15.6	-25.5
Difference:	1	-	-0.4	-9.7	-33.5

This confirms what we had previously suspected, that the additional protein-ligand hydrogen bond in the case of β-NAG seems to add little to the actual binding free energy, presumably due to the effects of solvation as described above, and despite much larger differences in the enthalpy and entropy which may reflect the differences between protein-ligand, protein-water, ligand-water and water-water H-bonds. Other studies in different systems and using different techniques give similar values, but the free energy may favour protein-ligand H-bonds when charged donors or acceptors are involved (Fersht et al.,1985; Bartlett & Marlowe,1987).

(ii) Triosephosphate Isomerase

In the above example, and in most other studies, we have been limited to studies of small substrate analogues or inhibitors rather than true substrates. This is either because the substrate is too complex or heterogeneous (as above), or more generally because the heat generated by the actual catalysed reaction that must take place in the presence of the full complement of substrates is usually sufficient to mask the effects of enzyme-substrate binding. We were, however, fortunate to find an exception in the case of triosephosphate isomerase (TIM) (Cooper & Jenkins,1973). This key enzyme of the glycolytic pathway is responsible for the rapid interconversion of dihydroxyacetone phosphate (DHAP) and D-glyceraldehyde-3-phosphate (GAP) and consists of a dimer of two apparently identical subunits with a total molecular weight of about 54,000. The overall equilibrium constant for the enzyme-catalysed isomerization reaction was known to be independent of temperature which implies that the enthalpy of the reaction is close to zero. This was confirmed by microcalorimetry ($\Delta H = 0 \pm 0.5$ kJ mol^{-1}). Consequently in this serendipitous situation it is possible to mix enzyme and substrate in the calorimeter and measure the binding process in the absence of any additional heat effects from the reaction itself.

What we measure, of course, is not the binding of any individual substrate but rather the average of the rapidly-established DHAP/GAP equilibrium mixture, but the differences are small in this case. An additional feature of interest with this enzyme is the availability of a transition state analogue, 3-phosphoglycollate (PGA), which is a powerful competitive inhibitor of TIM (Wolfenden,1970).

Binding parameters for both substrates and PGA have been determined by batch microcalorimetry and by kinetic techniques, with intriguing results. In the case of substrate (GAP) the K_{Diss} from calorimetry (0.14mM) is identical to the K_M from kinetics (0.15mM). The calorimetric enthalpy of binding per mole of enzyme is about -63 kJ mol^{-1}, compared to a Van't Hoff enthalpy from kinetics of around -32 kJ mol^{-1}. As noted earlier, the ratio between these two should be the number of binding sites per protein molecule, i.e. n=2 in this case as expected for the dimeric enzyme. This contrasts with the results found for the enthalpy of PGA binding where essentially identical values were found from calorimetry (-42 ± 1 kJ mol^{-1}) and kinetics (-41 ± 3 kJ mol^{-1}), indicating only one binding site on the dimeric TIM molecule for this competitive transition state analogue. This is puzzling and suggests that binding of PGA at one site is sufficient to inhibit substrate binding (or catalysis) at the other site on the opposing subunit. No mechanism for this long range interaction has yet been established, but the observations do illustrate the power of calorimetry in yielding data additional to just thermal parameters.

(iii) <u>Rhodopsin Photoenergetics</u>

The above examples have concentrated on the protein-ligand binding process which comprises the first step in any enzyme catalytic event. The next stage - the catalytic step itself - involves covalent changes in the substrate molecules mediated by the functional groups on the protein, and clearly involves major changes in interaction between the enzyme and substrate, together with possible transient perturbations of the protein structure as well. We would dearly like to monitor the energetics of this process, but there are severe technical difficulties. Firstly, the catalytic process is usually so fast that current calorimetric techniques would be unable to resolve them. Secondly, even if time resolution were improved or the process slowed down by use of lower temperatures, for example, it would still be difficult to synchronise events so that all molecules in the sample were in phase. The rhodopsin photoprocess has intrinsic properties which allow us to overcome at least some of these obstacles. Although not an enzyme, rhodopsin shows properties that should be common to all proteins and is, in any case, of such fundamental importance that it merits intensive study in its own right. Much of our work in recent years has been devoted to this, including development of new calorimetric techniques (photo-microcalorimetry) for the measurement of the energetics of light-activated biomolecular processes (Cooper & Converse,1976;Cooper 1979,1981,1982;Cooper et al.1986,1987, and references therein).

The primary processes of visual excitation take place in rhodopsin, a coloured membrane-bound protein of about 35,000 molecular weight containing one molecule of 11-cis retinal (vitamin A aldehyde) as chromophore. This chromophore is attached to a specific lysine side-chain via a Schiff base (imine) linkage of a kind also found in many enzyme mechanisms. After absorption of a single photon, rhodopsin undergoes a complex sequence of reactions involving both structural and chemical changes in the molecule. The details are rather too involved to dwell on here, but the chemical changes include, at different stages, isomerization of the retinal to the extended all-trans configuration and hydrolysis of the Schiff base linkage. These covalent changes in the chromophore also induce major conformational changes in the protein. We have determined the energetics of all these steps using a purpose-built calorimeter equipped with a special fibre-optics cell (Fig.3) to allow synchronised initiation of the reaction by monochromatic light, and taking advantage of the ability to halt the rhodopsin photoreaction at each stage by adjustment of temperature (down to liquid nitrogen temperatures for the very earliest stages) and pH, as appropriate.

A slight technical diversion is appropriate here. It is occasionally found that calorimetric experiments using different buffer solutions can give markedly different heat effects. That is, ΔH's obtained from identical samples at the same pH but in different buffers (Tris or phosphate, for example) may differ by upto 50 kJ mol^{-1}. This can be quite disconcerting and one must always be on the alert for possible buffer-specific side reactions. But the explanation here is quite simply that we are observing the buffer doing its proper job (Sturtevant,1962; Cooper & Converse,1976). Imagine a reaction which involves the uptake or release of a proton (H$^+$) and remember that calorimetry measures the totality of heat effects in a reaction. In a well-buffered system, any H$^+$ ions released will be immediately taken up by the buffer ions (and vice versa) as part of the pH buffering action. Now, protonation or ionisation reactions have quite large enthalpies, comparable to the non-covalent enthalpies that we normally study, which are different for different buffer species - hence the differences described above. Far from being a nuisance, this phenomenon is extremely useful in identifying and quantifying otherwise unsuspected protonation changes associated with the reaction of interest, and we routinely repeat experiments in a range of buffers to check for this. Indeed microcalorimetry is by far the most accurate way of detecting such processes in biological systems since the associated pH changes are usually too small to measure precisely.

To return now to rhodopsin: photoisomerization of the retinal chromophore is the very first event in the whole process and takes place within a few picoseconds, or less, with a massive increase in energy of the system - more than 60% of the photon energy is stored in the molecule at this stage, presumably as a result of the major strains exerted on the chromophore and the active site region of the protein by the gross change in shape of the retinal group. One functional consequence of this is that ground-state isomerization of retinal, which takes place quite readily in the free molecule, is actually inhibited by the protein as a result of the increased energy barrier imposed by the protein. In some respects rhodopsin is acting here as an "anti-enzyme", in the sense that we normally view catalysis occurring in enzymes as a result of a reduction in transition state barriers. But this property of rhodopsin, though at first sight surprising, is in fact essential for its purpose in facilitating the excited-state photoisomerization pathway, whilst inhibiting the ground-state dark reaction which would otherwise produce a large background noise on the visual response (Cooper, 1979). Subsequent steps in the process take place over gradually slower time scales, from nanoseconds to minutes. Initially the stored energy from the first stage dissipates in a sequence of exothermic steps as the conformational stresses and strains in the active site region relax throughout the protein. But there then follows an endothermic step ("metaI --> metaII"), coincident with the secondary

biochemical processes in the photoreceptor cell responsible for visual
signal transduction, which shows the peculiar buffer dependence in its
enthalpy mentioned above due to an uptake of a proton by the protein at
this stage. The endothermic nature of this stage suggests, along with
other indirect evidence, that covalent bond cleavage is taking place
here. This is in fact the stage at which hydrolysis of the retinal-
lysine imine bond occurs, as confirmed by recent Raman spectroscopy and
isotope labelling experiments (Cooper et al. 1987), and the proton
uptake is due to the anomalously low pK_A of the active site lysine thus
exposed. (This is another feature suspected to be common to Schiff base
enzymes.) The final stages of the rhodopsin photolysis sequence involve
migration and release of the now-unfettered all-trans retinal from the
protein.

This has been but a brief glimpse of what is really quite an
involved process of energy transduction and dissipation in a specific
protein molecule, and interested readers are referred to the original
publications for more details. But it does, together with the other
examples outlined above, illustrate the wealth of molecular detail, not
only on the energetics and thermodynamics, but also on other features
such as binding stoichiometries and protonation changes that can be
deduced from the careful application of microcalorimetric techniques to
biomolecular processes.

ACKNOWLEDGEMENTS

Much of the work reported here was done in collaboration with many
capable colleagues, including F.M.Jenkins, C.A.Converse, S.F.Dixon,
M.Tsuda, J.L.Robb and, currently, M.A.Nutley, B.Sanders and C.Johnson.
Financial support from the S.E.R.C. is gratefully acknowledged.

REFERENCES

Bartlett,P.A. & Marlowe,C.K.(1987) Science, 235:569.
Beddell,C.R.,Moult,J. & Phillips,D.C.(1970) in "Ciba Foundation
    Symposium on Molecular Properties of Drug Receptors", R.Porter &
    M.O'Connor, eds., Churchill, London, p.85.
Biltonen,R.L. & Langerman,N.(1979) Meth.Enzymol., 61:287.
Bjurulf,C. & Wadsö,I.(1972) Eur.J.Biochem., 31:95.
Blake,C.C.F., Johnson,L.N., Mair,G.A., North,A.C.T., Phillips,D.C. &
    Sarma,V.R.(1967) Proc.Roy.Soc.Ser.B, 167:378.
Cooper,A. & Jenkins,F.M.(1973) in "Protides of the Biological Fluids",
    H.Peeters,ed., Pergamon Press, Oxford, p.457.
Cooper,A.(1974) Biochemistry, 13:2853.
Cooper,A. & Converse,C.A.(1976) Biochemistry, 15:2970.
Cooper,A.(1979) Nature, 282:531.
Cooper,A.(1981) FEBS Lett., 123:324.
Cooper,A.(1982) Meth.Enzymol., 88:667.
Cooper,A.,Dixon,S.F. & Tsuda,M.(1986) Eur.Biophys.J., 13:195.
Cooper,A.,Dixon,S.F.,Nutley,M.A. & Robb,J.L.(1987) J.Am.Chem.Soc.
    109:7254.
Dahlquist,F.W. & Raftery,M.A.(1968) Biochemistry, 7:3269.
Fersht,A. et al.,(1984) Nature, 314:235.
Hu,C.Q. & Sturtevant,J.M. (1987) Biochemistry, 26:178.
Kauzmann,W.(1959) Adv.Protein Chem., 14:1.

Langerman,N. & Biltonen,R.L.(1979) Meth.Enzymol., 61:261.
Sturtevant,J.M.(1962)   in   "Experimental   Thermochemistry",   Vol.II,
     H.A.Skinner,ed., Interscience, New York, p.427.
Sturtevant,J.M.(1974) Annu.Rev.Biophys.Bioeng.,3:35.
Sturtevant,J.M.(1977) Proc.Nat.Acad.Sci., 74:2236.
Sturtevant,J.M.(1987) Annu.Rev.Phys.Chem., 38:463.
Wadsö,I.(1974) Pure Appl.Chem., 38:529.
Wadsö,I.(1983) Pure Appl.Chem., 55:515.
Wolfenden,R.(1970) Biochemistry, 9:3404.

STRUCTURE AND DYNAMICS OF PHOSPHOLIPID MEMBRANES FROM NANOSECONDS TO
SECONDS

Josef F. Holzwarth

Fritz-Haber-Institut der Max-Planck-Gesellschaft
Faradayweg 4-6, D-1000 Berlin 33, West Germany

ABSTRACT

The "Iodine-Laser Temperature Jump" (ILTJ) technique offers the
unique possibility to measure dynamic processes in membrane systems as
well as protein structures from nanoseconds to seconds, resulting in
relaxation amplitudes which are correlated with the enthalpy changes of
these often complicated systems. The new electron microscopic (EM)
technique of fast freezing ($<10^{-4}$ s) of thin lamellar preparations and
the use of a cryo EM allows the determination of structural details
without the need of contrast chemicals. ILTJ measurements from 10^{-9} s -
10^0 s are presented which cover the whole crystalline-fluid transition
of unilamellar vesicles (UVs) from dipalmitoylphosphatidylcholine (DPPC)
or dimyristoylphosphatidylcholine (DMPC). Especially tailored probe
lipids like acridineorangelecithin (AOL) and diphenylhexatrien-
palmitoylphosphatidylcholine (DPH PC) were used to confirm the turbidity
measurements by time resolved absorption and fluorescence anisotropy
changes. Five well separated relaxation signals of increasing
cooperativity could be observed. By reconstructing the equilibrium data
of the phase transition of membranes from the amplitudes of the kinetic
relaxations it could be proved that the dynamic processes represent the
whole crystalline-fluid transition. A model is presented which aims to
explain the kinetic results on a molecular basis. UVs with incorporated
cholesterol, gramicidin A and bacteriorhodopsin showed a strong shift of
the slower relaxation amplitudes (100 µs - 20 ms) towards the 10 µs time
range. We explain these results by the preference of intermediate states
of order of the annular lipids in the surrounding of functional units
like bacteriorhodopsin. We therefore call the relaxations τ_3 around 10
µs functional important movements FIMs after Frauenfelder.

INTRODUCTION

In contrast to a wide range of thermodynamic studies far fewer
kinetic investigations of well defined lipid membrane systems have been
carried out. These is mainly due to the lack of an experimental
technique which allows for kinetic measurements from 10^{-9} s to 10^0 s.
ESR /1/ and NMR /2/ studies as well as fluorescence polarization
lifetime measurements /3/ have been the favorite technqiues, but these
instruments only cover the nanosecond to picosecond time regime and
require models to obtain dynamic information. Infrared spectroscopy and

dynamic neutron scattering are limited to the pico- to femto-second time range /4/. Mössbauer spectroscopy might get into the 100 ns time domain. Pressure jump techniques /5/ as well as conventional joule-heating temperature jump /5/ were applied but either strong disturbing field effects or the release of pressure limited both methods to times longer than 500 μs /6/.

Our "Iodine-Laser Temperature-Jump" arrangement /7/ is the only instrument available which allows measurements from 10^{-9} s to 10^{0} s without producing unwanted physical or chemical effects /8/. Furthermore the relaxation amplitudes are linked to the enthalpy of the reaction under investigation by the well known Van't Hoff equation: $d\ln K/dT = \Delta H/RT^2$. If small temperature jumps (1-2°C) are applied this equation simplifies to $\Delta K/K = \Delta H \cdot \Delta T/RT^2$. Because of the so far unmatched time range as well as the flexibiliy in the detection parameters which are fluorescence-, absorption-, light scattering (turbidity)- and conductance changes we applied our ILTJ to investigate dynamic changes in lipid membranes and could establish a correlation with the ΔH of these dynamic processes. The ILTJ provides the chance to predict the most important time domain in respect of the ΔH of the process under investigation. The limited time resolution of most other instruments and the lack of a linear correlation between the measured signals and the ΔH of the dynamic processes of interest often led scientists to wrong conclusions about the imporant time ranges in respect of enthalpy changes (ΔH). This is especially true for the crystalline-fluid transition in lipid membranes and dynamic changes in proteins and enzymes. Theoreticians used mainly picosecond and femtosecond events to model such complex systems like membranes and proteins; this restricted procedure does not provide any information about the extremely important time range from 10^{-8} s to 10^{-1} s. In this time domain biology operates. We applied our ILTJ to fill the so far foggy area of dynamic processes in nano-, micro- and millisecond times. The main phase transition in pure lipid membranes and lipid preparations containing reconstituted polypeptides and proteins like the channel-forming gramicidin (GA) or the light driven proton pump bacteriorhodopsin (BR) were investigated to demonstrate the most important time ranges for dynamic changes. The kinetic results were compared with thermodynamic data either from turbidity/temperature, fluorescence/temperature, and absorption/ temperature or microcalorimetric (DSC) measurements. The recently developed cryo-electron microscopic technique (CEM) using thin lamellar preparations and avoiding any contrast chemicals was used to get structural information about the lipid preparations. To monitor the kinetic changes after a fast temperature jump ($\geq 10^{-9}$ s) we utilized mainly three different techniques: light scattering or turbidity from the whole solution, light absorption from especially tailored lipids and fluorescence anistropy changes observed through probe molecules. In this way we could rule out any specific misleading effects from one of the detection methods. Our target was to select the most important time areas between 10^{-9} s and 10^{0} s for dynamic changes in membrane systems and to compare those findings with thermodynamic measurements using the ΔH of reaction as measure. In this way it is possible to prove that all dynamic processes which contribute to ΔH are covered and to select the important ones in respect of enthalpy. A future result should be a time dependent model of interactions inside biological membranes. The kinetic technique of ILTJ used is not limited to membrane systems; it can be applied to any system which shows equilibrium concentrations of reactants and products not to different "relaxation equilibria" and posesses an enthalpy (ΔH) of reaction. Only purely entropic reactions cannot be observed.

To demonstrate the wide time range which can be covered by ILTJ
figure 1 shows the logarithmic timescale from femtoseconds to the stone
age. If one transfers the time window of ILTJ (10^{-9} - 10^{0} s) to longer
times it would be analogous to the time stretching from 1 s to human
life.

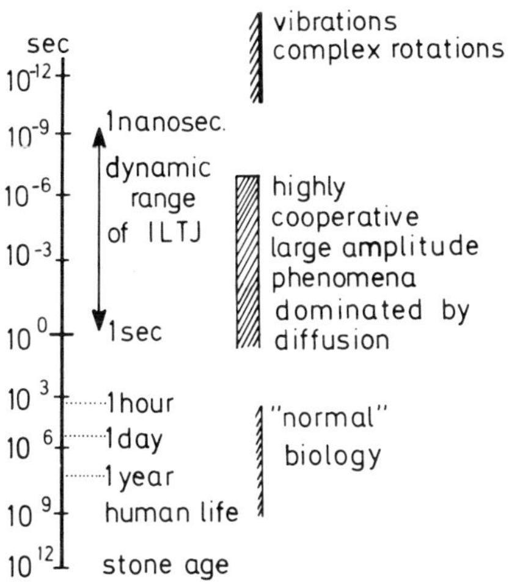

Fig. 1. Schematic timescale of events demonstrating the unmatched
timeresolution of the "Iodine-Laser Temperature-Jump" (ILTJ)
technique.

MATERIALS AND METHODS

<u>Materials</u>

The lipids dipalmitoylphosphatidylcholine, DPPC, dimyristoylphosphatidylcholine, DMPC, cholesterol, CHOL, as well as gramicidin A',
GA, were of the purest grade available from Fluka Switzerland, the
fluorescence probe molecules diphenylhexatrien, DPH, and trimethylamino-
diphenylhexatriene, TMA DPH, were supplied by Molecular Probes, Texas.
The specially tailored lipids 1{3[p-(6-phenyl-1,3,5-hexatrienyl)-phe-
nyl]-propionyl}-2-palmitoyl-3-phosphatidylcholine, DPH PC, was synthesized by E. Thomas, University of Salford, England and {2-[3,6-bis(di-
metyl-amino)-.10-acridino]ethyl}-(2,3-dipalmitoyl-D-L-1 glyceryl)-phos-
phat, AOL, was a gift of H. Zimmermann, University of Freiburg, West
Germany. All other chemicals were purchased from Merck, West Germany,
either p.a. or suprapure if available. Only triply distilled water from
quartz instruments was used and all preparations were handled with great
care to avoid any impurities to get into the samples.

<u>Vesicle Preparation</u>

The vesicles were prepared by a modified "injection method" /9,10/. 20-50 µM of lipid were dissolved in 1 ml of ethanol and slowly injected (20-40 min) into 10 ml of pure buffer at a temperature 10°C above the phase transition temperature, T_m, stirring the emulsion carefully. Afterwards the solution was dialized for 8 h against pure buffer to remove the alcohol /11/. The vesicle preparations were characterized by electron microscopy, light scattering as well as turbidity - absorption - or fluorescence anisotropy - temperature dependences. Differential scanning calorimetry, DSC, was used to measure the thermodynamic ΔH. Fluorescence and absorption labelling was achieved by adding a few µℓ of the probe dissolved in methanol to the unilamellar vesicles, UVs, giving the desired probe/lipid ratio.After an incubation time of 1 hour at a temperature above T_m under stirring,the methanol was completely removed as could be proved by spectroscopic evidence. DPHPC and AOL were incorporated /12/ into the UVs together with the lipids in the original ethanolic solution. Figure 2 gives the bond structure of lipid and probe molecules. CHOL was incorporated as described in reference /13/. Gramicidin was introduced together with the lipids in the original ethanolic solution /12/. Bacteriorhodopsin, BR, was reconstituted into DMPC membranes by N. Dencher, Freie Universität, Dept. Physics, Berlin, West Germany /14/. All preparations were checked by electron microscopy using either negative staining by uranyl acetate or the new CEM-technique described under methods. A size distribution of a typical sample is shown in figure 3.

<u>Methods</u>

<u>Iodine-Laser Temperature-Jump, ILTJ</u>. Since the pionieering work of Czerlinski and Eigen /15/ the "joule heating" temperature jump technique has been used extensively for kinetic studies in aqueous solutions, because of the temperature dependence of almost every chemical equilibrium in solution. But there is a serious problem with the joule heating T-jump. The temperature increase is created by the discharge of a capacitor through an electrolytic solution. This has the consequence that the sample is exposed to an electric field of 20-50 kV per cm; when the field decreases the temperature of the sample rises. In fact joule-heating is primarily a field jump technique with the decaying field energy being transfered into temperature energy by carrying ions through the solution. The rise time of the T-jump (time resolution of the instrument τ = R·C/2) is limited by the electrolytic content of the solution: high ionic strength (0.1 M) = time resolution of ca. 5 µs, low ionic strength (<0.01) = time resolution of ca. 50-100 µs, etc. Furthermore the high electric field which initiates the T-jump might cause serious physical or chemical effects in the sample under investigation; especially complex molecules like proteins or biopolymers like DNA are disturbed by the field in a way that reliable measurements are often not possible /16,17/. Membrane preparations like the ones which are used in our investigations cannot stand electric fields higher than 10 kV/cm without rupture of the bilayer.

To avoid the serious limitations of the joule-heating T-jump we have developed the ILTJ. This technique uses the photon emission of an iodine laser in the near IR at λ=1.315 µm to create a very fast temperature rise in aqueous solution. This occurs by photon absorption of rotational-vibrational states of water molecules. The absorption of water at 1.315 µm is such that very homogeneous T-jumps in layers up to 3 mm can be achieved /7/. Beside water no additional molecules or ions are necessary to achieve the T-jump. The rise time of the temperature is only governed by the emission time of the iodine laser which can be as short as 100 picoseconds. The severe problems caused by the initial electric field in joule heating are not existing as long as one does not exceed the power threshold of water (15 Gigawatt/cm^2 /18/). The long time limit is given by the decay time of the T-jump which is typically around 1 s in a volume of 500 µl in a quartz cuvette located in a brass-block to be thermostated. Table 1 summarizes the characteristica of our ILTJ. Applications other than the ones described in this article are found in the literature /19,20/. Kinetic measurements with the ILTJ to explain the first step of ligand binding to peroxidases /21/ and DNA-dye /22/ interaction were published recently: A schematic representation of

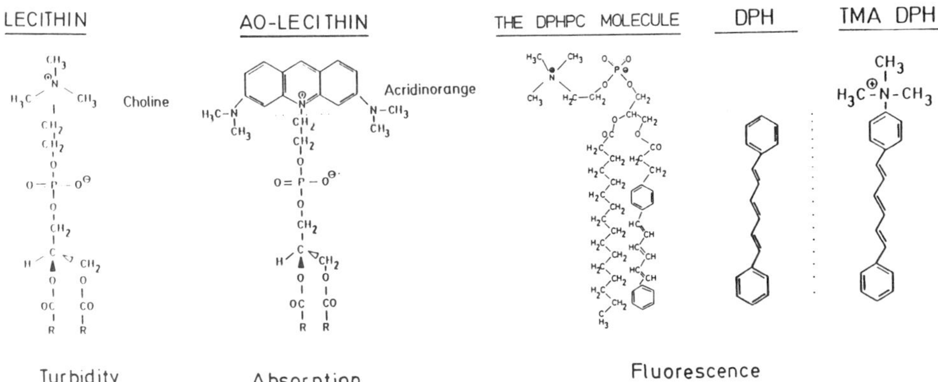

Fig. 2. Bond structure of lecithin and two especially tailored probe lipids like acridineorange lecithin (AOL) for absorption detection and diphenylhexatrien lecithin (DPHPC) for fluorescence measurements as well as two "free" probes diphenylhexatrien (DPH), and trimethylammoniumdiphenylhexatrien (TMADPH) for fluorescence anisotropy experiments.

a T-jump in DMPC vesicle containing solutions and an experimental arrangement of the ILTJ are drawn in figures 4 and 5. Figure 6 shows a signal in the most crucial time range of 1-100 µs were shockwaves could be monitored to prove the quality of our measurements. Such signals are impossible to be achieved with other instruments. To improve the signal to noise ratio we record all signals in transient digitizers like Tektronix 7912 AD, Biomation 8100, Tektronix 390 AD etc. A single T-jump can thus be monitored through 10752 channels set in such a way that the whole time scale from 10^{-9} s to 10^1 s is covered. By on-line connection to desk top computers like HP 9845 B or HP 9816 the signals can be sampled and relaxation times as well as their corresponding amplitudes can be calculated. We typically sample 9 signals, which improves the signal to noise by a factor of $\sqrt{9} = 3$. Our calculation program allows for the computation of the correct relaxation amplitudes for the different time windows, the sum of which is linearly correlated to the ΔH of reaction by the Van't Hoff equation (see Introduction). Direct comparison of kinetic results and the thermodynamic ΔH of reaction from DSC is therefore possible.

<u>Spectroscopic and DSC Measurements.</u> Turbidity transition curves were measured on a UV/VIS spectrophotometer (Perkin Elmer 555) at 300 nm or 360 nm. The temperature inside the cuvette (Helma, QS, Müllheim, West Germany) was scanned with a rate of 18° C/h using a Haake F3C thermostat equipped with the PG11 temperature control unit (Haake, Berlin, West Germany) and determined in control experiments with a digital thermometer.

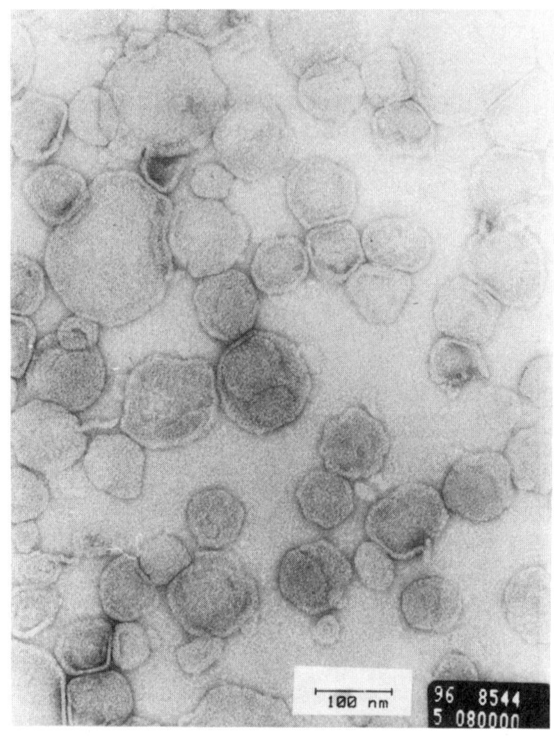

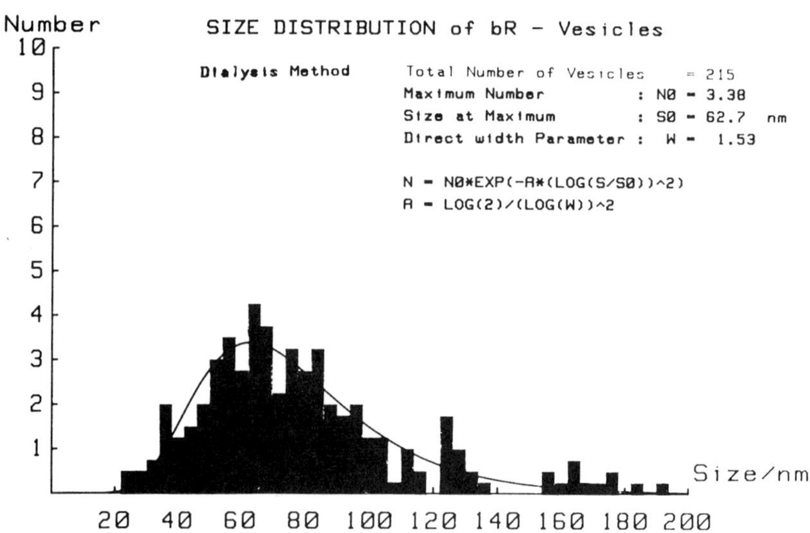

Fig. 3. Electron micrograph of vesicles containing bacteriorhodopsin
BR achieved by negative staining with uranylacetate showing
the unilamellar membranes of the vesicles (UVS) together with
the corresponding size distribution of the same preparation.

Table 1. Characteristic physical data of the ILTJ and two important
physical barriers for its application which are thermaliza-
tion time and dielectric breakdown of the solvent water.

1. <u>Laser</u>

Wavelength	:	$1.315\,\mu m \hat{=} 0.943\,eV \hat{=} 2.3 \times 10^{14}\,Hz$
Emission time	:	$\geq 150 \times 10^{-12}\,s$
Photon energy	:	$= 1.51 \times 10^{-19}\,J$
Photon number per pulse	:	$\sim 6.62 \times 10^{-18} \ \hat{=}\ 1\,J$

2. <u>Probe</u>

Heated volume H_2O	:	$140 \times 10^{-3}\,cm^3$
Temperature jump energy	:	$\Delta T = 1\,°C$ in $140\,\mu l$
	:	$E = 0.585\,J$
Number of photons	:	$N_P = 3.87 \times 10^{18}$
Number of H_2O molecules	:	$N_M = 4.69 \times 10^{21}$

3. <u>Thermalization</u>

Thermal conductivity	:	$k = 6.1 \times 10^3\,J/s\,cm\,°C\ (25\,°C)$
Heat quantity transported (Photon energy)	:	$Q = k \cdot area \cdot time \cdot \Delta C / distance$
Area = (distance)^2 (11×11 H_2O molecules of 3.1 Å)	:	$A = (31 \times 10^{-8})^2\,cm^2$
Thermalization time $\Delta\,°C = 1$	:	$t_{max} = 8 \times 10^{-11}\,s$

4. <u>Dielectric Breakdown of H_2O</u>

Power - threshold	:	$P_T = 1.5 \times 10^{10}\,W/cm^2$
Laser - power (1J in 150 ps)	:	$P_L = 0.7 \times 10^{10}\,W$

Differential scanning calorimetric, DSC, measurements were performed on a MC-1 microcalorimeter (Amherst, Mass., USA) using 0.9 ml of
solution for reference as well as the sample cell and a scan rate of
21.7° C/h. The enthalpy values, ΔH, were determined by referring the
transition curves to a calibration pulse and measuring the area of the
curves by paper weighting. Final results were obtained by averaging over
at least two different preparations.

<u>Cryo Electron Microscopy, CEM.</u> Electron microscopy is widely used
to determine structures. These structures are averages over a time which
depends on the fixation time of the preparation. If the structure of an
aqueous solution of membrane samples is monitored the structure being
monitored in the electron microscope is the one which is reached at the
time when the sample has a temperature at which the displacement of
atoms from their average position (diffusion) is stopped. Normally the
temperature of liquid nitrogen of ca. 100 K is sufficient to stabilize
membrane structures in aqueous solutions. In addition to this low
temperature the water of preparations, suitable for electron microscopy
of membranes in a CEM, has to be amorphous, because otherwise the
crystal structure of water could determine the structure seen on the EM
picture.

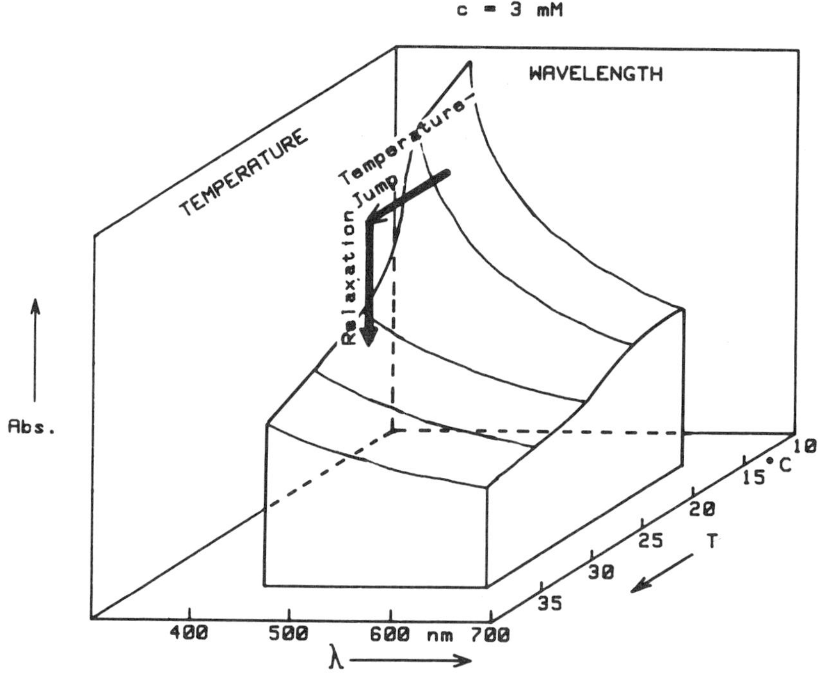

Fig. 4.   Three dimensional representation of a temperature-jump experiment in aqueous solutions of dimyristoylphosphatidyl-
Choline (DMPC) vesicles. The detection parameter to follow
the chemical relaxation given as Abs. could be a light
scattering (turbidity) - a light absorption - or a
fluorescence change.

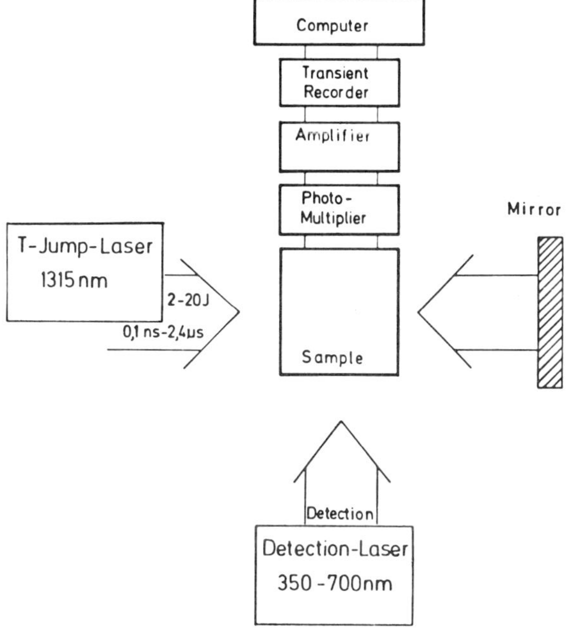

ADVANTAGES

1. Temperature-Jump, independent of additives, by photon absorption of vibrations of water.
2. No electric field disturbing the reactants.
3. Time resolution from picoseconds to seconds.
4. Sample volumes 10 µℓ - 100 µℓ.
5. Flexible detection: optical or conductance.

Fig. 5. Principle experimental arrangement of the ILTJ stating the advantages of this relaxation technique.

We used copper or carbon grids of 400 mesh (35 μm × 35 μm free space) to carry the aqueous solution which contained the UVs. By shooting the grid with a speed of ca. 4 m/s into liquid propane kept at a temperature of ca. 100 K we could stabilize lamellar layers of the aqueous solutions of 200-300 nm thickness in times shorter than 10^{-2} s, applying a cooling rate of 5×10^{4} K/s. The measured cooling rates as well as a micrograph of our UV preparations are shown in figures 7 and 8. Beside the extremely high cooling rates applied, the electron micrographs could be achieved without the use of any staining material taking advantage of the contrast between the lipid bilayer membrane and the surrounding water. The instrument used was the DEECO 250 Cryoelectron microscope of the Fritz-Haber-Institut built by Heide and Jäger.

The structural resolution is better than 3 nm as can be seen in figure 8 by the double lamellar vesicle in the left lower corner. The cooling of our preparation started at 301 K ca. 4° C above the phase transition temperature PT_m and the figure shows the perfect spherical shape of the vesicles. If cooling starts below PT_m a facetted surface is monitored. This proves that the cooling rate is faster than the dynamic processes causing the phase transition because otherwise the structures seen in the micrographs should be similar. At which point in time of the phase transition process we stop the diffusion and stabilize an intermediate structure is still not accurately known, but it is certainly between 10^{-4} and 10^{-2} s. We hope to get further insight into well time resolved structures by CEM in the near future.

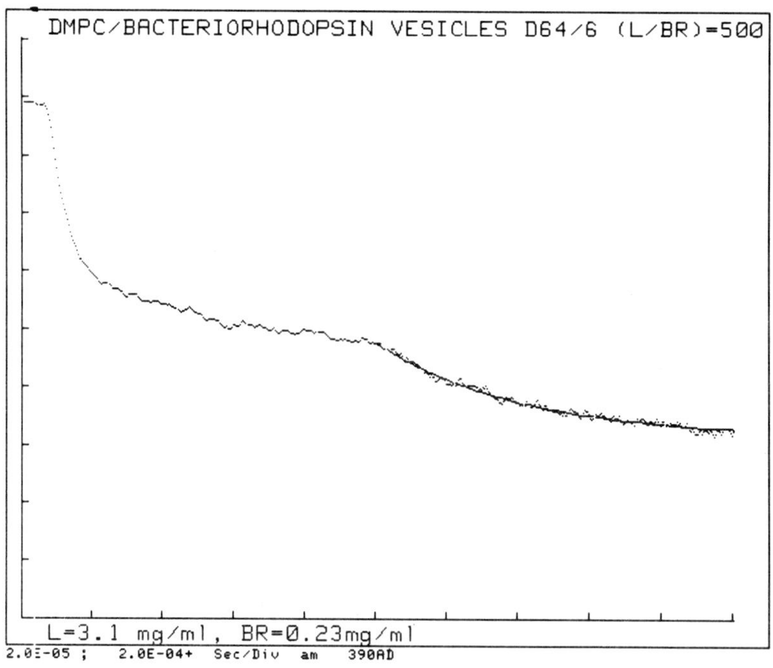

Fig. 6. Relaxation signal from unilamellar vesicles of DMPC containing 0.2 % bacteriorhodopsin showing τ_3, τ_4 and τ_5. The timebase is 2×10^{-5} for the first 5 divisions followed by 2×10^{-4} s for the last ones. The signal is a sample of 6 experiments and the solid line between division 5 and 10 is a computer fitting of relaxation τ_5.

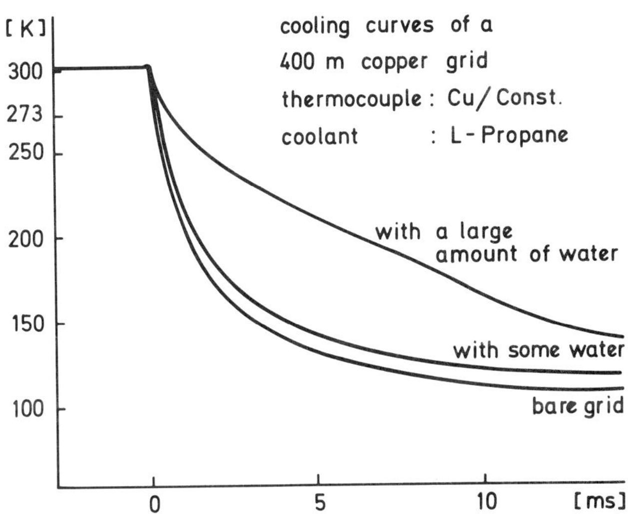

Fig. 7. Decrease of temperature measured through a thermocouple
fitted to the copper grids which are used for the cryoelec-
tronmicroscopic (CEM) experiments. Grid size 35 μm^2. Entry
velocity of the grid was 1.5 m/s. The figure shows the
cooling dependence on the amount of water on the grid.
Typical cooling curves used in our CEM are those similar to
some water.

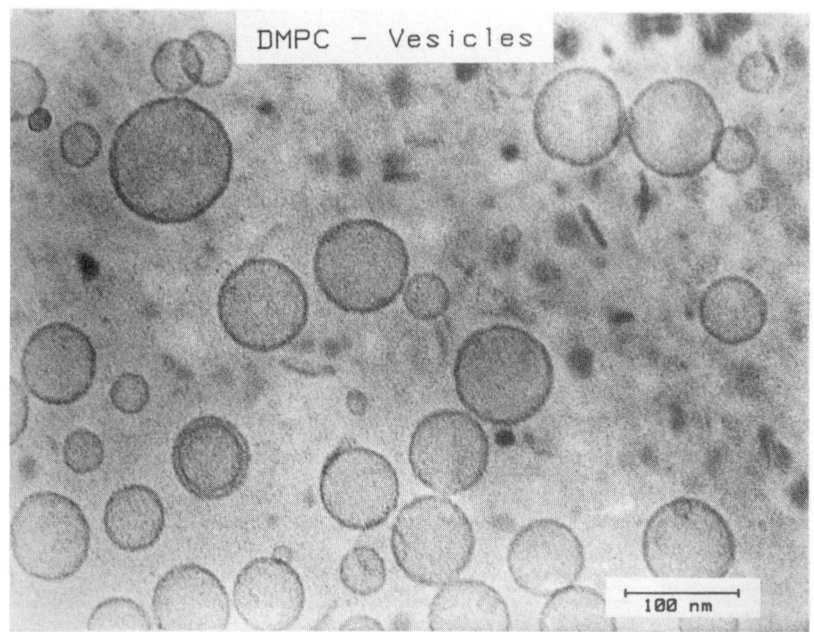

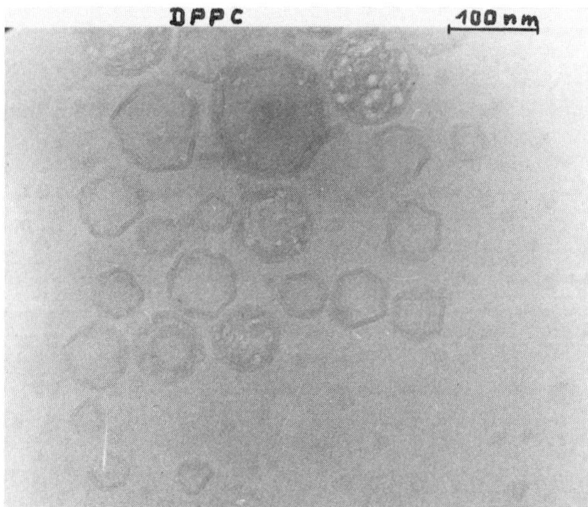

Fig. 8. Cryoelectronmicrographs of dimyristolyphosphatidylcholine
(DMPC) vesicle preparations showing the space resolution of
ca. 3 nm, the size distribution as well as the spherical
structure for DMPC vesicles cooled to 100 K in times of ca.
10^{-3} s from a temperature above the phase transition T=28 °C.
The dipalmitoylphosphatidylcholine (DPPC) vesicle pictures on
the lower electronmicrograph were achieved from a preparation
cooled to 100 K from a starting temperature well below the
phase transition showing a facetted surface structure. The
water surrounding the vesicles was amorphous at 100 K.

Preparations of unilamellar vesicles, UVs, of DPPC were measured statically as demonstrated in figure 9 using turbidity as the detection parameter for the phase transition T_m. The decrease of the signal with increasing wavelength is obvious; at wavelengths larger than 460 nm almost no signal is left. If we observe UVs containing the fluorescence probe,DPHPC, signals like in figure 10 are observed and figure 11 shows the equilibrium turbidity- and absorption-temperature dependences of UVs containing AOL as probe molecule. Figures 9, 10, 11 clearly demonstrate that the equilibrium temperature dependences of all three completely different observation methods give similar results, which proves that no misleading technique dependent parameter is involved.

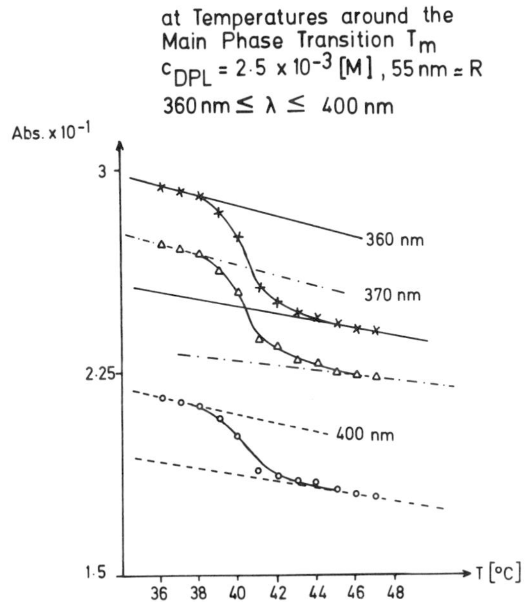

Fig. 9. Turbidity-temperature dependences of unilamellar vesicles of dipalmitoylphosphatidylcholine (DPL) showing the turbidity decrease associated with the phase transition as well as the decreasing signal with increasing wavelengths of observation. The size of the vesicles was ca. 100 nm. Centre of the phase transition T_m = 40.5 ± 0.2 °C.

Figure 12 shows the five well separated relaxations which could be observed after a temperature jump rising in 10^{-9} s by monitoring the decrease of turbidity at different temperatures. The magnitude of the T-jump was ca. 1° C to resolve the phase transition in respect of temperature. Relaxation times are indicated together with their corresponding relaxation amplitudes.

Figure 13 contains the kinetic results of similar preparations like in figure 12 exept that the detection parameter was the time resolved equilibrium fluorescence anisotropy at times longer than 10^{-6} s to make sure that the fluorescence lifetime of the probe molecule DPHPC is at least 100 times shorter. In this way a good time average for the "wobbling" of the probe molecule is achieved thus reporting about the space which is available for movements of lipid molecules at times longer than 10^{-6} s after the T-jump /12/. Figure 14 shows the experimental results for UVs with AOL incorporated to monitor absorption changes of this probe similar to figure 11, but in figure 14 the time

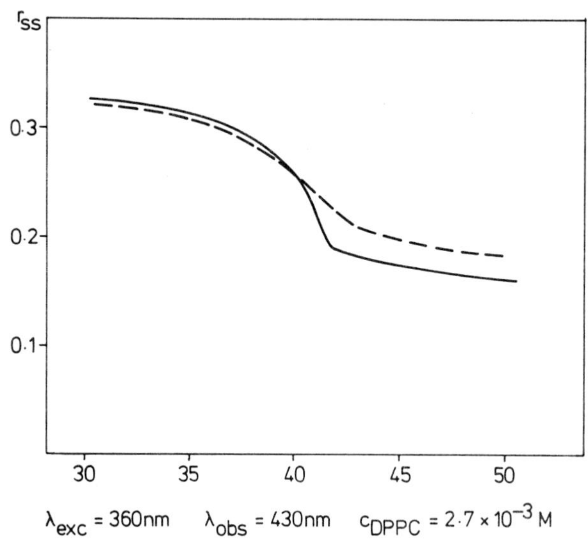

Fig. 10. Fluorescence anisotropy-temperature dependence of dipalmitoylphosphatidylcholine vesicles containing 0.25 % of the probe DPHPC solid line and an additional amount of 15 % of cholesterol (CHOL) dashed line. $r_{SS}=(I_\parallel-I_\perp)/(I_\parallel+2I_\perp)$. $I_\parallel$ is the fluorescence intensity from linearly polarised light of 360 ± 15 nm observed above 430 nm with the same polarisation. $I_\perp$ is the fraction of fluorescence with perpendicular polarisation compared to the excitation.

resolved changes after 10^{-6} s are drawn. That the probe molecules DPHPC and AOL cannot monitor the two fastest relaxations which are seen in figure 12 can be understood because lateral diffusion of lipid molecules is necessary for the absorption changes of AOL at 484 nm or the time resovled equlibrium fluorescence anisotropy from DPHPC being observed. The model of the dynamic changes of the five relaxations in figure 16 which represents the whole phase transition dynamics will make this understandable. In figure 15 the normalized amplitudes of the five relaxations from figure 12 are summarized for each temperature and the equilibrium turbidity/temperature dependence of the same preparation is

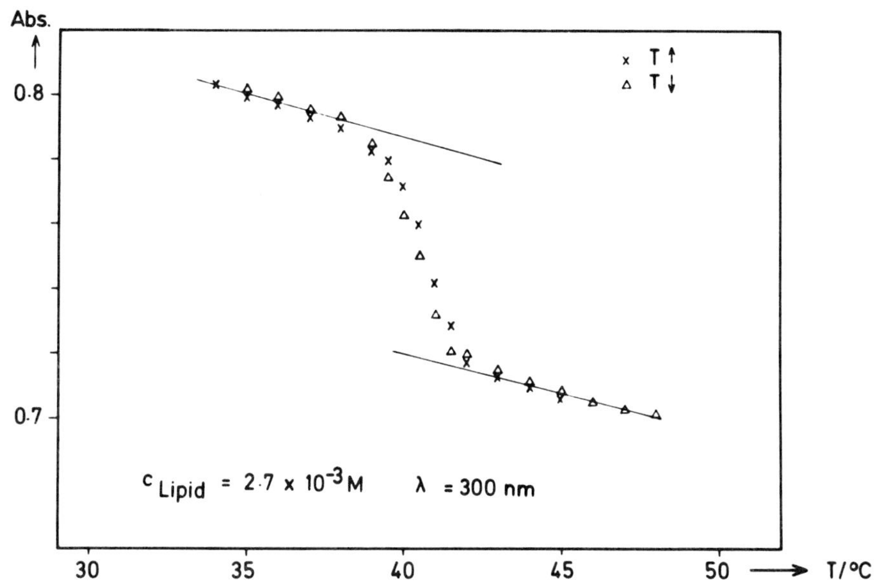

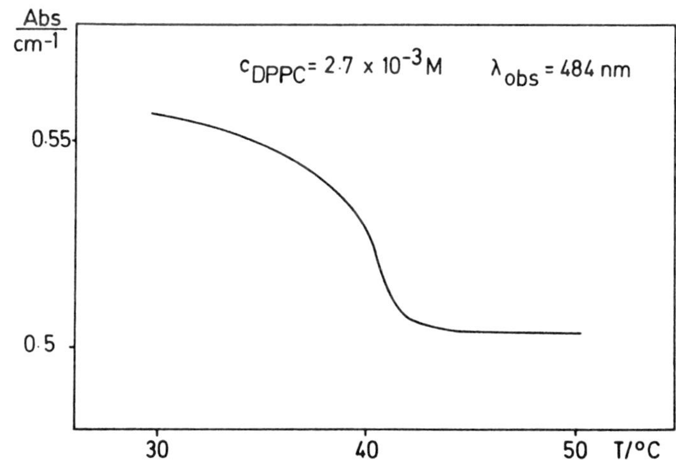

Fig. 11. Measurements of temperature-turbidity at 300 nm and temper-
ature-absorption at 484 nm representing the phase transition
of DPPC vesicles which contain ca. 1 % of the probe lipid
acridineorangelecithine (AOL). Size of the vesicles was ca.
100 nm. The observed signals prove that AOL is a good
indicator for the phase transition.

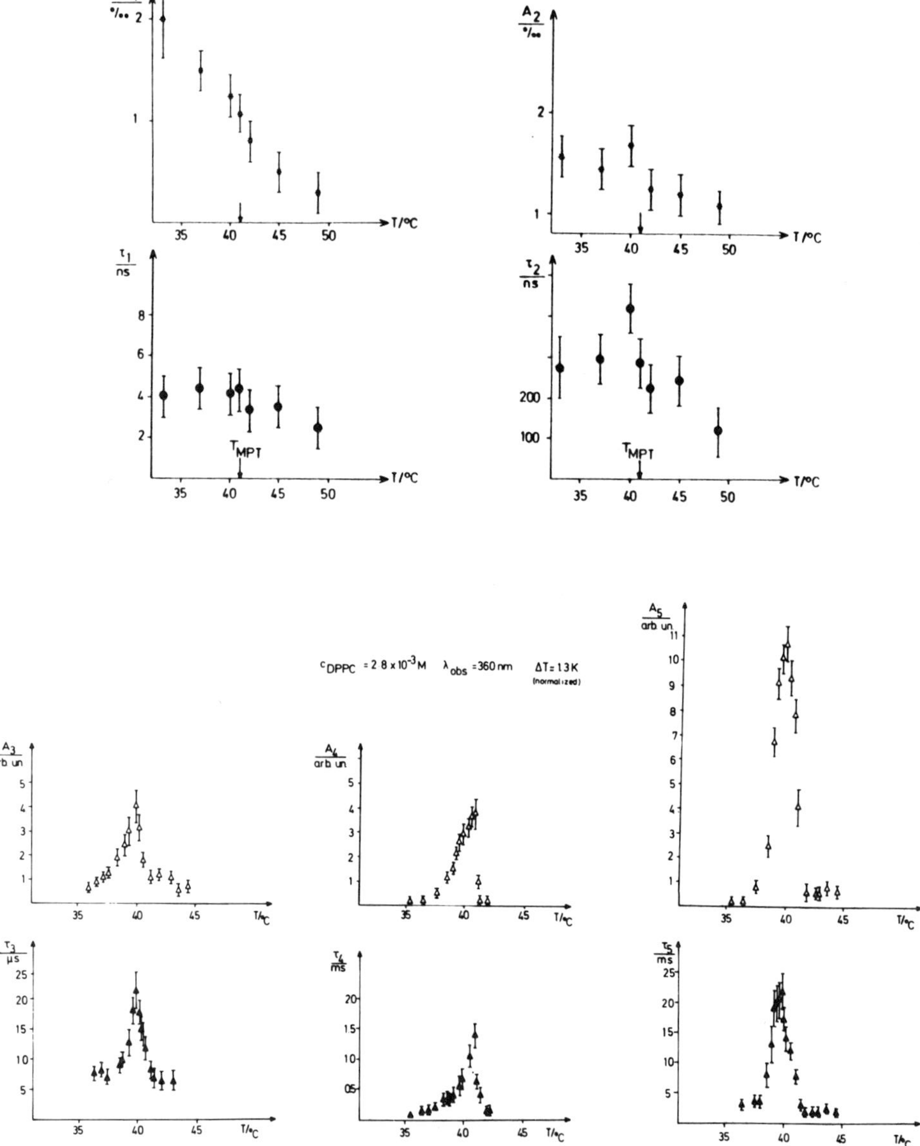

Fig. 12. Five relaxation signals observed after a 1 ns T-jump in
DPPC vesicle preparations showing a size distribution between 20 and 120 nm with a maximum around 45 nm. The five
relaxation times as well as their corresponding amplitudes
are included. The temperatures indicated are the start
point of T-jumps of ca. 1 °C. The maxima for the three
slower relaxations $\tau_3 \sim 25$ µs, $\tau_4 \sim 1.5$ ms and $\tau_5 \sim 25$ ms
demonstrate the strong cooperativity of the dynamic
processes which created those signals.

400 / 1

$C_{DPPC} = 2.7 \times 10^{-3}\,M$ $\lambda_{obs} > 395\,nm$ $\Delta T \sim 1K$

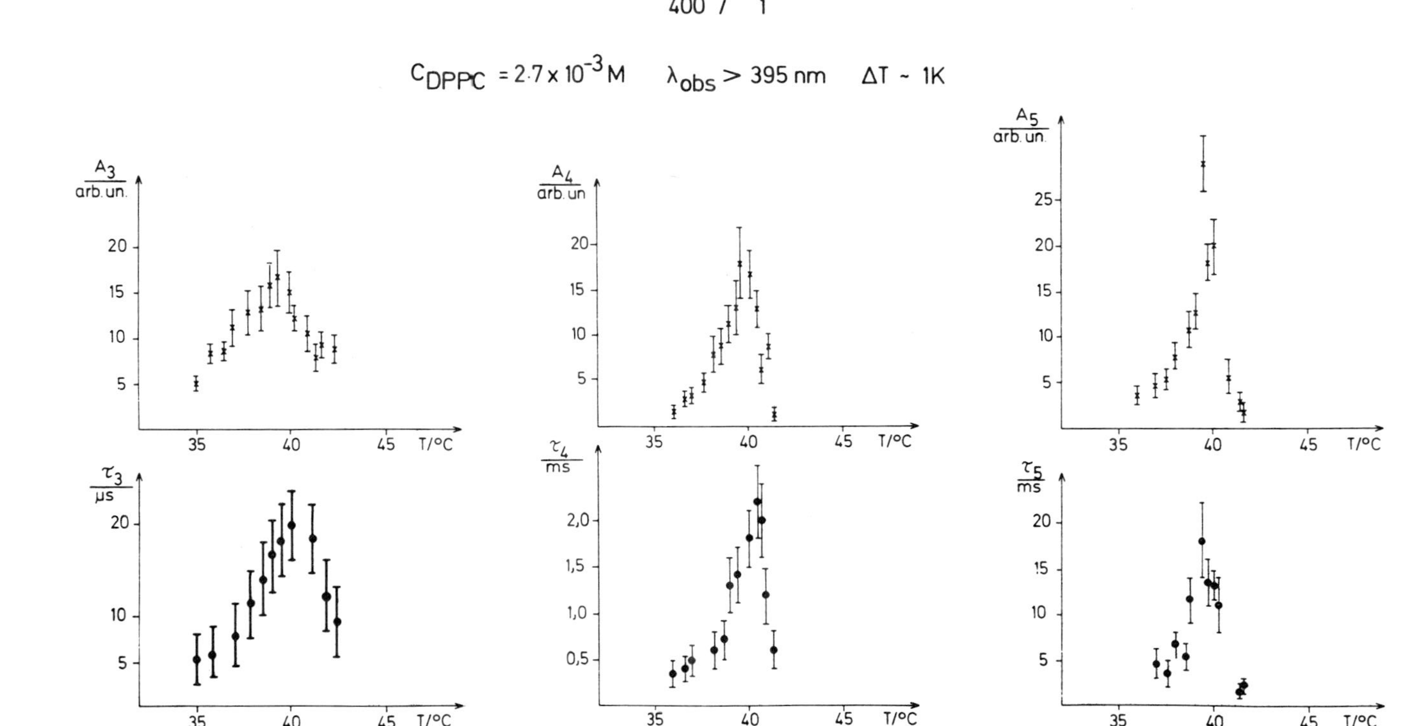

Fig. 13. Three relaxation signals produced by a T-jump of ca. 1 °C in vesicle preparations from DPPC containing 0.25 % of the fluorescence probe DPHPC observed as decrease of the time resolved equilibrium fluorescence anisotropy $r_{ss} = (I_\parallel - I_\perp)/(I_\parallel + 2I_\perp)$ in the temperature range of the phase transition.

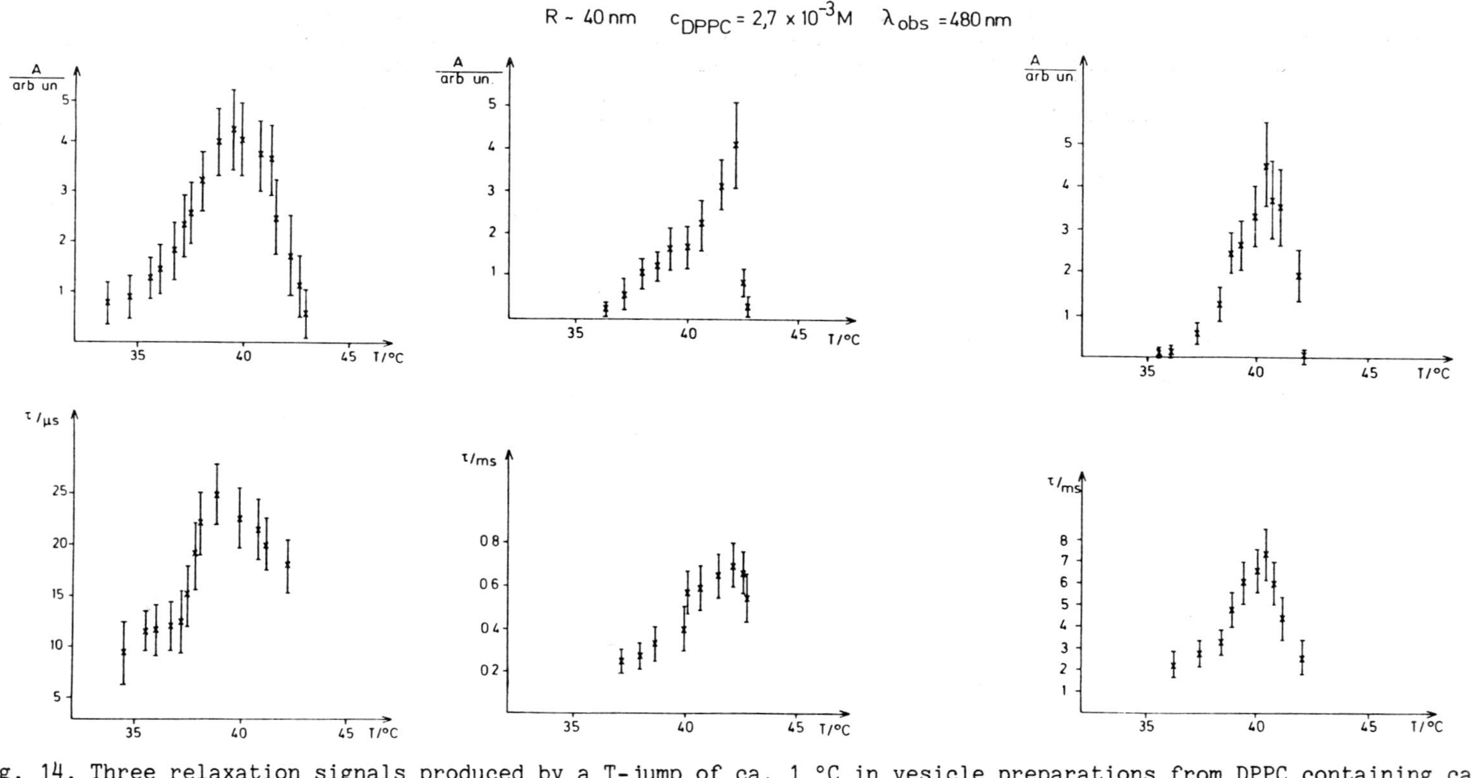

Fig. 14. Three relaxation signals produced by a T-jump of ca. 1 °C in vesicle preparations from DPPC containing ca. 1 % of the absorption probe lipid AOL observed as an absorption change at 480 ± 10 nm, showing the increasing mobility of the lipids.

included. The data of figure 15 demonstrate two important results:
(1) The sum of the five relaxation amplitudes allow the perfect reconstruction of the equilibrium measurements which proves that we monitored all important dynamic processes representing the phase transition
(2) The three slower relaxations contribute more than 75% to the ΔH of the phase transition because our relaxation amplitudes are linearly related to the ΔH of reaction as demonstrated in the introduction.

Summarizing the results of figures 12,13, 14 and 15 we are in a position to design a molecular model for the whole phase transition in lipid bilayer membranes, figure 16 contains the results. By changing the hydrocarbon chain length of the lipids, DPPC or DMPC, and the headgroup, choline or serine, it was possible to attribute the fastest process to the formation of simple rotational isomers (kinks) in single molecules because the 4 ns relaxation called τ_1 is neither sensitive to the hydrocarbon chain length nor the nature of the headgroups /23/. Relaxation 1 showed only very weak maxima near T_m of τ_1 and its corresponding amplitude A_1 which decreased steadily. Relaxation 2 around 300 ns showed a clear maximum at T_m in both τ_2 and A_2; it also changed with the nature of the headgroups.

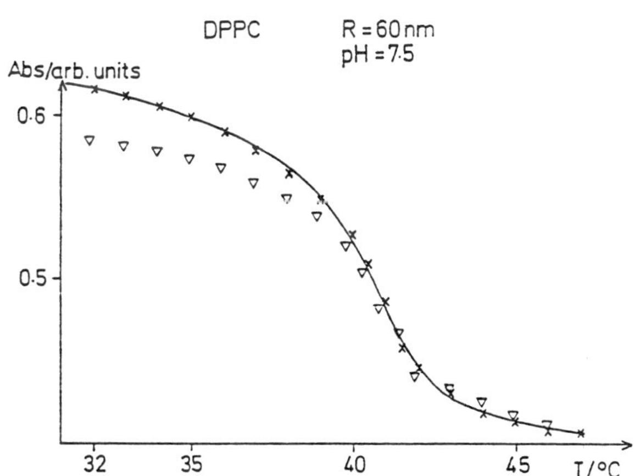

Fig. 15. Temperature-turbidity dependences of DPPC vesicles measured in a spectrophotometer at 360 nm solid line, and the sum of either the three slower relaxation amplitudes of figure 12, Δ, or the sum of all five relaxation amplitudes, X, at the temperatures indicated demonstrating the perfect reconstruction of the thermodynamic signal from the relaxation measurements.

The relaxation times τ_2 were measured at T_m as: 250 ns for DMPC, 270 ns for DPPS at pH 7.5, 320 ns for DPPS at pH 5.8 and 410 ns for DPPC. The maxima for A_2 at T_m are larger for DPPS vesicles than they are for DMPC or DPPC. Taking these results into account we explain relaxation 2 by the onset of the free rotation of the headgroups being no longer fixed in respect to the hexagonal lattice of the hydrocarbon chains of the lipids. Relaxation 3 around 10-20 µs is strongly cooperative; it showed large maxima of its relaxation time and its amplitude. The experiment with the absorption probe AOL in figure 14 proved that lateral diffusion has already started in this time range because we monitored the monomere-aggregate equilibrum of AOL which needs lateral diffusion for being established. We can attribute signal 3 to the formation of complex rotational isomers, gauches, in the hydrocarbon chains of the lipids so that the lateral diffusion is no longer hindered by the chains of the opposite monolayers and the whole membrane can expand in lateral direction.

At longer times processes 4 and 5 are observed. Their molecular explanation is more difficult than that of the faster phenomena. From the time scale of 100 µs to 30 ms the strong cooperativity and the influence of additives like cholesterol and peptides we conclude that relaxation 4 is due to the formation of clusters of lipids in a more

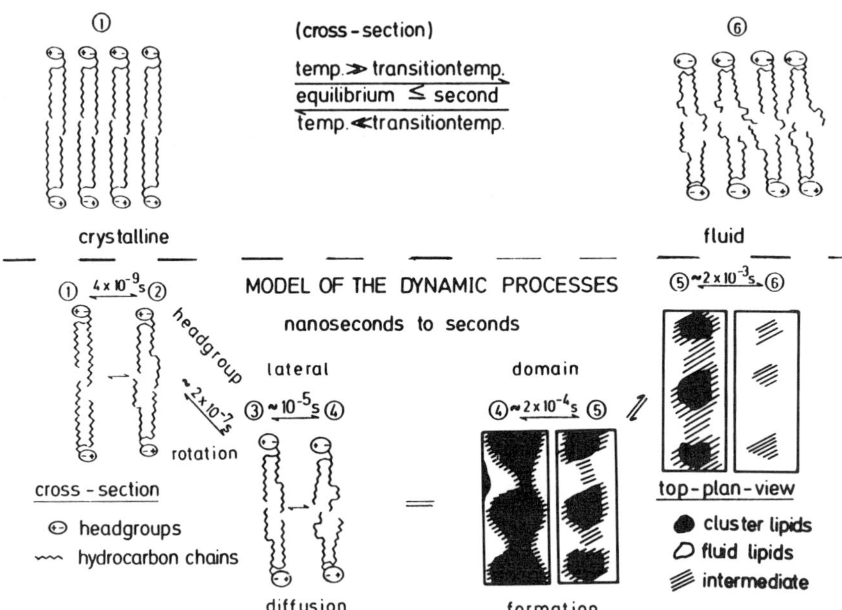

Fig. 16. Structural model for the five separated relaxation signals representing the whole phase transition dynamics in lipid membranes as measurd in DMPC vesicles. Kink formation followed by headgroup disordering, complex rotational isomer formation (gauche) connected with lateral diffusion of lipids, and the formation of crystalline domains as well as their disintegration are schematically drawn to cover the timerange from 1 ns to 0.1 s.

crystalline state surrounded by lipids of intermediate order and lipids
in the fluid state as figure 16 shows. This discontinuity in the state
of order of the lipids exists until we reach the 10-30 ms time regime
where relaxation 5 shows the disappearance of such discontinuities
leaving the bilayer in a fluid state. I would like to point out that the
time dependent molecular model structures for the 5 relaxations are only
existing for the duration of the relaxation times, which makes it very
hard to detect them by structural methods like electron microscopy or X-
ray diffraction. Theoretical models which used time resolved
fluorescence polarization lifetime measurements or NMR data as a basis
came to the wrong conclusion that the phase transition is completed
after 10^{-6} s which is in clear contradiction to our results and is
mainly due to the techniques which do not allow for the detection of
slower processes. Recent Monte Carlo simulations of the phase transition
dynamics by Mouritsen /24/ have shown, that our model is in very good
agreement with his results. Unfortunately Monte Carlo simulations do not
give any information about the time-scale in which the events are
occurring. Mouritsen could clearly demonstrate that at least 10
different states of order of the lipids from crystalline to fluid have
to be used to get reasonable results.

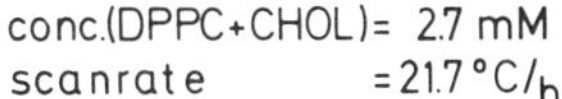

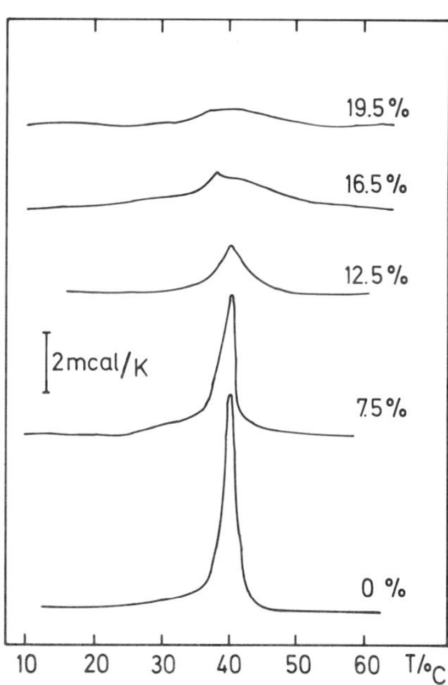

Fig. 17. Differential scanning calorimetric DSC measurements of DPPC
vesicles containing increasing amounts of cholesterol showing a decreasing ΔH as well as a broadening of the phase
transition enthalpy temperature dependence.

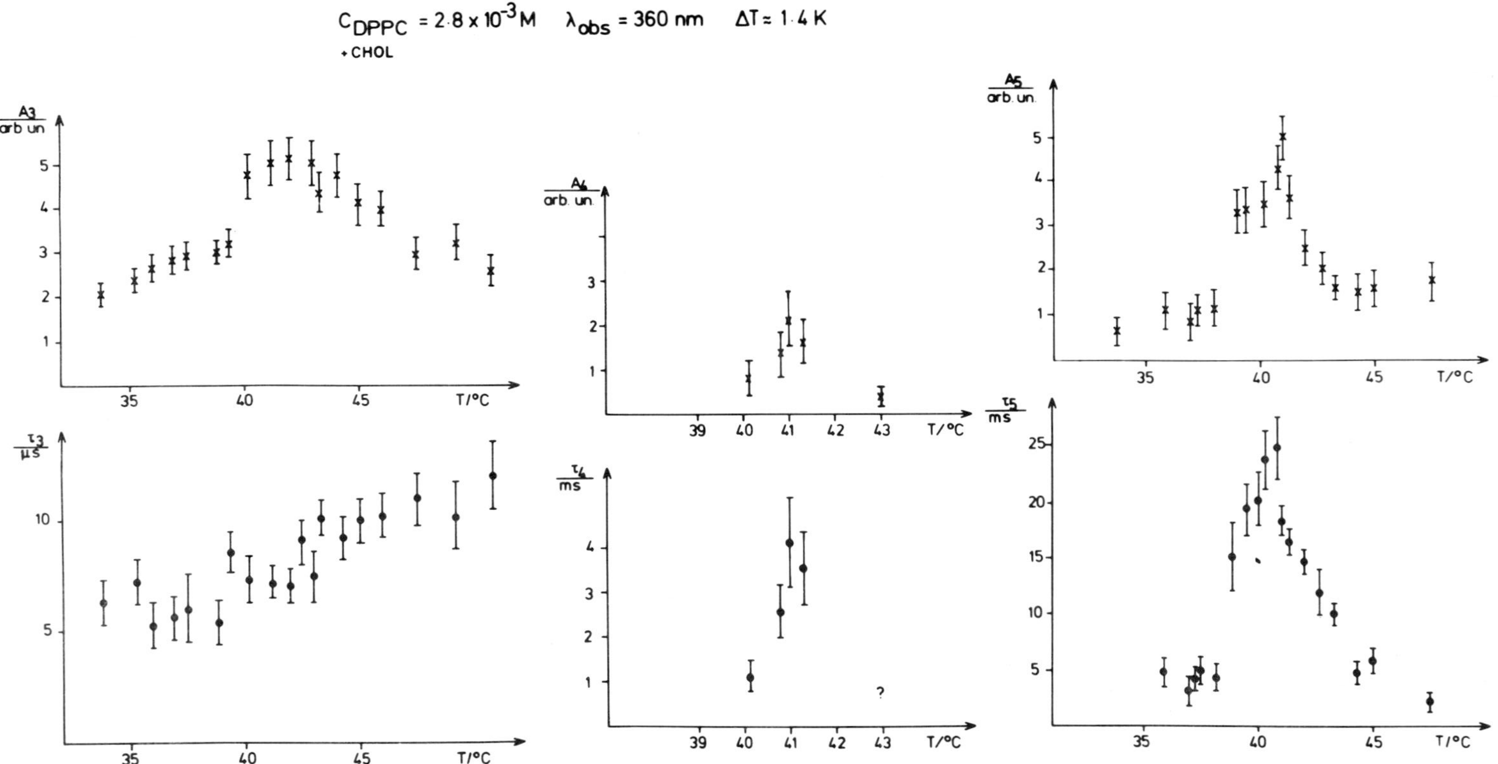

Fig. 18. Turbidity-temperature dependencies of the three slower relaxations between 1 µs and 0.1 s in DPPC vesicles containing 16.5 % of cholesterol and covering ~80 % of the transition enthalpy.

After we could design a molecular model for the main phase transition in lipid membranes we will investigate if the incorporation of functional units into the bilayer might cause characteristic alterations in the relaxation behavior. We concentrated on the energetically important time window from 10^{-6} to 10^{-1} s. The incorporation of cholesterol is the simplest modification of a lipid membrane. The equilibrium fluorescence anisotropy measurements of figure 10 show a small decrease of anistropy below the phase transition temperature, T_m, and a larger increase above T_m. The DSC measurements in figure 17 indicate a decrease in ΔH as well as a broadening of the phase transition towards higher temperatures with increasing CHOL content. Higher concentrations than 18% CHOL were not possible because otherwise the vesicle preparations became unstable.

The T-jump experiments summarized in figure 18 for the energetically important time range from 1 μs to 30 ms show characteristic differences to the pure lipid membrane. Relaxation signal 3 which is attributed to the formation of complex rotational isomers in the hydrocarbon chains of the lipids and the start of lateral diffusion of lipid molecues is weakened in its cooperativity and the temperature-relaxation time dependence is broadened and has lost its maximum. CHOL reduces the amplitude as well as the relaxation time of signal 4. Signal 5 shows a similar cooperativity as compared with the pure lipid membrane but a shoulder at higher temperatures is observed. This findings can be explained by the preference of CHOL to favour an intermediate state of the lipid molecules in its neighborhood, thus broadening relaxation 3, weakening relaxation 4 and creating additional relaxations at temperatures above T_m. The relaxation amplitudes of figure 18 can be reconstructed by the superposition of a narrow and a broad part as done in figure 19. The narrow component is connected with the change of free lipids and the broad component belongs to the hydrogen-bonded lipid-cholesterol units. A comparison of figures 19 and 17 demonstrates how well the sum of the amplitudes $\Sigma A = A_3 + A_4 + A_5$ agrees with the DSC data. This again proves that the amplitudes of our relaxations are linearly correlated with the ΔH of reaction.

In figure 20 we have summarized relaxation experiments of DPPC vesicles containing the channel forming polypeptide gramicidin, the lipid/GA ratio was 15/1. The most striking differences to the pure lipid membrane are the almost immeasurable relaxation 5 and the shoulder in relaxation 3 at temperatures below T_m as well as the broadening of τ_3. Relaxation 4 is not markedly altered.

We can conclude that GA shifts the relaxations into the 10 μs range and lowers the ΔH of the dynamic processes involved in the phase transition; the latter observation is in good agreement with DSC measurements on the same preparation included in figure 21. More details will be published /25/.

In the following part I will report results on the influence of the membrane protein bacteriorhodopsin on the phase transiton dynamics in DMPC bilayers. Integral membrane proteins like the light driven proton pump BR span the lipid bilayer matrices of biological membranes because they possess highly hydrophobic surfaces which are in contact with the alkane chains of annular lipids and hydrophilic areas exposed to the aqueous phase on both sides of the membrane. To study the biological activity of BR in detail, it is suitable to separate the protein from other components of the natural membrane, and to reconstitute it in a functionally active state in lipid bilayers.

Reconstituted BR in unilamellar DMPC vesicles is a good model system to investigate the protein itself as well as the interaction of annular lipids with the hydrophobic surface of BR. The protein can be kept either in an active state pumping protons by exposing it to photons of 560 nm wavelength or passive if no photons of a suitable wavelength are present. The related interactions can best be studied in the temperature range of the phase transition around 23° C. We investigated the relaxations of pure DMPC membranes and compared these results with analogous bilayers containing increasing amounts of functionally active BR. Figure 22 summarizes the kinetic data for a DMPC/BR ratio of 199/1 in the time range of 10^{-6} to 10^{-1} s; the full lines show the amplitude time dependences for pure DMPC vesicles after T-jumps of ca. 1° C which started at the temperatures indicated. The dotted lines are similar kinetic measurements for reconstituted BR in DMPC bilayers. As a result we could define the range of interaction between the protein BR and the annular lipids reaching as far as three to four lipid layers showing an exponential decay; this could be achieved by changing the DMPC/BR ratio. Furthermore we could observe that the two relaxation processes around 0.5 ms and 5-20 ms which are vey important in pure DMPC membranes are negligible in DMPC/BR preparations containing more than 1 mole % of BR.

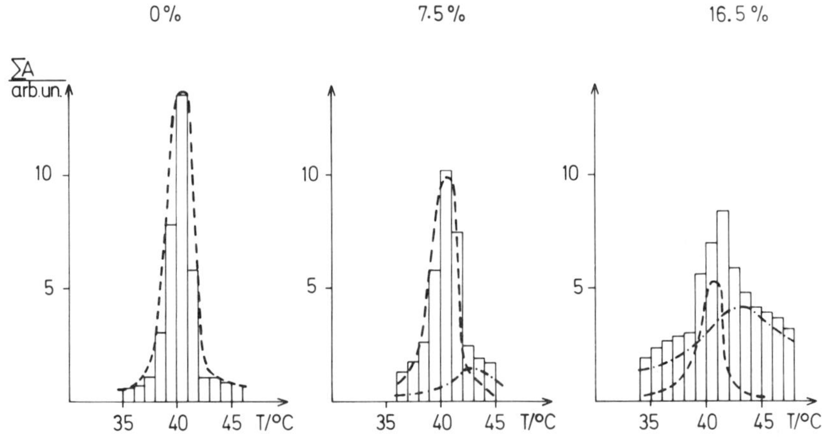

Fig. 19   Sum of the three relaxations between 10^{-6} s and 10^{-1} s from
figure 18 for the temperature intervals indicated and increasing concentrations of cholesterol. The summarised
amplitudes prove the good reconstruction of the DSC observations from kinetic data and allow for the splitting of
the signal into a component from DPPC lipids (narrow) and
DPPC-cholesterol complexes (broad).

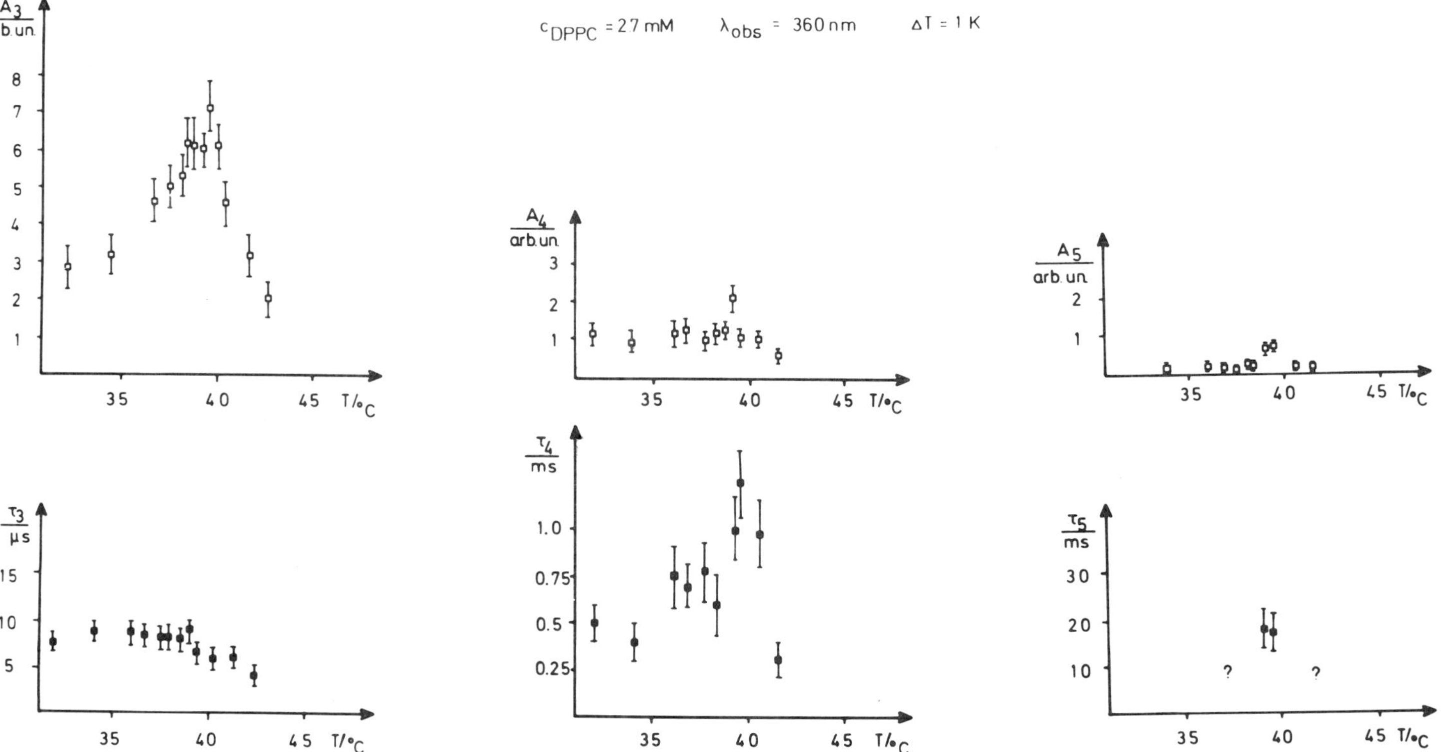

Fig. 20. Turbidity-temperature dependences for the three slower relaxations from 10^{-6} to 10^{-1} s of DPPC vesicles containing ~6.7 % of the channel forming polypeptide gramicidine A. The marked differences to pure DPPC vesicles in figures 12, 13 and 14 become clear.

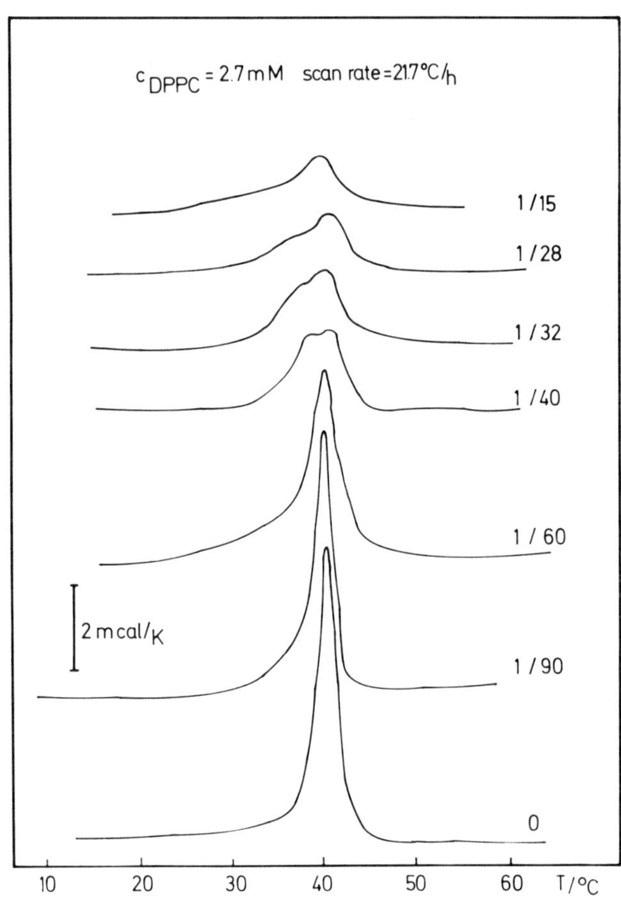

Fig. 21. Differential scanning calorimetric (DSC) measurement of vesicles of DPPC containing increasing amounts of gramicidin at ratios indicated showing the decrease and broadening of the phase transition enthalpy.

The relaxation process in the time range of 10 µs was strengthened by
BR. BR pumps protons through the DMPC bilayer if its chromophore retinal
is excited by photons of 560 nm. In the time range of milliseconds
strong spectral changes either at ca. 410 nm or 560 nm can be monitored
being due to the biological activity of BR. At 360 nm the turbidity
changes due to the increasing mobility of the lipids during the phase
transition can be observed. By monitoring spectral changes at 550 nm
(BR) and 360 nm (DMPC) simultaneously after a T-jump we can separate
dynamic processes due to the activity of BR from kinetic processes
resulting from DMPC. By comparing the most important BR and lipid
relaxations as done in figure 23 we observed that the lipid motions
around 10 µs and the functional activity of BR are strongly coupled. The
lipids have to undergo their phase transition before the activity of BR
changes. Keeping in mind that the Arrhenius dependences in figure 23 are
the sum of several complex equilibria they nevertheless demonstrate the
strong interaction between BR and the annular lipids as well as the
preference of the 10 µs time domain to create highly mobile lipids for
interaction with the protein surface. We therefore call the relaxation
around 5 µs a "functional important movement", FIM, (after Frauenfelder)
of the surface of the protein, BR; this is reflected in the fast motions
of the annular lipids.

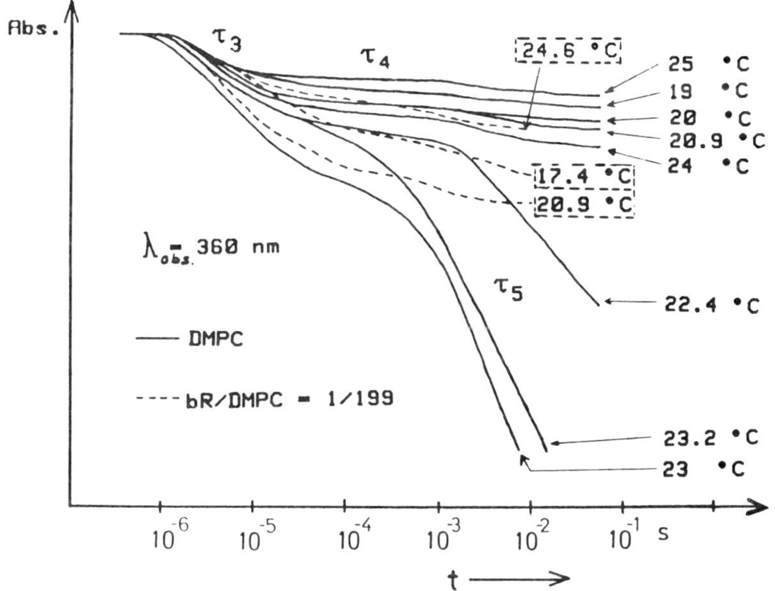

Fig. 22. Semilogarithmic plot of the turbidity temperature dependences in pure DMPC vesicles (solid lines) and 0.5 %
bacteriorhodopsin containing vesicles (dashed lines) after
fast temperature jumps of ca. 1 °C starting at the temperatures indicated. The strong difference between pure DMPC
and BR containing preparations is evident.

CONCLUSIONS

The highly versatile ILTJ technique is described which allows
kinetic measurements in times from 10^{-9} to 10^0 s by creating temperature
jumps in water without the need of additional ions or molecules. The
phase transition dynamics of lipid membranes were measured and a
molecular model explains the processes behind the 5 well separated
relaxations. A correlation between ΔH of reaction and the relaxation
amplitudes could be established which proves that the most important
dynamic changes are occurring in the micro- to milli-second time range.
The influence of cholesterol and gramicidin as well as bacteriorhodopsin
on the phase transition dynamics could be measured. By comparing all
kinetic data we can conclude that the 5 µs time domain is very important
for a favorable interaction of the lipid molecules with protein
surfaces. We call the underlying dynamic changes "functional important
movements", FIMs, of the protein surface. In the future we will try to
find further evidence for the biological importance of dynamic changes
in the microsecond time range, which we rate far more important for
biology than picosecond simulations.

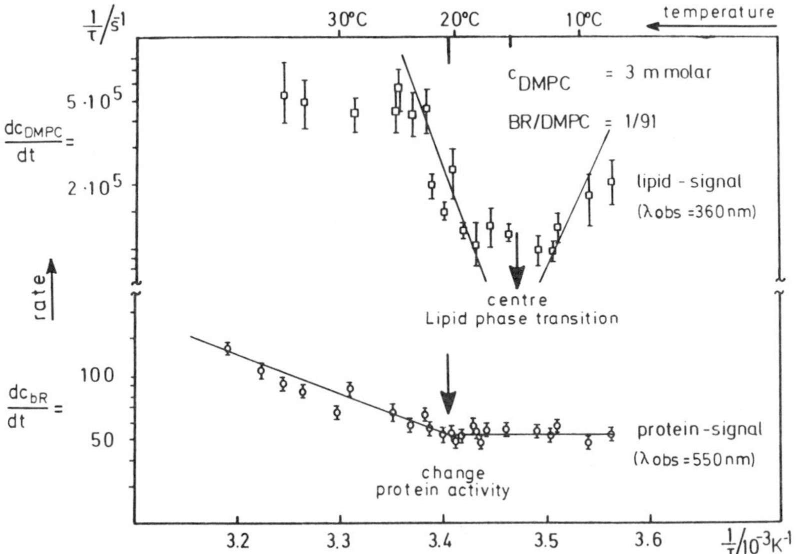

Fig. 23. Arrhenius dependences of the fast lipid relaxation near 5 µs
observed as a turbidity change at 360 nm and the slow
protein signal near 20 ms observed as an increasing absorption from the retinal chromophore inside bacteriorhodopsin.
After the lipids have increased their mobility, the protein
changes its activity.

ACKNOWLEDGEMENTS

I would like to thank my coworkers H. Wolff, A. Schmidt and W.
Frisch who developed the ILTJ technique together with me as well as A.
Genz and R. Groll who performed most of the experiments included in this
paper, last but not least I am thankful to the German Research
Foundation, DFG, for a grant, which made this investigation possible.

REFERENCES

1. D. Marsh, Electron Spin Resonance: Spin Labels, pp. 51-137 in: Membrane Spectroscopy, E. Grell ed., Springer Verlag Berlin,
2. R. J. Smith and E. Oldfield, Science 225:280 (1984).
3. L. Brand, J. R. Knutson, L. Davenport, J. M. Beechem, R. E. Dale, D. G. Walbridge and A. A. Kowalczyik, Time Resolved Fluorescence Spectroscopy, pp. 259-305, in: Spectroscopy and the Dynamics of Molecular Biological Systems, P. Bayley and R. E. Dale, eds., Academic Press, New York (1985).
4. a) F. Fillaux "Vibrational Spectroscopy", in: The Enzyme Catalysis Process, A. Cooper and J. Houber, eds., Plenum Pub. Corp., New York (1989).
    b) S. Cusack "Dynamic Neutron Scattering", in: ibid.
5. W. Knoche, Pressure Jump Methods, pp. 187-210, in: Investigation of Rates and Mechanisms of Reactions Part II, G. G. Hammes, ed., A. Wiley Interscience Publ., New York (1974).
6. J. F. Holzwarth, W. Frisch and B. Gruenewald, Fast Dynamic Processes in the Hydrocarbon Tail Region of Phospholipid Bilayers, in: Microemulsion, I. D. Robb, ed., Plenum Publ. Corp., New York (1982).
7. a) J. F. Holzwarth, Laser Temperature Jump pp. 47-59, in: Techniques and Applications of Fast Reactions in Solution, W. J. Gettins and E. Wyn-Jones, eds., D. Reidel Pub. Comp., Dordrecht Holland (1979),
    b) W. Frisch, A. Schmidt, R. Volk and J. F. Holzwarth, Laser T-Jump Arrangement with Time Resolution in the Second to Picosecond Range pp. 61-70, in: ibid (1979),
    c) J. F. Holzwarth, A. Schmidt, H. Wolff and R. Volk, J. Phys. Chem. 81:2300 (1977),
    d) see also ref. 19 and 20.
8. J. F. Holzwarth, V. Eck and A. Genz, Iodine Laser Temperature Jump: Relaxation Processes in Phospholipid Bilayers on the Picosecond to Millisecond Time Scale pp. 351-378, in: Spectroscopy and the Dynamics of Molecular Biological Systems, P. M. Bayley and R. E. Dale, eds., Academic Press, London (1985).
9. J. M. Kremer, M. W. Esker, C. Pathmamanocharan and C. Wiersema, Biochemistry 16:2932 (1977).
10. A. Genz and J. F. Holzwarth, Colloid+Polymer Sci. 263:484 (1985).
11. V. Eck and J. F. Holzwarth, Fast Dynamic Phenomena in Vesicles of Phospholipids During Phase Transitions, pp. 2059-2079, in: Surfactants in Solution, Vol. 3, K. L. Mittal and B. Lindman, eds., Plenum Pub. Corp, New York (1984).
12. A. Genz and J. F. Holzwarth, Eur. Biophys. J. 13:323 (1986).
13. A. Genz, J. F. Holzwarth and T. Y. Tsong, Biophys. J. 50:1043 (1986).
14. M. P. Heyn and N. A. Dencher, Reconstruction of Monomeric Bacteriorhodopsin into Phospholipid Vesicles, pp. 31-35, in: Methods in Enzymology, Vol. 88, L. Packer, ed., Acad. Press, New York (1982).
15. a) G. H. Czerlinski and M. Eigen, Z. Elektrochemie 63:652 (1959),
    b) C. F. Bernasconi, in: Relaxation Kinetics, C. F. Bernasoni, ed., Academic Press, New York (1976)
    c) D. H. Turner, Temperature Jump Methods, pp. 141-189, in: Investigation of Rates and Mechanisms of Reactions Part II, C. F. Bernasoni, ed., John Wiley+Sons, New. York (1986).
16. A. Dawson, J. Gormally, E. Wyn-Jones and J. F. Holzwarth, J. C. S. Chem. Comm. 386 (1981).

17. a) B. Marcandalli, C. Winzek and J. F. Holzwarth, Ber. Bunsenges.
     Phys. Chem. 88:368 (1984),
     b) B. Marcandalli, W. Knoche and J. F. Holzwarth, Gazetta Chimica
     Italiana 116:417 (1986),
     c) see also ref. 22.
18. a) W. C. Natzle, C. B. Moore, D. M. Goodall, W. Frisch and J. F.
     Holzwarth, J. Phys. Chem. 85:2882 (1981),
     b) D. M. Goodall, R. C. Greenhow, B. Knight, J. F. Holzwarth and W.
     Frisch, Single Photon Infrared Photochemistry: Wavelength and
     Temperature Dependence of the Quantum Yield for the Laser Induced
     Ionization of Water, pp. 561-568, in: Techniques and Applications
     of Fast Reactions in Solution, W. J. Gettins and E. Wyn-Jones, eds.
     D. Reidel Pub. Comp., Dordrecht Holland (1979).
19. J. J. Bannister, J. Gormally, J. F. Holzwarth and T. A. King, The
     Iodine Laser and Fast Reactions, pp. 227-233, in: Chemistry in
     Britain 20:227 (1984).
20. J. F. Holzwarth, Application of the Iodine Laser Temperature Jump
     in Biophysical Chemistry, pp. 94-125, in: Proceedings of the First
     International Workshop: Iodine Laser and Applications, B. Králiková
     and J. Krása, eds., Institute of Physics, Czechoslovac Academy of
     Sciences, 180:40 Prag 8 (1986).
21. J. F. Holzwarth, F. Meyer, M. Pickard and H. B. Dunford,
     Biochemistry 27:6628 (1988).
22. B. Marcandalli, G. Stange and J. F. Holzwarth, J. C. S. Faraday
     Trans. I, 84:2807 (1988).
23. B. Gruenewald, W. Frisch and J. F. Holzwarth, Biochim. Biophys.
     Acta 641:311 (1981).
24. L. Cruzeiro-Hanson and O. G. Mouritsen, Biochim. Biophys. Acta
     944:63 (1988).
25. A. Genz, T. Tsong and J. F. Holzwarth, Colloids and Surfaces, in
     preparation (1989).

THE IMPACT OF RECOMBINANT DNA TECHNIQUES ON THE STUDY OF ENZYMES

John R. Coggins

Department of Biochemistry
University of Glasgow
Glasgow G12 8QQ
Scotland

INTRODUCTION

Enzymologists have long dreamed of being able to change individual
amino acid residues in enzymes in order to study their contribution to
catalysis or substrate binding or regulation. Despite intensive research
efforts over the past 40 years this cannot yet be achieved either by the
total chemical synthesis of the enzyme protein, nor is it easy to achieve
by site-specific chemical modification. However the recent dramatic
advances in molecular biology have made it possible. Purposely modified
enzymes can now be prepared by combining the new techniques of
manipulating DNA in vitro with enzymes (so called recombinant DNA
techniques) with oligonucleotide chemistry. This has lead to the
development of the new science of enzyme engineering. Besides providing
a systematic means of manipulating enzyme structure recombinant DNA
techniques have also revolutionised enzymology in two other important
ways. Firstly it is now possible to produce large quantities of known
but previously inaccessible enzymes by the cloning and overexpression of
their genes. Secondly the misery has been taken out of the determination
of primary structure since sequencing the DNA of cloned genes is very
much faster than sequencing the enzymes as proteins.

The purpose of this chapter is to explain how many of the practical
aspects of enzymology have been radically changed by the application of
recombinant DNA technology.

STRATEGY AND TACTICS IN MODERN ENZYMOLOGY

The emergence of the science of molecular enzymology in the 1950's
and 1960's depended to a large extent on the study of enzymes such as
chymotrypsin and lysosyme that were readily available in pure form on the
gram scale. During the 1970's the focus of interest began to shift to
the study of enzymes that were present in living cells in only minute
quantities and could only be purified with great difficulty in microgram
amounts. However the techniques of protein chemistry, which were
necessary for primary structure analysis and the identification of

functional residues, still required milligram quantities of material and
the techniques of physical biochemistry, such as X-ray crystallography
and high field NMR spectroscopy, which are required to determine the
tertiary structure of proteins, usually required hundreds of times this
quantity for a total structural analysis.

The development of techniques for cloning genes and subsequently
over-expressing them offered salvation to enzymologists working on low
abundance enzymes. The overall experimental strategy which is now
employed for the study of such enzymes is summarised in Table 1.

Table 1.   Strategy for studying a low abundance enzyme

1.   Clone the gene encoding the enzyme
2.   Sequence the gene
3.   Work out a method for overexpressing the enzyme gene and
        accumulating large quantities of enzyme protein
4.   Develop a simple, large scale purification procedure for the
        enzyme
5.   Study the enzyme using all the usual techniques employed on
        readily available enzymes
6.   Determine the tertiary structure
7.   Use site-directed mutagenesis to study structure function
        relationships

This scenario is most readily followed for enzymes from prokaryotic
organisms where the genes are not split by non-coding sequences (introns)
and for which cloning and overexpression is generally straight forward.
In eukaryotic organisms enzyme genes almost always consist of a mosaic of
coding (exons) and non-coding sequences. In these species the initial
RNA copy (transcript) of the gene is processed to remove the non-coding
sequences and only the final mature messenger RNA (mRNA) has a sequence
corresponding to the protein sequence. It is this sequence that must be
copied into DNA and cloned and overexpressed.   Fortunately there is a
viral enzyme (reverse transcriptase) which is able to synthesise
complementary DNA (cDNA) copies of mRNA's. In eukaryotic species
therefore the first step in cloning an enzyme gene is to make an mRNA
preparation under conditions which maximise the synthesis of the desired
enzyme and then prepare cDNA from this material for cloning. This will
be discussed further later.

THE KEY ENABLING TECHNOLOGIES

The successful application of recombinant DNA techniques to the
study of enzymes has depended on a number of crucial methodological
advances in the fields of molecular biology, biochemistry and chemistry.
These are reviewed in this section.

<u>DNA enzymology</u>

One of most important crucial advances that opened up the
recombinant DNA field was an improved knowledge of DNA enzymology
(Kornberg, 1980) and subsequently the commercial availability of enzyme
preparations which allowed the facile manipulation of DNA in vitro
(Fersht, 1985). Briefly methods were developed for the replication (that

is copying) and repair of duplex DNA in the test tube using a proteolytic
fragment of the enzyme DNA polymerase I of <u>E. coli</u> (the Klenow enzyme).
It was also found that duplex DNA could be specifically cleaved with
enzymes. Of particular significance was the discovery of a class of
highly specific endonucleases (known as type II restriction enzymes) that
could cleave duplex DNA at rare and specific sites. These restriction
enzymes recognise particular nucleotide sequences (usually tetra or
hexanucleotide sequences) and scores of them are now available with
different sequence specificities. The fragments of DNA generated by
restriction enzyme digestion can be readily rejoined by another class of
enzymes, the DNA ligases. Using all these enzymes in the appropriate
order novel DNA molecules (recombinant molecules) can be assembled in the
test tube by "cutting out and sticking together" (restricting and
ligating) natural DNA fragments. It is also possible to substitute
natural fragments of DNA with chemically synthesised fragments (see
below).

<u>Oligonucleotide chemistry</u>

     Until recently the chemical synthesis of oligodeoxynucleotides was a
difficult and lengthy process and entirely the preserve of a small band
of dedicated synthetic chemists (Amarnath & Broom, 1977).   However
within the last 10 years methods have been developed for the rapid manual
synthesis of oligonucleotides (Itakura, 1982; Gait, 1984; Caruthers,
1987) and these are now used routinely by non-specialists. Moreover the
methods have been automated and "gene machines" are now commercially
available which are capable of coupling as many as 100 nucleotide units
per day (Hunkapillar et al., 1984).

     The solid phase method of synthesis is used. Suitably protected
mononucleotides are added sequentially to a nucleoside covalently
attached to an insoluble polymeric support [Figure 1] . The growing
oligonucleotide chain remains attached to the support throughout the
synthesis and the excess starting materials, coupling reagents and
by-products are removed by washing with solvents. The solid support is
generally silica which has been chemically modified in introduce amino
groups. The first nucleoside, protected on its 5' hydroxyl group with
the acid labile dimethoxytrityl group, is attached to these amino groups,
via its 3' hydroxyl group, using succinic acid so that the final linkage
between the solid support and the nucleoside is an ester linkage. The
chains are built up in a 3' to 5' direction. The synthesis cycle
consists of deprotecting the 5'-OH of the first nucleoside and coupling
on a 5'-protected deoxynucleoside derivative. Generally phosphoramidites
are used with tetrazole as the activating agent and a phosphite triester
linkage is formed. Any unreacted 5'-OH (usually 1 to 2% of the material)
is then blocked (capped) with acetic anhydride and the phosphite triester
linkage oxidised to a phosphate triester linkage. At every stage
unreacted materials and by-products are washed away with solvent and then
the synthesis cycle is repeated as many times as necessary to complete
the assembly of the desired oligodeoxynucleotide. At the end of the
synthesis the oligonucleotide is fully deprotected, hydrolyzed from the
support, and purified to homogeneity by polyacrylamide gel
electrophoresis. This electrophoretic step completely resolves the full
length oligonucleotide product from the shorter capped sequences
resulting from the small amount of incomplete coupling at each step.

     Synthetic oligonucleotides are used as probes for gene cloning, as
reagents for sequencing and for site-directed mutagenesis, and as
fragments from which complete synthetic genes can be assembled.

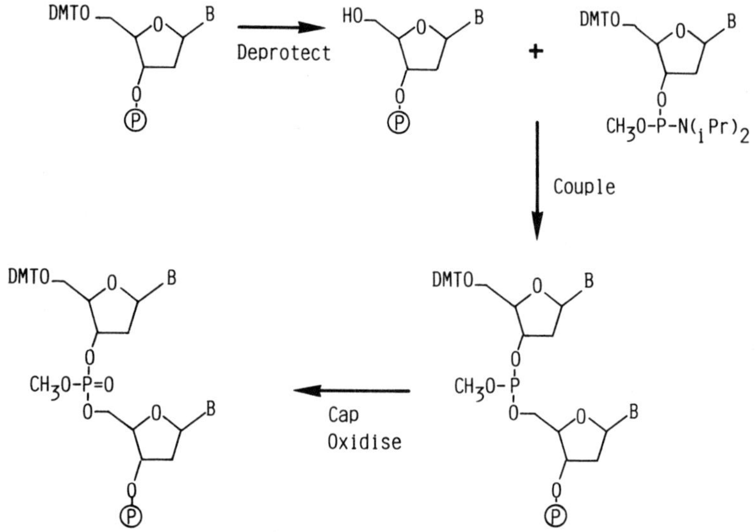

DMTO–[sugar]–B →(Deprotect) HO–[sugar]–B + DMTO–[sugar]–B, $CH_3O-\overset{\cdot}{P}-N(_iPr)_2$

(Couple)

DMTO–[sugar]–B, $CH_3O-P=O$ ... ←(Cap, Oxidise) DMTO–[sugar]–B, CH_3O-P ...

Figure 1. Steps in the synthesis of a dinucleotide by the solid phase
method. DMT:dimethoxytrityl protecting group; B:appropriately protected
base (adenine, thymine, guanine or cytosine); P: silica or controlled
pore glass support; Ipr:isopropyl. (Adapted from Caruthers, 1987).

<u>DNA Sequencing</u>

Although several methods for the rapid sequencing of DNA have been
developed (Gilbert, 1981; Sanger, 1981) one method, the Sanger "dideoxy"
or chain termination method (Sanger, Nicklen & Coulson, 1977) has become
predominant. This method can be used to determine sequences at the rate
of many kilobases per week in experienced hands. As a result it is now
possible, providing that the cloned gene or a suitable cDNA clone is
available, to determine the DNA coding sequence of an enzyme and deduce
its amino acid sequence in a few weeks and in favourable cases even in a
few days. In recent years protein sequencing techniques have also
greatly improved both in speed and in the much smaller amounts of
material required (see Walsh, 1987) but they still cannot match the
rapidity of DNA sequencing.

CLONING

The cloning of a gene encoding a particular enzyme from a
prokaryotic organism such as <u>Escherichia coli</u> is now relatively straight
forward (Old & Primrose, 1985; Glover, 1985; Berger & Kimmel, 1987).
First a gene bank or gene library must be constructed. DNA from the
organism is fragmented by partial digestion with a restriction enzyme to
produce a series of fragments some of which are likely to contain the
required gene. These DNA fragments are inserted into a vector which is a
double stranded DNA molecule that can replicate after foreign DNA
fragments have been inserted into it. The vector (generally a plasmid
or a bacteriophage) has to be prepared for the insertion of the DNA
fragments by digestion with an appropriate restriction enzyme. After the

DNA fragments have been ligated into the vector, the mixture of
recombinant vector molecules is mixed with a suitable host organism such
as E. coli; the vector DNA enters (transforms) the host and is
replicated. Conditions are chosen so that only one vector molecule
enters each host cell and the resulting "library" therefore consists of a
heterogeneous population of host cells containing different recombinant
vectors. The library is plated out so that single colonies can be
isolated which contain host cells carrying only one version of the
recombinant vector. Then these colonies are screened for the desired
gene or gene product.

When preparing gene libraries it is most important to ensure that a
fully representative set of DNA fragments is inserted into the vector;
it is also important to realise that only a few of the many thousands of
different restriction fragments inserted (cloned) into a vector will
contain the desired gene. This means that a sensitive screening method
is essential to identify those colonies containing the desired gene.

The simplest way to screen a gene library for the gene of a
prokaryotic enzyme is to use a complementation assay. This is
conveniently illustrated by work from my laboratory in Glasgow on the
cloning of the genes for the enzymes of the shikimate pathway of E.
coli. E. coli mutants were available which lacked each of the enzymes of
the pathway. These mutants, referred to as auxotrophs, were unable to
grow on minimal medium but would grow if the minimal medium was
supplemented with the various end products of the pathway or if they were
transformed with plasmids (vectors) containing the missing genes. By
transforming the appropriate mutants with recombinant plasmids containing
fragments of E. coli DNA and selecting for growth on minimal medium we
were, over a three year period, able to clone the genes encoding six of
the seven enzymes of this pathway (see Duncan, Edwards & Coggins, 1987
and Anton & Coggins, 1988 and references therein).

If there is no simple complementation assay available to screen the
gene library more laborious techniques must be used. If an antibody is
available to detect the enzyme of interest then this can be used to
screen the library for the production of enzyme protein (Huynh et al.,
1985; Helfman & Hughes, 1987; Mierendorf et al., 1987). Alternatively
a DNA hybridisation assay can be used (Berger & Kimmel, 1987). This
involves using a radio-labelled DNA fragment, which is complementary to
the desired gene, to screen DNA derived from single colonies in the gene
library. The problem is to obtain a satisfactory probe. If any protein
sequence is available, for example from microscale protein sequencing of
the purified enzyme, then a short oligonucleotide probe (preferably at
least 20 nucleotides long) corresponding to the protein sequence can be
designed and synthesized (Hunkapillar et al., 1984; Lathe, 1985; Wallace
& Miyada, 1987). The degeneracy of the genetic code and ignorance about
codon usage in many species means that it is usually impossible to design
a unique probe and generally a mixture of labelled nucleotides must be
used for the initial screening. Alternatively a fragment of gene
encoding the protein from a closely related species can be used as a
probe. Such heterologous probes work well if the species divide is very
narrow but very rarely work between distantly related species.

The cloning of eukaryotic genes is generally more complicated than
the cloning of prokaryotic genes because of the mosaic structure of thir
genes which has already been mentioned. Eukaryotic gene libraries are
therefore generally made with cDNA (and not genomic DNA); this allows
screening methods which rely on expression of the enzyme such as antibody

detection and complementation to be used. cDNA clones are also suitable
for the construction of strains that overexpress the enzyme. For very
low abundance, eukaryotic enzymes for which antibodies are not available,
the most direct method of obtaining clones is to purify a few micrograms
of the enzyme using modern high performance chromatographic methods and
then to take advantage of the new supersensitive automated protein
sequencers (Hunkapillar et al., 1986) and obtain 10 or more residues of
protein sequence. It will then be possible to design and synthesise an
oligonucleotide probe for screening the cDNA library. The possible
experimental strategies for cloning the gene for a low abundance
eukaryotic enzyme are listed in Table 2.

     Table 2.   Some possible strategies for obtaining the gene of a low
                     abundance eukaryotic enzyme.

1.   Purify the enzyme and use microsequencing to determine the
     N-terminal sequence and/or some internal sequence. Then design and
     make a synthetic oligonucleotide probe and screen a cDNA library.
2.   Purify the enzyme and raise an antibody and use this to screen an
     expression library.
3.   Use complementation; it will almost certainly be necessary to put
     the eukaryotic cDNA fragments behind a very strong promoter for this
     to work (referred to as using a "forced" expression library).
4.   Use a heterologous probe; this approach works well if the species
     divide is very narrow (for example between two higher plant species)
     but very rarely works between distantly related species.
5.   Synthesize the gene (the protein sequence must be known).

ACHIEVING OVEREXPRESSION

     The achievement of overexpression is not always straight forward.
E. coli genes can frequently be overexpressed in E. coli and yeast genes
in yeast but the overexpression of an enzyme gene in an organism other
than the original host (heterologous expression) is sometimes very
difficult to achieve. The enzyme may be produced in poor yield or in a
denatured form (Marston, 1987) and problems can also arise for enzymes
that require substantial post-translational modification. Sometimes it
is useful to overexpress the desired enzyme as a fusion protein by
placing its gene in the correct reading frame within the coding sequence
of a readily overexpressed protein such as the beta-galactosidase of E.
coli (Marston, 1987). It is then necessary to have a strategy for
releasing the desired enzyme from the fusion protein, for example by
limited proteolysis with a suitably specific protease. It is also
sometimes possible to arrange that the over-expressed enzyme is secreted
and this can greatly simplify purification (for a good example of this
see Gardell et al., 1985).

     Our own work on the shikimate pathway enzymes of E. coli (see for
example Anton & Coggins, 1988) illustrates very clearly the advantages of
achieving overexpression. Typically these enzymes are present in
wild-type cells at the level of 50 to 250µg per 20g wet weight of cells.
Purifications of 2,000- to 20,000-fold are required to give homogeneous
enzyme and the overall yields are typically in the range of 10 to 20%.
Hence 50µg is a good yield from 20g of cells and frequently only a few
micrograms of material are obtained from a preparation. 100-fold
overexpression, which can be very simply obtained by placing the gene in
a high copy number plasmid (Duncan et al., 1984), makes 5mg quantities of

418

enzyme available. At this level of expression up to 1% of the cellular
protein is the desired enzyme. By using specially designed
overexpression vectors the yield can easily be raised by another factor
of 10 (Anton & Coggins, 1988). These improved yields have revolutionised
our approach to the study of these enzymes.

SITE-DIRECTED MUTAGENESIS

     Before the advent of recombinant DNA techniques there were two
approaches for obtaining altered enzymes so that structure function
relationships could be explored. The genetic approach was restricted to
the isolation of phenotypically selectable mutants. The mutants were
produced by random mutagenesis of the whole organism with radiation or
chemical mutagens. The alternative approach employed by protein chemists
involved chemical modification with either group specific (Means &
Feeney, 1971) or site specific reagents (Shaw, 1970). The aim in this
case was to chemically alter specific amino acid side chains; it was
frequently impossible to restrict the changes to a single site, many side
chains could not be altered because they were chemically inert and it was
generally not possible to alter one naturally occurring side chain into
another. Neither approach allowed the enzymologist to alter specifically
and totally a single amino acid of his or her choice. The availability
of enzyme genes has totally altered this situation. It is now possible
to change deliberately single codons and therefore single amino acid side
chains and to express the altered gene and isolate and characterise the
altered enzyme.

     Two completely specific methods are available. One, which will not
be considered in detail here, involves chemical synthesis of the gene
(Rossi & Zoller, 1987). Mutagenesis via gene synthesis can be
accomplished either by total gene synthesis or by replacement of a
limited segment of the gene by a synthetic oligonucleotide fragment that
codes for the desired mutation. The mutant gene fragment is "stitched"
into the gene by using the appropriate combination of restriction enzymes
and DNA ligase.

     The more commonly used approach for site-specific mutagenesis is to
use the method of oligonucleotide-directed mutagenesis (Smith, 1982;
Rossi & Zoller, 1987). The mutagen is a synthetic
oligodeoxyribonucleotide and the method relies on the stability of short
oligodeoxyribonucleotide-DNA duplexes containing one or more mismatches.
The target gene must first be inserted into the single standed circular
DNA of a bacteriophage (virus) such as M13. An oligonucleotide, usually
20 to 30 nucleotides long, which is complementary to the target gene in
the region of interest except for a limited mismatch at the intended site
of mutation, is then synthesised. The synthetic oligonucleotide is
hybridised to its complementary sequence in the cloned gene [see Figure
2] and then the mutant gene is synthesised enzymically (using the Klenow
fragment of DNA polymerase and DNA ligase) from nucleotide triphosphates
by extending the synthetic primer which contains the mutation site and
using the wild-type gene as a template [Figure 2]. The completed double
stranded DNA contains a strand carrying the wild-type gene and a strand
carrying the mutant gene. It is used to transform a suitable bacterial
host and produce a mixture of progeny some of which contain only the
mutant gene and some only the original wild-type gene. Bacterial cells
containing phage carrying the mutant gene can be identified readily since
their DNA will hybridise to the oligonucleotide used for mutagenesis at
higher temperatures (because there will be no mismatches) than will the
DNA from cells containing the wild-type gene (Smith, 1982).

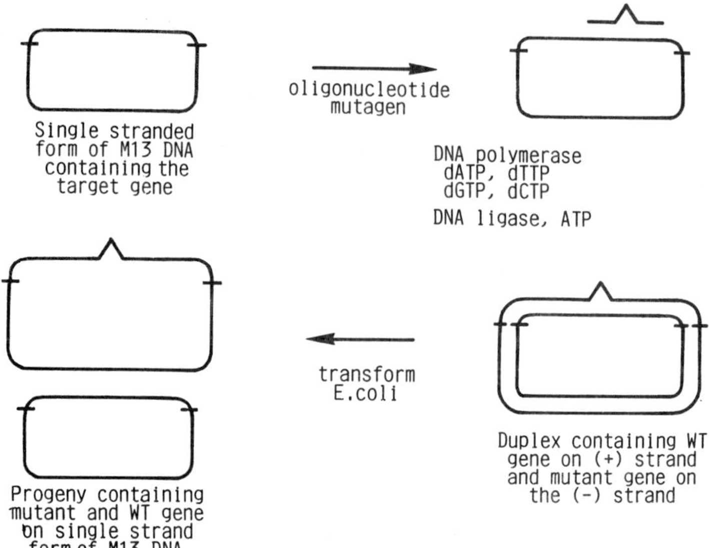

Figure 2.   Scheme for the construction of a mutant gene by
oligonucleotide-directed mutagenesis. A synthetic oligonucleotide
containing one or more mismatched bases is annealed to the target gene
which has been cloned into the single standed vector M13.   The
oligonucleotide is extended with nucleotides and enzymes to produce
closed circular double stranded DNA which contains wild-type (+) strand
and mutant (-) strand. This duplex is used to infect (transform) E. coli
and gives rise to mutant and wild-type progeny. The mutant progeny are
selected by hybridisation with the oligonucleotide used as the mutagen
and used for the production of mutant protein.

This method allows mutant genes to be isolated quickly and
reliably. It is also very versatile since it can not only be used to
introduce single point mutations, but also multiple point mutations
(because of the greater number of mismatches longer oligonucleotides are
needed for this) as well as insertions and deletions [see Figure 3].

Despite its elegant simplicity there are some practical problems and
difficulties which the enzymologist using site-directed mutagenesis
should take into consideration. Firstly it is always necessary to check
mutant genes by sequencing them completely. This will ensure that no
other changes have been accidently introduced and that the mutation
itself has not been repaired by the host organism. Although the
achievement of multiple mutations looks easy on paper it is often very
difficult to introduce several mutations simultaneously. Large genes (5
kilobases or longer corresponding to a protein subunit size of 150,000 or
greater) are difficult to handle in M13 and different single-stranded
vectors such as those of the pEMBL family may have to be used to handle
them (Dente et al., 1985). It is essential to express the mutant gene in
cells that lack the wild-type gene as it is sometimes difficult to

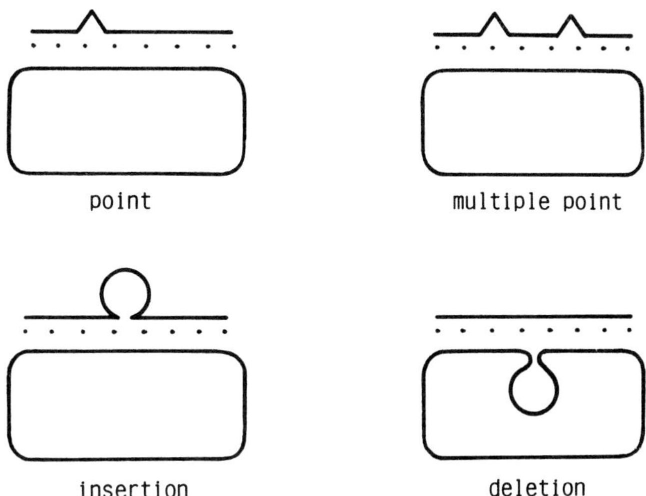

Figure 3. Scheme for obtaining point mutations, multiple point mutations, insertions and deletions by oligonucleotide-directed mutagenesis in M13.

separate the mutant and wild-type enzyme and the proper characterisation of the mutant enzyme in the presence of the wild-type enzyme may be impossible. Sometimes even a single amino acid change may affect folding and as a result it may not be possible to isolate some mutant enzymes. It should also be recognised that one good "mutant maker" can keep several enzymologists very busy.

Site-directed mutagenesis also suffers from two important limitations: it is restricted to the 20 naturally occurring protein amino acids and it does not provide a means of modifying the polypeptide backbone.

OTHER METHODOLOGICAL ADVANCES

Rapid advances in other non-molecular biological areas have also contributed significantly to the new enzymology. Purification techniques have dramatically improved over the past 10 years with the wide spread application of affinity chromatography (Dean et al., 1985) and with the introduction of high performance chromatography for protein purification (Burgess, 1987). The methods of protein chemistry have become much more sensitive particularly with the introduction of new super sensitive sequencers and fab mass spectrometers for the direct sequencing of proteins (Hunkapillar et al., 1984; Walsh, 1987). The use of synchrotron radiation and introduction of new detectors has also facilitated structure determination by X-ray crystallography (Hendrickson, 1987) and recently very high field nmr spectroscopy has been used directly to determine the tertiary structure of some smaller enzymes (Markley, 1987; Redfield in this volume).

PROTEIN ENGINEERING

The coming together of all these techniques has lead to the birth of
a new subject "protein engineering". This term has come to be used only
since recombinant DNA techniques have been used to manipulate protein
structure. The use of the term frequently carries the implication that
proteins can only be engineered by recombinant DNA techniques and that at
the present time proteins are being engineered routinely (Wetzel, 1986).
This is an oversimplification and I think it is useful to consider the
pragmatic definition of protein engineering put forward by Wetzel (1986)
in the first issue of the journal "Protein Engineering". He defined
"engineering" as the "art or science of making practical application of
the knowledge of the pure sciences" and continued "how can we engineer
proteins before we understand the fundamental rules by which primary
sequence determines higher structure, and by which structure, at all
levels, determines function?" Clearly a strict definition of protein
engineering based on a conventional definition of engineering is too
exclusive at this stage in the development of the subject. We must
instead be pragmatic and look at the sorts of experiment that can be
carried out with our current level of knowledge. Following this argument
Wetzel (1986) has recognised four areas which can reasonably be
considered as protein engineering:

1.   Work taking advantage of sequence information only, for eample
     domain shuffling.
2.   Screening randomly generated mutants.
3.   Experiments to elucidate and quantitate structure-function
     relationships (analytical protein engineering).
4.   Experiments in which a protein of improved features is confidently
     synthesised from a design based on well understood structure
     function relationships (true protein engineering).

It is the last two of these areas which are of particular interest
to enzymologists.

ENZYME ENGINEERING

The advent of the science of protein engineering has at last
provided enzymologists with a method to alter the amino acid sequence of
an enzyme at will. The essential requirements for enzyme engineering are
outlined in Table 3.

Table 3.   The essential requirements for enzyme engineering

1.   Cloned gene
2.   Over producing strain
3.   Simple, efficient purification procedure
4.   Sequence
5.   Access to synthetic oligonucleotides
6.   Strategy for site-directed mutagenesis
7.   Knowledge of the tertiary structure and mechanism of the
        enzyme
8.   A clearly defined question

The first 6 requirements are obvious; the requirement for knowledge
of the tertiary structure and the mechanism is essential if changes are
to be made in a rational manner. The final requirement recognises that

enzyme engineering involves a considerable experimental input and can
only be justified when there is a significant question to be answered.

At the present time the principal application of protein engineering
in enzymology is in the study of the basis of enzyme catalysis (Blow et
al., 1986; Fersht et al., 1986; Leatherbarrow & Fersht, 1986; 1987;
Fersht, 1987; Shaw, 1987; Wells & Estell, 1988). As will be discussed in
the next chapter the comparison of engineered and wild-type enzymes is
providing a wealth of information about catalysis. However it must be
recognised that the proper kinetic characterisation of the altered
enzymes is essential if interpretable results are to be obtained.
Absolute rate contants, that is rate constants independent of enzyme
concentration, must be measured (Fersht, 1987). This is because enzymes
can rarely be isolated in such a way that all the protein present is
fully active. This makes the comparison of two different enzymes (mutant
and wild-type or two different mutants) impossible unless the
concentration of active sites is known. This difficulty, of measuring
the number of functional active sites, was solved almost 30 years ago for
chemically modified enzymes by the introduction of the "all or none"
assay (Ray & Koshland, 1960) and lead to the development of active site
titrants (for a modern discussion of the problem see Fersht, 1985). When
characterising mutant enzymes kinetically, either enzyme titrants must be
used, or else pre-steady state rate constants, which can be made
independent of enzyme concentration, must be measured (Fersht, 1987).
Both active site titration and the application of pre-steady state
kinetics require that an intermediate accumulates during the reaction as
is the case with trypsin (Rutter et al., 1987) and tyrosyl-tRNA
synthetase (Fersht & Leatherbarrow, 1987) (see next chapter).

SUMMARY

The application of recombinant DNA techniques to the study of
enzymes has lead to the development of the new science of enzyme
engineering. The enzymologists dream of a method for the facile
modification of enzyme structure has been realised by the technique of
site-specific mutagenesis. It is now possible to tailor and re-design
existing enzymes by mutating their genes and expressing the mutant
enzymes. It is even possible, in principle, to synthesize new enzymes
via the synthesis and expression of their genes. The field is exciting,
especially as a tool for basic research on enzymes, and in time it will
undoubtedly lead to the design and construction of enzymes with enhanced
properties and the ability to catalyse novel reactions.

REFERENCES

Anton, I.A. & Coggins, J.R., 1988, Sequencing and overexpression of the
Escherichia coli aroE gene encoding shikimate dehydrogenase Biochem. J.,
249:319.

Berger, S.L. & Kimmel, A.R., 1987, "Guide to Molecular Cloning
Techniques" (Methods in Enzymology, Volume 152), Academic Press, New York.

Blow, D.M., Fersht, A.R. & Winter, G., 1986, "Design, Construction and
Properties of Novel Protein Molecules" Royal Society, London.

Burgess, R.R., 1987, Protein purification, in "Protein Engineering",
Oxender, D.L. & Fox, C.F. (eds.), Liss, New York p. 71.

Caruthers, M.H., 1987, DNA synthesis as an aid to protein modification and design, in "Protein Engineering", Oxender, D.L. & Fox, C.F. (eds.), Liss, New York p. 65.

Dean, P.D.G., Johnson, W.S. & Middle, F.A., 1985, "Affinity Chromatography, A Practical Approach", IRL Press, Oxford.

Dente, L., Sollazzo, M., Baldari, C., Cesareni, G. & Cortese, R., 1985, The pEMBL family of single-stranded vectors, in "DNA Cloning, A Practical Approach", Volume I, Glover, D.M. (ed.), IRL Press, Oxford p. 101.

Duncan, K., Lewendon, A. & Coggins, J.R., 1984, The purification of 5-enolpyruvylshikimate 3-phosphate synthase from an overproducing strain of _Escherichia_ _coli_ _FEBS Letters_ 165:121.

Duncan, K., Edwards, R.M. & Coggins, J.R., 1987, The pentafunctional _arom_ enzyme of _Saccharomyces_ _cerevisiae_ is a mosaic of monofunctional domains _Biochem. J._ 246:375.

Fersht, A.R., 1985, "Enzyme Structure and Mechanism" (2nd ed.), Freeman, San Francisco.

Fersht, A.R., 1987, Kinetic aspects of purposely modified proteins, in "Protein Engineering", Oxender, D.L. & Fox, C.F. (eds.), Liss, New York p. 225.

Fersht, A.R. & Leatherbarrow, R.J., 1987, Structure and activity of tyrosyl-t-RNA synthetase, in "Protein Engineering", Oxender, D.L. & Fox, C.F. (eds.), Liss, New York p. 269.

Fersht, A.R., Leatherbarrow, R.J. & Wells, T.N.C., 1986, Binding energy and catalysis: a lesson from protein engineering of tyrosyl-tRNA synthetase, _Trends. Biochem. Sci._, 11:321.

Gait, M.J., 1984, "Oligonucleotide Synthesis, A Practical Approach", IRL Press, Oxford.

Gardell, S.J., Craik, C.S., Hilvert, D., Urdea, M.S. & Rutter, W.J., 1985, Site-directed mutagenesis shows that tyrosine-248 of carboxypeptidase A does not play a crucial role in catalysis, _Nature_ 317:551.

Glover, D.M., 1985, "DNA Cloning, A Practical Approach, Volumes I & II", IRL Press, Oxford.

Helfman, D.M. & Hughes, S.H., 1987, Use of antibodies to screen cDNA expression libraries prepared in plasmid vectors, in "Guide to Molecular Cloning Techniques", Berger, S.L. & Kimmel, A.R. (eds.) (Methods in Enzymology Volume 142), Academic Press, New York p. 451.

Hendrickson, W.A., 1987, X-ray diffraction, in "Protein Engineering", Oxender, D.L. & Fox, C.F. (eds.), Liss, New York p. 5.

Hunkapillar, M., Kent, S., Caruthers, M.H., Dreyer, W., Firca, J., Griffen, C., Horvath, S., Hunkapillar, T., Tempst, P. & Hood, L., 1984, A microchemical facility for the analysis and synthesis of genes and proteins, _Nature_ 310:105.

Huynh, T.V., Young, R.A. & Davis, R.W., 1985, Constructing and screening
cDNA libraries in gt10 and gt11, in "DNA Cloning, A Practical
Approach", Volume I, Glover, D.M. (ed.), IRL Press, Oxford p. 49.

Itakura, K. 1982, Chemical synthesis of genes, Trends. Biochem. Sci.,
7:442.

Kornberg, A., 1980, "DNA Replication", Freeman, San Francisco.

Lathe, R., 1985, Synthetic oligonucleotide probes deduced from amino acid
sequence data, theoretical and practical considerations, J. Mol. Biol.
183:1.

Leatherbarrow, R.J. & Fersht, A.R., 1986, Protein Engineering 1:7.

Leatherbarrow, R.J. & Fersht, A.R., 1987, Use of protein engineering to
study enzyme mechanisms' in "Enzyme Mechanisms", Page, M.I. & Williams,
A. (eds), The Royal Society of Chemistry, London, p78.

Markley, J.L., 1987, One- and two-dimensional nmr spectroscopic
investigations of the consequences of amino acid replacements in
proteins, in "Protein Engineering", Oxender, D.L. & Fox, C.F. (eds.),
Liss, New York p. 15.

Marston, F.A.O., 1986, The purification of eukaryotic polypeptides
synthesized in Escherichia coli, Biochem. J. 240:1.

Mathews, B.W., 1987, Structural aspects of purposely modified proteins,
in "Protein Engineering", Oxender, D.L. & Fox, C.F. (eds.), Liss, New
York p. 221.

Means, G.E. & Feeney, R.E., 1971, "Chemical Modification of Proteins",
Holden Day, San Francisco.

Mierendorf, R.C., Percy, C. & Young, R.A., 1987, Gene isolation by
screening gt11 libraries with antibodies, in "Guide to Molecular
Cloning Techniques", Berger, S.L. & Kimmel, A.R. (eds.) (Methods in
Enzymology Volume 142), Academic Press, New York p. 458.

Old, R.W. & Primrose, S.B., 1985, "Principles of Gene Manipulation, An
Introduction to Genetic Engineering", Blackwell, Oxford.

Oxender, D.L. & Fox, C.F., 1987, "Protein Engineering", Liss, New York.

Ray, W.J. & Koshland, D.E., 1960, Comparative structural studies on
phosphoglucomutase and chymotrypsin, Brookhaven Symposia 13:135.

Rossi, J. & Zoller, M., 1987, Site-specific and regionally directed
mutagenesis of protein-encoding sequences, in "Protein Engineering",
Oxender, D.L. & Fox, C.F. (eds.), Liss, New York p. 51.

Rutter, W.J., Gardell, S.J., Rocniak, S., Hilvert, D., Sprang, S.,
Fletterick, R.J. & Craik, C.S., 1987, Redesigning proteins by genetic
engineering, in "Protein Engineering", Oxender, D.L. & Fox, C.F. (eds.),
Liss, New York p. 257.

Shaw, E., 1970, Chemical modification by active-site-directed reagents,
in "The Enzymes", 3rd ed., Volume 1, Boyer, P.D. (ed.), Academic Press,
New York p. 91.

Shaw, W.V., 1987, Protein Engineering: the design, synthesis and characterisation of factitious proteins' Biochem. J., 246:1.

Smith, M., 1982, Site directed mutagenesis, Trends. Biochem. Sci., 7:440.

Wells, J.A. & Estell, D.A., 1988, Subtilisin - an enzyme designed to be engineered, Trends. Biochem. Sci., 13:291.

Wallace, R.B. & Miyada, C.G., 1987, Oligonucleotide probes for the screening of recombinant DNA libraries, in "Guide to Molecular Cloning Techniques", Berger, S.L. & Kimmel, A.R. (eds.) (Methods in Enzymology Volume 142), Academic Press, New York p. 432.

Walsh, K.A., 1987, "Methods In Protein Sequence Analysis:1986", Humana Press, Clifton.

Wetzel,R., 1986, What is protein engineering, Protein Engineering, 1:3.

SITE DIRECTED MUTAGENESIS AS A TOOL TO STUDY ENZYME CATALYSIS

John R. Coggins

Department of Biochemistry
University of Glasgow
Glasgow G12 8QQ
Scotland

INTRODUCTION

Although the use of site-directed mutagenesis to engineer proteins, which was introduced in the preceding chapter, is less than 10 years old there have already been many publications describing the application of this technique to the study of enzymes. In this chapter I shall give some examples to illustrate how enzymologists are using site-diected mutagenesis to study enzyme catalysis and at the same time discuss the advantages and the limitations of the technique. Further examples may be found in the several excellent reviews that have appeared recently (Blow et al., 1986; Fersht et al., 1986a; Leatherbarrow & Fersht, 1986; 1987; Oxender & Fox, 1987; Shaw, 1987; Wells & Estell, 1988).

THE POTENTIAL AND THE LIMITATIONS OF THE TECHNIQUE

The practical aspects of achieving mutagenesis using specific oligonucleotides as mutagens were described in the preceding chapter. Providing the cloned gene and a suitable expression system are available the preparation of an enzyme which has been specifically altered at one or more sites can now be carried out routinely (Rossi & Zoller, 1987). The most critical aspect of the work is the very careful kinetic and structural comparison of the wild-type enzyme with the various mutants (Fersht, 1987). The first requirement is to have adequate structural and mechanistic information about the enzyme in question. Ideally the 3d-structure of the enzyme should be known. The general approach is to make single amino acid changes that will result in the deletion of a single interaction, such as a hydrogen bond, with the substrate or the product or with some putative "intermediate" on the catalytic pathway. It is generally assumed that making single amino acid changes of this kind will only have a very localised effect and that the mutant enzyme will fold up and attain an overall structure not significantly different from the structure of the wild-type enzyme. The only difference between the wild-type and mutant enzymes should be at the changed side chain. It is important to realise that this is an assumption which needs to be verified for each mutant. Absolute proof would require that the

3d-structure of each mutant enzyme should be solved and compared with the wild-type enzyme but generally this is too ambitious an undertaking. Often before attempting mutagenesis the proposed change is modelled using energy minimisation and molecular dynamics simulations to establish that it will not be too disruptive (Karplus, 1987). However in every case some experimental evidence about the structure of the mutant protein should also be sought. One possibility is to establish that the chemical reactivity of neighboring side chains is unaltered. Alternatively a physical chemical method such as circular dichroism, or fluorescence or, if sufficient material is available, nmr spectroscopy (Markley, 1987), should be used to establish that the basic conformation of the enzyme has not been changed by mutagenesis.

Site-directed mutagenesis is most useful as a tool for studying the contributions of individual amino acid side chains to the catalytic process. Interactions that have arisen through evolution are removed to produce less efficient enzymes. These are then characterised kinetically to build up a picture of the effect of adding side chains to the less 'evolved' enzymes. Mutations that result in completely inactive enzyme are best avoided since interpretation may be difficult. The lack of activity may be due to the removal of an essential catalytic group, such as the active centre serine of a serine protease, but it might also be due to the inability of the mutant enzyme to fold into its proper tertiary structure. Generally most information is obtained by comparing a series of mutants all of which have at least some activity.

IDENTIFYING CATALYTIC RESIDUES

The most obvious application of site-directed mutagenesis is to investigate the importance of individual catalytic residues in enzymes. I shall use as examples the work of Rutter and Craik and their colleagues (Rutter et al., 1987) on Tyr-248 of carboxypeptidase A and Asp-102 of trypsin. Despite all the mechanistic and crystallographic work on carboxypeptidase A (Lipscomb, 1980) and on the serine proteases (Stroud et al., 1975; Fersht, 1985) the catalytic roles of both these residues have remained the subject of much speculation.

Carboxypeptidase A

Carboxypeptidase A (CPA) is an exopeptidase which catalyses the hydrolysis of aromatic and branched chain aliphatic amino acids from the C-terminus of proteins and peptides. It also catalyses the hydrolysis of certain esters of aromatic alcohols with N-acylamino acids. Its structure and mechanism have been extensively studied (reviewed in Lipscomb, 1980). The structural work indicates that only two amino acid side chains, those of Glu-270 and Tyr-148, and the active centre zinc ion are near enough to the cleaved peptide bond to be involved in catalysis as shown in Figure 1 (Hartsuck & Lipscomb, 1971.)

Two possible mechanisms have been proposed for the cleavage of peptide and ester substrates: one involves direct nucleophilic attack of Glu-270 and the formation of a mixed anhydride intermediate while the other involves water as the nucleophile with Glu-270 acting as a general base catalyst (Hartsuck & Lipscomb, 1971) (see Figure 2).

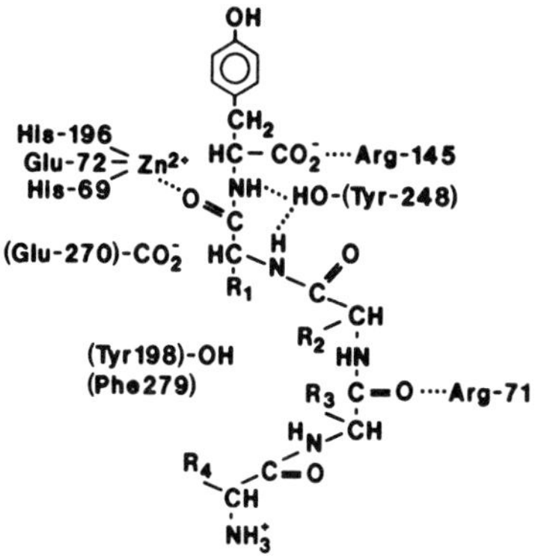

Figure 1.   The proposed mode of binding of a tetrapeptide substrate to
the active site of carboxypeptidase A (adapted from Hartsuck
& Lipscomb, 1971).

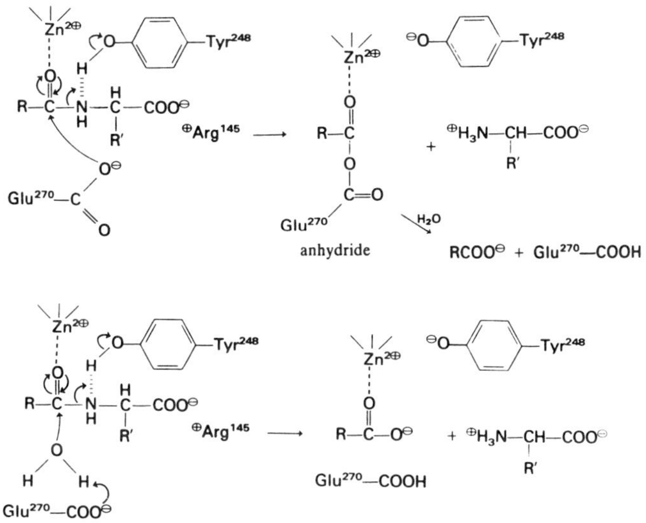

Figure 2.   Possible mechanisms for the action of carboxypeptidase A
(adapted from Walsh, 1979). The 'anhydride' mechanism
involving direct attack by Glu-270 on the scissile peptide
bond is shown above and the mechanism in which Glu-270
functions as a general base to increase the nucleophilicity
of a water molecule is shown below.

For both mechanisms the postulated role of Tyr-248 is to be a
general acid catalyst contributing a proton to the incipient amine
generated by peptide bond cleavage (Lipscomb, 1980) (see Figure 2). This
role for Tyr-248 had been suggested by crystallographic studies which
showed that the hydroxyl of Tyr-248 is far from the active centre in the
free enzyme but in the presence of substrate moves to be within hydrogen
bonding distance of the cleaved peptide bond (Hartsuck & Lipscomb, 1971;
Lipscomb, 1980).

The role of the phenolic hydroxyl group of carboxypeptidase A was
investigated by replacing Tyr-248 with phenylalanine, in other words the
hydroxyl group was deleted (Gardell et al., 1985; Rutter et al., 1987).
The mutagenesis was carried out on rat CPA since a cDNA encoding this
enzyme had previously been isolated (Gardell et al., 1985). The rat
enzyme has a very similar amino acid sequence to the bovine enzyme (78%
of the residues are identical) which had been used in all the earlier
mechanistic and structural work and all the residues postulated to have a
role in catalysis are conserved between the two species. Expression of
the wild-type and the mutant enzyme was carried out in yeast using the
alpha factor system for expression and secretion (Gardell et al., 1985).
Contrary to expectation the kinetic characteristics of the wild-type
enzyme and the mutant (Tyr248Phe) towards peptide substrates were very
similar (see Table 1) (Gardell et al., 1985). In particular the k_{cat}
values for the hydrolysis of the simple peptide substrate (Bz-Gly-Phe-OH)
and the simple ester substrate (Bz-Gly-OPhe) were essentially unaltered
by the mutation while the k_{cat} for a second peptide substrate
(Cbz-Gly-Gly-Phe-OH) was reduced to 46% of the value found for the
wild-type enzyme.

Clearly the loss of the hydroxyl group of Tyr-248 does not have a
significant effect on catalysis; its phenolic hydroxyl group cannot
therefore play an essential role as the general acid catalyst required
for peptide bond hydrolysis.   The mutation does however lead to a
significant increase in K_m for the two peptide substrates and this
suggests that Tyr-248 has a role in substrate binding (Gardell et al.,
1985).

Table 1.   Comparison of the kinetic parameters of carboxypeptidase A
             and carboxypeptidase A Tyr248Phe (from Gardell et al., 1985).

Substrate	Kinetic parameter	Wild-type (Tyr248)	Mutant (Phe248)
Bz-Gly-Phe (peptide)	k_{cat}	18	21
	K_m	39	199
	(k_{cat}/K_m)	(0.46)	(0.11)
Cbz-Gly-Gly-Phe (peptide)	k_{cat}	52	23
	K_m	27	194
	(k_{cat}/K_m)	(1.9)	(0.12)
Bz-Gly-OPhe (ester)	k_{cat}	1200	1140
	K_m	96	136
	(k_{cat}/K_m)	(12.5)	(8.4)

[Units: k_{cat} seconds^{-1} (s^{-1}); K_m micromolar (μM)]

This work, which was the first example of the use of site-directed
mutagenesis to prove that a particular amino acid side chain was not
required for catalysis, is a very good example of the power of protein
engineering as an analytical tool for establishing the role of individual
groups in enzyme catalysis. The final stage of the work, the
confirmation of the 3d-structure of the mutant enzyme, has not yet been
reported.

Trypsin is an endopeptidase which cleaves peptide chains on the C-terminal side of lysine and arginine residues. It is a member of the extensively studied serine protease family and is structurally and mechanistically very closely related to chymotrypsin and elastase (reviewed in Stroud et al., 1975). Early chemical modification experiments established that the active site residue, Ser-195, was hyper-reactive and became acylated during catalysis (see Figure 3). Further chemical experiments and crystallography established the close proximity of the side chains of His-57 and Asp-102 to the active site serine and these two residues are thought to play a crucial role in enhancing the nucleophilicity of Ser-195. The His-57 side chain must be deprotonated for the enzyme to be active and it clearly acts as a general base assisting in the removal of the proton from Ser-195 and subsequently donates this proton to the leaving group (the full mechanism is shown in Figure 3). The precise role of the side chain of Asp-102 which occurs in its deprotonated form in the active enzyme has been a matter of considerable argument. It is thought to share a proton with the imidazole of His-57 which would be expected to increase the basicity

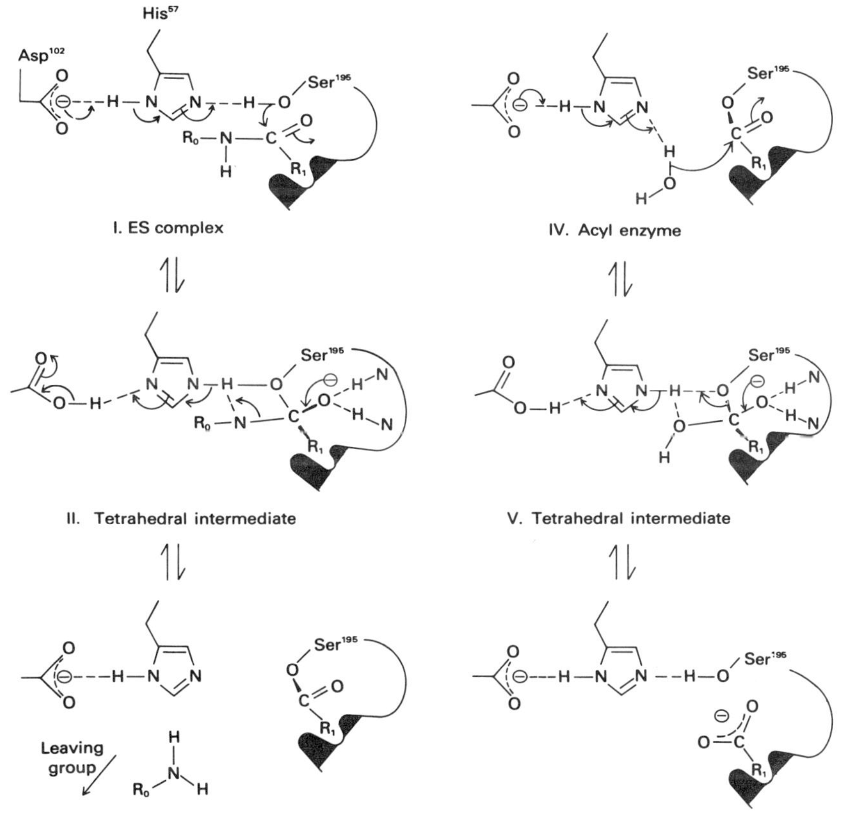

Figure 3. Mechanism of action of trypsin and the related mammalian serine proteases (from Walsh, 1979; adapted from Stroud et al., 1975).

of this residue and orientate it so as to enhance the nucleophilicity of
Ser-195. To investigate the role of Asp-102 of trypsin Rutter and Craik
and their co-workers (Rutter et al., 1987) have changed it to an
asparagine.

As with carboxypeptidase A all the structural and mechanistic work on
trypsin had been carried out on the bovine enzyme but the mutagenesis was
carried out on a full length copy of the rat trypsinogen coding sequence
(trypsinogen is the inactive precursor from which trypsin can be prepared
by limited proteolysis). Rat trypsin is 74% identical to the bovine
enzyme and all the active site residues are conserved. Both wild-type
and mutant (Asp102Asn) rat trypsin were purified from suitable stable
cell lines that had been designed to secrete rat trypsinogen. Kinetic
characterisation of the mutant and wild-type enzymes showed that k_{cat}
at neutral pH had been reduced 5000-fold and the pK_a of the active
centre histidine had been reduced from 6.8 to 5.3. The reactivity of the
active centre Ser-195 towards diisopropylfluorophosphate was found to
have decreased 10,000-fold but, in contrast, the reactivity of the active
centre His-57 towards the specific alkylating reagent Tos-Lys-CH_2Cl was
only reduced 5-fold. This suggested that the imidazole side chain of
His-57 was still properly positioned in the active site of the mutant
enzyme. The primary effect of mutation seemed to be on the
nucleophilicity of Ser-195 which was consistent with the idea that the
entire catalytic triad, with its precise network of hydrogen bonds, was
required for catalysis. To throw further light on this the mutant enzyme
was crystallised in the presence of the competitive inhibitor benzamidine

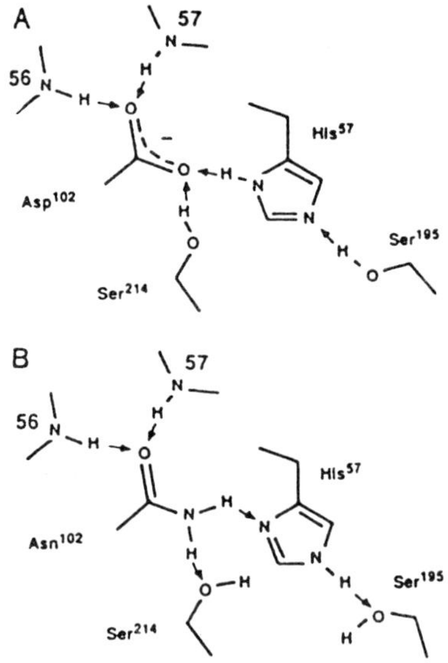

Figure 4.   (A) The hydrogen bond net work found in wild type trypsin
             compared with (B) the hydrogen bond network found in the
             mutant trypsin Asp102Asn. In the mutant trypsin the
             orientation of the hydrogen bond between His-57 and Ser-195
             is the reverse of that found in the wild type; this accounts
             for the low nucleophilicity of Ser-195 observed in the
             mutant enzyme (from Sprang et al., 1987).

and its structure determined. The results (Sprang et al., 1987; see Fig.4) showed that in the mutant enzyme the imidazole side chain of His-57 was hydrogen bonded to Asn-102 in such a way (Figure 4B) that it could no longer accept the hydrogen of Ser-195 and so could not function as a general base to activate this residue. In contrast the wild-type enzyme, which has aspartic acid at position 102 (Figure 4A), has the proper arrangement of hydrogen bonds to activate Ser-195.

MODIFYING SPECIFICITY

Site-directed mutagenesis is also being used as a tool to study and modify enzyme specificity. I shall again use the work of Rutter and Craik and their colleagues on trypsin as an example of what can be achieved (Rutter et al., 1987).

As mentioned above trypsin cleaves peptide chains preferentially at lysine and arginine residues. The specificity site is a pocket which has a wide entrance flanked by residues Gly-216 and Gly-226 and at the bottom of the pocket there is the negatively charged residue Asp-189 (Stroud et al., 1975). Arginine side chains fit nicely into the pocket and form a cyclic network of hydrogen bonds with Asp-189. The shorter lysine side chains interact with Asp-189 via a water molecule; this water molecule is displaced when the bulkier arginine side chain occupies the pocket. The structural basis of the substrate specificity of trypsin has been probed by mutating the two glycine residues at the entrance of the specificity pocket to alanine residues (Craik et al., 1985). Small changes would be expected in the tertiary structure of the trypsin-substrate complex if methyl groups are added in this way to perturb the specificity pocket. Such changes would be expected to affect substrate alignment and therefore the kinetic parameters of the mutant enzyme. Modeling studies suggested that the mutant trypsin Gly216Ala should show relatively better activity for arginine- as compared to lysine-containing substrates because the water molecule that is presumably displaced by the methyl group of Ala-216 does not take part in arginine binding. In contrast it was predicted that the mutant Gly226Ala should show a preference for lysine-containing substrates because the methyl group of Ala-226 will conflict with the binding of arginine side chains to Asp-189.

The mutants trypsin-216Ala, trypsin-226Ala and trypsin-216Ala, 226Ala were made and characterised kinetically (Craik et al., 1985; Rutter et al., 1987; Table 2). As with the studies on Asp102 of trypsin described above the mutagenesis was carried out on the rat enzyme. The results showed that the addition of a single methyl group at either position 216 or 226 greatly reduced the ability of trypsin to bind either lysine- or arginine-containing substrates; the K_m's were greatly increased. However the catalytic activity was altered in a discriminatory manner since trypsin-216Ala worked relatively better, as predicted, on arginine substrates and trypsin-226Ala worked better, also as predicted, on lysine substrates. The double mutant, trypsin-216Ala,226Ala, with its greatly constricted binding pocket showed very low catalytic activity.

An attempt has also been made to alter the substrate specificity of trypsin completely by changing Asp-189 into a lysine residue (Graf et al., 1987). The mutant enzyme, trypsin-189Lys showed no activity towards lysine- and arginine-containing substrates. It had been hoped that the mutant enzyme would show some activity towards acidic substrates. This

was not the case. Computer modeling suggested that the newly introduced
Lys-189 side chain might lie outside the substrate binding pocket which
was therefore expected to be simply a hydrophobic cavity analogous to
that found in chymotrypsin. The observation that the mutant enzyme
showed some low chymotrypsin activity was consistent with this prediction.

Table 2.   Comparison of the kinetic parameters for various trypsin
            mutants modified at the entrance of the specificity site.

Enzyme	Substrate type	K_m (μM)	k_{cat} (min^{-1})	$\dfrac{k_{cat}}{K_m}$	Arg/Lys[a]
Wild-type	Arg	14	1400	100	11
	Lys	140	1300	9	
Ala226Gly (revertant)	Arg	12	1700	150	12
	Lys	120	1500	13	
Gly216Ala	Arg	390	1000	2.6	29
	Lys	3900	340	0.09	
Gly226Ala	Arg	480	13	0.027	0.57
	Lys	3700	170	0.047	
Gly216Ala, Gly226Ala (double mutant)	Arg	220	1.1	0.0051	2.8
	Lys	330	0.6	0.0018	

[a] Ratio of k_{cat}/K_m values for arginine and lysine substrates; data
taken from Craik et al., 1985.

These experiments involving mutation of the specificity site of
trypsin have provided new information about how this enzyme recognises
its substrate. They have not yet resulted in a very active enzyme with
altered specificity. To date site-directed mutagenesis has been most
useful as an analytical tool for establishing which amino acid side
chains are involved in substrate recognition rather than as method for
the 'useful' alteration of enzyme specificity.

UNDERSTANDING TRANSITION STATE STABILISATION

The distinctive feature of enzyme catalysis compared with solution
catalysis is that the enzyme specifically binds its substrate and can use
binding energy to enhance catalytic rate. Pauling was the first to
suggest that an enzyme should be complementary in structure to the
transition state of the substrate rather than to the substrate itself so
that the enzyme would tend to deform the substrate into the transition
state (Pauling, 1948). The current view on the utilization of binding
energy in catalysis supports the concept of enzyme-transition state
complementarity but does not demand that the substrate is distorted by

the enzyme. The presence of binding sites on the enzyme that form better
bonds with the transition state of the substrate than with the unreacted
substrate is sufficient to increase the turnover number of the enzyme
(k_{cat}). Various lines of experimental evidence support the idea that
there is differential binding of transition state and substrate leading
to transition state stabilisation, for example, structure activity
studies with modified substrates (Fersht, 1985), X-ray diffraction
studies on enzymes such as the serine proteases and lysosyme (Fersht,
1985) and studies on the binding of transition state analogues to enzymes
(Wolfenden & Frick, 1987.)

Site-directed mutagenesis has proved to be a particularly valuable
tool for studying the stabilisation of transition states by enzymes since
very often the stabilisation involves specific hydrogen bonds and the
technique allows these to be specifically removed and the effect on
catalysis measured. It is thus possible to examine directly the
catalytic role of the binding energy associated with individual groups on
an enzyme. This can be simply illustrated by the work of Wells and his
collaborators on subtilisin (Wells et al., 1986; Wells & Estell, 1988).

Subtilisin is a bacterial serine protease mechanistically and, at
least as far as its active site is concerned, structurally very similar
to the trypsin family of mammalian serine proteases (Robertus et al.,
1972).

Table 3.   The catalytic residues of the serine proteases.

Enzyme	Equivalent catalytic triad residues		
Subtilisin	Asp-32	His-64	Ser-221
Trypsin Chymotrypsin Elastase	Asp-102	His-57	Ser-195

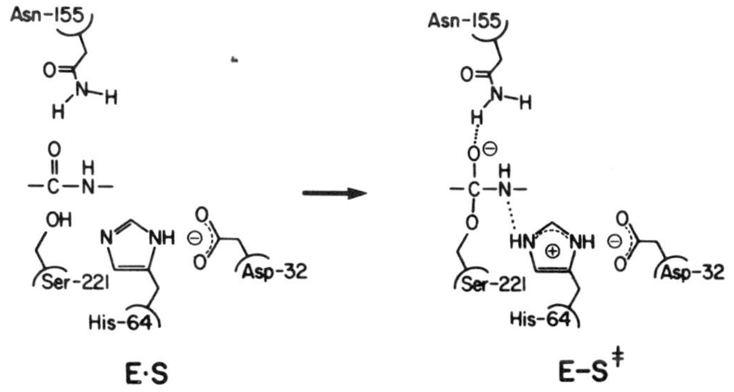

Figure 5.   The first step of the mechanism of subtilisin showing the
             formation of the 'oxyanion' transition state; the transition
             state is stabilised not only by the hydrogen bond to the
             side chain of Asn-152, which is shown, but also by a
             hydrogen bond to the main chain amide of Ser-221 (Robertus
             et al., 1972; Wells et al., 1986).

Subtilisin has a similar mechanism to trypsin (see Figure 3); the transition state on the way to acylenzyme formation is an 'oxyanion' which is believed to be stabilised by hydrogen bonding to the side chain amide of asparagine-155 and the main chain amide of serine-221 (Robertus et al., 1972; the interaction with Asn-155 is shown in Figure 5). In comparison, in chymotrypsin and the other mammalian serine proteases, both stabilising interactions involve the main chain amides of Ser-195 and Gly-193. It is not possible to alter these stabilising interactions in the mammalian enzymes since site-directed mutagenesis can only alter amino acid side chains; the backbone cannot be changed. However it is a simple matter to alter the side chain of Asn-152 of subtilisin especially since subtilisin is a bacterial enzyme and its gene isolation and manipulation have proved particularly straight forward (Wells & Estell, 1988). A number of mutations have been made at this position and the kinetic contants for the purified mutant enzymes are given in Table 3.

Table 4.   Kinetic constants for wild-type subtilisin and various mutants altered at amino acid 155.

Amino acid 155	Asn (wild-type)	Thr	Gln	Asp	His
k_{cat} (s^{-1})	50	0.02	0.06	0.02	0.2
10^{-4} K_m (M)	1.4	2	0.3	0.3	0.2
10^{-3} k_{cat}/K_m	360	0.1	2	0.8	9
$\Delta\Delta G^{\neq}$ (kJ mol^{-1})	0	20	13	15	9.2

$\Delta\Delta G^{\neq}$ is calculated from the relation

$$\Delta\Delta G^{\neq} = -RT\ln \frac{(k_{cat}/K_m)_{mutant}}{(k_{cat}/K_m)_{Asn-155}}$$

All the mutations dramatically reduce k_{cat} but scarcely affect K_m. Clearly the replacement of the Asn-152 side chain is extremely detrimental to the enzyme. The kinetic data indicate that the effect of mutating Asn-155 on the energy differerence between the transition state (E-S) and the ground state (E + S) is to cause a loss of from 9 to 20 kJ/mol stabilisation energy. These data are entirely consistent with Asn-155 having a role in stabilising the 'oxyanion' transition state.

To date the most elegant application of site-directed mutagenesis to the study of the role of transition state stabilisation in enzyme catalysis has been the work of Fersht and Winter and their collaborators on tyrosyl-tRNA synthetase (for reviews of this work see Fersht et al., 1986a; 1986b; Fersht & Leatherbarrow,1987)

The reaction catalysed is a two step process:

$$E + Tyr + ATP = E.Tyr\text{-}AMP + PP_i$$

$$E.Tyr\text{-}AMP + tRNA^{Tyr} = E + Tyr\text{-}tRNA^{Tyr} + AMP$$

The _Bacillus_ _stearothermophilus_ enzyme has been extensively studied;
it is very stable; extensive molecular biological work has been carried
out and excellent overproducing strains are available. The 3-d structure
of the enzyme and of the enzyme-tyrosine and the enzyme-tyrosyladenylate
complexes have been determined by Blow and his collaborators (Rubin &
Blow, 1981; Monteilhet at al., 1984). The enzyme is an ideal one for
study by site-directed mutagenesis since there are at least 9 hydrogen
bonds involved in substrate binding (Figure 6) and it was therefore not
surprising that it was the first enzyme of known 3-d structure to be
studied by site-directed mutagenesis (Winter et al., 1982). Extensive
kinetic work has been done on the wild-type enzyme; the amount of active
enzyme can be readily determined by active site titration and the rate
constants for both activation and transfer can be readily determined by
stopped flow and quenched flow methods (Fersht, 1985). Much attention
has been focussed on the first step, the formation of tyrosyl adenylate.
This involves the attack of a fully ionised nucleophile (R-CO$_2^-$) on
an activated compound (MgATP) and there is a good leaving group (MgPP).
The classical enzymatic processes of acid-base or covalent catalysis do
not apply. The reaction, which involves simple in line displacement, is
catalysed simply as a result of transition state stabilisation. Modeling
studies indicate that the transition state ressembles the
pentaco-ordinate intermediate in the reaction scheme (Figure 7).

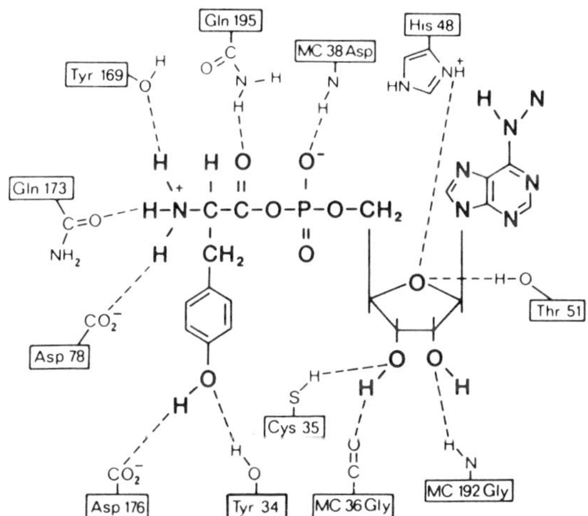

Figure 6.   The active site of tyrosyl-tRNA synthetase showing the
possible hydrogen bond interactions between tyrosyl
adenylate and the enzyme (from Fersht et al., 1986).

The roles of the two side chains (His-45 and Thr-40) involved in
stabilising the transition state have been evaluated by mutation and full
kinetic analysis. Changing His-45 to glycine removes the imidazole chain
that makes a hydrogen bond with an oxygen in the transition state (Figure
8). The effect of this mutation His45Gly is to reduce k_{cat} very
considerably (Table 5). Changing Thr-40 to alanine removes the hydrogen
bond between the hydroxyl group of the side chain and the transition
state. The effect is also to lower k_{cat} (Table 5). The changes
involving Thr-40 and His-45 increase the energy level of the transition
state but do not effect the energy levels of the enzyme-substrate complex.

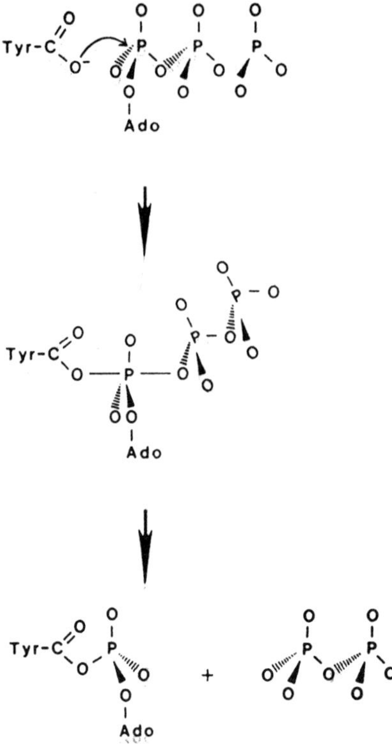

Figure 7.   Chemical mechanism for the formation of tyrosyl adenylate
(from Leatherbarrow et al., 1985).

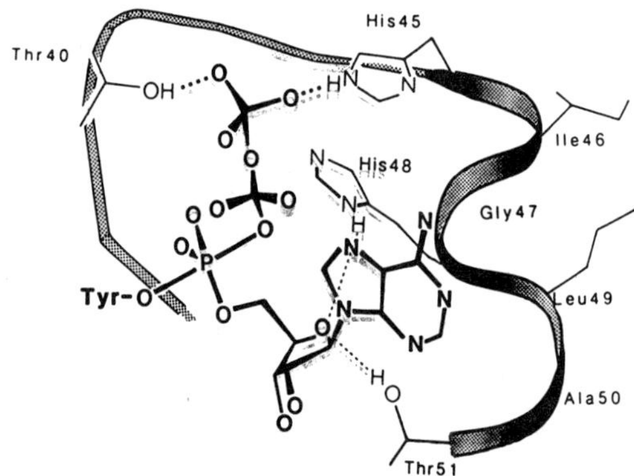

Figure 8.   Model of the pentaco-ordinate transition state for the
formation of tyrosyl adenylate built into the active site of
tyrosyl-tRNA synthetase (from Fersht et al., 1986b).

For a model intramolecular reaction between ATP and tyrosine a
reaction rate of 10^{-3} s^{-1} has been estimated (Leatherbarrow et al.,
1985). This is similar to the rate observed with the least active mutant
and suggests that the reaction in this case is just a simple
intramolecular reaction. The wild-type enzyme has a turnover

Table 5.   Pre-steady state kinetic parameters for the formation of
            tyrosyl adenylate by mutants of tyrosyl-tRNA synthetase
            (from Leatherbarrow, Fersht & Winter, 1985)

Enzyme	k_{cat} (s^{-1})	K_S for Tyr (μM)	K_S for ATP (μM)
Wild-type	38	12	4.7
His45Gly	0.16	10	1.2
Thr40Ala	0.0055	8	3.8
T40A, H45G	0.00012	4.5	1.1

number of 38 s^{-1} which is 4 X 10^4 faster than the solution model.
This acceleration must be due to transition state stabilisation
presumably mediated as shown in Figure 9 (Leatherbarrow et al., 1985).

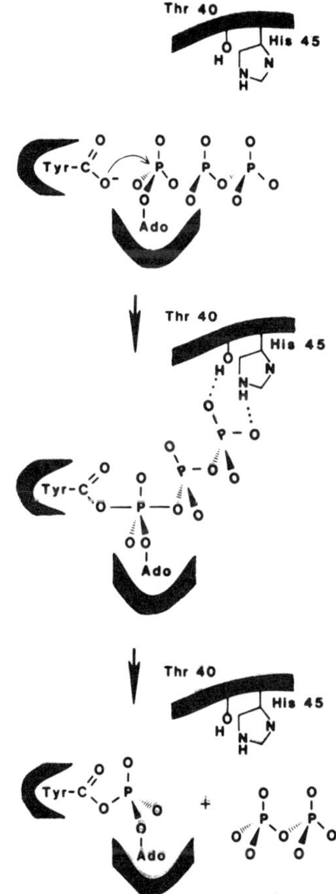

Figure 9.   Possible mechanism for the stabilisation of the
               transition state for tyrosyl adenylate formation
               by residues Thr-40 and His-45 of tyrosyl-tRNA
               synthetase (from Fersht & Leatherbarrow, 1987).

SUMMARY

Site-directed mutagenesis is an invaluable tool which is allowing
enzymologists to elucidate enzyme mechanisms and study enzyme specificity
very much more easily than in the past. The work described in this
chapter on tyrosyl-tRNA synthetase, the serine proteases and
carboxypeptidase A illustrates what can already be achieved at the level
of analysing mechanisms. It also indicates very clearly that the
methodology will now allow the rational alteration of enzyme function.
Already the first examples of such rational manipulations are being
reported. Good examples are the work of Fersht's group on altering the
pH-activity profile of subtilisin by the modification of surface charge
(Thomas et al., 1985; Russell & Fersht, 1987) and the work of Holbrook
and Clarke and their collaborators on the manipulation of the specificity
of lactate dehydrogenase (Clarke et al., 1989). Site-directed
mutagenesis is clearly taking enzymology into a new and exciting phase.

REFERENCES

Blow, D.M., Fersht, A.R. & Winter, G., 1986, "Design, Construction and
Properties of Novel Protein Molecules", Royal Society, London.

Clarke, A.R., Atkinson, T. & Holbrook, J.J., 1989, From analysis to
synthesis: new ligand binding sites on the lactate dehydrogenase
framework, Trends. Biochem. Sci. 14:101.

Craik, C.S., Largman, C., Fletcher, T., Roczniak, S., Barr, P.J.,
Fletterick, R. & Rutter, W.J., 1985, Redesigning trypsin: alteration of
the substrate specificity, Science 228:291.

Fersht, A.R., 1985, "Enzyme Structure and Mechanism" (2nd ed.), Freeman,
San Francisco.

Fersht, A.R., 1987, Kinetic aspects of purposely modified proteins, in
"Protein Engineering", Oxender, D.L. & Fox, C.F. (eds.), Liss, New York
p. 225.

Fersht, A.R. & Leatherbarrow, R.J., 1987, Structure and activity of
tyrosyl-t-RNA synthetase, in "Protein Engineering", Oxender, D.L. & Fox,
C.F. (eds.), Liss, New York p. 269.

Fersht, A.R., Leatherbarrow, R.J. & Wells, T.N.C., 1986a, Binding energy
and catalysis: a lesson from protein engineering of tyrosyl-tRNA
synthetase, Trends. Biochem. Sci. 11:321.

Fersht, A.R., Leatherbarrow, R.J. & Wells, T.N.C., 1986b, Structure and
activity of tyrosyl-tRNA synthetase: the hydrogen bond in catalysis and
specificity, Phil. Trans. R. Soc. Lond. A 317:305.

Gardell, S.J., Craik, C.S., Hilvert, D., Urdea, M.S. & Rutter, W.J.,
1985, Site-directed mutagenesis shows that tyrosine-248 of
carboxypeptidase A does not play a crucial role in catalysis, Nature
317:551.

Graf, L., Craik, C.S., Patthy, A., Roczniak, S., Fletterick, R.J. &
Rutter, W.J., 1987, Selective alteration of substrate specificity by
replacement of aspartic acid-189 with lysine in the binding pocket of
trypsin, Biochemistry 26:2616.

Hartsuck, J.A. & Lipscomb, W.N., 1971, Carboxypeptidase A, in "The Enzymes", 3rd edn., Volume III, Boyer, P.D. (ed.), Academic Press, New York p. 1.

Karplus, M., 1987, The prediction and analysis of mutant structures, in "Protein Engineering", Oxender, D.L. & Fox, C.F. (eds.), Liss, New York p. 35.

Leatherbarrow, R.J. & Fersht, A.R., 1986, Protein Engineering 1:7.

Leatherbarrow, R.J., Fersht, A.R. & Winter, G., 1985, Transition-state stabilisation in the mechanism of tyrosyl-tRNA synthetase revealed by protein engineering, Proc. Natl. Acad. Sci. USA 82:7840.

Leatherbarrow, R.J. & Fersht, A.R., 1987, Use of protein engineering to study enzyme mechanisms', in "Enzyme Mechanisms", Page, M.I. & Williams, A. (eds), The Royal Society of Chemistry, London, p. 78.

Lipscomb, W.N., 1980, Carboxypeptidase A mechanisms, Proc. Natl. Acad. Sci. USA 77:3875.

Markley, J.L., 1987, One- and two-dimensional nmr spectroscopic investigations of the consequences of amino acid replacements in proteins, in "Protein Engineering", Oxender, D.L. & Fox, C.F. (eds.), Liss, New York p. 15.

Monteilhet, C., Blow, D.M. & Brick, P., 1984, Interaction of crystalline tyrosyl-tRNA synthetase with adenosine, adenosine monophosphate, adenosine triphosphate and pyrophosphate in the presence of tyrosinol, J. Mol. Biol. 173:477.

Oxender, D.L. & Fox, C.F., 1987, "Protein Engineering", Liss, New York.

Pauling, L., 1948, Nature of forces between large molecules of biological interest, Nature 161:707.

Robertus, J.D., Kraut, J., Alden, R.A. & Birktoft, J.J., 1972, Subtilisin: a stereochemical mechanism involving transition-state stabilisation, Biochemistry 11:4293.

Rossi, J. & Zoller, M., 1987, Site-specific and regionally directed mutagenesis of protein-encoding sequences, in "Protein Engineering", Oxender, D.L. & Fox, C.F. (eds.), Liss, New York p. 51.

Rubin, J. & Blow, D.M., 1981, Amino acid activation in crystalline tyrosyl-t-RNA synthetase from Bacillus stearothermophilus, J. Mol. Biol. 145:489.

Russell, A.J. & Fersht, A.R., 1987, Rational modification of enzyme catalysis by engineering surface charge, Nature 328:496.

Rutter, W.J., Gardell, S.J., Rocniak, S., Hilvert, D., Sprang, S., Fletterick, R.J. & Craik, C.S., 1987, Redesigning proteins by genetic engineering, in "Protein Engineering", Oxender, D.L. & Fox, C.F. (eds.), Liss, New York p. 257.

Shaw, W.V., 1987, Protein Engineering: the design, synthesis and characterisation of factitious proteins' Biochem. J. 246:1.

Sprang, S., Standing, T., Fletterick, R.J., Stroud, R.M., Finer-Moore, J., Xuong, N-H, Hamlin, R., Rutter, W.J. & Craik, C.S., 1987, The three-dimensional structure of Asn^{102} mutant of trypsin: role of Asp^{102} in serine protease catalysis, Science 237:905.

Stroud, R.M., Krieger, M., Koeppe, R.E., Kossiakoff, A.A. & Chambers, J.L., 1975, Structure function relationships in the serine proteases, in "Proteases and Biological Control", Reich, E., Rifkin, D.B. & Shaw, E., eds., Cold Spring Harbor Laboratory, New York, p. 13.

Thomas, P.G., Russell, A.J. & Fersht, A.R., 1985, Tailoring the pH-dependence of enzyme catalysis using protein engineering, Nature 318:375.

Walsh, C., 1979, "Enzymatic Reaction Mechanisms", Freeman, San Francisco.

Wells, J.A., Cunningham, B.C., Graycar, T.P. & Estell, D.A., 1986, Importance of hydrogen-bond formation in stabilizing the transition state of subtilisin, Phil. Trans. R. Soc. Lond. A317:415.

Wells, J.A. & Estell, D.A., 1988, Subtilisin - an enzyme designed to be engineered, Trends. Biochem. Sci. 13:291.

Wells, T.N.C. & Fersht, A.R., 1985, Hydrogen bonding in enzymatic catalysis analysed by protein engineering, Nature 316:656.

Winter, G., Fersht, A.R., Wilkinson, A.J., Zoller, M. & Smith, M., 1982, Redesigning enzyme structure by site-directed mutagenesis: tyrosyl tRNA synthetase and ATP binding, Nature 299:756.

Wolfenden, R. & Frick, L., 1987, Transition state affinity and the design of enzyme inhibitors, in "Enzyme Mechanisms", Page, M.I. & Williams, A. (eds), The Royal Society of Chemistry, London, p. 97.

ENZYMES AS CATALYSTS IN ORGANIC SYNTHESIS

Stanley M. Roberts

Department of Chemistry
University of Exeter
Exeter, Devon, England

1. INTRODUCTION AND BACKGROUND INFORMATION

The ability of the enzymes that are present in micro-organisms to
catalyse chemical reactions has been known for thousands of years; indeed
the fermentation of sugars into alcohol using yeast is described in the
early scriptures.[1]

Today there are many hundreds of scriptures (now called research papers)
describing transformations that can be accomplished by fungi and other micro-
organisms.[2] The vast majority of these transformations involve hydrolysis
reactions (*e.g.* esters or amides), the oxidation of alcohols, the reduction
of carbonyl compounds (*e.g.* aldehydes or ketones) or hydroxylation reactions.
The whole cell system sometimes catalyses a cascade of reactions (*e.g.* in
the fermentation of a sugar) or a single-step biotransformation. The substrate can be a natural product or a substance foreign to the organism.

In recent years single step biotransformations involving unnatural
substances have increasingly been investigated using enzymes isolated from
micro-organisms, plants, and animals. Thus in the area of biotransformations there has been a distinct move away from the use of whole cell
systems towards the use of purified enzymes.

The use of an isolated enzyme to perform a chosen conversion is often
preferred for a number of reasons. First it circumvents the need to obtain/
store or culture micro-organisms. Secondly it avoids the loss of material
due to side reactions involving different metabolic pathways in the whole
cell system. Thirdly it can avoid problems due to poor diffusion of
substrates across biological membranes and to the potential cytotoxicity
of the materials.

Not only are some single step biotransformations best achieved by using isolated enzymes but sequential processes involving the same transformation on closely related substances can be performed by a purified enzyme. Thus the degradation of protein using proteases is important to the detergent manufacturers. The use of glycosidases in the food industry and the employment of lipases in the transesterification of fats are further examples of the utility of enzymes. In fact, the bulk of the enzymes that are produced in the world today are used in large scale conversions of this type.

Whether it is appropriate to use an isolated, partially purified enzyme or a whole-cell system for a particular operation depends upon the biotransformation in question. Hydrolase enzymes such as lipases, esterases, proteases and amidases are readily available and easy to use, and represent the method of choice for performing hydrolysis reactions. The utilization of other enzymes such as dehydrogenase enzymes (oxidoreductases) is not as simple an exercise due to inherent problems such as co-factor recycling (*vide infra*). Thus the use of bakers' yeast for the reduction of ketones, diketones and ketoesters is the preferred method in many instances. Examples of the use of enzymes for the oxidation of ostensibly non-activated carbon centres are difficult to find; by comparison there are numerous examples of the use of whole cell systems in this regard. The regioselective hydroxylation of the steroid nucleus has been thoroughly studied and many noteworthy transformations have been documented.[3]

Perusal of the recent literature (1983-date) suggest that over 90% of the current work in the area of biotransformations is involved in assessing conversions of un-natural substrates using oxidoreductases and hydrolases. (The two classes of enzymes are being researched to an equal extent). This emphasis and this trend is reflected in this report.

The chief advantages that can be gained from the use of enzymes (and whole-cell systems) in the synthesis of new organic compounds are as follows. First, enzymes operate under mild conditions of temperature and pH, conditions which sensitive molecules can often tolerate. Secondly, regioselective functionalization of a non-activated carbon centre is a property displayed by some enzymes that is particularly difficult to emulate using conventional chemical methods. Finally, and perhaps most importantly, reactions catalysed by enzymes are often highly stereoselective; the synthesis of optically active products from racemic mixtures and non-chiral substrates is a noteworthy feature of many reactions catalysed by enzymes.

The preparation of chiral organic molecules using enzyme catalysed reactions has created a lot of excitement amongst the scientists working in this area since the products formed from the biotransformations are often useful as starting materials for the synthesis of more complex structures of interest to scientists in the Pharmaceutical, Agrochemical, and Perfumery Industries.[4]   The preparation of optically active compounds by means of biotransformations will form the under-pinning theme of the remainder of this review which will concentrate on oxidoreductases (enzymes that catalyse oxidation and reduction reactions involving addition and removal of hydrogen atom equivalents or the insertion of an oxygen atom between two atoms), hydrolases (enzymes that catalyse the hydrolysis of esters and amides) and miscellaneous enzymes covering a range of other biotransformations.

2. BIOTRANSFORMATIONS LEADING TO OPTICALLY ACTIVE MATERIALS

2.1 Oxidoreductases

The central rôle played by the carbonyl group in synthetic organic chemistry has ensured that enzyme controlled reductions of this moiety have been studied extensively.   Practically speaking, the reduction of a ketone to an alcohol using bakers' yeast involves a very simple procedure.   For example, only a handful of bakers' yeast and table sugar in tap-water is needed to reduce 40 g of ethyl acetoacetate (ethyl 3-oxobutanoate) to the (*S*)-alcohol (1) with high selectivity (85–95% e.e.) and good yield (59–76%).[5]   The equally useful ethyl 3(*R*)-hydroxybutanoate (2) is produced industrially by microbial β-hydroxylation of n-butyric acid and can be obtained from cells of *Zoogloea ramigera*.[6]

Substituents can be tolerated at the 2-position and the 4-position of the 3-oxobutanoate system.   For example ethyl 4-*tert*-butoxy-3-oxobutanoate gave the hydroxy-ester (3) in 70% yield and 97% e.e.[7]   The absolute configuration and optical purity of the reduction products depend on the nature of the 4-substituent and the alcohol portion of the ester moiety.   For example yeast reduction of ethyl 4-chloro-3-oxobutanoate gave an excess of the 3(*S*)-alcohol (4) while the corresponding octyl ester gave optically pure octyl 4-chloro-3(*R*)-hydroxybutanoate (5).   The involvement of two different dehydrogenase enzymes seems likely.[8]   Readers unfamiliar to this area may

(1) R^1 = OH , R^2 = H

(2) R^1 = H , R^2 = OH

(3)

have already formed the impression that the product arising from a yeast
reduction of a β-keto ester is a total lottery. This is not the case.
The major product can usually be predicted from the generalised situation
illustrated in Figure 1.[9] Thus, as predicted by this model, octyl 3-oxo-
pentanoate gave the 3(*S*)-alcohol (7) with high selectivity (> 95% e.e.).[10]
Reduction of octyl 2-methyl-3-oxobutanoate with yeast gave octyl 3(*S*)-
hydroxy-2-(*R*)-methylbutanoate (8) (82% yield).[11]

Yeast reductions of 3-oxobutanoates can be modified using immobilised
systems. In particular polyurethane immobilisation can improve or reverse
the stereoselectivity of the reduction.[12] Other micro-organisms behave
differently towards β-keto-esters. *Thermoanaerobium brockii* reduced methyl
4-chloro-3-oxobutanoate to the 3(*R*)-hydroxy-compound (6) with good
selectivity.[13]

The cyclic β-keto-esters (9) are reduced to the corresponding *cis*-
hydroxy-esters in exquisite chemical and optical yields using *Mucor racemosus*
or *M. circinelloides* while the corresponding *trans*-hydroxy-esters are
obtained using *Rhizopus arrhizus*.[14]

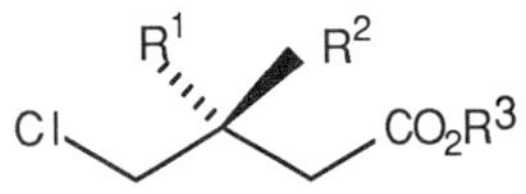

(4) R^1 = H, R^2 = OH, R^3 = C_2H_5

(5) R^1 = OH, R^2 = H, R^3 = C_8H_{17}

(6) R^1 = OH, R^2 = H, R^3 = CH_3

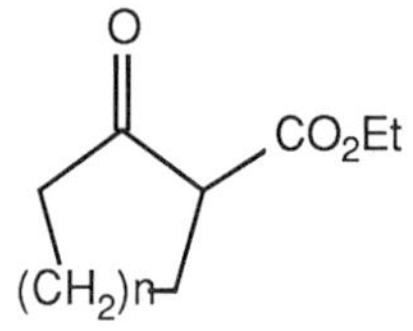

n = 1,2

(9)

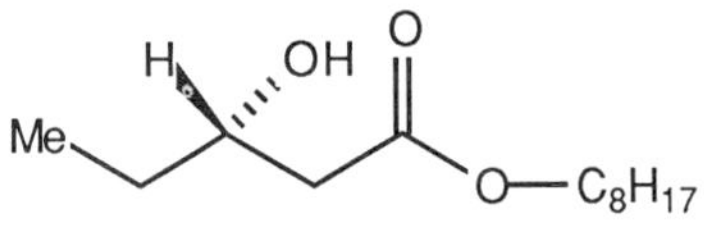

(7)

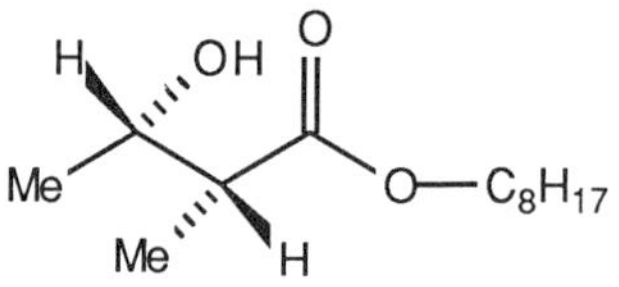

(8)

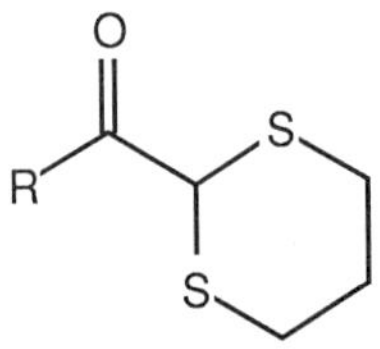

(10)

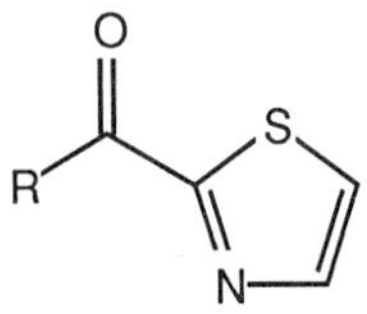

(11)

446

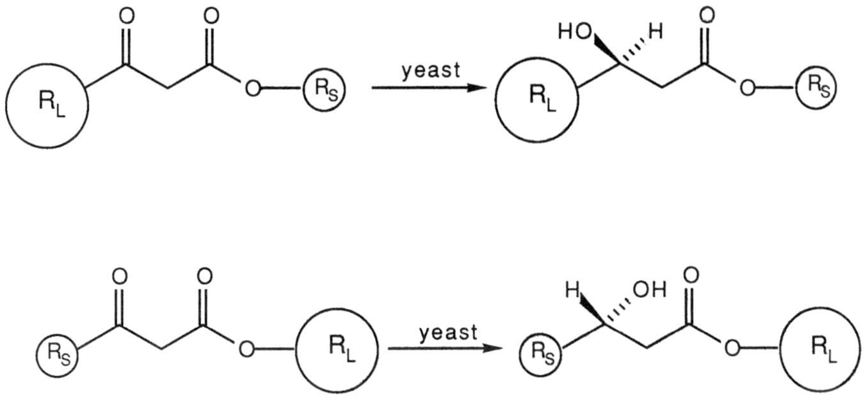

Figure 1

Figure 2

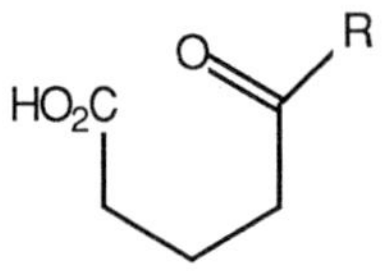

(12)

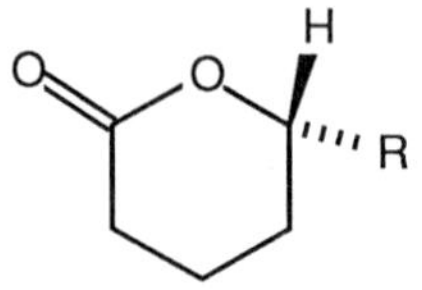

(13)

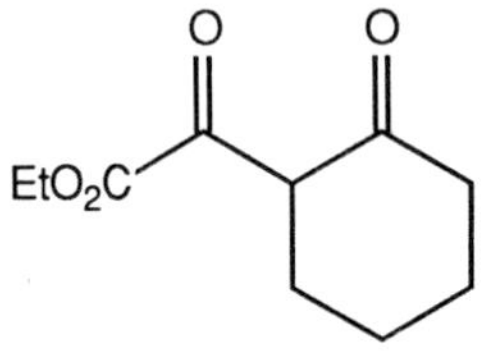

(14)

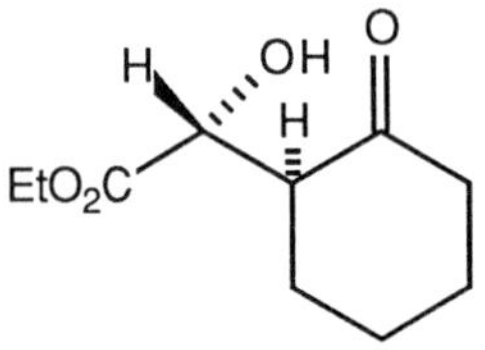

(15)

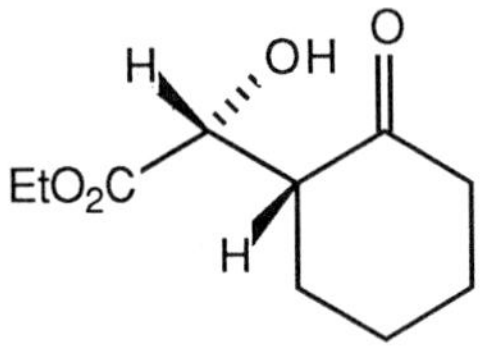

(16)

Many other types of carbonyl compound have been reduced to give
optically active products using oxidoreductase enzymes in whole cell preparations of yeasts and other micro-organisms. Even with the armoury of
chiral reducing agents available to the organic chemist,[15] the cheapness
and practical simplicity of yeast reductions have meant that this method is
often used to prepare optically active (S)-alcohols from the corresponding
ketones. For example 13(S)-hydroxy-9Z,11E-octadecadienoic acid has been
prepared by reduction of the corresponding ketone with yeasts.[16]

Various protected L-glyceraldehydes have been obtained by bakers'
yeast reduction of the ketones (10) and (11); the products possess the
(S)-stereochemistry at the reduction site.[17] Bakers' yeast reduction of
the compound (10, R = $CH_2(CH_2)_2CO_2Me$) gave a useful synthon for leukotriene-
B_4.[18] Prelog's rule can be used to predict the outcome of a reduction of
a simple ketone using yeast (Figure 2). Some recently reported yeast
catalysed reductions should suffice to further illustrate the dependability
of this rule; a range of keto-acids (12) have been converted into the
optically active lactones (13),[19] and the ketone (14) is reduced to the
optically pure alcohols (15) and (16) in 57% yield.[20] The diketone (17)
gave the (S),(S)-diol (18) with very high selectivity (82% yield, 97% e.e.).[21]

448

Similarly benzil (19) gave the $(S),(S)$-diol (40% yield; 94% e.e.) with the
yeast *Saccharomyces uvarum* (the enantiomer was obtained [55% yield; 96%
e.e.] using the micro-organism *Rhodotorula mucilaginosa*).[22]

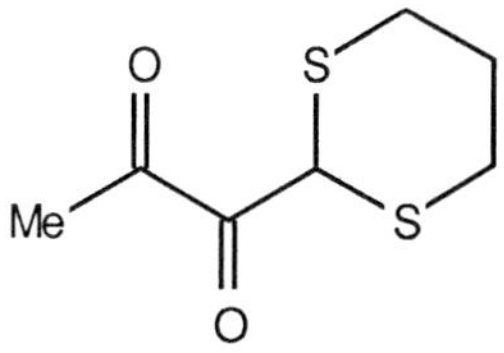

(17) 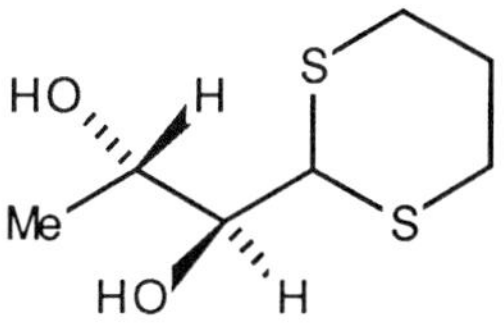   (18)   (19)

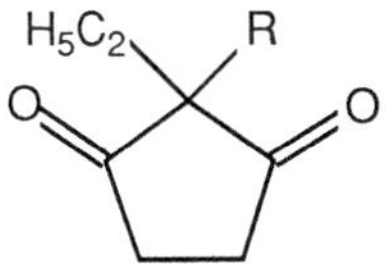

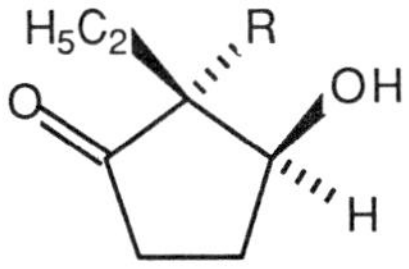

(20)   R = CH₂CH₂COCH₂CH₂SPh   (21)

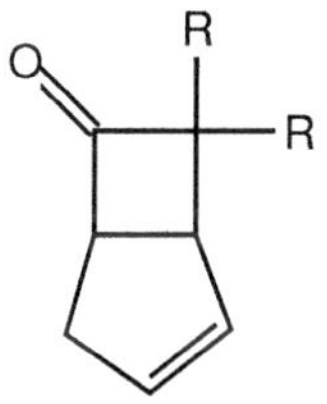

(22) R = H   (23)   (25)

(24) R = Me

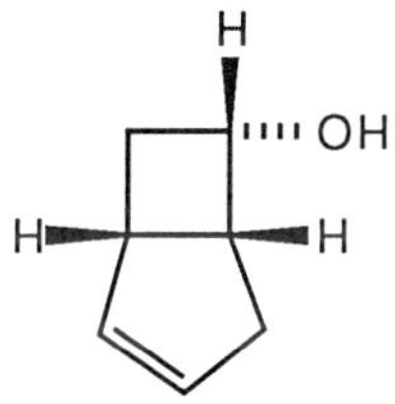

Finally the reduction of prochiral cyclopentanediones using yeast to
give optically active building blocks for the preparation of steroids has
been widely exploited. A typical example of this beautiful strategy
[(20) → (21)] has been reported recently.[23]

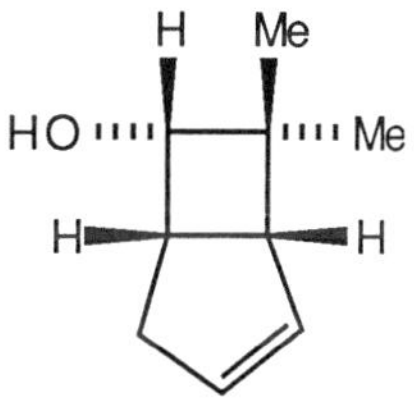

Hopefully, the scope for carrying out reductions of ketones using micro-
organisms has been demonstrated. Problems can occur. Obviously there are
a multitude of other processes occurring within the cell which may cause

unwanted side-reactions in the process under investigation.   Furthermore,
as illustrated in one example above, yeasts often express more than one
oxidoreductase and the enzymes frequently show differing selectivities;[24]
the competing enzymes may show different catalytic abilities under different
culture conditions.   Quite frequently the micro-organism is prone to
product inhibition (even the wines and spirits industry has failed to find
a yeast capable of surviving in a solution containing more than 14%
ethanol!).   For these reasons a lot of work has concentrated on using cell-
free, partially purified oxidoreductase enzymes.   When using such isolated
enzymes, the reaction mixture should mimic the intracellular environment of
the cell through the use of pH buffers;   frequently oxygen must be excluded.
More importantly a way must be found to achieve the regeneration of enzyme
co-factors.   The most important of these co-factors is nicotinamide adenine
dinucleotide (NAD^+) or its phosphorylated counterpart ($NADP^+$) and the
corresponding reduced forms (NADH, NADPH).   The development of new regeneration techniques is at the leading edge of current research, but several
highly workable techniques are available.[25]   The simplest procedure is to
use a sacrificial alcohol (such as ethanol) as the source of hydrogen atoms.
In early work horse liver alcohol dehydrogenase (HLAD) was the favoured
enzyme and this protein was used to enantiospecifically reduce a simple
ketone;   reduction of the NAD^+ so formed was accomplished in a coupled
reaction with ethanol ((a) = (b) = HLAD, Figure 3).   More recently a double
enzyme system of 3α,20β-hydroxysteroid dehydrogenase ((a) Figure 3) and HLAD
or yeast alcohol dehydrogenase ((b) Figure 3) has been used.[26]

A broad range of ketones can be reduced by using different oxidoreductases (Figure 4).

In many cases the enantioselectivity of the oxidoreductase is highly
predictable.   HLAD follows the Prelog rule of hydride addition or removal
from the *re* face of the ketone (Figure 2).

A range of oxidoreductase enzymes including those mentioned in Figure
4 are available from commercial sources.   HLAD has been investigated
extensively[27] and a start has been made on determining the stereoselectivity
of reductions using the dehydrogenase enzyme from *T. brockii*.[28]   For
example the racemic bicycloheptenone (22) gave the 6(*S*)-alcohol (23)
[> 95% e.e.] and recovered optically active ketone.[29]   Another useful
dehydrogenase enzyme is 3α,20β-hydroxysteroid dehydrogenase;   reduction
of the ketone (24) with this protein gave the 6(*S*)-bicycloheptenol (25)
(> 95% e.e.) and recovered optically active ketone.[29]   Glycerol dehydrogenases from *Enterobacter aerogenes* and *Cellulomonas sp.* have been used to
reduce various 2-hydroxy-ketones.   For example, 1-hydroxypropan-2-one and

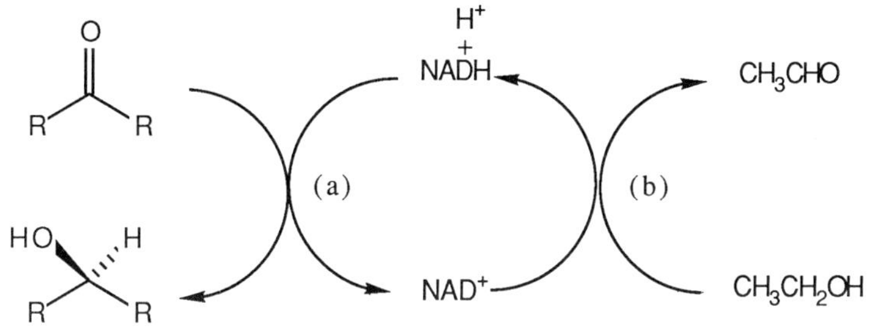

Figure 3

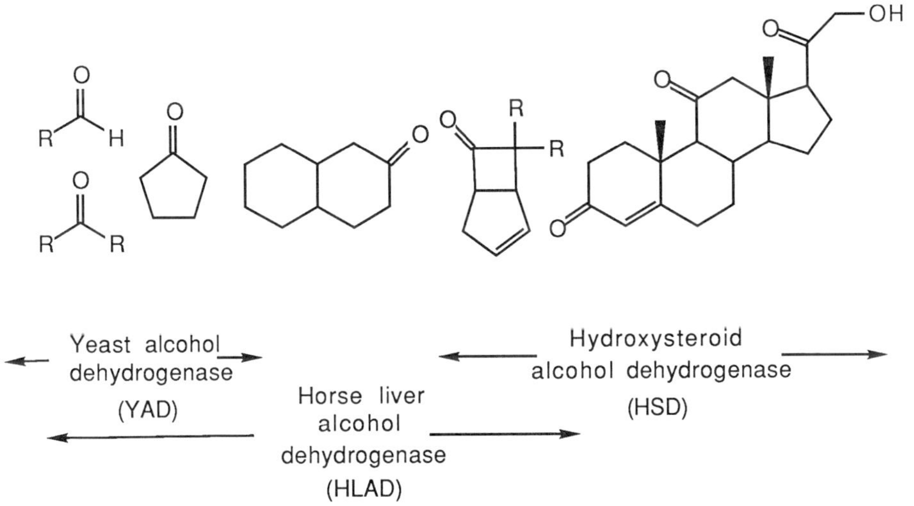

Figure 4

1-hydroxybutan-2-one gave the corresponding (R)-diols (e.e. 95-98%).
Comparable results were obtained using whole cell yeast fermentations.[30]

There are two classes of enzymes that catalyse the reduction of double
bonds; hydrogenases use molecular hydrogen as a co-factor[31] whilst enoate
reductases (which are found in some yeasts) use NADH.[32]   An example of the
former type of reaction is the reduction of the allene (26) into the β,γ-
unsaturated acid (27) using *Clostridium sp.* and hydrogen gas; yeast
reduction of the enedione (28) to give the optically active diketone (29)
is an example in the second category (note that the enedione was reduced
on a 13 kg scale!).

Two more examples from the recent literature emphasise the potential
importance of this under-exploited area.   Reduction of the chloroketones
(30) with bakers' yeast gave the (S)-chloroketones selectively (44-84% e.e.)
Further reduction gave the diastereoisomers 3(S)-chloro- and 3(R)-chloro-
alkan-2(S)-ols with very high selectivity (*ca.* 98% e.e.).[33]   Finally
geraniol (31) was reduced to (R)-citronellol (32) (98% e.e.) using bakers'
yeast.[34]

The oxidation of alcohols to aldehydes and ketones by micro-organisms
and isolated enzymes has not been widely used in synthesis. The work by
Jones using HLAD is a major contribution to this area: the oxidation of a
meso-diol (33) by HLAD proceeds with *pro*-S selectivity to give the hydroxy-
aldehyde (34) (which is further oxidised *in situ* to give the cyclic δ-
lactone).[27,35]

The replacement of a hydrogen atom by a hydroxyl group at a centre
remote from other functionality is practically impossible with conventional
chemical reagents. In contrast a range of enzymes is available that will
hydroxylate virtually any site in a steroid with complete regiospecificity.[3]
(The conversion of the steroid (35) into cortisol (36) is a classic example).
Much the same is true regarding the hydroxylation of terpenes[36] and other
alicyclic compounds.[37] Smaller molecules can be hydroxylated in a very
selective fashion. For example *iso*-butyric acid yields 3-hydroxy-2(*S*)-
methylpropionic acid on treatment with *Pseudomonas putida*. Finally the
norbornane derivative (37) is selectively hydroxylated to give the useful
hydroxy acid (38) in 57% yield and 85% e.e. using *Aspergillus awamori*.[38]

2.2 <u>Hydrolases</u>

2.2.1 <u>Esterases and Lipases</u>

The amount of research into the production of optically active compounds
using esterases and lipases has increased almost exponentially over the last
ten years. Many studies have been carried out into the behaviour of these
hydrolytic enzymes under various conditions, with a view to examining changes
in stereoselectivity. One of the pioneers in the field is Professor J.B.
Jones; in one case the Canadian group have studied the effect of co-solvents
and temperature on the enantioselectivity of pig liver esterase (PLE)
catalysed hydrolysis of 3-substituted dimethyl glutarates (39).[39] Thus
in 20% aqueous methanol and at -10 $^{\circ}$C, significant improvements were
obtained. The enantiomeric excess obtained with the 3-methyl diester
((39), R_1 = Me, R_2 = H), increased from 79% to 97%. With small substituents

(33)   (34)

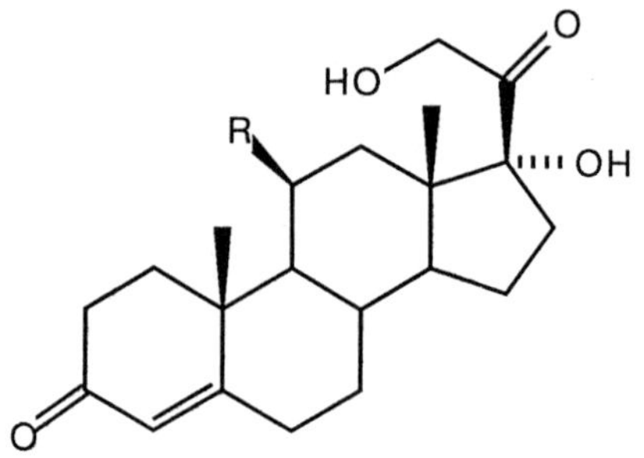

(35) R = H

(36) R = OH

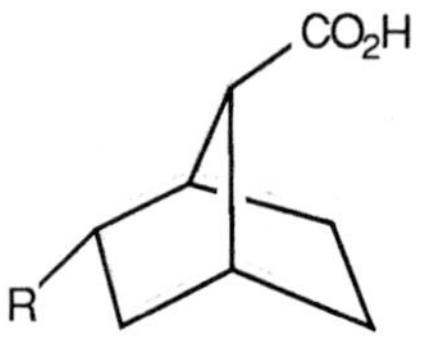

(37) R = H
(38) R = OH

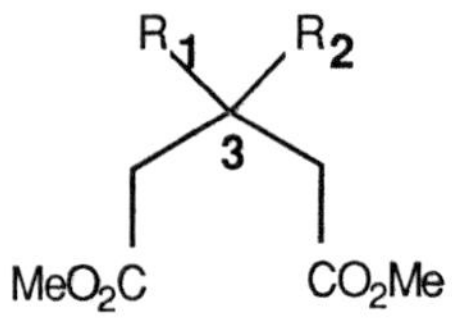

(39)

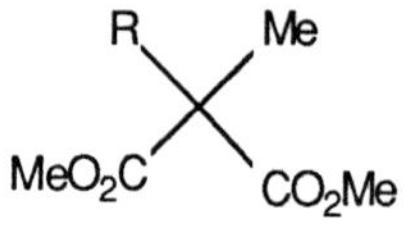

(40)

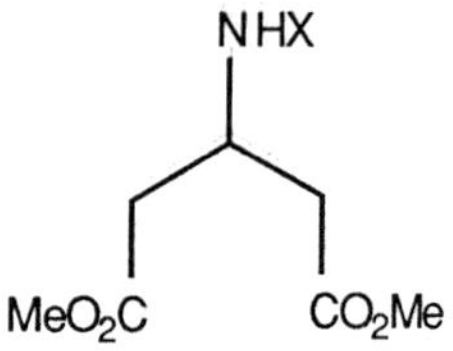

(41)

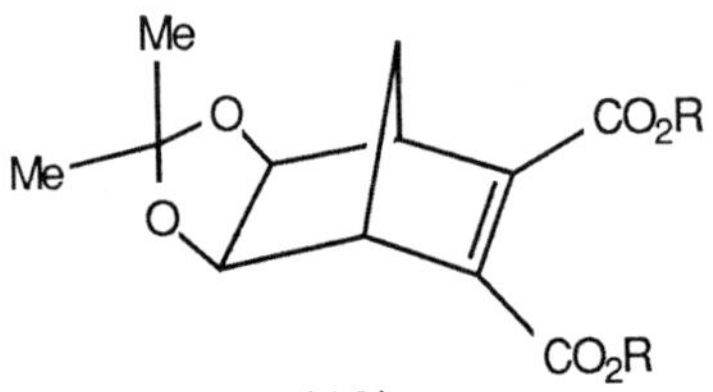

(42)

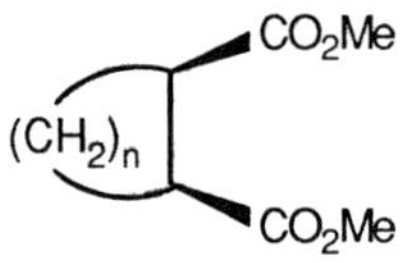

(43)

RCH(NHCOMe)CO$_2$Et

(44)

at C-3, hydrolysis favoured the *pro-S* ester function whereas *pro-R* selectivity was observed with large C-3 substituents. These results were interpreted in terms of an active site model containing a preferred binding site for small substituents and a less-preferred binding site for large substituents. Björkling *et al*[40] carried out a similar study on dialkylated malonate diesters (40) in DMSO-water mixtures. Marked increases in stereo-selectivity were observed (up to 50% DMSO), accompanied by a lowering (by approximately one order of magnitude) of enzyme activity. When R in diesters (40) was small (Et, $Me(CH_2)_6$) the (*S*)-half ester was formed selectively, but with large groups, the (*R*)-half ester was formed selectively. With several substrates there was a marked tendency for the proportion of (*R*)-isomer to increase with increasing DMSO concentration. Ohno and his colleagues[41] have studied the hydrolysis of derivatives of dimethyl 3-amino-glutarates (41). The same trend in relation to the bulk of the C-3 sub-stituent (NHX in (41)) was observed as in the work of Jones *et al*:[39] the half ester formed had the (*R*)-configuration with smaller substituents, changing to the (*S*)-configuration with increasing steric bulk. These results were again interpreted in terms of the binding site geometry. Derivatives of malic, aspartic and glutamic acids were also studied, as were esters of the tricyclic *meso*-system (42). A significant variation in the rate of hydrolysis was observed on changing the alcohol component R. The rate was highest with R = Me, falling markedly as R was changed to Et, *n*-Pr, *i*-Pr and *n*-Bu. Ohno has accomplished a number of elegant natural product synthesis using synthons derived from enzyme catalysed reactions.[41]

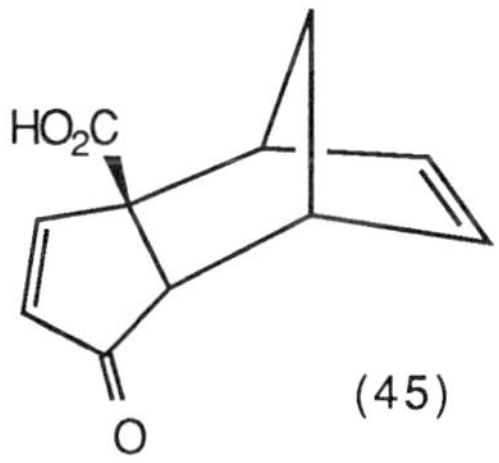

(45)

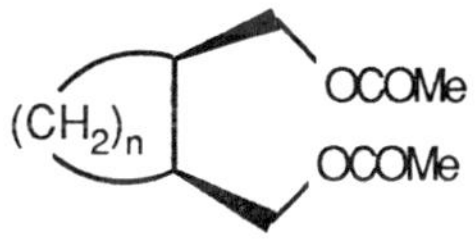

(46)

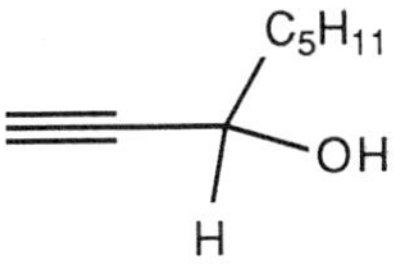

(47)

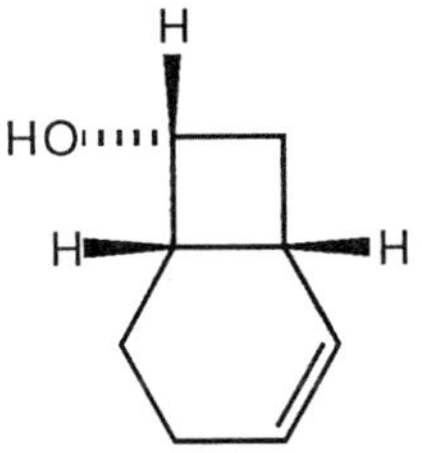

(48)

The resolution of cyclic *meso*-diesters of the type (43) using pig
liver esterase and porcine pancreatic lipase has been the subject of the
attention of Jones[42] and others.[43]   The beauty of this strategy is that,
using an enzyme catalysed reaction, it is possible to get a very high yield
of optically pure monoester.

It has also proved to be useful to hydrolyse racemic α-amino-esters
(44) to furnish optically acid amino-acid derivatives and recovered
optically active ester.[44]   Similarly PLE catalysed hydrolysis of the
corresponding racemic ethyl ester was the method of choice for the synthesis
of the complex monocarboxylic acid (45).[45]

Enzyme catalysed hydrolysis reactions have also been used to afford
useful optically active alcohols.   Cyclic diacetates of the type (46)
provide useful optically active mono-ols.[46]   Simple alcohols such as
oct-1-yn-3-ol (47) can be obtained in optically active form using whole
cell[47] or enzyme catalysed reactions.[48]   More complex alcohols such as
the bicyclo[4.2.0]octenol (48)[49] and the bicyclo[2.2.1]heptenol (49)[50]
can be prepared very conveniently by hydrolysis of the corresponding acetate
using enzymes such as *Muchor miehei* lipase.

Other esterases have been investigated to ascertain if different
selectivities are observed when compared to commercially available esterases
and lipases.   For example the acetyl cholinesterase from electric eel has
been used to prepare 3(R),5(S)-3-acetoxy-5-hydroxycyclopentene.[51]

(49)   (50)   (51)

R = alkyl , aryl , CH$_2$OH , CH$_2$CO$_2$H , (CH$_2$)$_2$NH$_2$

(52)   (53)

PLE and a number of fungal lipases have been shown to catalyse enantioselective Michael additions to trifluoromethyl methacrylate (50), thus displaying a type of activity quite different from the normal hydrolysis
reaction.[52]   However it is the area of esterification and the use of enzymes
in low water systems that has recently attracted a lot of interest and
activity.   The use of a yeast enzyme to catalyse the stereoselective esterification of bromopropionic acid in a solvent system of butanol, hexane and
a trace of water is an illustration of work from Klibanov's pioneering
studies.[53]   Many other substrates and enzymes have now been investigated[54]
some on a multi-gram scale.   The use of immobilised enzymes has been
particularly important in this field.

Ramos Tombo *et al* have described an elegant use of porcine pancreatic
lipase to convert 1,3-propanediol derivatives (51) into (*S*)- and (*R*)-
monoacetates by, respectively, enantioselective hydrolysis of the diacetates
and enantioselective trans-esterification with methyl acetate as donor of the
acetyl group.[55]   Enantiomeric excesses were highest when R = $CH_2=CH(CH_2)_2-$
and C_6H_5- (> 90% for both enantiomers).

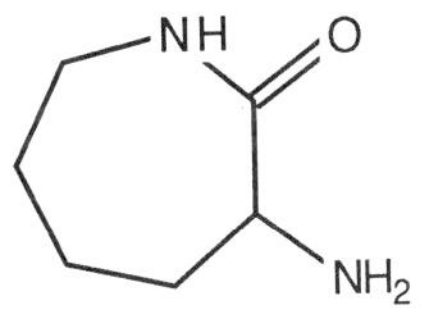

(54)

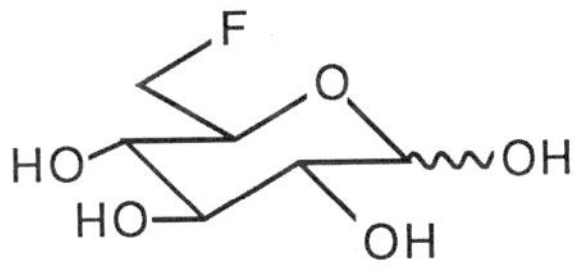

(55)

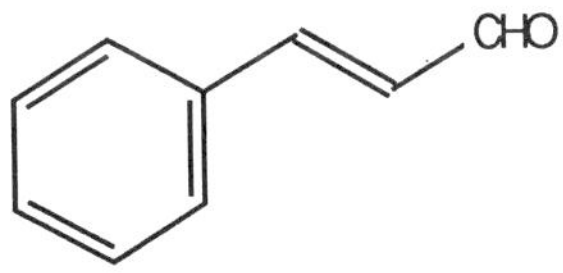

(56)

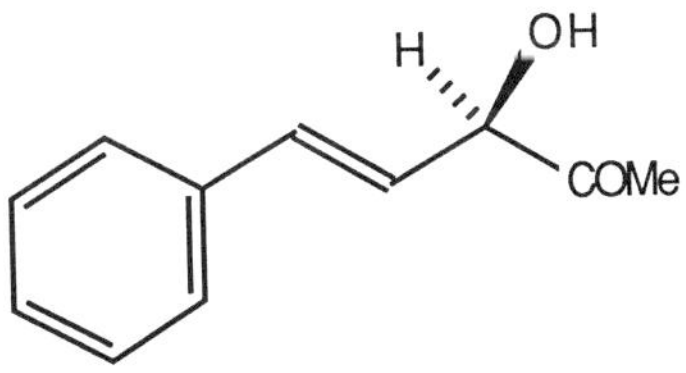

(57)

2.2.2 Amidases

The conversion of fermented penicillins into 6-aminopenicillanic acid
using an amidase from *E. coli* is a commercially important process.   The
amidase has been purified and immobilised.[56]   Acylases such as hog kidney
acylase have been widely employed for the hydrolysis of amide bonds in

racemic substrates to give optically active products. A range of racemic
N-acetyl amino acids afford L-amino acids (52) and D-N-acetylamino acids on
treatment with an acylase.[57] Similarly hydantoinase-catalysed hydrolyses
of substrates of type (53) offer an alternative route to resolved α-amino
acids. D-Hydantoinase is commonly employed; this produces D-carbamates
which are readily converted into the D-amino acid.[58] Important compounds
produced in the latter way include D-phenylglycine and D-*p*-hydroxyphenyl-
glycine for use in the synthesis of semi-synthetic penicillins.

Cyclic amides can also be resolved. For example δ-aminocaprolactam
(54) is hydrolysed to L-lysine (leaving D-δ-aminocaprolactam for recycling)
using various yeasts and bacteria.

Proteases and lipases have been utilized, with great success, for the
coupling of two amino-acid derivatives or peptide fragments. Tripeptides
consisting of three different amino-acids can be obtained in one step using
papain as the catalyst.[59]

3. MISCELLANEOUS REACTIONS CATALYSED BY ENZYMES

The epoxidation of alkenes using oxygenase enzymes is a transformation
that has been little explored. The conversion of oct-1-ene into (*R*)-1,2-
epoxyoctane (70% e.e.) using *Pseudomonas oleovoranes*[60] gives an indication
of the potential importance of this methodology.

Enzymes that catalyse aldol reactions are being investigated. For
example fructose bisphosphate aldolase catalyses a stereospecific aldol
reaction between dihydroxyacetone phosphate and a number of aldehydes to
give ketose-1-phosphates which upon removal of the phosphate groups by
hydrolysis (catalysed by an acid phosphatase) are converted stereospecific-
ally to aldose derivatives (catalysed by glucose isomerase). These combined
enzymatic processes allow the preparation of compounds such as 6-desoxy-6-
fluoro-glucose (55) on a 4-20 mmol scale.[61]

Yeast transforms aldehydes like cinnamaldehyde (56) to the corresponding
hydroxyketone (57) and then the diol (58) utilising acetaldehyde generated
in situ.[62]

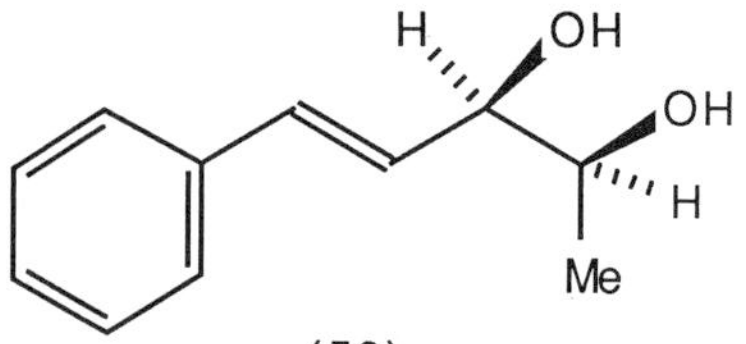

(58)

The enzyme oxynitrilase is obtained from almonds and catalyses the stereospecific addition of cyanide to a broad range of aldehydes. The optically active cyanohydrins (59) can be converted into α-hydroxy carboxylic acids, α-hydroxy ketones and β-aminols.[63]

Enzymes and whole cell systems are available that catalyse the hydrolysis of nitriles. For example the micro-organism *Brevibacterium sp* R 312 possesses a nitrile hydratase and an amidase with wide substrate selectivities. Thus highly functionalised nitriles such as α-alkoxy-α-amino nitriles can be hydrolysed to the corresponding acid in good yield.[64]

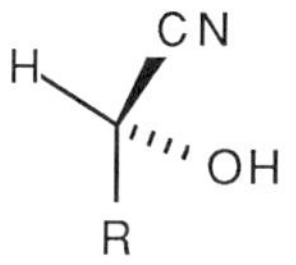

(59)

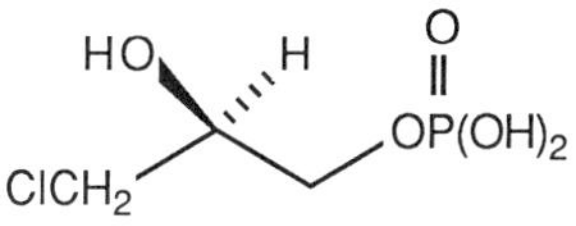

(60)

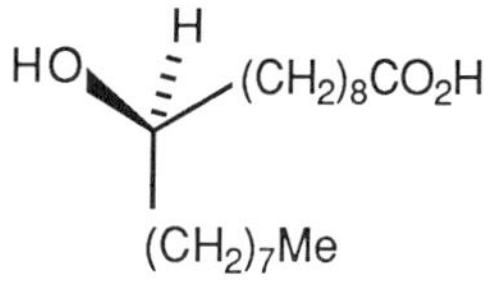

(61)

(62)

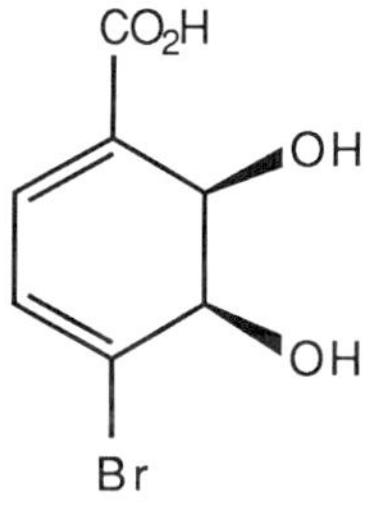

(63)

Glycerol kinase from *Saccharomyces cerevisiae* has been utilised to phosphorylate a number of glycerol analogues. 3-Chloropropane-1,2-diol gave the phosphate (60). The phosphorylating agent, adenosine triphosphate (ATP), was regenerated *in situ* using phosphoenolpyruvate and the enzyme pyruvate kinase.[65]

Phospholipase D from *Streptomyces sp.* catalysed the transfer of phosphatidyl residues from phosphatidylcholines to the 5'-hydroxy group of nucleosides using a two-phase system. A variety of 5'-phosphatidyl-nucleosides were prepared in high yields using this strategy.[66]

Optically active sulphoxides have been prepared from the corresponding sulphides by incubation with the micro-organism *Corynebacterium equi* or *Mortierella isabellina.* Yields are often variable but enantiomeric excesses of the chiral sulphoxides can be very high.[67]

Highly selective *anti*-additions to carbon-carbon double bonds are
controlled by a number of enzymes.   The substrates are generally α.β-
unsaturated carboxylic acids.   For example aspartase catalyses the addition
of ammonia to fumaric acid to give L-aspartic acid;   similarly the action
of fumarase gives rise to L-threochloromalic acid from chlorofumaric acid.[68]
Additions to unactivated double bonds are also known, for instance oleic
acids can be converted into 10(R)-hydroxystearic acid (61).

A particularly interesting biotransformation is the conversion of
benzene into *cis*-3,5-cyclohexadiene-1,2-diol.[69]   Recent work has shown that
many benzene derivatives can be processed by the relevant micro-organism,
Pseudomonas putida, to give optically active polysubstituted cyclohexadienes.
The conversion of the benzoic acid derivative (62) into the diene (63) is an
example from the current research in this area.[70]

4. FUTURE POSSIBILITIES IN THE AREA OF BIOTRANSFORMATIONS

The number of applications of the use of enzymes such as esterases,
lipases, and dehydrogenases for the preparation of useful quantities of
un-natural fine chemicals will continue to increase rapidly.

Enzymes able to catalyse other transformations will be the focus of
much attention;   the synthetic utility of enzymes that catalyse carbon-
carbon bond formation (*e.g.* the aldolases) will be explored in some detail.

The employment of genetically manipulated micro-organisms to accomplish
useful multi-stage processes will be increasingly in evidence.   The modification of enzymes by site-directed mutagenesis in order to extend the range
of the natural catalytic properties is well established but more dramatic
changes in the catalytic properties of such proteins can be expected.
Indeed, a start has been made on this new adventure.   Thus three amino-acid
substitutions have been made in the lactate dehydrogenase from *Bacillus
stearothermophilus*.   The wild type enzyme has a catalytic specificity for
pyruvate over oxaloacetate of *ca.* 1000 whereas the triple mutant has a
reversed specificity for oxaloacetate over pyruvate of 500.[71]

The possibility of using of antibodies as catalysts will be explored
further.   The idea behind this work is that antibodies raised to a stable
mimic of a transition state of the reaction under investigation should have
the capacity to stabilize, to some extent, the appropriate transition
state.[72]

REFERENCES

1. For an interesting historical review see M.K. Turner in 'The Chemical
    Industry' (*ed.* C.A. Heaton), Blackie, Glasgow and London, 1986,
    p. 284.
2. For full reviews see J.B. Jones, *Tetrahedron*, 1986, *42*, 3351 and
    references therein.
3. R.A. Johnson, 'Oxygenations with Micro-organisms' in 'Oxidation in
    Organic Chemistry' *ed.* W.S. Trahanovsky, Academic Press, New York,
    1978, Volume C; W. Charney and H.L. Herzog, 'Microbial Trans-
    formations of Steroids', Academic Press, New York, 1967, p. 26.
4. S. Butt and S.M. Roberts, *Nat. Prod. Reports*, 1986, *3*, 489.
5. B. Wipf, E. Kupfer, R. Bertazzi, and H.G.W. Leuenberger, *Helv. Chim.
    Acta*, 1983, *66*, 485; see also J. Ehrler, F. Giovannini,
    B. Lamatsch, and D. Seebach, *Chimia*, 1986, *40*, 172.
6. K. Mori and H. Watanabe, *Tetrahedron*, 1981, *37*, 1341; *idem. ibid*,
    1986, *42*, 295.   Professor Mori has synthesized a wide range of
    natural products from optically active 3-hydroxyalkanoates.
7. D. Seebach and M. Eberle, *Synthesis*, 1986, 37.
8. C. Fuganti, P. Grasselli, P. Casati, and M. Carmeno, *Tetrahedron
    Letters*, 1985, *26*, 101; W.-R. Shieh, A.S. Gopalan, and C.J. Sih,
    J. Am. Chem. Soc., 1985, *107*, 2993.
9. C.J. Sih and C.-S. Chen, *Ang. Chem. Int. Ed. Engl.*, 1984, *23*, 570.
10. P. De Shong, M.-T. Lin, and J.J. Ferez, *Tetrahedron Letters*, 1986,
    27, 2091.
11. K. Nakamura, T. Miyai, K. Nozaki, K. Ushio, S. Oka, and A. Ohno,
    Tetrahedron Letters, 1986, *27*, 3155.
12. K. Nakamura, M. Higaki, K. Ushio, S. Oka, and A. Ohno, *Tetrahedron
    Letters*, 1985, *26*, 4213.
13. D. Seebach, M.F. Züger, F. Giovannini, B. Sonnleitner, and A. Fiechter,
    Angew. Chem. Int. Ed. Engl., 1984, *23*, 151.
14. D. Buisson and R. Azerad, *Tetrahedron Letters*, 1986, *27*, 2631.
15. *e.g.* the Noyori Reagents, see R. Noyori, T. Ohkuma, M. Kitamura,
    H. Takaya, N. Sayo, H. Kumobayashi and S. Akutagawa, *J. Am. Chem.
    Soc.*, 1987, *109*, 5856.
16. H. Suemune, N. Hayashi, K. Funakoshi, H. Akita, T. Oishi, and K. Sakai,
    Chem. Pharm. Bull., 1985, *33*, 2168.
17. G. Guanti, L. Barfi, and E. Narisano, *Tetrahedron Letters*, 1986, *27*,
    3547; T. Fujisawa, E. Kojima, T. Itoh, and T. Sato, *Chem. Letters*,
    1985, 1751; Y. Takaishi, Y.-L. Yang, D.D. Tullio, and C.J. Sih,
    Tetrahedron Letters, 1982, *23*, 5489.
18. C.-Q. Han, D. Di Tullio, Y.-F. Wang, and C.J. Sih, *J. Org. Chem.*,
    1986, *51*, 1253.
19. M. Utaka, H. Watabu, and A. Takeda, *Chem. Letters*, 1985, 1475.
20. S. Tsuboi, E. Nishiyama, M. Utaka, and A. Takeda, *Tetrahedron Letters*,
    1986, *27*, 1915.
21. T. Fujisawa, E. Kojima, T. Itoh, and T. Sato, *Tetrahedron Letters*,
    1985, *26*, 6089.
22. D. Buisson, S. El Baba, and R. Azerad, *Tetrahedron Letters*, 1986, *27*,
    4453.
23. W.M. Dai and W.-S. Zhou, *Tetrahedron*, 1985, *41*, 4475; see also
    D.W. Brooks, H. Mazdiyarni, and P.G. Grothaus, *J. Org. Chem.*, 1987,
    52, 3223.
24. see, for example, C. Fuganti, P. Grasselli, P. Casati, and M. Carmeno,
    Tetrahedron Letters, 1985, *26*, 101.
25. R. Bowen and S.Y.R. Pugh, *Chem. Ind. (London)*, 1985, *10*, 323.

26. J. Leaver, T.C.C. Gartenmann, S.M. Roberts, and M.K. Turner in
     'Biocatalysis in Organic Media' (*eds*. C. Laane, J. Tramper, and
     M.D. Lilley), Elsevier, Amsterdam, 1987, 411.
27. J.B. Jones in 'Enzymes in Organic Synthesis' (*eds*. R. Porter and S.
     Clark), Ciba Foundation Symposium 111, Pitman, 1985, p. 3.
28. E. Keinar, E.F. Hafeli, K.K. Seth and R. Laned, *J. Am. Chem. Soc.*,
     1986, - , 162; for an example of a reduction using growing
     T. brockii see A. Belan, J. Bolte, A. Fauve, J.G. Gourcy and
     H. Veschambre, *J. Org. Chem.*, 1987, *52*, 256.
29. S. Butt, H.G. Davies, M. J. Dawson, G.C. Lawrence, J. Leaver, S.M.
     Roberts, M.K. Turner, B.J. Wakefield, W.F. Wall, and J.A. Winders,
     Tetrahedron Letters, 1985, *26*, 5077; S.M. Roberts, *Chem. and Ind.*,
     1988, 384.
30. L.G. Lee and G.M. Whitesides, *J. Org. Chem.*, 1986, *51*, 25.
31. B. Rambeck and H. Simon, *Angew. Chem. Int. Ed. Engl.*, 1974, *13*, 609.
32. H.G.W. Leuenberger, W. Boguth, E. Widner, and R. Zell, *Helv. Chim.
     Acta*, 1979, *62*, 455; H. Simon, H. Gunther, J. Bader, and W.
     Tischer, *Angew. Chem. Int. Ed. Engl.*, 1981, *20*, 861; for a recent
     example of a bakers' yeast reduction of an α,β-unsaturated aldehyde
     see *Tetrahedron Letters*, 1988, *29*, 2197.
33. M. Utaka, S. Konishi, and A. Takeda, *Tetrahedron Letters*, 1986, *27*,
     4737.
34. P. Gramatica, P. Manitto, D. Monti, and G. Speranza, *Tetrahedron*,
     1986, *42*, 6687; see also P. Gramatica, G. Speranza, P. Manitto,
     and D. Monti, *ibid*, 1987, *43*, 4481.
35. K.P. Lok, I.J. Jakovac, and J.B. Jones, *J. Am. Chem. Soc.*, 1985,
     107, 2521; see also G.L. Lemière, J.A. Lepoivre and F.C.
     Alderweireldt, *Tetrahedron Letters*, 1985, *26*, 4527: the difficulties
     in recycling NAD(P)$^+$ have been discussed, L.G. Lee and G.M.
     Whitesides, *ibid*, 1985, *107*, 6999.
36. Y. Khandelwal, N.J. de Souza, S. Chatterjee, B.N. Ganguli, and R.H.
     Rupp, *Tetrahedron Letters*, 1987, *28*, 4089.
37. J.-D. Fourneron, A. Archelas, B. Vigne, and R. Furstoss, *Tetrahedron*,
     1987, 43, 2273; V. Lamare, J.D. Fourneron, R. Furstoss, C. Ehret,
     and B. Corbier, *Tetrahedron Letters*, 1987, *28*, 6269.
38. Y. Yamazaki and H. Maeda, *Tetrahedron Letters*, 1985, *26*, 4775.
39. C.J. Francis and J.B. Jones, *J. Chem. Soc.*, *Chem. Commun.*, 1984, 579;
     L.K.P. Lau, R.A.H.F. Hui, and J.B. Jones, *J. Org. Chem.*, 1986,
     51, 2047.
40. F. Björkling, J. Boutelje, S. Gatenbeck, K. Hult, T. Norin, and
     P. Szmulik, *Tetrahedron*, 1985, *41*, 1347; F. Björkling,
     J. Boutelje, S. Gatenbeck, K. Hult, and T. Norin, *Tetrahedron Letters*,
     1985, *26*, 4957.
41. M. Kurihara, K. Kamiyama, S. Kobayashi, and M. Ohno, *Tetrahedron Letters*,
     1985, *26*, 5831.
42. G. Sabbioni and J.B. Jones, *J. Org. Chem.*, 1987, *52*, 4565.
43. H.-J. Gais and K.L. Lukas, *Angew. Chem. Int. Ed. Engl.*, 1984, *23*, 142.
44. B.I. Glanzer, K. Faber, and H. Griengl, *Tetrahedron*, 1987,*43*, 771.
45. A.J.H. Klunder, F.J.C. van Gastel, and B. Zwanenburg, *Tetrahedron
     Letters*, 1988, *29*, 2697.
46. H. Hemmerle and H.-J. Gais, *Tetrahedron Letters*, 1987, *28*,
     3471.
47. B.I. Glänzer, K. Faber, and H. Griengl, *Tetrahedron*, 1987, *43*, 5791.
48. C. Chan, P.B. Cox, and S.M. Roberts, *J. Chem. Soc.*, *Chem. Commun.*,
     1988,
49. I.C. Cotterill, H. Finch, D.P. Reynolds, S.M. Roberts, H.S. Rzepa,
     K.M. Short, A.M.Z. Slawin, C.J. Wallis, and D.J. Williams,
     J. Chem. Soc., *Chem. Commun.*, 1988, 470.
50. G. Eichberger, G. Penn, K. Faber, and H. Griengl, *Tetrahedron Letters*,
     1986, *27*, 2843.

51. D.R. Deardorff, A.J. Matthews, D.S. McMeekin, and C.L. Craney, *Tetrahedron Letters*, 1986, *27*, 1255.

52. In addition trifluoroethanol reacts selectively with various α,β-unsaturated ketones and esters to give the γ-hydroxyl carbonyl compounds, T. Kitazume and N. Ishikawa, *Chem. Letters*, 1984, 1815.

53. G. Kirchner, M.P. Scollar, and A.M. Klibanov, *J. Am. Chem. Soc.*, 1985, *107*, 7072.

54. P.E. Sonnet, *J. Org. Chem.*, 1987, *51*, 3477.

55. G.M. Ramos Tombo, H.-P. Schär, X. Fernandex, and O. Ghisalba, *Tetrahedron Letters*, 1986, *27*, 5707.

56. T.A. Savage in 'Biotechnology of Industrial Antibiotics', *ed.* E.J. Vandamme, Marcel Dekker, New York, 1984, p. 171.

57. I. Chibata in 'Asymmetric Reactions and Processes in Chemistry' (*eds.* E.L. Eliel and S. Otsuka), *Am. Chem. Soc.*, Washington, 1982; see also C. Wandrey in 'Enzymes as Catalysts in Organic Synthesis' (*ed.* M. Schneider), D. Reidel, Dordrecht, 1986, p. 263.

58. H. Yamada in 'Enzyme Engineering' (*eds.* I. Chibata, S. Fukui, and L.B. Wingard), Vol. 6, Plenum Press, New York, 1982, p. 97; T. Fukumura, *Agric. Biol. Chem.*, 1976, *40*, 1687, 1695; *idem*, *ibid*, 1977, *41*, 1327.

59. C.F. Barbas III and C.-H. Wong, *Tetrahedron Letters*, 1988, *29*, 2907; E.K. Bratovanova, I.B. Stoineva, and D.D. Petkov, *Tetrahedron*, 1988, *44*, 3633; A. A. Ferjancic, A. Puigserver, and H. Gaertner, *Biotech. Letters*, 1988, *10*, 101.

60. M.-J. de Smet, B. Witholt, and H. Wynberg, *J. Org. Chem.*, 1981, *46*, 3128; A.G. Katopodis, K. Wimalasena, J. Lee, and S.W. May, *J. Am. Chem. Soc.*, 1984, *106*, 7928.

61. C.-H. Wong and G.M. Whitesides, *J. Org. Chem.*, 1983, *48*, 3199; J.R. Durrwachter, D.G. Drueckhammer, K. Nozaki, H.M. Swears, and C.-H. Wong, *J. Am. Chem. Soc.*, 1986, *108*, 7812.

62. C. Fuganti, P. Grasselli, and S. Servi, *J. Chem. Soc., Perkin Trans. I*, 1983, 241; C. Fuganti, P. Grasselli, S. Servi, F. Spreafico, and C. Zirolfi, *J. Org. Chem.*, 1984, *49*, 4087.

63. E. Hochuli, *Helv. Chim. Acta*, 1983, *66*, 489; L. Jaenicke and J. Preun, *Eur. J. Biochem.*, 1984, *138*, 319.

64. Y. Vo-Quang, D. Marais, L. Vo-Quang, F. Le Goffic, A. Thiery, M. Maestracci, A. Arnaud, and P. Galzy, *Tetrahedron Letters*, 1987, *28*, 4057.

65. D.C. Crans and G.M. Whitesides, *J. Am. Chem. Soc.*, 1985, *107*, 7008, 7019.

66. S. Shuto, S. Ueda, S. Imamura, K. Fukukawa, A. Matsuda, and T. Ueda, *Tetrahedron Letters*, 1987, *28*, 199.

67. H. Ohta, Y. Okamoto, and G. Tsuchihashi, *Chem. Letters*, 1984, 205; H.L. Holland, H. Popperl, R.W. Ninniss, and P.C. Chenchaiah, *Can. J. Chem.*, 1985, *63*, 1118.

68. M.A. Findeis and G.M. Whitesides, *J. Org. Chem.*, 1987, *52*, 2838.

69. S.V. Ley, F. Sternfeld, and S.C. Taylor, *Tetrahedron Letters*, 1987, *28*, 225.

70. S.J.C. Taylor, D.W. Ribbons, A.M.Z. Slawin, D.A. Widdowson, and D.J. Williams, *Tetrahedron Letters*, 1987, *28*, 6391.

71. A.R. Clarke, C.J. Smith, K.W. Hart, H.W. Wilks, W.N. Chia, T.V. Lee, J.J. Birktoft, L.J. Banaszak, B.A. Barstow, T. Atkinson, and J.J. Holbrook, *Biochem. Biophys. Res. Commun.*, 1987, *148*, 15.

72. A.D. Napper, S.J. Benkovic, A. Tramontano, and R.A. Lerner, *Science*, 1987, *237*, 1041.

ENZYMES IN NON-AQUEOUS SYSTEMS

Marcel Waks

Unité Associée 586 du CNRS
Université René Descartes
75270 Paris Cedex 06, France

INTRODUCTION

Enzymes active in non-aqueous systems have been also designated as
"dry enzymes"[1]. The notion does not imply the absence of any water at
all in the reaction medium, however, since enzymatic activity requires
at least 0.2 g of water per gram of protein. This amount of water,
determined in sequential rehydration experiments on solid vitrified
samples of lysozyme by Careri et al[2], is tightly bound to the surface of
the protein and can be considered to be a different solvent from bulk
water. Indeed its physicochemical properties are comparable to
"biological water" [3] also present at membrane interfaces and which plays
a key role in the control of many important biological processes.

Yet, under such unusual experimental conditions where an organic
solvent can constitute more than 98% v/v of the reaction system, many
enzymes, usually studied in water, display conventional Michaelian
activity and an increased stability toward extreme temperatures. They
can yield completely new catalytic reactions, moreover, and thus they
are not "hibernating enzymes" [4].

Previously, organic solvents have been utilized extensively in
studies of macromolecules, especially of enzymes. The aim of these
studies was to dissolve the protein in a protic solvent or in a mixture
with water. This type of experiment, leading in most cases to a rapid
inactivation of the catalytic activity, has been discussed by Singer[5].
However after appropriate chemical modifications, some enzymes become
soluble in organic solvents and retain their activity[6,7]. The ability
of some organic solvents to stabilize proteins has been related to
mechanisms both of preferential hydration and of effects of the
hydrocarbon solvent on the water structure. The former phenomenon
reflects the inability of hydrocarbon additives to interact with the
protein itself, which excludes the solvent from its surface[8].

It seems also that enzymes can function in a non-aqueous
environment without being dissolved in the organic fluid. As pointed out
by Klibanov and his associates[9], this approach allows the use of water-
insoluble substrates as well as the easy recovery of the products, and
it takes advantage of the possible reuse of the enzymes. A different
course has been followed by the laboratories of Martinek[10] and Luisi[11].

In their experiments, the enzymes are solubilized in optically transparent water droplets, separated from the bulk hydrocarbon solvent by a monolayer of surfactant. Such an organized system is composed of reverse micelles which are dynamic entities of sizes comparable to proteins. They exhibit the processes of collision and exchange of their contents on a microsecond time scale. They are thus able to react either with aqueous solutions of entrapped substrates, or with substrates soluble only within the surrounding organic solution. Moreover, because of the peculiar physicochemical properties of the encapsulated water[12], some substrates hardly soluble in bulk water become soluble in the water pool of reverse micelles. In addition we have shown that membrane proteins and among them especially those only soluble in mixtures of organic solvents[13] can be inserted into the amphipathic and aqueous microenvironment of reverse micelles.

Thus non aqueous systems can provide versatile opportunities in the field of enzymology by yielding products of entirely new biotechnological processes[14], as foreseen by Singer in the sixties[5].

1) CHARACTERISTICS OF NON-AQUEOUS SYSTEMS (N.A.S.)

Diverse systems allow an enzyme to function actively in the presence of a minimal amount of water and provide the ability to solubilize effectively hydrophobic substrates and products in the bulk organic solvent. Although close to zero, the water content of the system is crucial, minor changes causing large differences in enzymatic activity. In most cases, the actual amount of bound water is barely sufficient to cover the macromolecular surface with a monolayer, assuming a homogeneous distribution of water molecules. Precise control of the water content of the system constitutes, therefore, a predominent factor in the choice of the type of system to use. The satisfaction of this and similar requirements in a given system may be contradictory and must be assayed by trial and error to obtain vigorous enzymatic activity.

There is general agreement that, whatever the system, the organic solvent should be water immiscible. For the choice of the most appropriate solvent, Laane et al[15] have proposed the use of a polarity index: log P. It designates the logarithm of the partition coefficient P of a given compound in the octanol-water two phase system. This approach has proven to be successful for enzymatic reactions involving solid enzymes[16] and enzymes in reverse micelles[15,17], or enzymes adsorbed on glass beads. In addition, the resistance of biocatalysts to heat denaturation also seems to be related to log P of the organic solvent[18]. According to Laane this index is linked to the ability of a given solvent to "distort" the protein-water interactions by making contact with it. Solvents displaying a log P > 4 seem to favor biocatalysis. Klibanov[9] suggests that this mechanism is related instead to the ability of the solvent to strip the protein of its essential, tightly bound water, by partitioning it from the enzyme to the solvent.

Suspension of enzymes in non-aqueous solvents are inhomogeneous two phase systems. Some of their advantages have been discussed above. In contrast, although a micellar solution is composed of a continuous organic phase, a surfactant, sometimes a cosurfactant, water molecules and the protein, it constitutes a homogeneous single phase system in the macroscopic sense. Such a system comprises an interphase constituted by the hydrocarbon tails of the surfactant in contact with the organic fluid and an interface with water at the contact of the polar head groups of the surfactant which can be modified by a cosurfactant (in general a water immiscible aliphatic alcool). Thus there exists a simple compartmentalization with an inside and an outside. A substrate will be able to partition between the encased water, the interphase and the organic phase, accumulating in a specific region depending on the polarity. Generally, the interphase or the organic solvent will be favored for hydrophobic solutes. Parameters such as ionization constants and oxidation reduction potentials will also be modified. The distribution of the products into the different phases is one of the factors which can lead to displacement of the reaction equilibrium for example by using α-chymotrypsin for peptide synthesis[19].

2) WHY ENZYMES IN N.A.S.

From what has been discussed in the preceeding section one might suspect that there is a considerable potential for the use of enzymes in N.A.S., for a number of the limitations of conventional aqueous solutions can been avoided. The net binding energy of a substrate to an enzyme is the difference between the energies of the molecules when free in solution and when joined in the enzyme-substrate complex. One way to change the substrate specificity, therefore, is to replace water by another reaction medium, thereby changing the energetic balance. In some cases an alteration or a reversal of substrate specificity can be obtained in this way[20,21].

Because of the aforementioned solubility properties, high local concentrations of substrates and products can be obtained and as a consequence reasonable rates and extents of conversion. Moreover, in micellar solutions a very large area (100 m^2 per milliliter of micellar solution) for exchange of reactants at the interface with the continuous organic phase is available. The mass transfer to and from the biocatalyst is not rate limiting. The physical parameters affecting the microenvironment of the enzyme can be controlled by the key parameter w_o which is defined as the molar ratio of water to surfactant. W_o determines the size of the micelles as well as the microviscosity, polarity, dielectric constant and activity of the encased water. Reversal of reaction equilibria can be obtained at low w_o values[22]. Thus by controlling these experimental conditions, it should be possible to fix the intermolecular geometry and to orient and to stabilize the reaction partners.

Stability in terms of resistance to denaturation at high temperatures has been shown for several enzymes in N.A.S. Among them are porcine lipase[23], chymotrypsin[18], and terpene cyclase[24]. Douzou[25], on the other hand has successfully employed micellar solutions in cryoenzymology experiments at temperatures as low as -36°C, where encased water does not freeze.

Finally it should be emphasized that in addition to the solubility properties of substrates and products in N.A.S., water-insoluble integral membrane proteins can be incorporated into reverse micelles[13] at reasonable concentrations and with a high degree of periodicity in structure, the amphipathic character of these microassemblies probably preventing the protein from unfolding. This opportunity could be exploited for the practical use of in view of catalytic reactions involving enzymes functionning in a membrane-bound state and possibly reacting with lipophilic substrates. Moreover, these enzymes can be specifically modified by reactants soluble either in the water pool or in the bulk organic solvent, reverse micelles functionning then as microreactors.

3-1 Dry enzymes in organic solvents

Knowledge of the precise state of solid proteins in hydrocarbon solvents is at present scarce[26]. Zaks and Klibanov[16] propose interpretations of their kinetic data and make comparisons with the observations reported by Rupley et al[27] and Finney and Poole[4] in sequential rehydration experiments on solid dry lysozyme. Such comparisons require, however, some caveats. There are obvious differences between a dry enzyme in a vacuum and the same enzyme suspended in a solvent of a given polarity, viscosity and dielectric constant. In the first case no interactions take place except for those with a few water molecules and with other protein molecules. In the second case, the protein is in contact with a hydrocarbon solvent which can interact with it. This appears to affect the catalytic properties of chymotrypsin in the presence of a constant amount of water but in solvents of differing dielectric constant[16].

One of the most intriguing differences between the reports on rehydration experiments on dry enzymes[27] and those of Zaks and Klibanov[16] concerns the minimal amount of water required for enzymatic activity. While the former investigators have not observed any measurable catalytic activity with less than 0.2 g of water per g of protein (i.e. 20% w/w), corresponding to about 200 molecules of H_2O per molecule of protein, the latter authors report optimal transesterification by chymotrypsin in an organic solvent with much less water i.e. 50 molecules per protein molecule provided that N-acetyl-phenylalanine is also present, although in neither case are solution conditions or monolayer coverage of the enzyme required. Caution must therefore must be exercised in the interpretation and comparison of experimental data. It is ironic that in N.A.S., water is one of the most critical parameters in initiating enzymatic activity.

In this connection, one important finding of Zaks and Klibanov[16] is that the ionization state of proteins in aqueous solutions is conserved after freeze-drying and subsequent transfer into a hydrocarbon solvent. For example, transesterification reactions catalysed by chymotrypsin in octane occur optimally when the protein is lyophilized from aqueous buffer at pH 7.8. This pH is also optimal for hydrolytic activity. The same type of result was also obtained with porcine lipase. Furthermore it has been shown that the state of ionization in solid hen yolk phosvitin is not altered from the state in the aqueous solution from which it is prepared[28]. As far as chymotrypsin is concerned, moreover, the active ionic conformation existing at pH 7.8 happens also to be the most stable in organic solvents at both ambient and elevated temperature. It should be noted that the improved stability of dry enzymes in N.A.S. as a function of temperature has been interpreted in terms of the crucial roles of water and pH in thermoinactivation processes[29]. Obviously the scarcity of water in the type of system described herein prevents it from playing its normal role and renders ineffective at least some denaturation mechanisms.

It has been established[4,27] that dry enzymes display marked structural rigidity. Finley and Poole[4] claim that at least 10% (w/w) of water appears necessary before any loosening of the protein structure and before any ionization can take place. The same conclusion has been

reached for the enzymes dispersed in non aqueous solvents[16]. In the case
of chymotrypsin enzyme activity is not obtained until the readdition to
the suspension 0.1% (v/v) water corresponding to over 1000 water
molecules per protein molecule. Lyophilization from an aqueous buffer
containing 0.25% (w/w) N-acetyl-phenylalanine (i.e. 50 molecules per
protein molecule) has the same effect. The acylated amino acid may act
in two ways: its binding to the active site would stabilize the native
conformation[16]. It is also possible for the small molecule to serve as a
plasticizer, in the same manner as water.

Protein rigidity is also suggested by the restricted enzymatic
specificity of porcine lipase in organic solvents[23], which is probably
due to increased constraints on the active site. Zaks and Klibanov
conclude that enzymes in the solid state can be "locked" into active
structures if properly treated prior to removal of the aqueous medium.
The conformation does not appear to change after suspension of the
enzyme in a hydrocarbon solvent, in marked contrast to the behavior of
enzymes dissolved in water-miscible protic solvents. In the former case
the enzyme appears to remain frozen in a native-like conformation
by intramolecular interactions because of insurmountable kinetic
barriers[16].

3-2 <u>Proteins / enzymes entrapped in reverse micelles</u>

Upon transfer of proteins from either an aqueous solution or from
a dry powder into the water interior of reverse micelles several
situations may arise according to the particular forces involved. i) The
overall conformation of the protein is not measurably affected or only
slightly changed. This seems to be the case for many water-soluble
enzymes of high molecular weight such as liver alcohol dehydrogenase,
and for enzymes of low molecular weight containing enough disulfide
bridges to stabilize them in a given conformation. ii) There is a marked
conformational change upon incorporation of the enzyme. This behavior
has been observed with membrane proteins released from the membrane and
inserted into the reverse micellar system designated then as membrane
mimetic[30]. The myelin basic protein (MBP), pI = 10.6, for example,
carries charges complementary to those of the sulfonates of the
surfactant polar head groups in reverse micelles of sodium bis (2-
ethylhexyl) sulfosuccinate (AOT). This type of structure remains
unaffected by additional water. iii) A third type of protein or peptide
structure can undergo a refolding process upon incorporation into
reverse micelles containing small amounts of water, but with large
amounts of added water they unfold, probably as a result of competition
for hydrogen bonding between the peptide chain and the water dipoles.
This effect has been illustrated by ACTH (1-24), a basic peptide active
at interfaces[31], and by lysozyme, a basic enzyme. In the latter case,
however, the conformation can be stabilized in the presence of a
specific inhibitor[32].

Because of the optical transparency of reverse micelles the
behavior and organization of proteins in these systems is now better
understood compared to enzyme powder dispersions in organic solvents.
Although we have studied in detail from both a conformational and
dynamic point of view model peptides and proteins that lack enzymatic
activity, all things being equal, there are no reasons why enzymes
should exhibit entirely different behavior under the same conditions.

4) ORIGIN OF THE CONFORMATIONAL CHANGES OBSERVED IN REVERSE MICELLES

The possible conformational changes occuring upon uptake of proteins by reverse micelles arise from at least two major types of mechanism: electrostatic interactions and hydrogen-bonding in entrapped water. The first originates from attractions and repulsions between charged protein side-chains and the sulfonates of the AOT polar head groups. In the case of myelin basic protein this is strongly suggested by the observation that the final folded structure observed in reverse micelles is totally unaffected by the amount of water present as determined by circular dichroism. Moreover, replacing AOT by a non ionic surfactant (Nikkol) weakens the electrostatic interactions with a partial loss of the α-helix content of MBP as a consequence[30].

This interpretation is confirmed by fluorescence studies of MBP in reverse micelles, for upon excitation at 290 nm a blue shift of the sole Trp is observed from 350 nm in aqueous solution to 335 nm in reverse micelles. This wavelength remains unchanged with increasing amounts of water and indicates a shielding of the Trp from it. The influence of electrostatic interactions is supported also by the fluorescence emission maximum (341 nm) in Nikkol micelles at w_o = 5.6. Thus, there is a correlation between the absence of charged head groups, an increased tryptophan accessibility to solvent and an unfolding of MBP. The firm anchoring of MBP at the micellar interface of AOT micelles represents, then, the driving force reponsible for the conformational change observed.

The unusual physical properties of entrapped water[12] constitute another important factor responsible for conformational changes of proteins. At low water content, intramolecular bonding within the protein will be favored, as a consequence of the fact that the water in the micelles is largely unavailable to solvate the protein, owing to its stronger interactions with the surfactant. At high w_o values water in excess of the amount needed to hydrate the surfactant is present and can solvate the protein, making peptide backbone-water interactions more prominent. Such observations were first reported by Seno et al[33] with basic homopolyaminoacids.

We have investigated the respective importance of charges and of the state of micellar water in the conformational changes of synthetic adrenocorticotropin hormone (ACTH) peptides containing residues (1-24) and (5-10) at various w_o values in comparison with the small water soluble molecule, N-acetyl-tryptophanamide (NATA). For NATA the wavelength of maximum fluorescence emission varies continuously as a function of w_o, which indicates a progressively more polar environment of the chromophore. The change in polarity arises from changes in the physical properties of the entrapped water, such as the dielectric constant and viscosity. The behavior of the basic ACTH (1-24) (6 positive charges) differs from that of the shorter ACTH (5-10) (2 positive charges). The latter peptide behaves as NATA, with continuously varying wavelength of maximum fluorescence emission from 336 nm to 345 nm. ACTH (1-24), in contrast, undergoes a variation from 334 nm at w_o = 2.0 to 338 nm at w_o = 7, where all the water is still tightly bound. The maximum emission wavelength then remains invariant between w_o = 7.0 and 22.4, indicating that the environment of the chromophore is shielded from additional free water due to the attractive interactions between the peptide and the micelle inner wall[31].

At the same time ACTH (1-24) undergoes a dramatic conformational change from a β-pleated sheet to an unfolded structure as a function of

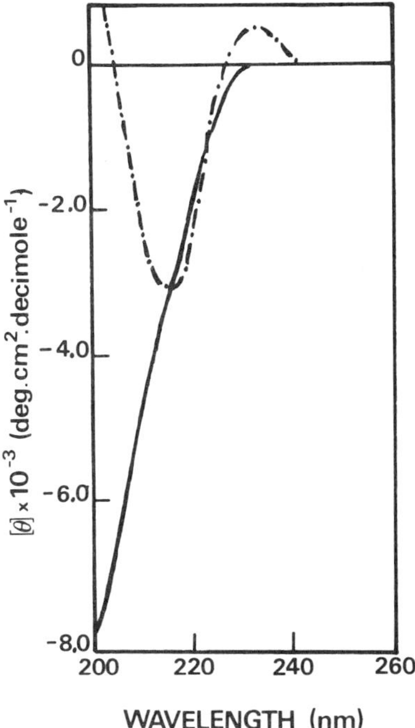

Fig. 1. Far ultraviolet circular dichroism spectra of ACTH (1-24) in micellar solutions at 25 mM AOT and at w_o values 3.5 (·—·) and 22.4 (——). Adapted from Ref. 31.

the presence of unbound water, in which interaction of the peptide chain with mobile water dipoles become possible (Fig.1). This occurs even though the part of the peptide containing the tryptophan remain anchored to the inner wall. These examples point to the respective roles of the charges compared to the properties of immobilized water in stabilizing the protein conformation. In conclusion, electrostatic interactions may not prevent an extensive unfolding of short polypeptide chains as in ACTH (1-24), whereas in larger proteins such as MBP a more subtle balance between the stabilizing forces and unfolding must exist. For example initially, the electrostatic forces may induce a conformational change in the macromolecule while intramolecular hydrogen bonds could stabilize and subsequently orient it.

5) DYNAMICS OF PROTEINS IN REVERSE MICELLES

The dynamics of protein molecules studied by time resolved fluorescence techniques yields interesting information about the organisation of the macromolecules within micelles. These techniques demonstrate that the blue shift of the tryptophan fluoresence emission maximum is accompanied by severe restriction of the side chain rotational motion upon incorporation of MBP into reverse micelles, in striking contrast to its motion in water. Furthermore, the rotational motion of the fluorophore becomes faster with increased w_o values. Thus it appears that increasing the water content of micelles does not measurably modify the overall protein conformation but only the internal protein dynamics[30].

We have studied[31] in detail the Trp internal dynamics and overall peptide rotational diffusion of the basic peptide ACTH (1-24) and of the shorter peptide ACTH (5-10) in comparison with neutral glucagon. The results may yield on a more limited scale a clue to what happens with larger proteins. Upon transfer of all these peptides from aqueous solution into reverse micelles, a severe slowing down of the peptide internal dynamics and a strong limitation of the angle of Trp rotation is observed. The variation of the long correlation time of the Trp fluorescence anisotropy decay as a function of the water content of reverse micelles is depicted in Fig.2. The overall tumbling of all peptides is slowed in reverse micelles as compared to water.

For the overall rotational motion of a rigid body entrapped in reverse micelles, two limiting cases can be considered according to their size. At low w_0 values, corresponding to small-size micelles, the microscopic viscosity is considerably higher than in bulk water[11]. The solute is not allowed to rotate freely within the micelle, and the long correlation time of the anisotropy decay corresponds to the rotational diffusion of the micelle itself. At high w_0 values, the rotational correlation time of the micelles is larger since their volumes increase. However, the microscopic viscosity inside the micelle decreases at the same time, and tumbling of the solute within the micellar water pool starts to occur. In this case, the long correlation time is a combination of both rotational processes: the rotational diffusion of the peptide within the micelle and the rotational diffusion of the micelle itself.

The strength of coupling between the two rotational motions is governed not only by the interactions between the solute and the surfactant, but also by interactions with embedded water. In this

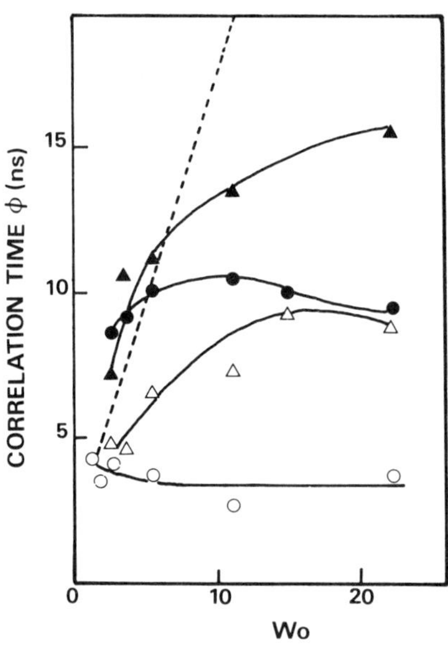

Fig. 2. Variation of the long correlation time in nanoseconds of the Trp fluorescence anisotropy decay in peptides inserted into reverse micelles as a function of w_0, (▲) ACTH (1-24), (●) glucagon, (△) ACTH (5-10), (O) NATA, (—––) variation of the rotational correlation time of the micelles calculated from the Stokes-Einstein relation. Adapted from Ref. 31.

respect, ACTH (1-24) interacts more strongly than glucagon and ACTH (5-10) with the AOT sulfonate groups and therefore displays a stronger coupling. The influence of the peptide-water interactions on the coupling mechanism is suggested by the fact that despite their differences in molecular weight, ACTH (5-10) (MW: 830) and glucagon (MW: 3.500) display similar overall tumbling at high w_o values. It can be concluded that both electrostatic interactions of the side-chains and hydrogen bonding with water and probably also with the surfactant head groups are crucial not only for the localization and conformational distribution but also for the dynamics of peptides and proteins with respect to the micellar interface.

CONCLUSION

Among the questions arising from experiments carried out with enzymes in water restricted environments some have received operational answers which open new biological opportunities. From a different perspective, however, many fundamental questions related to protein thermodynamics and kinetics remain still to be elucidated. For example when Levinthal[1] asks "Is a dry enzyme frozen in an active state from which denaturation, in the unfolding sense, is virtually impossible or at least requires much greater energy input than is normally observed in water?", the results of the experiments described herein strongly suggest that the enzymes so far investigated exist frozen in an apparently active state. Frozen structures, indeed, have been previously described in aqueous protein-protein associations[34] that did not involve enzyme-substrate complexes or transition states. In the latter case, however, the enzyme must free itself of product and be restored to its initial conformation after activity. So far we have no compelling evidence about how these events occur in dry enzymes suspended in organic solvents.

We lack, therefore, a large body of significant information about essential reaction mechanisms in N.A.S. : what is the direct effect of water immiscible organic solvents on the peptide chain itself and by what mechanisms can they interfere with enzymatic activity? How much rigidity is acceptable for a biocatalyst to function or how little water is required for induce the necessary amount of flexibility of it in N.A.S.? Similarly, in reverse micelles where spectroscopic methods disclose severe barriers to protein dynamical motions, the implications of these effects for enzymatic activity are not yet understood. Obviously these questions need to be investigated further. A point worth noting is that while aqueous solutions of surfactant and bulk hydrocarbon solvents may act as powerful denaturing agents for enzymes, they can also maintain or even restore a native-like folded structure under experimental conditions where the same agents exist within an organized membrane mimetic system[35].

ACKNOWLEDGEMENTS

M.W wishes to thank Dr. Peter Kahn for invaluable help and constant encouragements and Dr. C.Nicot for many insightful discussions. The excellent editorial work of M.Vacher is gratefully acknowledged.

This work was supported in part by grants from I.N.S.E.R.M. and C.N.R.S.

REFERENCES

1. C. Levinthal, Dry enzymes. Proteins: Struct. Funct. Genetics. 1: 2-3, (1986).

2. G. Careri, E. Gratton, P. H. Yang, and J. A. Rupley, Correlation of IR spectroscopic, heat capacity, diamagnetic susceptibility and enzymatic measurements on lysozyme powder. Nature., 284: 572-573, (1980).

3. R. Cooke, and I. D. Kuntz, The properties of water in biological systems. Annu.Rev.Biophys.Bioeng., 3: 95-126, (1974).

4. J. L. Finney, and P. L. Poole, Protein hydration and enzyme activity: the role of hydration-induced conformation and dynamic changes in the activity of lysozyme. Comments Mol.Cell.Biophys., 2: 129-151, (1984).

5. S. J. Singer, The properties of proteins in nonaqueous solvents. in "Advances in protein chemistry." Jr. C. B. Anfinsen, M. L. Anson, K. Bailey, and J.T.Edsall, eds., New York: Academic Press. 17: 1-68, (1962).

6. T. Yoshimoto, A. Ritani, K. Ohwada, K. Takahashi, Y.Kodera, A. Matsushima, Y. Saito, and Y. Inada, Polyethylene glycol derivative-modified cholesterol oxidase soluble and active in benzene. Biochem.Biophy.Res.Comm., 148: 876-882, (1987).

7. H. F. Gaertner, and A. J. Puigserver, Peptide synthesis catalyzed by polyethylene glycol-modified chymotrypsin in organic solvents. Proteins: Struct. Funct. Genetics, 3: 130-137, (1988).

8. L. Y. Lee, and J. Lee, Thermal stability of proteins in the presence of polyethylene glycols. Biochemistry., 26: 7813-7819, (1987).

9. A. Zaks, and A. M. Klibanov, Enzyme-catalysed processes in organic solvents. Proc.Natl.Acad.Sc. USA., 82: 3192-3196, (1985).

10. K. Martinek, A. V. Levashov, N. Klyachko, Y. L. Khmelnitski, I. V. Berezin, Micellar enzymology., Eur.J.Biochem., 155: 453-468, (1986).

11. P. L. Luisi, L. Magid, Solubilization of enzymes and nucleic acids in hydrocarbon micellar solutions. Crit.Rev.Biochem., 20: 409-474, (1986).

12. P. L. Luisi, M. Giomini, M. P. Pileni, and B. H. Robinson, Reverse micelles as hosts for proteins and small molecules., Biochim. Biophys. Acta., 947: 209-246, (1988).

13. A. Delahodde, M. Vacher, C. Nicot, and M. Waks, Solubilization and insertion into reverse micelles of the major myelin transmembrane proteolipid. FEBS Lett., 172: 343-347, (1984).

14. A. L. Margolin, and A. M. Klibanov, Peptide synthesis catalyzed by lipases in anhydrous organic solvents. J. Am. Chem. Soc., 109: 3802-3804, (1987).

15. C. Laane, S. Boeren, K. Vos, C. Veeger, Rules for optimization of biocatalysis in organic solvents. Biotechnol. Bioeng.,30: 81-87, (1987).

16. A. Zaks, and A. M. Klibanov, Enzymatic catalysis in nonaqueous solvents. J.Biol.Chem., 263: 3194-3201, (1988).

17. C. Laane, and C. Veeger, On the design of reversed micellar media for the enzymatic synthesis of apolar compounds. in " Methods in Enzymology" Immobilized Enzymes and Cells. K. Mosbach, ed., 136: 216-229, (1987).

18. M. Reslov, P. Adlercreutz, B. Mattiasson, Organic solvents for bio-organic synthesis. Optimization of parameters for a chymotrypsin catalysed process. Appl.Microbiol.Biotechnol., 26: 1-8, (1987).

19. P. Lüthi, and P. L. Luisi, Enzymatic synthesis of hydrocarbon-soluble peptides with reverse micelles. J. Am. Chem. Soc., 106: 7285-7286. (1984).

20. A. Zaks, and A. M. Klibanov, Substrate specificity of enzymes in organic solvents vs. Water is reversed. J.Am.Chem.Soc., 108: 2767- 2768, (1986).

21. K. Martinek, Yu. L. Khmel'nitskii, A. V. Levashov, and I.V. Berezin Substrate specificity of alcohol dehydrogenase in a colloidal solution of water in an organic solvent. Dokl. Akad. Nauk SSSR, 263: 737-741, (1982).

22. K. Martinek, Yu. L. Khmel'nitskii, A. V. Levashov, N. L. Klyachko, A. N. Semenov, and I. V. Berezin, Biocatalysis and regulation of chemical equilibrium in a system of inverted micelles of a surfactant in an organic solvent. Dokl.Akad.Nauk SSSR, 256: 1423-1426, (1981).

23. A. Zaks, and A. M. Klibanov, Enzymatic catalysis in organic media at 100°C. Science., 224: 1249-1251, (1984).

24. C. J. Wheeler, and R. Croteau, Terpene cyclase catalysis in organic solvent/minimal water media: demonstration and optimization of (+) α-Pinene cyclase activity. Arch.Biochem.Biophys., 248: 429-434, (1986).

25. P. Douzou, E. Keh, and C. Balny, Cryoenzymology in aqueous media: micellar solubilized water clusters. Proc.Natl.Acad.Sci. USA. 76: 681-684, (1979).

26. F. M. Wasacz, J. M. Olinger, R. J. Jakobsen, Fourrier transform infrared studies of proteins using nonaqueous solvents. Effect of methanol and ethylene glycol on albumin and immunoglobulin G. Biochemistry, 26: 1464-1470, (1987).

27. J. A. Rupley, E. Gratton,and G. Careri, Water and globular proteins Trends in Biochem. Sci., 8: 12-22, (1983).

28. B. Prescott, V. Renugopalakrishnan, M. J. Glimcher, A. Bhushan, and G. J. Thomas, Jr., A Raman spectroscopy study of hen egg yolk phosvitin: structures in solution and in the solid state. Biochemistry, 25: 2792-2798, (1986).

29. T. J. Ahern and A. M. Klibanov, The mechanism of irreversible enzyme inactivation at 100°C. Science, 228: 1280-1284, (1985).

30. C. Nicot, M. Vacher, M. Vincent, J. Gallay, and M. Waks, Membrane proteins in reverse micelles: myelin basic protein in membrane-mimetic environment. Biochemistry, 24: 7024-7032, (1985).

31. J. Gallay, M. Vincent, C. Nicot, and M. Waks, Conformational aspects and rotational dynamics of synthetic adrenocorticotropin - (1-24) and glucagon in reverse micelles. Biochemistry, 26: 5738-5747, (1987).

32. D. Steinmann, H. Jackle, and P. L. Luisi, A comparative study of lyysozyme conformation in various reverse micellar systems. Biopolymers, 25: 1133-1156, (1986).

33. M. Seno, H. Moritomi, Y. Kuroyanagi, K.Iwamoto, and G. Ebert, Conformational studies of basic poly(α-amino acid)s in reversed micelles. Colloid Polymer Sci., 262: 727-733, (1984).

34. M. Waks, and S. Beychok, Induced conformational states in human apo-hemoglobin on binding of haptoglobin 1-1. Effect of added heme as a probe of frozen structures. Biochemistry, 13: 15-22, (1974).

35. J. H. Fendler, Enzymes and membrane mimetic systems. In "Membrane Mimetic Chemistry" J. Wiley and Sons eds, New York. 235-292, (1982).

ADDITIONAL SEMINARS AND STUDENT WORKSHOPS

E.Gratton: Structural fluctuations and enzyme catalysis.

J.R.Knutson: Membrane phase transition - a fluorescence study.

G.Weber: Role of pressure on protein dynamics as seen by fluorescence
    spectroscopy.

WORKSHOP I: Fluorescence relaxation techniques.

WORKSHOP II: Molecular modulation.

WORKSHOP III: Proteins in industry and biotechnology.

POSTERS

A.Bianchi,P.Grigolini,R.Mannella,V.Palleschi & M.R.Torquati (Italy).
     First passage time problems in the presence of colored noise.

B.M.Britt & W.L.Peticolas (U.S.A.). The hydrolysis of a chromophoric
     ester substrate by carboxypeptidase-A appears to proceed via a
     cyclic metal-anhydride intermediate.

G.Cheron,G.Noat,J.Nari & J.Ricard (France). A transient kinetic study
     of β glucosidase from soybean cell wall.

J.C.Dewan (U.S.A.). Greatly reduced radiation damage in ribonuclease
     crystals mounted on glass fibers.

D.T.F.Dryden (U.K.). Resolution of the heterogeneous fluorescence of
     3-phosphoglycerate kinase.

M.T.Giudici-Orticoni,J.Buc,A.Cornish-Bowden & J.Ricard (France).
     Structural kinetics of fructose biphosphatase from spinach
     chloroplasts.

A.B.Hanley,C.Kwiatkowska & A.R.Mackie (U.K.). DNA manipulation in
     reverse micelles.

G.Hervé (France). Mechanisms of allosteric regulation in aspartate
     transcarbamylase.

R.Hilhorst & C.Veeger (Netherlands). Technological applications of
     enzymes in reversed micelles.

J.F.Holzwarth,F.Meyer,M.Pickard & H.B.Dunford (Germany). Mechanism
     of azide binding to chloro- and horseradish peroxidase: an iodine
     laser temperature jump investigation.

J.L.Houben & G.Tropiano (Italy). State coupling and state dynamics.

J.L.Houben & G.Tropiano (Italy). Fluorescence lifetime analysis,
     stochastic processes and noise.

C.M.L.Hutnik & A.G.Szabo (Canada). Protein conformation and dynamics:
     a time-resolved fluorescence spectroscopic study.

F.J.Jacob (Belgium). Site-directed mutagenesis of streptomyces albus G
     β-lactamase.

C.M.Johnson & N.C.Price (U.K.). Changes in the monomeric phosphoglycerate mutase from schizosaccharomyces pombe in the presence of
its cofactor.

V.Leloup (France). An application of enzymic hydrolysis in heterogeneous
medium: alpha-amylolysis of starch gels.

A.Matagne (Belgium). Kinetic studies of β-lactamases.

L.Mouawad,M.Desmadril,D.Perahia & J.M.Yon (France). The use of energy
minimization for the interpretation of fluorescence anisotropy
decay data.

P.R.V.Nayudu (Australia). Transphosphorylation of suitable alcohol by
intestinal alkaline phosphatase from ATP.

H.Ondag (Turkey). Effects of excess thymidine and methotrexate on
human peripheral blood and fibroblast cultures.

A.Perico (Italy). Local dynamics in biological macromolecules.

B.Pispisa (Italy). Theoretical models of noncovalent diastereomeric
electron-transfer adducts.

M.Pozsgay,P.R.Carey,C.Choma, & T.Lessard (Canada). Structure-function
studies of the insecticidal protein from bacillus thuringiensis.

G.D.Reinhart (U.S.A.). The role of coupling entropy in allosteric
response.

B.Sanders (U.K.). Dynamics and thermodynamics of genetically
engineered proteins.

S.Sorrentino,R.De Prisco & M.Libonati (Italy). Human seminal
ribonuclease. A potential marker in human prostatic carcinoma.

A.Tekin (Turkey). Localization of herpes simplex type I virus early
antigens in fibroblast cell culture by indirect immune fluorescence
technique.

M.S.Thorniley,Y.A.B.D.Wickramasinghe & P.Rolfe (U.K.). Non-invasive
in-vivo monitoring of changes in the redox state of cytochrome
oxidase and the oxygenation of haemoglobin by near infra-red
spectroscopy.

M.P.Weir,M.A.Chaplin,C.W.Dykes,D.M.Wallace & A.N.Hobden (U.K.).
Structure-activity relationships of recombinant human interleukin-2
missense mutants.

PARTICIPANTS (L: lecturer, D: co-director)

Aysar AKTIMUR, Ege University School of Medicine, Dept. of Physiology,
T-35100 Bornova, Izmir, Turkey.

Giuliano ALAGONA, Istituto di Chimica Quantistica ed Energetica
Molecolare - CNR, Via Risorgimento 35, I-56100 Pisa, Italy.

Aitor ALANA PARDO, Dept. of Biochemistry & Molecular Biology, University
of the Basque Country, P.O.Box 644, SP-48080 Bilbao, Spain.

Max BATENBURG, Unilever Research Laboratory, Postbus 114, NL-3130 AC
Vlaardingen, The Netherlands.

Astrid BIRK, Institut fur Physikalische-Chemie, Westfalische Wilhelms-
Universitat, Schlossplatz 4, D-4400 Munster, Germany.

B. Mark BRITT, Department of Chemistry, University of Oregon, Eugene,
Oregon 97403, U.S.A.

Mercedes CAMPILLO, Laboratorio de Bioestadistica y Epidemologia,
Universidad Autonoma de Barcelona, SP-08193 Bellaterra, Barcelona,
Spain.

Giorgio CARERI (L), Dipartimento di Fisica, Università di Roma "La
Sapienza", Piazzale A.Moro 5, I-00185 Roma, Italy.

Paul R. CAREY (L), Division of Biological Sciences, NRC Canada, Ottawa
K1A OR6, Canada.

Paolo CAVATORTA, Dipartimento di Fisica, Via delle Scienze, I-43100
Parma, Italy.

Guilene CHERON, Centre Biochimie Biologie Moléculaire - CNRS, B.P. 71,
31 Chemin Joseph Aiguier, F-13402 Marseilles, Cedex 9, France.

Christin CHOMA, Division of Biological Sciences, National Research
Council of Canada, Ottawa K1A OR6, Canada.

Pere CLAPES, Laboratory of Peptides, Dept. of Biological and Organic
Chemistry, Jorge Girona Saalgada 18-26, SP-08034 Barcelona, Spain.

John R. COGGINS (L), Biochemistry Department, University of Glasgow,
Glasgow G12 8QQ, Scotland, U.K.

Alan COOPER (L/D), Dept. of Chemistry, University of Glasgow, Glasgow
G12 8QQ, Scotland, U.K.

Maria Veiga CUNHA, Biochemistry Department, Microbiology Unit, Oxford University, South Parks Road, Oxford OX1 3QU, U.K.

Stephen CUSACK (L), Institut Laue Langevin BP 156X, Avenue des Martyres, F-38042 Grenoble Cedex, France.

Francesca CUTRUZZOLA, Centro di Biologia Molecolare - CNR, Università di Roma "La Sapienza", Piazzale A.Moro 5, I-00185 Roma, Italy.

Gunay DEMIREL, Dept. of Chemistry, Faculty of Arts and Sciences, Middle East Technical University, T-06351 Ankara, Turkey.

Kurt Dwaine DESHAYES, Dept. of Chemistry & Biochemistry, University of California, 405 Hilgard Avenue, Los Angeles, CA 90024-1569, U.S.A.

Michel DESMADRIL, Laboratoire d'Enzymologie Physico-chimique et Moléculaire, Bâtiment 433, Université de Paris-Sud, F-91405 Orsay, France.

John DEWAN, Chemistry Dept., New York University, 4 Washington Place, Room 514, New York, NY 10003, U.S.A.

Lorenzo DI BARI, Department of Chemistry, University of Pisa, Via Risorgimento 35, I-56100 Pisa, Italy.

David T.F. DRYDEN, Biochemistry Department, University of Newcastle, C. Cookson Bldg., Framlington Place, Newcastle-upon-Tyne NE2 4HH, U.K.

Francesco FARSACI, Istituto di Fisica, Università di Messina, Contrada Papardo, Salita Sperone 31, I-98010 Santa Agata, Messina, Italy.

Francois FILLAUX (L), Laboratoire de Spectroscopie Infrarouge et Raman, 2 Rue Henri Dunant, F-94320 Thiais, France.

Manfred FISCHER, Institut fur Physikalische-Chemie, Westfalische Wilhelms-Universitat, Schlossplatz 4, D-4400 Munster, Germany.

Teresa FONSECA (L), Chemistry Department, Colorado State University, Fort Collins, CO 80523, U.S.A.

Caterina GHIO (L), Istituto di Chimica Quantistica ed Energetica Molecolare - CNR, Via Risorgimento 35, I-56100 Pisa, Italy.

Rita GIORDANO, Istituto di Fisica, Università di Messina, Contrada Papardo, Salita Sperone 31, I-98010 Santa Agata, Messina, Italy.

Jesus GIRALDO, Laboratorio de Bioestadistica y Epidemologia, Universidad Autonoma de Barcelona, SP-08193 Bellaterra, Barcelona, Spain.

Marie Therese GIUDICI ORTICONI, Centre Biochimie Biologie Moleculaire - CNRS, B.P. 71, 31 Chemin Joseph Aiguier, F-13402 Marseilles, Cedex 9, France.

Antonino GRASSO, Istituto di Fisica, Università di Messina, Contrada Papardo, Salita Sperone 31, I-98010 Santa Agata, Messina, Italy.

Enrico GRATTON (L), Department of Physics, University of Illinois, 1110 West Green Street, Urbana-Champaign, IL 61801, U.S.A.

Anthony Brian HANLEY, AFRC Institute of Food Research, Colney Lane, Norwich, Norfolk NR4 7UA, U.K.

Guy HERVÉ, Laboratoire d'Enzymologie, Conseil National de la Recherche Scientifique, F-91190 Gif-sur-Yvette, France.

Riet HILHORST, Department of Biochemistry, Agricultural University, Dreijenlaan 3, NL-6703 HA Wageningen, The Netherlands.

Virginia HO, Lab. Nacional Engenheria Tecnologia Industrial ITI, Departamento Tecnologia Industrias Quimicas, P-2745 Queluz, Portugal.

Jozef F. HOLZWARTH (L), Fritz-Haber-Institut der Max-Planck Gesellschaft Faradayweg 4-6, D-1000 Berlin, Germany.

Julien L. HOUBEN (L/D), Istituto di Chimica Quantistica ed Energetica Molecolare - CNR, Via Risorgimento 35, I-56100 Pisa, Italy.

Cindy Mary-Lynn HUTNIK, Division of Biological Sciences, National Research Council of Canada, Ottawa K1A OR6, Canada.

James T. HYNES (L), Department of Chemistry, University of Colorado, Boulder, CO 80309-0215, U.S.A.

Francoise JACOB, Service de Microbiologie Appliquée, Bâtiment Chimie B6, Université de Liège Sart-Tilman, B-4000 Liège, Belgium.

Christopher Mark JOHNSON, Department of Chemistry, University of Glasgow, Glasgow G12 8QQ, Scotland, U.K.

Jay R. KNUTSON (L), Department of Health and Human Services, NIH, National Heart Lung & Blood Institute, Bethesda, MD 20892, U.S.A.

Alessandro LAMI (L), Istituto di Chimica Quantistica ed Energetica Molecolare - CNR, Via Risorgimento 35, I-56100 Pisa, Italy.

Valerie LELOUP, Institut National de la Recherche Agronomique, Rue de la Géraudière, F-44072 Nantes Cedex 03, France.

Franz MARK, Max Planck Institut fur Strahlenchemie, Stifstrasse 34-36, D-4330 Mulheim a.d. Ruhr, Germany.

Andre MATAGNE, Laboratoire d'Enzymologie, Bâtiment Chimie B6, Université de Liège Sart-Tilman, B-4000 Liège, Belgium.

Andrew MAZAR, University of Illinois, College of Medicine, 1853 W.Polk 312-CMW, Chicago, IL 60612, U.S.A.

Liliane MOUAWAD, Laboratoire d'Enzymologie Physico-chimique et Moleculaire, Bâtiment 433, Université de Paris-Sud, F-91405 Orsay, France.

Pagadala Ramachandra V. NAYUDU, Department of Zoology, Monash University, Clayton, Victoria 3168, Australia.

Paolo NEYROZ, Dipartimento di Chimica Biologica, Ospedale Maggiore, Viale Gramsci 14, I-43100 Parma, Italy.

Hulya ONDAG, Nenehatun cad. no 25/2 Kucukesat, Ankara, Turkey.

Ferhan OZADALI, Food Engineering Dept., Middle East University, 06531 Ankara, Turkey.

Fritz PARAK (L), Institut fur Physikalische-Chemie, Westfalische Wilhelms-Universitat Munster, Schlossplatz 4-7, D-4400 Munster, Germany.

Luis Marcelo PEREIRA, LNETI, National Laboratory of Industrial Research, Estrada das Palmeiras, Queluz de Baixo, P-2745 Queluz, Portugal.

Angelo PERICO, Centro di Studi Chimico-Fisici di Macromolecole Sintetiche e Naturali - CNR, Corso Europa 30, I-16132 Genova, Italy.

Basilio PISPISA, Dipartimento di Scienze e Tecniche Chimiche, Università Roma Torvergata, Via O.Raimondo, I-00173 Roma, Italy.

David PISTON, University of Illinois at Urbana, Dept. of Physics, 1110 W.Green Street, Urbana, IL 61801, U.S.A.

Marianne POZSGAY, Division of Biological Sciences, NRC-Canada, Sussex Drive, Ottawa K1A OR6, Canada.

Eric RAIMBAUD, Institut National de la Recherche Agronomique, Rue de la Géraudière, F-44072 Nantes Cedex 03, France.

Christina REDFIELD (L), Inorganic Chemistry Laboratory, Oxford University, South Parks Road, Oxford OX1 3QR, U.K.

Gregory D. REINHART, Dept. of Chemistry, 620 Parrington Oval, Room 208, Norman, Oklahoma 73019, U.S.A.

Torquato RISTORI, Istituto di Biofisica - CNR, Via San Lorenzo 26, I-56100 Pisa, Italy.

Stan M. ROBERTS (L), Department of Chemistry, University of Exeter, Stocker Road, Exeter EX4 4OD, U.K.

John A. RUPLEY (L), Department of Biochemistry, University of Arizona, Tucson, AZ 85721, U.S.A.

Nicole S. SAMPSON, Dept. of Chemistry, University of California, Berkeley, CA 94720, U.S.A.

Beth SANDERS, Department of Chemistry, University of Glasgow, Glasgow G12 8QQ, Scotland, U.K.

Ray SOMORJAI (L), Division of Biological Sciences, NRC Canada, Ottawa K1A OR6, Canada.

Salvatore SORRENTINO, Dept. of Organic and Biological Chemistry, University of Naples, Via Mezzocannone 16, I-80134 Napoli, Italy.

Arthur SZABO (L), Division of Biological Sciences, NRC Canada, Ottawa K1A OR6, Canada.

Ayfer TEKIN, Hacettepe University, Faculty of Medicine, Dept. of Medical Biology, Ankara, Turkey.

Maureen S. THORNILEY, Dept. of Postgraduate Medicine, University of Keele, Thonburrow Drive, Hartshill, Stoke-on-Trent ST4 7QB, U.K.

Maria Rita TORQUATI, Istituto Nazionale di Fisica Nucleare, Via Vecchia Livornese, S. Piero a Grado (Pisa), Italy.

Carlo Alberto VERACINI, Dept. of Chemistry, University of Pisa, Via Risorgimento 35, I-56100 Pisa, Italy.

Stefano VERONESI, IFAM-CNR, Via del Giardino 7, I-56100 Pisa, Italy.

Marcel WAKS (L), UA 586 CNRS - Biochimie, Rue des Saints Pères, F-75006 Paris Cedex 06, France.

Franco WANDERLINGH (L), Dipartimento di Fisica, Università di Messina, Via dei Verdi, I-98100 Messina, Italy.

Ulderico WANDERLINGH, Istituto di Fisica, Università di Messina, Contrada Papardo, Salita Sperone 31, I-98010 Santa Agata, Messina, Italy.

Arieh WARSHEL (L), Department of Chemistry, University of Southern California, University Park, Los Angeles, CA 90089-1062, U.S.A.

Herman C. WATSON (L), Department of Biochemistry, Bristol University, University Walk, Bristol BS8 1TD, U.K.

Gregorio WEBER (L), Biochemistry Department, University of Illinois, Urbana-Champaign, IL 61801, U.S.A.

Malcom Peter WEIR, Genetic Unit, Glaxo Research Group, Greenford Road, Middlesex UB6 OHE, U.K.

Ab initio methods, 345, 346
Absorption spectroscopy, 126, 176
Accessibility, 32, 133
Acoustic measurements, 250
Acrylamide
   fluorescence quenching by, 127
ACTH, 470
Activated complex, 335
Activation
   energy, 164, 285, 297, 300
      distribution of, 216
   entropy, 285
   free energy, 285, 300
Active site, 59-63, 237, 283, 302,
      305
Acyl enzyme, 15, 59, 173, 191, 431
Adiabatic mapping, 348
Alcohol dehydrogenase, 450
Aldol reactions, 458
Algorithms, 33, 37, 48
   minimization, 45
   folding, 41, 45
Allostery, 12, 19, 165
Alpha-helix, 14, 29, 44, 55, 60,
      145, 149
   dipole, 60, 133
AMBER, 46
Amidases, 457
Ammonia, 97
Amphipathic structures, 32
Anharmonicity, 98, 114, 166, 358
Anisotropic potential, 35
Annular lipids, 383, 405
Antibodies as catalysts, 460
Antigenic determinants, 32
Apoazurin, 130, 136
Arginine, 65, 434
Aromatic amino acids, 123-139
   ring flips, 103
Arrhenius expression, 164, 410
Artificial intelligence, 44
Aspartic acid, 60
Assymetric double well, 118
Atom transfer, 285, 289
Autocorrelation function, 258, 268

Average coordinates, 205
Avoided crossing, 277
Azurin, 132, 136

Bacteriorhodopsin, 218, 383, 405
Band
   intensity, 71, 85
   position, 71
   width, 71
Barrier, 35, 469
   crossing, 165, 286
   frequency, 288
   kinetic, 35, 165, 286, 288, 300,
      469
   proton transfer, 300
   recrossing, 286
Beer-Lambert law, 200
Beta-sheet, 29, 44, 55, 145, 149
Binding, 12, 143, 144, 153, 154,
      211, 231, 369
Biotransformation, 443-463
   optically active, 444, 445
Boltzmann probability, 160, 371
Bombesin, 134
Bond angle, 27
Borgis-Hynes theory, 294
Born-Oppenheimer approximation, 99,
      298
Bound water, 466
Boundary conditions, 38
BPTI, 108, 109, 111-114
Bragg law, 205
Breathing modes, 108, 169
Brownian diffusion, 19, 253
Brownian motion, 160, 169, 219
Brownian oscillator, 106, 215
Buffer effects, 379

Caesium
   fluorescence quenching by, 127
Calcium binding proteins, 131
Calmodulin, 131
Calorimetry, 369, 386
Canonical ensemble, 34, 47
Carbonium ion, 321

Carbonyl
   group, 182, 445
   polarization, 183
Carboxylic acids, 118, 227
Carboxypeptidase A, 428
Catalytic resdidues, 428
Catalytic triad, 323
Channels, 165, 405
Charge density, 27
Charge relay mechanism, 15, 323
Charge transfer reaction, 290
Chiral synthesis, 444, 445
Chironomus thummi thummi, 208
Cholesterol, 383
Chymotrypsin, 14, 65, 174, 323, 467
Circular dichroism, 74
Cloning, 414, 416
Cluster size, 240, 402
Coherence length, 204
Collectively moving segments, 216
Collision
   frequency, 286
   process, 17
Complementation assay, 417
Complexity, 8
Condensation process, 227, 237
Conformation space, 34, 48
Conformational activation, 187
Conformational change, 103, 165,
     197, 210, 469, 470
Conformational drift, 170
Conformational heterogeneity, 123,
     129, 133
Conformational motion, 46, 103
Conformational substates, 104, 114,
     198, 207, 217
Contact ion pair, 288
Continuous jump diffusion, 214
Cooperativity, 12, 19, 58, 165, 398
Correlation
   function, 72, 298
   length, 35, 251
   time, 119
Cosurfactant, 466
COSY, *see* NMR
Coupling, 39, 299, 473
   fluctuations, 295
   non-linear, 93-102
Crambin, 109
Critical exponents, 236, 243
Critical region, 235
Cysteine protease, 173, 183
Cytochrome C, 111
Cytotoxins, 131

Debye-Waller factor, 89, 94, 110,
     118, 205, 270
Dehydrogenase, 55, 450
Delocalised systems, 173, 181
Density of states, 104, 109

Dielectric
   constant, 31, 301, 309
   properties, 214, 235, 237
   relaxation, 69, 127, 133, 237
Differential scanning calorimetry,
     386, 390, 408
Differential solvation, 289
Diffusion
   coefficient, 253, 268
   continuous jump, 214
   lateral, 396, 402
   rotational, 131, 472
Diffusion-collision model, 43, 50
Diffusional motion, 203, 213
   processes, 123, 252
Diffusive regime, 287
Dihydroxyacetone phosphate, 293,
     301, 338, 345, 377
Dimyristoylphosphatidyl choline, 383
Dipalmitoylphosphatidyl choline, 383
Dipole moment, 31, 70, 290
   helix, 60, 133
Discrete jump model, 214
Dissociation constant, 143, 153
Distribution, 75, 130, 163
   of activation energy, 216
   Boltzmann, 160, 371
   of frequencies, 109, 111, 114
   Lorentzian energy, 200
   structural, 204, 473
   zero point, 206
Dithioester chromophore, 184
DNA, 109, 252
   enzymology, 414
   sequencing, 414, 416
Domain, 19, 42
   motion, 61, 65, 103, 165, 169
Doppler effect, 200
Double-well potential, 96, 118, 294
Downhill trajectory, 316
Dry enzymes, 465

Eigenvectors, 108, 110
Elastase, 14
Elastic peak, 105
Electron
   polarization, 183
   transfer, 291, 297
Electron microscopy, 383
   cryo-, 383, 390
Electrostatic
   effects, 305, 321, 328
   interaction, 13, 30, 301, 470
Embryo-nucleus mechanism, 50
Empirical valence bond formulation,
     305
Energetics, 369
Energy
   of activation, 216, 285, 300
   distribution, 163, 200

Hydration, 223-234, 465
  shell, 218, 230
Hydrogen bond, 7, 13, 28, 60, 289,
        295, 301, 370, 376, 470
  dynamics, 99, 118
Hydrogen exchange, 228
Hydrolases, 444, 453
Hydropathy, 44
Hydrophobic interaction, 31, 51, 370
Hydrophobicity scales, 32, 44
Hysteresis, 170

Indole, 134
Induced changes, 7
Inelastic neutron scattering, 87,
        103-122, 248, 254
Inhibitor, 143, 144, 153, 154
Interface, 466
Intermediates, 65, 231
  acyl, 15, 191
  enzyme-substrate, 12, 55, 173
  tetrahedral, 15, 174, 191
Internal conversion, 124
Interphase, 466
Intersystem crossing, 73, 124
Intramolecular bonding, 470
Intramolecular mobility, 218
Iodide
  fluorescence quenching by, 127
Iodine-laser temperature
        jump, 383
Ion pair association, 291
Ionic resonance structure, 320
Ionization potential, 27
Ising model, 236
Isomerization, 43, 284, 293, 301,
        379, 402
Isomer shift, 201
Isotope effects, 294, 300, 380

Joule heating, 384

Kinase, 15, 19, 55, 459
Kinetic barriers, 35, 165, 286, 288,
        300, 469
Kinetics, 12, 168
  Michaelis-Menten, 12, 66, 283,
        293, 465
Kramers theory, 278, 286
Kynurenine, 132

Lactate dehydrogenase, 440, 460
Lamb-Mössbauer factor, 201, 203
Langevin
  dipole model, 305, 308
  equation, 168, 169, 278, 287
Larmor frequency, 202
Lateral diffusion, 396, 402
Lattice models, 236
Lecithin, 218, 387
Light scattering, 251

Line broadening, 71, 203
Linear free energy relationship,
        311, 317
Linear response approximation, 318
Lipases, 444, 453
Lipids, 383
Lissajoux curve, 280
Local mode, 79, 83
Long-range
  connectivity, 6, 235
    interaction, 43
Lorentzian energy distribution, 200
Low-frequency vibration, 104, 166,
        376
Low-temperature Raman spectroscopy,
        190
Lysine, 63, 134, 378, 434
Lysozyme, 6, 65, 108, 109, 141-158,
        225, 235, 248, 254, 321,
        341, 376, 465

Macroscopic model, 308
Markov chain, 35
Markov process, 35, 47
Mean-square displacement, 116, 118,
        201, 205
Membrane-bound state, 467
Membrane mimetic, 469
Membrane dynamics, 383-412
Membrane proteins, 466
Mesoscopic system, 18, 163, 277
Methionine, 65
Methyl group, 318
  torsion, 112
Micellar interface, 470
Michaelis-Menten kinetics, 12, 66,
        283, 293, 465
Microcalorimetry, 369-381
Microcanonical ensemble, 37
Microreactors, 467
Microscopic
  model, 308
  viscosity, 472
Minimization algorithms, 33, 45
Molar extinction coefficients, 126
Molecular dynamics simulation, 14,
        18, 31, 36, 48, 104, 107,
        114, 162, 275
Molecular mechanics, 14, 345
Molten globule state, 43, 49
Monte Carlo simulation, 31, 34, 47,
        332, 403
Morse oscillator, 84, 360
Morse potential, 84, 95, 306
Mössbauer spectroscopy, 104, 106,
        115, 119, 197-221
Multi-photon ionization, 359
Multiple-minimum problem, 45
Mutant enzymes, 351
Mutase, 62
Myelin basic protein, 469